SELECTED WORKS OF

EBERHARD HOPF

WITH COMMENTARIES

Eberhard Hopf
1902–1983

We thank Professor Hopf's daughter, Barbara Hopf Offenhartz, for the use of this photograph of him taken in New Rochelle in 1948–49 during his visit with Richard Courant and colleagues.

SELECTED WORKS OF EBERHARD HOPF WITH COMMENTARIES

Cathleen S. Morawetz
James B. Serrin
Yakov G. Sinai
Editors

American Mathematical Society
Providence, Rhode Island

2000 *Mathematics Subject Classification.* Primary 30–XX, 34–XX, 35–XX, 37–XX; Secondary 34C23, 35Q30, 45–XX, 47–XX.

Library of Congress Cataloging-in-Publication Data

Hopf, Eberhard, 1902–
[Papers. Selections]
Selected works of Eberhard Hopf : with commentary / Cathleen S. Morawetz, James B. Serrin, Yakov G. Sinai, editors.
p. cm.
English and German.
Includes bibliographical references
ISBN 0-8218-2077-X (alk. paper)
1. Mathematical analysis. 2. Fluid dynamics. 3. Ergodic theory. I. Morawetz, Cathleen S. II. Serrin, James B., 1962– III. Sinai, Iakov Grigor′evich, 1935– IV. Title.

QA300.5 .H662 2002
515—dc21
2002033254

A complete list of acknowledgments can be found at the back of this publication.

∞ The paper used in this book is acid-free and falls within the guidelines established to ensure permanence and durability.
Visit the AMS home page at `http://www.ams.org/`

10 9 8 7 6 5 4 3 2 1 07 06 05 04 03 02

Contents

Foreword ix

Acknowledgments xi

Contributors xiii

Curriculum Vitae xv

Works of Eberhard Hopf xvii

Preface xxiii

Selecta with Commentaries

Part I: Papers on Bifurcation, Fluid Dynamics, Integral Equations and Partial Differential Equations
EDITED BY CATHLEEN S. MORAWETZ AND JAMES B. SERRIN 1

Elementare Bemerkungen über die Lösungen partieller Differentialgleichungen zweiter Ordnung vom elliptischen Typus 3

Commentary by James B. Serrin 9

A Remark on Linear Elliptic Differential Equations of Second Order 15

Commentary by James B. Serrin 19

Zum analytischen Charakter der Lösungen regulärer zweidimensionaler Variationsprobleme 21

Commentary by Hans Weinberger 31

Über eine Klasse singulärer Integralgleichungen (with Norbert Wiener) 33

Commentary by Harold Widom 47

Über den funktionalen, insbesondere den analytischen Charakter der Lösungen elliptischer Differentialgleichungen zweiter Ordnung 49

Commentary by Hans Weinberger 89

Abzweigung einer periodischen Lösung von einer stationären Lösung eines Differentialsystems 91

Commentary by Martin Golubitsky and Paul H. Rabinowitz 111

Repeated branching through loss of stability, an example 119

A mathematical example displaying features of turbulence 127

Commentaries by Roger Temam 147

On S. Bernstein's theorem on surfaces $z(x, y)$ of nonpositive curvature 151

Commentary by Louis Nirenberg 157

The Partial Differential Equation $u_t + uu_x = \mu u_{xx}$ 159

Commentary by Peter Lax 189

Über die Anfangswertaufgabe für die hydrodynamischen Grundgleichungen 193

Commentary by James B. Serrin 213

Hamilton's theory and generalized solutions of the Hamilton-Jacobi equation 221

Commentary by Cathleen S. Morawetz 269

Part II: Papers in Ergodic Theory
EDITED BY YAKOV G. SINAI 271

Statistik der geodätischen Linien in Mannigfaltigkeiten negativer Krümmung 273

Statistik der Lösungen geodätischer Probleme vom unstabilen Typus. II. 317

Commentaries by Ya. G. Sinai 337

Closed Surfaces without Conjugate Points 339

Commentary by Ya. G. Sinai 345

Statistical Hydromechanics and Functional Calculus 347

Commentary by Ya. G. Sinai 385

On the Ergodic Theorem for Positive Linear Operators 387

Commentary by Donald Ornstein 393

Credits 395

Foreword

In 1997 James Serrin, Yakov Sinai and I agreed to edit a selection of the papers of Eberhard Hopf for publication. Hopf (1902-1983) was a founding father of ergodic theory and produced many beautiful and now classical results in integral equations and partial differential equations. In fact so basic, for example, is his maximum principle that it is often used without reference to its author.

Born in Austria, trained in Germany, Hopf spent several years at the Harvard Observatory and at M.I.T., returned to a permanent professorship in Leipzig in 1936 (to both dismay and understanding in the mathematical community), moved to Munich in 1944 and was a visiting professor at New York University in 1947. The remainder of his professional life he was a professor at Indiana University.

Hopf was not a prolific writer but a very large fraction of his work remains at the core of the fields he worked in and he wrote with such elegance and clarity that they are of great use today. The editors hope that the commentaries written for each paper will assist the reader in understanding the texts themselves and in particular illuminate the works that have been written in German. One notes for example that the paper on Burger's equation has been cited 539 times according to the "Web of Science," but there are hardly any references to his very well-known and useful bifurcation results. However, the phrase "Wiener-Hopf" appears in 645 titles in the A.M.S. Math Reviews.

In ergodic theory Hopf invented a method of proving ergodicity in hyperbolic systems, which is known as Hopf's method and is used very often in the modern theory of chaotic systems. It is based upon the notion of Hopf chains. His beautiful ergodic theorem for positive operators opened a new branch of ergodic theory. This list would not be complete without mentioning Hopf equations describing the evolution of correlation functions of measures in functional spaces.

Acknowledgments

The editors would like to thank the many people who helped with this volume. The first effort was to collect reprints of the original papers, a task in which Andrew Gleason, Erhard Heinz, Stefan Hildebrandt, Jurgen and Lucy Moser, and Eberhard Zeidler all assisted. Philip Anselone and Herbert Kalf donated their collections. Peter Lax and Louis Nirenberg lent their old reprints, some going back to the mid-twenties.

The next stage was to obtain willing commentators. Here James Serrin and Yakov Sinai took the major burden. But we needed more help, and this has been provided by Marty Golubitsky, Peter Lax, Louis Nirenberg, Donald Ornstein, Roger Temam, Harold Widom, and Hans Weinberger to all of whom we would like to express our gratitude.

The staff of the Courant Institute library, notably the head librarian, Carol Hutchins, and her assistant, Michael Haag, were enormously useful with the bibliography and other data. The hefty job of miscellaneous typing, correcting, and collating was in the able hands of James Hyre.

We would like finally to express our profound thanks for the help (including the photograph) received from Professor Hopf's daughter, Dr. Barbara Hopf Obercrantz.

Contributors

Martin Golubitsky	Deparment of Mathematics, Univeristy of Houston
Peter D. Lax	Courant Institute, New York University
Cathleen Synge Morawetz	Courant Institute, New York University
Louis Nirenberg	Courant Institute, New York University
Donald S. Ornstein	Department of Mathematics, Stanford University
Paul H. Rabinowitz	Department of Mathematics, University of Wisconsin
James B. Serrin	Department of Mathematics, University of Minnesota
Yakov G. Sinai	Department of Mathematics, Princeton University
Roger Temam	University of Paris XI
Hans F. Weinberger	School of Mathematics, University of Minnesota
Harold Widom	Department of Mathematics, University of California, Santa Cruz

Curriculum Vitae

1902	Born in Salzburg, Austria, April 17, 1902, son of Friedrich Hopf, chocolate manufacturer. Died on July 24, 1983 in Indiana, USA.
1925	Ph.D. Berlin.
1929	Habilitation. Married fellow student Ilse Wolf, daughter of the musicologist, Johannes Wolf.
1929 - 1932	Instructor of Mathematics and Astronomy at the University of Berlin, Germany.
1930 - 1932	International Fellow of Mathematics at Harvard University.
1932 - 1936	Assistant Professor of Mathematics at the Massachusetts Institute of Technology. The Hopfs had one daughter, Barbara Hopf Offenhartz, born April 24, 1936.
1936 - 1944	Professor of Mathematics at The University of Leipzig Germany.
1938	Member of the Sächsische Akademie der Wissenschaften.
1939	Coeditor of the Grundlehren der Mathematischen Wissenschaften (Yellow Series).
1942	Drafted to serve in German Aeronautical Institute.
1944 - 1949	Professor of Mathematics at the University of Munich, Germany, succeeding C. Caratheodory.
1945	Elected member of the Bayrische Akademie der Wissenschaften.
1947 - 1948	Visiting Professor at New York University.
1948 - 1962	Professor of Mathematics at Indiana University.
1962 - 1983	Research Professor of Mathematics at Indiana University.
1971	Gibbs lecturer of American Mathematical Society.
1952 - 1983	Editor, Journal of Rational Mechanics and Analysis, which later became Journal of Mathematics and Mechanics.

Biographies:

"In memoriam Eberhard Hopf: 1902-1983." *Indiana Univ. Math. J.* **32** (1983), no. 6, i-ii.

BECKERT, HERBERT, "Obituary of Eberhard Hopf." *Akad. Wiss. Leipzig Jahresber. Akademie-Verlag, Berlin.* (1986) pp. 238-245.

DENKER, M., "Eberhard Hopf: 04-17-1902 to 07-24-1983." *Jahresber. Deutsch. Math.- Verein.* **92** (1990), no. 2, pp. 47-57.

ICHA, ANDRZEJ, "Eberhard Hopf (1902-1983)." *Nieuw Arch. Wisk. (4).* **12** (1994), no. 1-2, pp. 67-84.

Works of Eberhard Hopf

Books:

Mathematical Problems of Radiative Equilibrium. Cambridge Tract No. 31, Cambridge University Press (1934).

Ergodentheorie, Ergebnisse der Mathematik und ihrer Grenzgebiete. 5, H. 2., Berlin, Julius Springer (1938), (reprinted).

Publications:

"Über die Zusammenhänge zwischen gewissen höheren Differenzenquotienten reeller Funktionen einer reellen Variablen und deren Differenziebarkeitseigenschaften," *Inaugural-Dissertation*, Berlin (1926).

"Bemerkungen zum ersten Randwertproblem der Potentialtheorie im Raume," *Sitzungsberichte der Berliner Mathematischen Gesellschaft*, **26** (1926), 43-48.

"On the Integral Equation for Radiative Equilibrium," (with E. Freundlich and U. Wegner), *R. Astron. Soc. Monthly Notices* **88** (1927), 139-142.

"Eine Bemerkung zur Theorie der partiellen Differentialgleichungen zweiter Ordnung vom elliptischen Typus," *Sitzungsberichte der Berliner Mathmatischen Gesellschaft.* (1927), 11-112.

"Elementare Bemerkungen über die Lösungen partieller Differntialgleichungen zweiter Ordnung vom elliptischen Typus," *Sitzungsberichte Preussische Akad. Wiss. Berlin.* (1927), 147-152.

"Ein Analogon zu einem Mittelwertsatze von H. A. Schwarz bei komplexen Polynomen," *Jahresbericht der Deutschen Math.-Vereinigung.* **37** (1928), 249-251.

"Bemerkungen zu einem Satze von S. Bernstein aus der Theorie der elliptischen Differentialgleichungen," *Math. Zeitschr.* **29** (1928), 744-745.

"Zum Problem des Strahlungsgleichgewichts in den äusseren Schichten der Sterne," *Zeitschrift für Physik* **46** (1928), 374-382.

"Über Strahlungsgleichgewicht in den äusseren Schichten der Sterne II," *Zeitschrift für Physik* **49** (1928), 155-161.

"Über lineare Integralgleichungen mit positivem Kern," *Sitzungsberichte Preussische Akad. Wiss. Berlin* (1928), 233-245.

"Über die geschlossenen Bahnen in der Mondtheorie," *Sitzungsberichte Preussische Akad. Wiss. Berlin* (1929), 401-413.

"Zum Strahlungsgleichgewicht der Sternatmosphären," *Astronomische Nachrichten* **235** (1929), 97-104.

"Über eine Klasse singularer Integralgleichungen: Bemerkungen zur Methode von Hardy und Titchmarsh," *J. London Math. Soc.* **4** (1929), 23-27.

"Kurze Bemerkung über wiederholte Quadrateren," *Math. Zeitschr.* **20** (1929), 373-374.

"Zum analytischen Charakter der Lösungen regulärer zweidimensionaler Variationsprobleme," *Math. Zeitschr.* **30** (1929), 404-413.

"Remarks on the Schwarzchild-Milne Model of the Outer Layers of a Star," *Astronomical Soc. Monthly Notices* **90** (1930), 287-293.

"Zwei Sätze über den wahrscheinlichen Verlauf der Bewegungen dynamischer Systeme," *Math. Annalen* **103** (1930), 710-719.

"Über den funktionalen, insbesondere den analytischen Charakter der Lösungen elliptischer Differentialgleichungen zweiter Ordung," *Math. Zeitschr.* **34** (1931), 194-233.

"Über eine Klasse singularer Integralgleichungen," (with Norbert Wiener), *Sitzungsberichte Preussische Akad. Wiss., Phys.- Math. Kl. Berlin* (1931), 696-706.

"Kleine Bemerkung zur Theorie der elliptischen Differentialgleichungen," J. Reine Angew. Math. 165 (1931), 50-51.

"Über den Polytropenindex eines Sternmodells, I," *Zeitschr. Astrophys.* **1** (1931), 67-70.

"Über den Polytropenindex eines Sternmodells, II," *Zeitschr. Astrophys.* **3** (1931), 71-76.

"Über den Polytropenindex eines Sternmodells, III," *Zeitschr. Astrophys.* **3** (1931), 108-115.

"On Emden's Differential Equation," *R. Astron. Soc. Monthly Notices* **91** (1931), 653-663.

"Mathematisches zur Strahlungsgleichgewichtstheorie der Fixsternatmosphären," *Math. Zeitschr.* **33** (1931), 109-128.

"Kleine Bemerkung zur Theorie der elliptischen Differentialgleichungen," *J. Reine Angew. Math.* **165** (1931), 50-51.

"Über lineare Gruppen unitärer Operatoren im Zusammenhange mit den Bewegungen dynamischer Systeme," *Sitzungsberichte Preussische Akad. Wiss. Phys.-Math. Kl. Berlin* (1932), 182-190.

"Remarks on the Schwarzschild-Milne Model of the Outer Layers of a Star, II," *R. Astron. Soc. Monthly Notices* **92** (1932), 863-868.

"On Certain Integral Equations," *Quart. J. Math.* **3** (1932), 269-272.

"On the Time Average Theorem in Dynamics," *Proc. Nat. Acad. Sci. (USA)* **18** (1932), 93-100.

"Complete Transitivity and the Ergodic Principle," *Proc. Nat. Acad. Sci. (USA)* **18** (1932), 204-209.

"Proof of Gibbs' Hypothesis on the Tendency toward Statistical Equilibrium," *Proc. Nat. Acad. Sci. (USA)* **18** (1932), 333-340.

"Theory of Measure and Invariant Integrals," *Trans. Amer. Math. Soc.* **34** (1932), 373-393.

"Remarks on Causality and Probability," *J. Math. Phys.* **13** (1934), 4-9.

"On Causality, Statistics and Probability," *J. Math. Phys.* **13** (1934), 51-102.

"Remarks on Causality and Probability," *J. Math. Phys.* **14** (1935), 1-35.

"Absorption Lines and the Integral Equation of Radiative Equilibrium," *R. Astron. Soc. Monthly Notices* **96** (1936), 522-533.

"Über die Bedeutung der willkürlichen Funktionen für die Wahrscheinlichkeitstheorie," *Jahresbericht d. Deutschen Math. Vereinigung* **46** (1936), 179-195.

"Fuchsian Groups and Ergodic Theory," *Trans. Amer. Math. Soc.* **39** (1936), 299-314.

"Ein Verteilungsproblem bei dissipativen dynamischen Systemen," *Math. Annalen* **114** (1937), 161-186.

"Beweis des Mischungscharakters der geodatischen Strömung auf Flächen der Krümmung minus Eins und endlicher Oberfläche," *Sitzungsberichte Preussische Akad. Wiss. Phys.- Math. Kl. Berlin* (1938), 333-344.

"Statistische Probleme und Ergebnisse in der klassischen Mechanik," Contribution to "Le principe ergodique et les probabilités en chaines," *Actualités Scientifiques* **737** (1938), 5-16.

"Le principe ergodique et les probabilités en chaine," Paris: *Hermann & Cie.*, (1938), 68.

"Statistik der geodatischen Linien in Mannigfaltigkeiten negativer Krümmung," *Akad. Wiss. Leipzig* **91** (1939), 261-304.

"Statistik der Lösungen geodatischer Probleme vom unstabilen Typus II," *Math. Annalen* **117** (1940), 590-608.

"Randbemerkungen zu einigen Existenzsätzen der Differentialgeometrie," *J. Ber. Deutschen Math. Vereinigung* **49** (1940), 253-255.

"Ein allgemeiner Endlichkeitssatz der Hydrodynamik," *Math. Annalen* **117** (1941) 764-775.

"Abzweigung einer periodischen Lösung von einer stationären Lösung eines Differentialsystems," *Akad. Wiss. Math.- Nateiw. Kl. (Leipzig)* **95** no. 1, (1943), 3-22.

"Über eine Ungleichung der Ergodentheorie," Sitzungsberichte Akad. Wiss. (München) (1944), 171-176.

"Kennzeichnung der durch Punkttransformationen erzeugten linearen Funktionaloperationen," *Sitzungsberichte Akad. Wiss. (München)* (1944), 233-236.

"Über die Funktionalgleichungen der trigonometrischen und hyperbolischen Funktionen," *Sitzungsberichte Akad. Wiss. (München)* (1945-1946), 167-173.

"Über einige Eigenschaften von Kurvenintegralen und über die Äquivalenz von indefiniten mit definiten Variationsproblemen," (with Wilhelm Damkohler), *Math. Annalen* **120** (1947), 12-20.

"Closed surfaces without conjugate points," *Proc. Nat. Acad. Sci., (USA)* **34** (1948), 47-51.

"A Mathematical Example Displaying Features of Turbulence," *Comm. on Pure and Appl. Math.* **1** (1948), 303-322.

"A Theorem on the Accessibility of Boundary Parts of an Open Point Set," *Proc. Amer. Math. Soc.* **1** (1950), 76-79.

"On S. Bernstein's Theorem on Surfaces z(x,y) of Nonpositive Curvature," *Proc. Amer. Math. Soc.* **1** (1950), 80-85.

"The Partial Differential Equation $u_t + uu_x = \mu u_{xx}$," *Comm. on Pure and Appl. Math.* **3** (1950), 201-230.

"Über die Anfangswertaufgabe für die hydrodynamischen Grundgleichungen," *Math. Nachr.* **4** (1951), 213-231.

"Statistical Hydromechanics and Functional Calculus," *J. Rational Mech. Anal.* **1** (1952), 87-123.

"A Remark on Linear Elliptic Differential Equations of Second Order," *Proc. Amer. Math. Soc.* **3** (1952), 791-793.

"Remarks on the Preceding Paper by D. Gilbarg," *J. Rational Mech. Anal.* **1** (1952), 419-424.

"On an Inequality for Minimal Surfaces $z = z(x, y)$," *J. Rational Mech. Anal.* **2** (1953), 519-522.

"On Certain Special Solutions of the M-equation of Statistical Hydrodynamics," (with E. W. Titt), *J. Rational Mech. Anal.* **2** (1953), 587-591.

"Correction to the Paper 'On an inequality for minimal surfaces $z = (x, y)$'," *J. Rational Mech. Anal.* **2** (1953), 801-802.

"The General Temporally Discrete Markoff Process," *J. Rational Mech. Anal.* **1** (1954), 13-45.

"Die Harnacksche Ungleichung für positive harmonische Funktionen," *Math. Zeitschr.* **63** (1955), 156-157.

"Repeated Branching through Loss of Stability, an Example," *Proc. of the Conference on Differential Equations at the University of Maryland* (1955) 49-56.

"On the Application of Functional Calculus to the Statistical Theory of Turbulence," *Proc. of Symposia in Appl. Math.*, vol. VII, Applied Probability, A.M.S. (1957), 41-50.

"The Temporal Behavior of a Wave Packet," *Duke Math. J.* **24** (1957), 477-480.

"Zur Kennzeichnung der Euklidischen Norm," *Math. Zeitschrift* **72** (1959), 76-81.

"On the ergodic theorem for positive linear operators," *J. Reine Angew. Math.* **205** (1960), 101-106.

"Some topics of ergodic theory," *C.I.M.E. Roma*, Instituto Matematico dell'-Universitá (1960), 1-64.

"Remarks on the functional-analytic approach to turbulence," *Proc. of Symposia in Appl. Math.* **13** (1962), Hydrodynamic Instability, 157-163.

"An inequality for positive linear integral operators," *J. Math. & Mech.* **12** (1963), 683-692.

"Remarks on 'An inequality for positive linear integral operators'," *J. Math. & Mech.* **12** (1963), 889-892.

"Hamilton's theory and generalized solutions of the Hamilton-Jacobi equation" (with E.D. Conway), *J. Math. & Mech.* **13** (1964), 939-986.

"Generalized solutions of non-linear equations of first order," *J. Math. & Mech.* **14** (1965), 951-973.

"On the right weak solution of the Cauchy problem for a quasilinear equation of first order," *J. Math. & Mech.* **19** (1969), 483-487.

"Ergodic theory and the geodesic flow on surfaces of constant negative curvature," *Bull. Amer. Math. Soc.* **17** (1971), 863-877.

Preface

There are two main parts to this volume. The first part, selected by J. Serrin, includes classic papers in analysis and fluid dynamics and the second part, selected by Ya. G. Sinai, in ergodic theory. In the first part, the papers are grouped together according to topic and ordered chronologically by the first paper of the group. It seems more sensible to have the reader find, for example, the two papers on the maximum principle for elliptic equations together although they were written many years apart.

Much of Hopf's early work was in astronomy. None of these papers have been included, but it is worth noting that the famous Wiener-Hopf result was undoubtedly connected to Hopf's work at the Harvard Observatory in the early thirties. Sinai has heard that the Russian physicist, Landau, once said that in mathematics there were some things that could not have been invented by physicists, notably the Wiener-Hopf method.

Several short biographies of Hopf have been written and have been listed at the end of his curriculum vitae.

Part I

Papers on Bifurcation, Fluid Dynamics, Integral Equations and Partial Differential Equations

Edited by
Cathleen S. Morawetz and James B. Serrin

Elementare Bemerkungen über die Lösungen partieller Differentialgleichungen zweiter Ordnung vom elliptischen Typus.

Von Dr. E. Hopf
in Berlin-Friedenau.

(Vorgelegt von Hrn. Schmidt.)

É. Picard[1], S. Bernstein[2], L. Lichtenstein[3] haben gezeigt, daß die Lösungen gewisser allgemeiner Differentialgleichungen vom elliptischen Typus Eigenschaften besitzen, welche an einen bekannten Satz über die Extrema harmonischer Funktionen erinnern. Während die genannten Autoren dabei von tiefer liegenden besonderen Darstellungen der Lösungen ausgehen, geben wir in dieser Note eine gänzlich elementare Begründung und Weiterführung jener Resultate.

I. Es sei $L(u) = \sum\limits_{1,1}^{n,n} a_{\nu\mu} u_{x_\nu x_\mu} + \sum\limits_{1}^{n} b_\nu u_{x_\nu}$ ein linearer partieller Differentialausdruck zweiter Ordnung von elliptischem Typus. Die $a_{\nu\mu}$ und b_ν seien stetige Funktionen des Punktes $P = (x_1, x_2, \cdots, x_n)$ eines n-dimensionalen Gebietes T und die quadratische Form $\sum\limits_{1,1}^{n,n} a_{\nu\mu}(P)\lambda_\nu\lambda_\mu$ sei in T überall positiv definit.

Wir beweisen dann den

Satz. Genügt die in T mit stetigen partiellen Ableitungen zweiter Ordnung versehene Funktion $u(P)$ dort überall der Ungleichung $L(u) \geqq 0$ und ist, wenn P_0 ein fester Punkt von T ist, $u(P) \leqq u(P_0)$ überall in T, so ist in T $u(P) \equiv u(P_0)$. Dasselbe folgt, wenn $L(u) \leqq 0$, $u(P) \geqq u(P_0)$ in T ist.

Beweis: Der Durchsichtigkeit halber beschränken wir uns auf den ebenen Fall, $x_1 = x$, $x_2 = y$,

$$L(u) = Au_{xx} + 2Bu_{xy} + Cu_{yy} + Du_x + Eu_y; \quad AC - B^2 > 0, \quad A > 0 \text{ in } T.$$

Im allgemeinen Falle verläuft der Beweis genau ebenso. Es sei also etwa

$$L(u) \geqq 0 \text{ in } T, \tag{1}$$

[1] Vgl. É. Picard, Traité d'Analyse II, 2. Aufl. Paris 1905. p. 29—30.

[2] Vgl. S. Bernstein, Math. Annalen 59 (1904), p. 69.

[3] Vgl. L. Lichtenstein, Palermo Rend. 33 (1912), p. 211; Math. Zeitschr. 20 (1924), p. 205 bis 206.

$u(P_0) = M$, sonst $u \leqq M$ in T. Wäre nun nicht $u \equiv M$, so kämen wir, wie folgt, zum Widerspruch. Es gäbe dann, wie leicht zu sehen, einen nebst Rand in T gelegenen Kreis, auf dessen Rande irgendwo, etwa in P_1,

$$u(P_1) = M \tag{2}$$

wäre, während in seinem Innern stets $u < M$ ausfiele. Ist K ein abgeschlossener, diesen Kreis in P_1 von innen berührender kleinerer Kreis — sein Radius sei R —, so wäre überall in K, ausgenommen seinen Randpunkt P_1, $u < M$. Endlich sei K_1 ein ganz in T gelegener Kreis um P_1 als Mittelpunkt und mit einem Radius R_1,

$$R_1 < R. \tag{3}$$

Der Rand von K_1 besteht nun aus dem zu K gehörigen (abgeschlossenen) Bogen S_i und dem außerhalb K liegenden Bogen S_a. Auf S_i wäre dann $u < M$, somit auch $u \leqq M - \varepsilon$ bei passendem $\varepsilon > 0$. In Summa wäre

$$u \leqq M - \varepsilon \text{ auf } S_i, \quad u \leqq M \text{ auf } S_a. \tag{4}$$

Wir wählen jetzt den Mittelpunkt von K zum Koordinatenursprung und betrachten die Funktion

$$h(P) = e^{-\alpha r^2} - e^{-\alpha R^2}; \quad r^2 = x^2 + y^2, \quad \alpha > 0.$$

Die Rechnung ergibt

$$e^{\alpha r^2} \cdot L(h) = 4\alpha^2(Ax^2 + 2Bxy + Cy^2) - 2\alpha(A + C + Dx + Ey).$$

Im Innern und auf dem Rande von K_1 ist wegen (3) stets $Ax^2 + 2Bxy + Cy^2 > 0$, somit auch $\geqq$ const. > 0. Wird die Zahl α groß genug angesetzt, so ist offensichtlich

$$L(h) > 0 \text{ in } K_1. \tag{5}$$

Ferner ist

$$h < 0 \text{ auf } S_a, \quad h(P_1) = 0. \tag{6}$$

Wir setzen schließlich

$$v(P) = u(P) + \delta \cdot h(P), \quad \delta > 0$$

mit so klein gewähltem δ, daß — man beachte (4)! — $v < M$ auf S_i ausfällt. Wegen (4) und (6) ist auch auf S_a $v < M$. Folglich wäre auf dem ganzen Rande von K_1 $v < M$, in seinem Mittelpunkte nach (2) und (6) $v = M$. Demnach hätte v irgendwo im Innern von K_1 sein Maximum. Dies ist aber unmöglich, da in K_1 v nach (1) und (5) der Ungleichung $L(v) > 0$ genügt; an einer Stelle, wo v ein Maximum hätte, müßte nach den Regeln der Differentialrechnung $L(v) = Av_{xx} + 2Bv_{xy} + Cv_{yy} \leqq 0$ sein[1].

[1] An dieser Stelle wäre $v_{xx}\lambda^2 + 2v_{xy}\lambda\mu + v_{yy}\mu^2 \leqq 0$ für beliebige λ, μ. Hier wende man etwa den algebraischen Satz an: Sind $\sum_{1,1}^{n,n} a_{ik}\lambda_i\lambda_k$ und $\sum_{1,1}^{n,n} b_{ik}\lambda_i\lambda_k$ zwei niemals negative quadratische Formen, so ist $\sum_{1,1}^{n,n} a_{ik} \cdot b_{ik} \geqq 0$. Vgl. L. Fejér, Math. Zeitschr. 1 (1918), p. 70—79. Ursprünglich finden sich Überlegungen dieser Art bei A. Paraf, Ann. de la Fac. d. Sc. d. Toulouse 6 (1892), insbes. p. 49—50.

Ist speziell $L(u) = 0$ in T, so kann, wenn u nicht konstant ist, keine der Ungleichungen $u(P) \leqq u(P_0)$, $u(P) \geqq u(P_0)$ in T gelten[1].

II. $\phi(x, y, u, p, q, r, s, t)$ sei eine für alle Punkte (x, y) eines ebenen Gebietes T und für beliebige $u, p, \cdots, t$ stetige Funktion ihrer acht Argumente. Setzt man, wie üblich, in ϕ

$$u = u(x, y), \quad p = u_x(x, y), \cdots, \quad s = u_{xy}(x, y), \quad t = u_{yy}(x, y) \quad (7)$$

ein, so ist die Gleichung $\phi = 0$ eine partielle Differentialgleichung zweiter Ordnung für $u(x, y)$. Wir setzen voraus, daß die Funktion ϕ in der Form

$$\phi = \phi_1 r + 2\phi_2 s + \phi_3 t + \phi_4 p + \phi_5 q + \phi_6 u \quad (8)$$

darstellbar ist, wo die ϕ_ν stetige Funktionen von $x, y, u, \cdots, t$ sind. Eine derartige Darstellung wird dann im allgemeinen auf verschiedene Weise möglich sein. Eine in T reguläre, d. h. dort mit stetigen partiellen Ableitungen zweiter Ordnung versehene Lösung $u(x, y)$ von $\phi = 0$ heiße eine in T elliptische Lösung von $\phi = 0$, wenn bei passender Darstellung (8) der Funktion ϕ und nach Einsetzen von (7) in die ϕ_ν

$$\phi_1\phi_3 - \phi_2^2 > 0, \quad \phi_1 > 0 \quad (8a)$$

überall in T ausfällt.

Erhält man nach Einsetzen von (7) etwa $\phi_1 = A(x, y)$, $\phi_2 = B(x, y)$, $\cdots$, $\phi_6 = F(x, y)$, so ist die in T elliptische Lösung von $\phi = 0$ auch eine in T reguläre Lösung der linearen Differentialgleichung

$$L(u) + Fu = Au_{xx} + 2Bu_{xy} + Cu_{yy} + Du_x + Eu_y + Fu = 0;$$
$$AC - B^2 > 0, A > 0 \quad (9)$$

vom elliptischen Typus mit in T stetigen Koeffizienten.

Wir beweisen dann unter den obigen Voraussetzungen die folgenden Sätze:

Satz 1. Ist für eine in T elliptische Lösung u von $\phi = 0$ nach Einsetzen von (7) $\phi_6 = 0$ überall in T, so erreicht sie, wenn sie keine Konstante ist, nirgends in T ihre obere oder untere Grenze.

Satz 2. Ist für eine in T elliptische Lösung u von $\phi = 0$ nach Einsetzen von (7) $\phi_6 \leqq 0$ in T, so hat u in keinem Punkte von T ein positives Maximum oder negatives Minimum, es sei denn, daß u in der Umgebung eines solchen Punktes konstant ist.

Satz 3. Eine in T elliptische Lösung von $\phi = 0$ kann an einer Stelle, wo sie verschwindet, kein Extremum haben, es sei denn, daß sie in der Umgebung dieser Stelle identisch verschwindet.

Nach dem oben Gesagten können wir uns beim Beweise auf Lösungen einer linearen Differentialgleichung (9) beschränken.

Satz 1. folgt unmittelbar aus dem Hilfssatze.

---

[1] Übrigens bleiben die obigen Überlegungen unverändert, wenn man nur Beschränktheit der $A, B, \cdots, E$ voraussetzt. Dann muß aber das Bestehen einer Ungleichung $A\lambda^2 + 2B\lambda\mu + C\mu^2 \geq \text{const}\,(\lambda^2 + \mu^2) > 0$ für alle λ, μ und überall in T gefordert werden.

Beweis von Satz 2.: Hat u im Punkte P_0 von T etwa ein positives Maximum, so ist in einem passenden Kreise K_0 um P_0 als Mittelpunkt $0 < u \leqq u(P_0)$, somit nach (9) $L(u) = -Fu \geqq 0$. Unser Hilfssatz, angewandt auf den Kreis K_0, lehrt unmittelbar, daß dann $u \equiv u(P_0)$ in K_0 sein muß.

Beweis von Satz 3.: Es sei $u(P_0) = 0$ und etwa $u(P) \leqq 0$ in einem Kreise K_0 um P_0 als Mittelpunkt. Wir setzen dann $u(x, y) = e^{\alpha x} \cdot v(x, y)$. v ist dann eine in T reguläre Lösung einer Gleichung

$$L(u) + Fu = e^{\alpha x}(L^*(v) + F^* v) = 0\,;\; L^*(v) = A v_{xx} + 2B v_{xy} + C v_{yy} + D^* v_x + E^* v_y$$

mit stetigen Koeffizienten. Dabei ist $F^* = e^{-\alpha x}(L(e^{\alpha x}) + F e^{\alpha x}) = \alpha^2 A + \alpha D + F$. Wird die Zahl α genügend groß gewählt, so ist in K_0 offenbar $F^* > 0$. Nun ist $v(P_0) = 0$, $v \leqq 0$ in K_0, somit $L^*(v) = -F^* v \geqq 0$. Nach dem Hilfssatze ist dann $v \equiv 0$, also $u \equiv 0$ in K_0.

III. Wir gehen zum Schluß noch auf engere Voraussetzungen ein, unter denen die allgemeinen Voraussetzungen des vorigen Abschnitts stets erfüllt sind. Die obigen Benennungen beibehaltend, setzen wir voraus

a) die Null ist in T eine Lösung der Differentialgleichung $\phi = 0$;

b) ϕ besitzt partielle Ableitungen $\phi_u, \phi_p, \cdots, \phi_s, \phi_t$, die für alle in Betracht kommenden Werte von $x, \cdots, t$ stetige Funktionen sind;

c) es ist $4\phi_r \phi_t - \phi_s^2 > 0$, $\phi_r > 0$ an jeder Stelle $(x, y, u, 0, 0, r, s, t)$, (x, y) in T.

Wir schreiben dann

$$\begin{aligned}\phi = {} & \phi(x, y, u, p, q, r, s, t) - \phi(x, y, u, 0, 0, r, s, t) \\ & + \phi(x, y, u, 0, 0, r, s, t) - \phi(x, y, u, 0, 0, 0, 0, 0) \\ & + \phi(x, y, u, 0, 0, 0, 0, 0) - \phi(x, y, 0, 0, 0, 0, 0, 0)\end{aligned}$$

und wenden auf jede dieser drei Differenzen eine bekannte Formel der Analysis an[1]. Wir erhalten dann ϕ in der Gestalt (8), wobei

$$\left.\begin{aligned} \phi_1 &= \int_0^1 \phi_r(x, y, u, 0, 0, \xi r, \xi s, \xi t)\, d\xi, \\ 2\phi_2 &= \int_0^1 \phi_s(x, y, u, 0, 0, \xi r, \xi s, \xi t)\, d\xi, \\ \phi_3 &= \int_0^1 \phi_t(x, y, u, 0, 0, \xi r, \xi s, \xi t)\, d\xi, \\ & \cdots\cdots\cdots\cdots\cdots \\ \phi_6 &= \int_0^1 \phi_u(x, y, \xi u, 0, 0, 0, 0, 0)\, d\xi \end{aligned}\right\} \qquad (10)$$

---

[1] Besitzt $f(x_1, x_2, \cdots, x_n)$ an allen in Betracht kommenden Stellen stetige partielle Ableitungen erster Ordnung, so ist

$$f(x_1, x_2, \cdots, x_n) - f(0, 0, \cdots, 0) = x_1 \cdot \int_0^1 f_{x_1}(\xi x_1, \xi x_2, \cdots, \xi x_n)\, d\xi + \cdots$$
$$\cdots + x_n \cdot \int_0^1 f_{x_n}(\xi x_1, \xi x_2, \cdots, \xi x_n)\, d\xi.$$

ist. Nach c) ist die quadratische Form $\phi_r \lambda^2 + \phi_s \lambda\mu + \phi_t \mu^2$ stets positiv definit, nach den ersten drei Formeln (10) also auch die Form $\phi_1 \lambda^2 + 2\phi_2 \lambda\mu + \phi_3 \mu^2$. Somit ist die Bedingung (8a) stets erfüllt. Jede in T reguläre Lösung von $\phi = 0$ ist dann eine in T elliptische Lösung. Unter diesen Voraussetzungen gilt

Satz 1'. Ist jede Konstante in T Lösung der Differentialgleichung $\phi = 0$, so besitzt eine beliebige in T reguläre Lösung dieser Gleichung die in Satz 1. angegebene Eigenschaft.

Satz 2'. Ist überall in T für jeden Wert von u $\phi_u(x, y, u, 0, 0, 0, 0, 0) \leqq 0$[1], so gilt für jede in T reguläre Lösung der Gleichung $\phi = 0$ die Aussage des Satzes 2.

Zu Satz 1': Ist jede Konstante Lösung, so ist nach (10) $\phi_6 \equiv 0$ in T.

Zu Satz 2': Nach (10) ist dann stets $\phi_6 \leqq 0$.

Ersetzen wir die Voraussetzung c) durch die etwas weitere:

c) Es ist $4\phi_r \phi_t - \phi_s^2 > 0$, $\phi_r > 0$ an jeder Stelle $(x, y, 0, 0, 0, r, s, t)$, (x, y) in T, so setzen wir diesmal

$$\begin{aligned}\phi = {} & \phi(x, y, u, p, q, r, s, t) - \phi(x, y, u, 0, 0, r, s, t)\\ & + \phi(x, y, u, 0, 0, r, s, t) - \phi(x, y, 0, 0, 0, r, s, t)\\ & + \phi(x, y, 0, 0, 0, r, s, t) - \phi(x, y, 0, 0, 0, 0, 0, 0).\end{aligned}$$

Nach obigem Muster folgt, daß jede in T reguläre Lösung von $\phi = 0$ von elliptischem Charakter ist. Es gilt dann unter diesen Voraussetzungen

Satz 3'. Jede in T reguläre Lösung der Gleichung $\phi = 0$ besitzt die in Satz 3. angegebene Eigenschaft.

Insbesondere gelten diese Sätze, wenn die Differentialgleichung $\phi = 0$ schlechthin vom elliptischen Typus ist, d. h. wenn die Bedingung c) für alle Werte von $x, y, u, p, \cdots, t$, (x, y) in T, erfüllt ist. Läßt man die durchgängig gemachte Voraussetzung, daß die Null eine Lösung ist, fallen, so sind die obigen Sätze nicht mehr richtig, sie gelten dann jedoch für die Differenz irgend zweier regulärer Lösungen von $\phi = 0$; denn sind v und w irgend zwei derartige Lösungen in T, so setze man $u = v - w$, $u_x = p$, $u_y = q, \cdots, u_{yy} = t$,

$$\begin{aligned}\psi(x, y, u, p, \cdots, t) = {} & \phi(x, y, u + w(x, y), \quad p + w_x(x, y), \cdots, \quad t + w_{yy}(x, y))\\ & - \phi(x, y, w(x, y), \; w_x(x, y), \cdots, \; w_{yy}(x, y)).\end{aligned}$$

$u(x, y)$ ist dann eine in T reguläre Lösung der Gleichung $\psi = 0$, wobei die Funktion ψ wieder unter die obigen Betrachtungen fällt. Satz 2'. enthält dann

---

[1] Allgemeiner kann man wegen $u\phi_6 = \phi(x, y, u, 0, 0, 0, 0, 0)$ voraussetzen, daß

$$\frac{1}{u}\,\phi(x, y, u, 0, 0, 0, 0, 0) \leqq 0$$

ist.

speziell einen bekannten Unitätssatz[1]; Satz 3'. besagt, daß irgend zwei Lösungsflächen sich nirgends berühren[2] können, ohne in der Umgebung des Berührungspunktes zusammenzufallen.

Übrigens übertragen sich obige Überlegungen mit Leichtigkeit in völlig entsprechender Weise auf den Fall von mehr als zwei unabhängigen Variablen.

---

[1] Ist für alle Werte der acht Variablen $\varphi_u \leqq 0$, so folgt aus der Gleichheit zweier in T regulärer Lösungen auf dem Rande von T ihre Identität in T. Dieser Satz spielt in den Untersuchungen von S. Bernstein über die erste Randwertaufgabe bei der Gleichung $\varphi = 0$ eine wichtige Rolle. Vgl. den Bericht von L. Lichtenstein in d. Enzykl. d. Math. Wissensch. Bd. II 3, Heft 8, insbesondere p. 1328—1329.

[2] Es ist hier eine in der Umgebung des Berührungspunktes einseitige Berührung gemeint.

Ausgegeben am 3. August.

Berlin, gedruckt in der Reichsdruckerei.

Elementare Bemerkungen über die Lösungen partieller Differntialgleichungen zweiter Ordnung vom elliptischen Typus, *Sitzungsberichte Preussiche Akad. Wiss.* **Berlin (1927), 147-152.**

Commentary

James B. Serrin

Eberhard Hopf's great paper on the maximum principle "Elementare Bemerkungen über die Lösungen partieller Differentialgleichungen zweiter Ordnug vom elliptischen Typus" is the keystone work on this prime result of elliptic partial differential equations. It has the beauty and elegance of a Mozart symphony, the light of a Vermeer painting. Only a fraction more than five pages in length, it contains seminal ideas which are still fresh after 75 years.

The maximum principle for harmonic and subharmonic functions was known to Gauss on the basis of the mean value theorem (1839); an extension to elliptic inequalities however remained open until the twentieth century. Bernstein (1904), Picard (1905), Lichtenstein (1912, 1924) then obtained various results by difficult means, as well as use of regularity conditions for the coefficients of the highest order terms. It was Hopf's genius to see that a "gänzlich elementare Begründen" could be given. The comparison technique he invented for this purpose is essentially so transparent that it has generated an enormous number of important applications in many further directions.

Here is Hopf's theorem in its main form:

Let $u(x)$, $x = (x_1, \ldots, x_n)$, be a C^2 function which satisfies the differential inequality

$$Lu \equiv \sum_{i,j} a_{ij} \frac{\partial^2 u}{\partial x_i \partial x_j} + \sum_i b_i \frac{\partial u}{\partial x_i} \geq 0$$

in a domain Ω, where the (symmetric) matrix $a_{ij} = a_{ij}(x)$ is locally uniformly positive definite in Ω and the coefficients a_{ij}, $b_i = b_i(x)$ are locally bounded.

If u takes a maximum value M in Ω, then $u \equiv M$ in Ω.

Hopf's proof (Section I), now a classic of the subject, is reproduced in the monographs of Protter and Weinberger (1967) and Gilbarg and Trudinger (1978), and in many other texts as well, particularly the second volume of Courant and Hilbert (1937 and 1962). [The hypothesis that u is of class C^2 is essential for the theorem, though not always strictly noted in presentations of the result. For maximum principles when u is not of class C^2, and even possibly only measurable, see e.g. Littman (1959).]

Hopf next observes (Section II) that one can allow the coefficients to depend on the solution u itself, provided that when they are evaluated along a solution the resulting functions $a_{ij}(x), b_i(x)$ satisfy the conditions of the main theorem. This allows him to deal explicitly with nonlinear as well as linear equations, a point not always noted by later authors.

In the same section he then notices two important corollaries (Sätze 2,3) dealing with the differential inquality $Lu + cu \geq 0$: First, for the case $c = c(x) \leq 0$ and a positive maximum, and second, when there is an extremum $M = 0$ with c being bounded but not necessarily non-positive.[1]

Because Hopf's formulation of these results is somewhat obscure, the main conclusions are worth restating here, which we do in terms of the operator L.

Theorem 1. *Let $u(x)$ be a C^2 function satisfying the differential inequality*

$$Lu + cu \geq 0 \quad (\leq 0)$$

in a domain Ω, where the coefficients of L satisfy the previous conditions, and $c = c(x)$ is a non-positive function on Ω. If u takes a positive maximum (negative minimum) value M in Ω, then $u \equiv M$.

Theorem 2. *Let the hypotheses of Theorem 1 hold, except that one now assumes only that the function c is locally bounded. If u takes on a vanishing maximum (minimum) value $M = 0$ in Ω, then $u \equiv 0$.*

The real depth of Hopf's nonlinear analysis shows up only in Section III, though the presentation is seriously obscured by the restriction to exact equations, as well as to the case where one of the solutions in question is assumed to vanish identically ("engere Voraussetzungen" according to Hopf). Accordingly we shall again restate the results, in slightly greater generality and in more usual notation.

Theorem 3. (Touching Lemma). *Let u, v be $C^2(\Omega)$ solutions of the nonlinear differential inequalities*

$$F(x, u, Du, D^2u) \geq 0 \qquad F(x, v, Dv, D^2v) \leq 0,$$

where F is of class C^1 in the variables u, Du, D^2u (notation obvious). Suppose also that the matrix

$$Q_{ij} \equiv \frac{\partial F}{\partial(D^2_{ij}u)}(x, u, Du, \theta D^2u + (1-\theta)D^2v)$$

is locally uniformly positive definite in Ω for all $\theta \in [0, 1]$. If $u \leq v$ in Ω and $u = v$ at some point x_0 in Ω, then $u \equiv v$ in Ω.

The terms u, Du in Q can be replaced by v, Dv.

Proof. Essentially following Hopf's proof of Satz $3'$, we write

[1] The latter possibility is not mentioned by Gilbarg and Trudinger. Moreover, Courant and Hilbert in their formulation of Satz 2 do not include the crucial restriction to a positive maximum.

$$\begin{aligned}
0 &\geq F(x,v,Dv,D^2v) - F(x,u,Du,D^2u) \\
&= F(x,u,Du,D^2v) - F(x,u,Du,D^2u) \\
&\quad +F(x,u,Dv,D^2v) - F(x,u,Du,D^2v) \\
&\quad +F(x,v,Dv,D^2v) - F(x,u,Dv,D^2v) \\
&= \sum a_{ij}D^2_{ij}(v-u) + \sum b_i D_i(v-u) + c(v-u) \\
&= L(v-u) + c(v-u),
\end{aligned}$$

where, for some values θ, θ_1, $\theta_3 \in [0,1]$ we have

$$\begin{aligned}
a_{ij} &= \frac{\partial F}{\partial (D^2_{i,j}u})(x,u,Du,\theta D^2v + (1-\theta)D^2u) = Q_{ij} \\
b_i &= \frac{\partial F}{\partial D_i u}(x,v,\theta_1 Dv + (1-\theta_1)Du, D^2u) \\
c &= \frac{\partial F}{\partial u}(x,\theta_2 v + (1-\theta_2)u, Dv, D^2v).
\end{aligned}$$

Clearly a_{ij}, b_i, c are locally bounded while a_{ij} is locally uniformly positive definite on Ω. Since by assumption $v - u \geq 0$ and $(v-u)(x_0) = 0$, it now follows from Theorem 2 that $v \equiv u$ in Ω.

To obtain the final conclusion of the theorem, one proceeds in the same way, though starting from the alternative decomposition

$$\begin{aligned}
0 &\geq F(x,v,Dv,D^2v) - F(x,u,Du,D^2u) \\
&= F(x,v,Dv,D^2v) - F(x,v,Dv,D^2u) \\
&\quad +F(x,v,Dv,D^2u) - F(x,v,Du,D^2u) \\
&\quad +F(x,v,Du,D^2u) - F(x,u,Du,D^2u).
\end{aligned}$$

The next result (essentially Satz $2'$ in a more general context and formulation) is stated as a comparison result, rather than a maximum principle, this being the underlying content of Hopf's result.

Theorem 4. (Comparison Lemma). *Let u, v be $C^2(\Omega)\cap C(\bar{\Omega})$ solutions of the nonlinear differential inequalities given in Theorem 3. Suppose that the matrix $Q = Q_{i,j}$ is positive definite in Ω and that*

$$\Psi = \frac{\partial F}{\partial u}(x,w,Dv,D^2v) \leq 0$$

for all functions $w \geq v$ (or simply for all functions w on Ω). Suppose $u \leq v$ on $\partial\Omega$. Then $u \leq v$ in Ω.

The terms u, Du in Q can be replaced by v, Dv if at the same time the terms Dv, D^2v in Ψ are replaced by Du, D^2u and the condition $w \geq v$ is replaced by $w \leq u$.

Proof. Suppose for contradiction that the conclusion $v - u \geq 0$ in Ω fails.

Then there will be a subdomain Ω' of Ω in which $v - u \leq 0$ but is not identically constant, and in which also $v - u$ takes on a negative minimum M. As in the proof of Theorem 3 one obtains

$$L(v-u)+c(v-u)\leq 0,$$

while by hypothesis $c \leq 0$ in Ω'. Hence by Theorem 1 we get $v-u\equiv M$ in Ω', a contradiction.

The final conclusion is obtained from the alternative decomposition in the proof of Theorem 3.

Using other decompositions, one can obtain various related results, e.g. Protter and Weinberger, Theorem 31 of Chapter 2.

A direct consequence of Theorem 4 is a uniqueness theorem for the Dirichlet problem for the nonlinear equation $F(x,u,Du,D^2u)=0$, a fact mentioned by Hopf in his final paragraph, though not explicitly formulated by him. Since the result is important, and a precise formulation in fact not immediate from Hopf's analysis, it is worth stating a definite result here.

Theorem 5. *Let u and v be C^2 solutions of the nonlinear equation*

$$F(x,u,Du,D^2u)=0$$

in a domain Ω, with $u=v$ on $\partial\Omega$. Suppose Q is positive definite in Ω for all $\theta\in[0,1]$, and $\Psi\leq 0$ in Ω for all functions w. Then $u\equiv v$.

This is an immediate corollary of Theorem 4, the main result being used to establish that $u\leq v$ and the final part of the theorem to get $v\leq u$. Here it is crucial that $\Psi\leq 0$ *for all functions w*.

It is surprising that the matrix Q in the hypothesis of Theorem 5 is, insofar as its second and third arguments are concerned, to be evaluated solely on the functions u and Du, *without any symmetric reference to v and Dv.*

Indeed specializing Theorem 5 to quasilinear equations, we find that for the equation

$$A_{ij}(x,Du)D^2_{ij}u+B(x,u,Du)=0$$

a sufficient condition for uniqueness is that the matrix $Q_{ij}=A_{ij}(x,Du)$ *needs to be positive definite (i.e. the equation needs to be elliptic) only when evaluated for either one (!) of the solutions u or v*, provided that $B(x,u,p)$ is, say, a non-increasing function of u for arbitrary arguments x,p. As far as the present author is aware, this last result (essentially due to Hopf, though not explicitly mentioned or stated by him) has not been previously noted.

This result applies at once to the quasilinear operator

$$F=(1+|\nabla u|^2)\Delta u-\sum\frac{\partial u}{\partial x_i}\frac{\partial u}{\partial x_j}\frac{\partial^2 u}{\partial x_i\partial x_j}$$

(mean curvature), since clearly

$$Q_{i.j}=(1+|\nabla u|^2)I_{i.j}-D_iuD_ju$$

is positive definite for all values of its arguments. Here of course there is no need to use the full strength of Theorem 5. On the other hand, if we consider the Dirichlet problem

$$(1+|\nabla u|^2)\Delta u-2\sum\frac{\partial u}{\partial x_i}\frac{\partial u}{\partial x_j}\frac{\partial^2 u}{\partial x_i\partial x_j}=0$$

in Ω, with $u = 0$ on $\partial\Omega$, then the matrix Q is *not* positive definite for arbitrary arguments D^2u. Nevertheless $Q = I$ for the function $u \equiv 0$, whence it follows that this function is the *unique* solution of the stated Dirichlet problem. (I wish to thank Prof. Weinberger for noting this example.)

A second and more subtle example, worth mentioning in the immediate context of Hopf's results, is the elementary Monge-Ampere equation

$$\frac{\partial^2 u}{\partial x^2}\frac{\partial^2 u}{\partial y^2} - \left(\frac{\partial^2 u}{\partial x \partial y}\right)^2 = f(x, y).$$

Here one checks that

$$Q_{i.j}\xi_i\xi_j = \frac{\partial^2 u}{\partial y^2}\xi_1^2 - 2\frac{\partial^2 u}{\partial x \partial y}\xi_1\xi_2 + \frac{\partial^2 u}{\partial x^2}\xi_2^2.$$

The discriminant of Q is then

$$\frac{\partial^2 u}{\partial y^2}\frac{\partial^2 u}{\partial x^2} - \left(\frac{\partial^2 u}{\partial x \partial y}\right)^2,$$

which is precisely $f(x, y)$ when evaluated at a solution u.

Suppose in particular that $f > 0$. It is easy to see then, that *any solution u is either everywhere strictly convex or everywhere strictly concave.* From this, one can check without difficulty that *if u and v are two convex solutions then Q is positive definite for the arguments $D^2_{i.j}(\theta u + (1 - \theta)v)$.*

Hence the Dirichlet problem for the elementary Monge-Ampere equation above has at most one convex solution. On the other hand, if u and v are concave solutions, then $-u$ and $-v$ are convex solutions and so, similarly, the Dirichlet problem can have at most one concave solution; altogether then *the problem can have at most two solutions.* This result is a special case of a theorem of Rellich (1933); see Courant and Hilbert (1962), page 324.

Other related maximum and comparison principles are discussed in Protter and Weinberger, Notes to Chapter 2, and in Gilbarg and Trudinger, Chapter 9, to which the reader is strongly referred; see also the references cited in Walter (1967), page 314.

Hopf's proof technique, as noted above, leads to other results of fundamental interest, particularly the celebrated Boundary Point Lemma (see the following paper of Eberhard Hopf in this selection) and a Harnack principle for elliptic equations having two independent variables; for this last result, see a paper of the present author (1956), reproduced by both Protter and Weinberger and Gilbarg and Trudinger. A nonlinear version of the Harnack principle in two variables has also been given recently by P. Pucci and the present author (2001).

Still another application of the comparison method, to the Phragmén-Lindelöf theorem, was given by Gilbarg(1954) and Hopf (1954).

References

BERNSTEIN, S.N., "Sur la généralisation du problème de Dirichlet," *Math Annalen*, **69** (1910), pp. 82-136.

COURANT, R., AND D. HILBERT, *Methods of Mathematical Physics*, volume 2. New York: Interscience Publishers, Inc., 1962.

GAUSS, C.F., *Allgemeine Theorie des Erdmagnetismus. Beobachtungen des magnetischen Vereins im Jahre, 1838.* Leipzig (1839). Also in Werke (collected works), **5**, p. 129.

GILBARG, D., AND N. TRUDINGER, *Elliptic Partial Differential Equations of Second Order.* Berlin: Springer-Verlag, (1983).

GILBARG, D., "The Phragman-Lindelöf theorem for elliptic ; artial differential equations," *J. Rational Mech. Anal.* (1952), pp. 411-417.

HOPF, E., "Remarks on the preceding paper by D. Gilba·g," *J. Rational Mech. Anal.* **1** (1952), pp. 419-424.

LICHTENSTEIN, L., "Beiträge zur Theorie der linearen partiellen Differentialgleichungen zweiter Ordnung vom elliptischen Typus." *Rend. Circ. Mat. Palermo,* **33** (1912), pp. 201-211.

LICHTENSTEIN, L., "Neue Beiträge zur Theorie der linearen partiellen Differentialgleichungen zweiter Ordnung vom elliptischen Typus." *Math. Zeitschrift,* **20** (1924), pp. 194-212.

LITTMAN, W., "A strong maximum principle for weakly L-subharmonic functions." *J. Math. Mech.* **8** (1959), pp. 761-770.

PICARD, E., *Traité d'Analyse.* Paris: Gauthier-Villars, 1905, volume 2.

PROTTER, M.H., AND H. F. WEINBERGER. *Maximum Principles in Differential Equations.* New York: Prentice-Hall, Inc., 1967.

PUCCI, P., AND J. SERRIN, "The Harnack inequality in R^2 for quasilinear elliptic equations." *J. d'Analyse Mat'ematique,* **85** (2001), pp. 307-321.

RELLICH, E., "Zur ersten Randwertaufgabe bei Monge-Ampérschen Differentialgleichungen vom elliptischen Typus." *Math. Annalen,* **107** (1933), pp.505-513.

WALTER, W., *Differential and Integral Inequalities.* Berlin: Springer-Verlag, 1964, 1970.

Reprinted from the
PROCEEDINGS OF THE AMERICAN MATHEMATICAL SOCIETY
Vol. 3, No. 5, pp. 791-793
October, 1952

A REMARK ON LINEAR ELLIPTIC DIFFERENTIAL EQUATIONS OF SECOND ORDER

EBERHARD HOPF

Consider a linear partial differential expression

$$L(u) = \sum_{i,k} a_{ik}(x) \frac{\partial u}{\partial x_i \partial x_k} + \sum_i b_i(x) \frac{\partial u}{\partial x_i}$$

with no term $c(x)u$. The coefficients a_{ik} and b_i are suppose to be continuous in an open connected set R of x-space, $x=(x_1, \cdots, x_n)$. Let x^0 denote a point on the boundary of R which has the property that R contains the interior of a hypersphere $|x-x^*|<r_0$ with x^0 on its boundary. Suppose that the coefficients are continuous at $x=x^0$ also. Let, finally, L be elliptic in $R+x^0$ such that the quadratic form

$$\sum a_{ik}(x)\lambda_i\lambda_k$$

is positive definite in each point of $R+x^0$.

This note contains a simple proof of the following:

THEOREM. *Suppose that $u=u(x)$ is of class C'' in R and that $u \geqq 0$, $L(u) \leqq 0$ in R. If the limit value of u at $x=x^0$ is zero, then either the normal derivative du/dn at $x=x^0$, understood as the limit inferior of $\Delta u/\Delta n$, is >0 or $u\equiv 0$ in R.*

Special cases of the theorem have been known for a long time. It contains, in particular, the fact that Green's function of L has a positive normal derivative along the boundary if the boundary is sufficiently smooth.

To prove the theorem we note first that $u \geqq 0$ in R and $u(x^0)=0$ trivially implies $du/dn \geqq 0$. The hypotheses that $u \geqq 0$ in R and $L(u) \leqq 0$ in R imply that either $u>0$ in R or $u\equiv 0$ in R. This follows from

Received by the editors February 18, 1952.

the sharp maximum-minimum-theorem.[1] It suffices to prove that $du/dn>0$ at x^0 if $u>0$ in R. Consider the sphere mentioned in the hypothesis. It may be assumed that its boundary has no other point in common with the boundary of R than x^0. Otherwise a second sphere which is internally tangent to the first one at $x=x^0$ would satisfy this condition. We choose its center as origin of the coordinate-system and we set $r=|x|$. Consider the closed spherical shell S: $r_0/2 \leqq r \leqq r_0$ where r_0 denotes the radius of the sphere.[2] Obviously u is continuous in S, and

$$\text{(1)} \qquad \begin{aligned} u &\geqq 0 \quad \text{on} \quad r = r_0, \\ u &= 0 \quad \text{at the point } x^0 \text{ of} \quad r = r_0, \\ u &> 0 \quad \text{on} \quad r = r_0/2. \end{aligned}$$

In my proof of the extremum-theorem I considered the auxiliary function

$$h(x) = e^{-ar^2} - e^{-ar_0^2}.$$

It has the property that $h>0$, $r<r_0$, and that

$$\text{(2)} \qquad L(h) > 0, \qquad r_0/2 < r < r_0,$$

if the constant a is chosen sufficiently large. The reader can easily verify this fact himself if he uses the ellipticity of L and the continuity of the coefficients in the closed region S. h is of class C'' in S, and

$$\text{(3)} \qquad h = 0 \quad \text{on} \quad r = r_0.$$

The function

$$v = u - \epsilon h, \qquad \epsilon > 0,$$

is of class C'' in the interior of S and continuous in S. Moreover, by (1) and (3),

$$\text{(4)} \qquad v \geqq 0 \quad \text{on} \quad r = r_0.$$

---

[1] E. Hopf, *Elementare Bemerkungen über die Lösungen partieller Differentialgleichungen zweiter Ordnung vom elliptischen Typus*, Sitzungsberichte der Berliner Akademie der Wissenschaften vol. 19 (1927) pp. 147–152.

[2] I owe the idea of using this type of region to my colleague D. Gilbarg who used it in a special case in order to prove the uniqueness of free boundary flow under more general conditions than considered hitherto. He considers a special differential equation $L(u)=0$ and uses a special solution h, $L(h)=0$, as an auxiliary function. See his paper *Uniqueness of axially symmetric flows with free boundaries*, Journal of Rational Mechanics and Analysis vol. 1 (1952) pp. 309–320, in particular pp. 314–315.

If the constant ϵ is chosen sufficiently small, then, by the third property (1),

$$v \geqq 0 \quad \text{also} \quad \text{on} \quad r = r_0/2. \tag{5}$$

By hypothesis, $L(u) \leqq 0$ in S, and by (2),

$$L(v) < 0, \qquad r_0/2 < r < r_0. \tag{6}$$

(4), (5), and (6) imply that $v \geqq 0$ holds in the whole of S. This follows again from the sharp extremum theorem or, this time more simply, from the more elementary fact that v cannot have a negative minimum in the interior of S. But $v \geqq 0$ in S and $v = 0$ at $x = x^0$ (see (3) and the second property (1)) imply that

$$\frac{dv}{dn} = \frac{du}{dn} - \epsilon \frac{dh}{dn} \geqq 0.$$

dh/dn is evidently >0. Hence $du/dn > 0$, q.e.d.

INDIANA UNIVERSITY

A Remark on Linear Elliptic Differential Equations of Second Order, *Proc. Amer. Math. Soc.* **3 (1952), 791-793.**

Commentary

James B. Serrin

In 1952 Eberhard Hopf observed that the comparison method used in Section I of his 1927 paper enables one similarly – though independent of the Maximum Principle itself – to prove the following Boundary Point Lemma:

Let $u(x)$ be a C^2 solution of the differential inequality $Lu \geq 0$ in a domain Ω, where the linear operator L satisfies the conditions stated in the preceding commentary for Hopf's 1927 maximum principle paper. Let x_0 be a point of $\partial\Omega$, which is also on the boundary of an open ball K contained in Ω

Assume finally that u is continous in $\Omega \cup x_0$ and has an inward directional derivative $\partial u/\partial \nu$ at x_0 in some direction ν pointing into K. Then if $u < M$ in Ω and $u = M$ at x_0, it follows that $\partial u/\partial \nu < 0$.

Several remarks need to be added. First, assuming the conclusion of the Boundary Point Lemma, one may use this result to obtain a direct proof of the Maximum Principle itself (see Protter and Weinberger, or Gilbarg and Trudinger).

Second, by using the Maximum Principle, one can get the following generalization of the Boundary Point Lemma:

If the condition $u < M$ in Ω is replaced by $u \leq M$ in Ω, then the conclusion is that either

$$u \equiv M \text{ in } \Omega \qquad \text{or} \qquad \partial u/\partial \nu < 0. \tag{0.1}$$

Third, it is obviously not necessary to have explicit directional derivatives at x_0, appropriate difference quotients can equally suffice.

Clearly, the Boundary Point Lemma can be used to develop nonlinear results as in Hopf's 1927 paper, and so to obtain comparison and uniqueness theorems for nonlinear equations with nonlinear boundary conditions (involving only first order derivatives). Here Neumann and Robin boundary conditions are obvious special cases. This leaves the Boundary Point Lemma standing second only in importance to the Maximum Principle itself, and shows Hopf once more standing at the forefront of creative mathematics.

Finally, it should be noted that the Boundary Point Lemma was independently obtained by the great Russian mathematician Olga Oleinik, also in 1952, her proof likewise being based on Hopf's comparison method.

REFERENCES

GILBARG, D., AND N. TRUDINGER, *Elliptic Partial Differential Equations of Second Order.* Berlin: Springer-Verlag, 1983.

HOPF, E., "Elementare Bemerkungen über die Lösungen partieller Differentialgleichungen zweiter Ordnung vom elliptischen Typus," *Sitzungsber, d. Preuss. Akad. d. Wiss.*, **19** (1927), pp. 147-152.

OLEINIK, O.A., "On properties of some boundary problems for equations of elliptic type." *Math. Sbornik*, N.S. 30 (72) (1952), pp. 695-702.

PROTTER, M.H., AND H. F. WEINBERGER. *Maximum Principles in Differential Equations.* New York: Prentice-Hall, Inc., 1967.

Zum analytischen Charakter der Lösungen regulärer zweidimensionaler Variationsprobleme.

Von

Eberhard Hopf in Berlin-Dahlem.

Die zweimal stetig differenzierbaren Lösungen $u = u(x, y)$ des regulären zweidimensionalen Variationsproblems

$$(1) \qquad \delta \iint_B F(x, y, u, p, q)\, dx\, dy = 0; \quad F''_{pp} F''_{qq} - (F''_{pq})^2 > 0$$

sind bei analytisch von x, y, u, p, q abhängendem F bekanntlich analytische Funktionen von x, y (Hilbertsches Theorem); denn unter diesen Voraussetzungen genügt $u(x, y)$ der analytischen elliptischen Differentialgleichung zweiter Ordnung

$$\frac{d}{dx} F'_p + \frac{d}{dy} F'_q - F'_u \equiv F''_{pp} u''_{xx} + 2 F''_{pq} u''_{xy} + F''_{qq} u''_{yy} + \ldots = 0$$

(Lagrangesche Gleichung des Variationsproblems). Wie Herr L. Lichtenstein [1]) gezeigt hat, sind deren zweimal stetig derivierbare Lösungen notwendig dreimal stetig derivierbar, somit nach dem Satz vom analytischen Charakter der (dreimal stetig derivierbaren) Lösungen [2]) elliptischer Differentialgleichungen notwendig analytisch. Die Voraussetzung der zweimaligen stetigen Derivierbarkeit kann durch geringere ersetzt werden. Ich beweise im folgenden auf eine Anregung von Herrn Lichtenstein hin den schärferen

Satz. *Jede Lösung $u = u(x, y)$ von (1), deren erste Ableitungen p, q einer Hölder-Bedingung genügen, ist analytisch,*

---

[1]) Vgl. L. Lichtenstein, Über den analytischen Charakter der Lösungen regulärer zweidimensionaler Variationsprobleme, Bull. de l'Ac. des Science de Cracovie 1913, S. 915–941.

[2]) Vgl. den Encyklopädieartikel von Herrn L. Lichtenstein, Neuere Entwicklung der Theorie partieller Differentialgleichungen zweiter Ordnung vom elliptischen Typus (II C 12), S. 1320–1324.

indem ich zeige, daß jede derartige Lösung zweimal stetig differenzierbar ist. In speziellen Fällen reicht die Voraussetzung der bloßen Stetigkeit von p, q aus, etwa wenn F ein Polynom zweiten Grades in p, q ist, oder, wie Herr T. Radó gezeigt hat, für $F = \sqrt{1+p^2+q^2}$. [3])

Es bedeutet eine für die folgenden Überlegungen nur unwesentliche Einschränkung, wenn aus Durchsichtigkeitsgründen F zum Beweise als unabhängig von x, y, u angenommen wird. Jede einmal stetig derivierbare Lösung $u(x, y)$ von

$$\text{(1a)} \qquad \delta \iint_B F(p, q)\, dx\, dy = 0; \quad F''_{pp} F''_{qq} - (F''_{pq})^2 > 0$$

genügt dann bei passender Hilfsfunktion $v(x, y)$ im Innern des Bereiches B dem System simultaner elliptischer Differentialgleichungen erster Ordnung

$$\text{(1b)} \qquad v'_x = -F'_q(u'_x, u'_y), \qquad v'_y = F'_p(u'_x, u'_y), \text{ [4])}$$

das bei zweimaliger Derivierbarkeit von u identisch mit der Lagrange-Gleichung zweiter Ordnung ist. Es genügt also, den obigen Satz für die Lösungen u, v von (1b) zu beweisen.

Es darf vielleicht noch bemerkt werden, daß die im folgenden entwickelte Methode — auf sie hoffe ich bei späterer Gelegenheit zurückzukommen — von den Existenzsätzen der Theorie der elliptischen Differentialgleichungen (Grundlösung, Greensche Funktion) keinerlei Gebrauch macht.

§ 1.

Systeme linearer elliptischer Differentialgleichungen erster Ordnung. Integralgleichungen für die ersten Ableitungen der Lösungen.

$$L(u, v) \equiv a(x, y)\, u'_x + b(x, y)\, u'_y - v'_y,$$
$$M(u, v) \equiv b(x, y)\, u'_x + c(x, y)\, u_y + v'_x,$$
$$a(x, y)\, c(x, y) - b^2(x, y) > 0, \; a(x, y) > 0$$

sei ein *elliptisches* System linearer Differentialausdrücke erster Ordnung Unser Ziel ist die Aufstellung gewisser Integralgleichungen für die ersten Ableitungen u'_x, u'_y der Lösungen u, v des elliptischen Gleichungssystems

$$L(u, v) = f, \qquad M(u, v) = g.$$

---

[3]) Vgl. loc. cit [1]), S. 936–941. Für die Analytizität der mit einer stetigen Normale versehenen Minimalflächen gibt Herr T. Radó einen kurzen Beweis; vgl. Math. Zeitschr. **24** (1925), S. 321–327.

[4]) Vgl. A. Haar, Über die Variation der Doppelintegrale, Crelles Journal **149** (1919), S. 1–18.

1. Wir betrachten vorerst den Fall **konstanter** Koeffizienten a, b, c. Setzt man dann

$$R(u,v) = a u'_x + b u'_y + b v'_x + c v'_y$$
$$S(u,v) = v'_x - u'_y,$$

so gilt, wie leicht zu sehen, die (Greensche) Identität

$$(2)\quad \begin{aligned} &\iint\limits_B [u^* L(u,v) + v^* M(u,v) + u R(u^*,v^*) + v S(u^*,v^*)]\, dx\, dy \\ &= \int\limits_C [v u^* - u(b u^* + c v^*)]\, dx + [v v^* + u(a u^* + b v^*)]\, dy \end{aligned}$$

für vier willkürliche, im Bereiche B und auf seinem (positiv durchlaufenen) Rande C einmal stetig derivierbare Funktionen u, v, u^*, v^*.

Die Funktionen

$$\Omega(x,y;\xi,\eta) = a(y-\eta)^2 - 2b(y-\eta)(x-\xi) + c(x-\xi)^2$$
$$\Gamma(x,y;\xi,\eta) = -\frac{\log \Omega(x,y;\xi,\eta)}{4\pi\sqrt{ac-b^2}}$$

besitzen bekanntlich wegen des elliptischen Charakters folgende einfache Eigenschaften. Die quadratische Form Ω ist positiv definit; bei passender Konstante $\omega > 0$ ist daher $\Omega \geqq \omega r^2$, [5])

$$r^2 = (x-\xi)^2 + (y-\eta)^2.$$

Wie leicht zu sehen, gelten Ungleichungen der Form

$$(3)\qquad |\Gamma'_\xi|, |\Gamma'_\eta| < \gamma_1 r^{-1}; \quad |\Gamma''_{\xi\xi}|, |\Gamma''_{\xi\eta}|, |\Gamma''_{\eta\eta}| < \gamma_2 r^{-2}. \text{ [5])}$$

Ferner ist

$$(3\text{a})\quad R(\Gamma'_\xi, \Gamma'_\eta) = a\Gamma''_{\xi\xi} + 2b\Gamma''_{\xi\eta} + c\Gamma''_{\eta\eta} = 0,\quad S(\Gamma'_\xi, \Gamma'_\eta) = \Gamma''_{\xi\eta} - \Gamma''_{\xi\eta} = 0.$$

Für jedes im Bereiche B und auf dessen Rande C einmal stetig derivierbares Funktionenpaar u, v gilt in jedem inneren Punkte (x, y) von B die Integralformel

$$(4)\quad u(x,y) = \begin{cases} \iint\limits_B [L(u,v)\Gamma'_\xi + M(u,v)\Gamma'_\eta]\, d\xi\, d\eta \\ -\int\limits_C [v\Gamma'_\xi - u(b\Gamma'_\xi + c\Gamma'_\eta)]\, d\xi + [v\Gamma'_\eta + u(a\Gamma'_\xi + b\Gamma'_\eta)]\, d\eta. \end{cases}$$

(Im Integranden ist $u = u(\xi,\eta)$, $v = v(\xi,\eta)$ zu setzen.) Man weiß, daß das Bereichintegral absolut konvergiert, da der Integrand wegen (3) an

[5]) Bekanntlich ist ω gleich der kleinsten Wurzel der charakteristischen Gleichung $\begin{vmatrix} a-\omega & b \\ b & c-\omega \end{vmatrix} = 0$; sie ist sicher $\geqq \frac{ac-b^2}{a+c}$. Betreffs der Konstanten γ_1 und γ_2 ist für später die Bemerkung von Wichtigkeit, daß sie nur von einer oberen Schranke für a, $|b|$, c und einer unteren Schranke für die Diskriminante $ac - b^2$ abhängen.

der singulären Stelle $\xi = x$, $\eta = y$ höchstens von erster Ordnung unendlich wird. Die Formel (4) ergibt sich in geläufiger Weise, indem man (2) — man ersetze dort die Buchstaben x, y durch ξ, η — mit

$$u^*(\xi, \eta) = -\Gamma'_\xi, \qquad v^* = -\Gamma'_\eta$$

auf den Bereich $B - (\varrho)$ anwendet, unter (ϱ) das Innere der die singuläre Stelle (x, y) ausschließenden Ellipse $\Omega = \varrho^2$ verstanden und nachträglich diese Ellipse auf den Punkt (x, y) zusammenzieht $(\varrho \to 0)$.

2. Wir gehen nun zum allgemeinen Fall variabler Koeffizienten über. Die Elliptizitätsbedingung $a(x, y)\, c(x, y) - b^2(x, y) > 0$ und $a(x, y) > 0$ sei in einem Bereiche B und auf seinem Rande C erfüllt. Wir verlangen im folgenden nur, daß a, b, c, f, g in B einer Hölder-Bedingung, etwa vom Exponenten α, $0 < \alpha \leqq 1$,

$$|a(x, y) - a(\xi, \eta)| \leqq \text{konst.}\, r^\alpha, \ldots, \qquad 0 < \alpha \leqq 1,$$

genügen.

Wir schreiben dann das Gleichungssystem

$$L(u, v) = f, \qquad M(u, v) = g$$

in der Form[5a)]

$$L_0(u, v) = a_0 u'_x + b_0 u'_y - v'_y = f - (a - a_0) u'_x - (b - b_0) u'_y,$$

$$M_0(u, v) = b_0 u'_x + c_0 u'_y + v'_x = g - (b - b_0) u'_x - (c - c_0) u'_y,$$

$$a_0 = a(x_0, y_0), \ldots,$$

unter (x_0, y_0) einen inneren Punkt von B verstanden, und wenden hierauf die Formel (4) mit $a = a_0, \ldots, L = L_0$, $M = M_0$ an. Dann ergibt sich

$$u(x, y) = \iint_B (f\Gamma'_\xi + g\Gamma'_\eta)\, d\xi\, d\eta$$

$$(5) \quad - \iint_B \{[(a - a_0)\Gamma'_\xi + (b - b_0)\Gamma'_\eta]\, u'_\xi + [(b - b_0)\Gamma'_\xi + (c - c_0)\Gamma'_\eta]\, u'_\eta\}\, d\xi\, d\eta$$

$$- \int_C [v\Gamma'_\xi - u(b_0\Gamma'_\xi + c_0\Gamma'_\eta)]\, d\xi + [v\Gamma'_\eta + u(a_0\Gamma'_\xi + b_0\Gamma'_\eta)]\, d\eta\,.$$

Unser Ziel ist hier die Bildung der ersten Ableitungen u'_x, u'_x. Daß beim Randintegral in jedem inneren Punkte (x, y) von B unter dem Integralzeichen differenziert werden darf, ist selbstverständlich. Betreffs der Bereichintegrale bemerken wir zunächst, daß

$$\iint_B (f\Gamma'_\xi + g\Gamma'_\eta)\, d\xi\, d\eta$$

$$(5\text{a}) \quad = \iint_B [(f - f_0)\Gamma'_\xi + (g - g_0)\Gamma'_\eta]\, d\xi\, d\eta + f_0 \int_C \Gamma\, d\eta - g_0 \int_C \Gamma\, d\xi\,,$$

$$f_0 = f(x_0, y_0), \qquad g_0 = g(x_0, y_0)$$

5a) Einer ähnlichen Zerlegung hatte sich bei anderer Gelegenheit bereits Herr A. Korn bedient. Vgl. Schwarz-Festschrift 1914, S. 215–229.

ist. Die einzelnen Terme in den Integranden der Bereichintegrale sind nunmehr im wesentlichen von gleicher Gestalt. Differentiation unter dem Integralzeichen wird hier im allgemeinen nicht gestattet sein, da dann die Integranden, z. B. $[a(\xi,\eta)-a(x_0,y_0)]\,u'_\xi\,\Gamma''_{\xi x}$, für $\xi=x$, $\eta=y$ im allgemeinen von zweiter Ordnung unendlich werden. Jedoch ist sie an der Stelle $x=x_0$, $y=y_0$ erlaubt, da dann nach der Differentiation der Integrand als Funktion von x, y, ξ, η für $\xi=x$, $\eta=y$ höchstens von $(2-\alpha)$-ter Ordnung unendlich wird (a, b, c, f, g sind Hölder-stetig!) [6]. Führen wir die Differentiation $\left.\frac{\partial}{\partial x}\right|_{\substack{x=x_0\\ y=y_0}}$ aus, so erhalten wir mit Rücksicht auf die Beziehungen $\Gamma'_x=-\Gamma'_\xi$, $\Gamma'_{\xi x}=-\Gamma'_{\xi\xi}$ usw. aus (5) und (5a) — wir gestatten uns, statt x_0, y_0 nachträglich wieder x, y zu schreiben — die Relationen

$$(6)\quad \left\{\begin{aligned} &\underline{u'_x(x,y)=B_1(x,y)+B_2(x,y)+C(x,y),}\\ &B_1(x,y)=\iint_B[u'_\xi H(x,y;\xi,\eta)+u'_\eta G(x,y;\xi,\eta)]\,d\xi\,d\eta,\\ &B_2(x,y)=-\iint_B\{[f(\xi,\eta)-f(x,y)]\,\Gamma''_{\xi\xi}+[g(\xi,\eta)-g(x,y)]\,\Gamma''_{\xi\eta}\}\,d\xi\,d\eta,\\ &C(x,y)\;=\int_C[v\Gamma''_{\xi\xi}-u(b\Gamma''_{\xi\xi}+c\Gamma''_{\xi\eta})+g\Gamma'_\xi]\,d\xi\\ &\qquad\qquad+[v\Gamma''_{\xi\eta}+u(a\Gamma''_{\xi\xi}+b\Gamma''_{\xi\eta})-f\Gamma'_\xi]\,d\eta,\ ^{7)} \end{aligned}\right.$$

mit den Abkürzungen

$$(6a)\quad \begin{aligned} H(x,y;\xi,\eta)&=[a(\xi,\eta)-a(x,y)]\,\Gamma''_{\xi\xi}+[b(\xi,\eta)-b(x,y)]\,\Gamma''_{\xi\eta},\\ G(x,y;\xi,\eta)&=[b(\xi,\eta)-b(x,y)]\,\Gamma''_{\xi\xi}+[c(\xi,\eta)-c(x,y)]\,\Gamma''_{\xi\eta}, \end{aligned}$$

$$\Gamma(x,y;\xi,\eta)=-\frac{\log[c(x,y)(x-\xi)^2-2b(x,y)(x-\xi)(y-\eta)+a(x,y)(y-\eta)^2]}{4\pi\sqrt{a(x,y)\,c(x,y)-b^2(x,y)}}.$$

Eine ganz analoge Formel ergibt sich natürlich für u'_y. Ersichtlich erhält man die Formel für u'_y, indem man in (6) und (6a) Γ'_ξ durch Γ'_η, $\Gamma''_{\xi\xi}$ durch $\Gamma''_{\xi\eta}$, $\Gamma''_{\xi\eta}$ durch $\Gamma''_{\eta\eta}$ ersetzt. Diese beiden Formeln bilden offenbar ein System von zwei linearen Integralgleichungen für die zwei Funktionen u'_x und u'_y, wenn u, v irgendwelche in B einmal stetig derivierbare Lösungen des elliptischen Systems $L(u,v)=f$, $M(u,v)=g$ bedeuten [8].

---

[6]) Natürlich ist das nur eine Plausibilitätsbetrachtung. Zur Vermeidung störender Zwischenbetrachtungen haben wir den strengen Beweis für die Derivierbarkeit unter dem $\int$-Zeichen unterdrückt; er ist indessen leicht nach potentialtheoretischem Muster zu erbringen.

[7]) In dem Randintegral ist genauer $a=a(x,y), \ldots, g=g(x,y)$ zu schreiben.

[8]) Zu diesen bemerkenswerten Integralgleichungen gibt es Analoga bei allen übrigen linearen elliptischen Problemen (Systeme höherer Ordnung, eine einzige

§ 2.

Anwendung auf die Bestimmung von Schranken für die Beträge der ersten Ableitungen der Lösungen.

Die Voraussetzungen des vorigen Paragraphen seien beibehalten. Mit $H_\alpha[a], H_\alpha[b], \ldots, H_\alpha[g]$ seien die „Hölder-Konstanten" von $a, b, \ldots, g$, d.h. die oberen Grenzen von $r^{-\alpha}|a(x,y)-a(\xi,\eta)|, \ldots, r^{-\alpha}|g(x,y)-g(\xi,\eta)|$ in B bezeichnet. Wir wenden die Integralgleichungen (6), (6a) usw. an auf den Beweis einiger Ungleichungen bezüglich der Lösungen u, v des Gleichungssystems

$$L(u,v)=f, \qquad M(u,v)=g.$$

Bezeichnet man mit $[u]$, $[v]$, $[f]$, $[g]$ die oberen Grenzen von $|u|$, $|v|$, $|f|$, $|g|$ im Bereiche B, so gilt das

Lemma. Es gibt drei positive Konstanten $\beta_1, \beta_2, \beta_3$ von der Beschaffenheit, daß für alle im Bereiche B einmal stetig derivierbaren Lösungen u, v von $L=f$, $M=g$

$$|u_x'|,\ |u_y'| \leqq \beta_1 R^{-1}([u]+[v]) + \beta_2([f]+[g]) + \beta_3 R^{\alpha}(H_\alpha[f]+H_\alpha[g])$$

bei beliebigem $R \leqq d$ gilt, unter d den Abstand des Punktes (x,y) vom Rande C von B verstanden; $\beta_1, \beta_2, \beta_3$ können so gewählt werden, daß die Ungleichungen bestehen bleiben, wenn die Koeffizientenfunktionen a, b, c in genügend engen Grenzen stetig abgeändert werden, jedoch so, daß dabei die Hölder-Konstanten $H_\alpha[a]$, $H_\alpha[b]$, $H_\alpha[c]$ unter einer festen Grenze liegen.

Beweis. Ersetzt man die Koeffizienten $a(x,y)$, $b(x,y)$, $c(x,y)$ durch irgendwelche neue, von ihnen in ganz B genügend wenig differierende, so werden aus Stetigkeitsgründen die Ungleichungen (3) mit festen Konstanten γ bestehen bleiben (man beachte die Fußnote [5])). Bleiben überdies bei der stetigen Abänderung der Koeffizienten deren Hölder-Konstanten durchweg unter einer festen Grenze, so gelten nach (6a) in B Ungleichungen der Form

$$(7) \qquad |H(x,y;\xi,\eta)|,\ |G(x,y;\xi,\eta)| < \text{konst.}\, r^{\alpha-2}.$$

Differentialgleichung beliebiger Ordnung); sie sind in jedem Falle Integralgleichungen für die höchsten vorkommenden Ableitungen. Auch der Weg, auf dem man zu ihnen gelangt, ist in allen Fällen derselbe (Übergang von konstanten zu variablen Koeffizienten, Differentiation bis zur höchsten in Frage kommenden Ordnung in der oben geschilderten Art und Weise). Sie scheinen zur Behandlung der feineren Fragen der Theorie — Eigenschaften der Lösungen unter geringsten Stetigkeitsvoraussetzungen — besonders geeignet zu sein. Ich hoffe darauf noch bei späterer Gelegenheit zurückzukommen.

Infolgedessen werden alle folgenden, lediglich auf (3) und (7) basierenden Abschätzungen gültig bleiben, wenn a, b, c den obigen Bedingungen gemäß variiert werden.

Um irgendeinen vorgegebenen inneren Punkt von B als Zentrum denke man sich eine ganz im Innern von B gelegene, sonst beliebige Kreisscheibe K_0, etwa mit dem Radius R_0. Offenbar ist das Lemma bewiesen, wenn wir die zu beweisende Ungleichung im Mittelpunkt von K_0 und mit $R = R_0$ als gültig erkannt haben. Wir beschränken uns auf die Punkte von K_0 und verstehen jetzt unter d den Abstand eines Punktes (ξ, η) von der Peripherie von K_0. Da nun u'_x und u'_y auf der abgeschlossenen Kreisscheibe K_0 stetig sind, wird das Maximum

$$(8) \qquad M = \operatorname*{Max}_{(\xi,\eta) \text{ in } K_0} [d\,|u'_\xi|,\ d\,|u'_\eta|]$$

in einem inneren Punkte (x, y) von K_0 erreicht; d_0 sei der Abstand dieses nunmehr festgehaltenen Punktes von der Peripherie. Wir betrachten eine neue Kreisscheibe $r \leqq R_1$ mit $r^2 = (\xi - x)^2 + (\eta - y)^2$, $R_1 < d_0$; sie ist offenbar ganz in K_0 gelegen. Da nun für jeden Punkt (ξ, η) dieser Kreisscheibe $R_1 + d \geqq r + d \geqq d_0$ gilt, ist wegen (8)

$$(8\text{a}) \qquad |u'_\xi|,\ |u'_\eta| \leqq \frac{M}{d_0 - R_1} \text{ für } r \leqq R_1; \quad \operatorname{Max}\{|u'_x(x, y)|,\ |u'_y(x, y)|\} = \frac{M}{d_0}.$$

Wir wenden jetzt die Integralformeln (6), (6a) und die analogen Formeln für u'_y auf die Kreisscheibe $r \leqq R_1$ an. Zunächst ergibt sich im Punkte (x, y) wegen (8a)

$$(9) \qquad M \leqq d_0 \{|B_1(x, y)| + |B_2(x, y)| + |C(x, y)|\}.$$

Für $|B_1|$ ergibt sich nun nach (7) und (8a)

$$(9\text{a}) \qquad |B_1(x, y)| \leqq \text{konst.} \frac{M}{d_0 - R_1} \cdot \iint\limits_{r \leqq R_1} r^{\alpha - 2}\, d\xi\, d\eta = \text{konst.} \frac{M R_1^\alpha}{d_0 - R_1}.$$

Weiter erhält man wegen (3) nach gleichem Muster

$$(9\text{b}) \qquad |B_2(x, y)| \leqq \text{konst.} (H_\alpha[f] + H_\alpha[g]) R_1^\alpha.$$

Was endlich das Peripherieintegral $C(x, y)$ betrifft, so ist auf dem Integrationswege $r = R_1$, $\xi = x + R_1 \cos\varphi$, $\eta = y + R_1 \sin\varphi$, somit nach (3)

$$(9\text{c}) \qquad |C(x, y)| \leqq \text{konst.} R_1^{-1}([u] + [v]) + \text{konst.} ([f] + [g]).$$

Setzt man noch $R_1 = \vartheta d_0$ $(\vartheta < 1)$, so folgt aus (9), (9a), (9b), (9c) und wegen $R_1 < d_0 \leqq R_0$

$$(10) \qquad M \leqq \text{konst.} \frac{\vartheta^\alpha}{1 - \vartheta} M + \text{konst.} R_0^{1+\alpha} (H_\alpha[f] + H_\alpha[g]) + \text{konst.} \vartheta^{-1}([u] + [v]) + \text{konst.} R_0([f] + [g]).$$

Hier kann man der Zahl ϑ einen solchen (festen) Wert erteilen, daß der erste Summand der rechten Seite $\leqq \frac{1}{2} M$ ausfällt. Durch Hinüberschaffen auf die linke Seite ergibt sich somit eine Ungleichung für M, die formal mit (10) übereinstimmt, wenn man dort rechts den ersten Summanden wegläßt. Bedenkt man endlich, daß im Mittelpunkte von K_0

$$|u'_x|, |u'_y| \leqq \frac{M}{R_0}$$

nach (8) ist, so ergibt sich in diesem Mittelpunkte die zu beweisende Ungleichung[8a]). Die schließlichen Konstanten β_ν hängen offenbar nur von den γ_ν in (3) und der konst. in (7) ab.

§ 3.

Beweis des Satzes über die Lösungen von (1).

Es sei $F(p, q)$ an der Stelle $p = p_0$, $q = q_0$ analytisch; u, v sei ein Lösungspaar des elliptischen Gleichungssystems (1b) mit $u'_x(x_0, y_0) = p_0$, $u'_y(x_0, y) = q_0$; die Funktionen u'_x, u'_y mögen in der Umgebung von x_0, y_0 einer Hölder-Bedingung mit dem Exponenten $\alpha \leqq 1$ genügen. Es ist dann zu zeigen, daß in der Umgebung dieses Punktes $u''_{xx}, u''_{xy}, u''_{yy}$ existieren und stetig sind. Es seien K_0 und K_1 zwei hinreichend kleine Kreisscheiben um x_0, y_0 als Zentrum mit

$$K_1 < K_0.$$

1. Wir zeigen zuerst, daß u'_x, u'_y in K_1 beschränkte Differenzenquotienten haben; zur Abkürzung werde

$$U_h(x, y) = \frac{u(x+h, y) - u(x, y)}{h}, \quad V_h(x, y) = \frac{v(x+h, y) - v(x, y)}{h} \tag{11}$$

gesetzt. Wir zeigen zunächst, daß U_h, V_h einem *linearen* elliptischen Gleichungssystem genügen. Es ist nämlich

$$\begin{aligned} a_h(x, y) \frac{\partial U_h}{\partial x} + b_h(x, y) \frac{\partial U_h}{\partial y} - \frac{\partial V_h}{\partial y} &= 0, \\ b_h(x, y) \frac{\partial U_h}{\partial x} + c_h(x, y) \frac{\partial U_h}{\partial y} + \frac{\partial V_h}{\partial x} &= 0, \end{aligned} \tag{12}$$

mit

$$\begin{aligned} a_h(x, y) &= \int_0^1 F''_{pp}(\vartheta p_2 + (1 - \vartheta) p_1, \vartheta q_2 + (1 - \vartheta) q_1)\, d\vartheta, \\ b_h(x, y) &= \int_0^1 F''_{pq}(\vartheta p_2 + (1 - \vartheta) p_1, \vartheta q_2 + (1 - \vartheta) q_1)\, d\vartheta, \\ c_h(x, y) &= \int_0^1 F''_{qq}(\vartheta p_2 + (1 - \vartheta) p_1, \vartheta q_2 + (1 - \vartheta) q_1)\, d\vartheta \end{aligned} \tag{12a}$$

[8a]) Bezüglich dieser Abschätzungsweise durch „Hinüberschaffen" vgl. auch M. Gevrey, Ann. de l'École Normale **35** (1918), S. 129–190, insbes. S. 145–148.

und den Abkürzungen

$$(12\text{b})\qquad \begin{aligned} p_1 &= u'_x(x, y), & p_2 &= u'_x(x+h, y), \\ q_1 &= u'_y(x, y), & q_2 &= u'_y(x+h, y). \end{aligned}$$

Zum Beweise ersetze man in (1b) x durch $x+h$ und ziehe die beiden so erhaltenen Gleichungen von den Gleichungen (1b) ab; mit Rücksicht auf die Abkürzungen (12b) ergibt sich dann

$$h\frac{\partial V_h}{\partial y} = F'_p(p_2, q_2) - F'_p(p_1, q_1) = \varphi(1) - \varphi(0) = \int_0^1 \varphi'(\vartheta)\, d\vartheta;$$

$$\varphi(\vartheta) = F'_p(\vartheta p_2 + (1-\vartheta)p_1, \vartheta q_2 + (1-\vartheta) q_1).$$

Mit Rücksicht auf die Abkürzungen (12a) ist nun

$$\int_0^1 \varphi'(\vartheta)\, d\vartheta = (p_2 - p_1) a_h + (q_2 - q_1) b_h,$$

woraus wegen $p_2 - p_1 = h\frac{\partial U_h}{\partial x}$, $q_2 - q_1 = h\frac{\partial U_h}{\partial y}$ die erste der Gleichungen (12) — die zweite ergibt sich natürlich analog — unmittelbar folgt.

Das System (12) ist formal für $h = 0$ vom elliptischen Typ im Kreise K_0. Zufolge der Voraussetzung werden die Hölder-Konstanten H_α der Funktionen a_h, b_h, c_h (bezogen auf den Bereich K_0) für alle hinreichend kleinen $|h|$ unter einer festen Schranke gelegen sein. Infolgedessen ist das Lemma von § 2 auf alle Punkte x, y des kleineren Kreises K_1 anwendbar — man nehme für R etwa die Differenz der Radien von K_0 und K_1 — mit $f = g = 0$. Dann ergibt sich alles in allem, daß u'_x und u'_y in K_1 einer Lipschitz-Bedingung genügen.

2. Wir wenden nun auf die Gleichungen (12) die Integralrelationen (6), (6a) an,

$$(13)\qquad \begin{aligned} \frac{\partial U_h}{\partial x} &= B_1(x, y; h) + C(x, y; h), \\ B_1(x, y; h) &= \iint\limits_{K_1} \left(\frac{\partial U_h}{\partial \xi} H_h + \frac{\partial U_h}{\partial \eta} G_h\right) d\xi\, d\eta, \\ C(x, y; h) &= \oint [V_h \Gamma''_{\xi\xi} - U_h(b_h \Gamma''_{\xi\xi} + c_h \Gamma''_{\xi\eta})\, d\xi \\ &\quad + [V_h \Gamma''_{\xi\eta} + U_h(a_h \Gamma''_{\xi\xi} + b_h \Gamma''_{\xi\eta})]\, d\eta, \\ \Gamma &= \Gamma_h(x, y; \xi, \eta). \end{aligned}$$

Wegen des obigen Resultates existieren nach einem bekannten Lebesgueschen Satze $u''_{\xi\xi} = \lim\limits_{h=0} \frac{\partial U_h}{\partial \xi}$, $u''_{\xi\eta} = \lim\limits_{h=0} \frac{\partial U_h}{\partial \eta}$, ... fast überall in K_1. Die zweiten Ableitungen u'' sind ferner beschränkt in K_1. Nun werden $\frac{\partial^2 \Gamma_h}{\partial \xi^2}, \frac{\partial^2 \Gamma_h}{\partial \xi \partial \eta}, \ldots, H_h, G_h$

für alle hinreichend kleinen $|h|$ festen Ungleichungen der Form (3) und (7) genügen. Infolgedessen ist der Integrand von B_1 dem Betrage nach für alle hinreichend kleinen $|h|$ unter einer festen summierbaren Funktion konst. $r^{\alpha-2}$ gelegen. Ferner konvergiert der Integrand für $h \to 0$ fast überall in K_1. Man erhält daher nach einem bekannten Lebesgueschen Satze

$$\lim_{h=0} B_1(x, y; h) = B_1(x, y; 0) = \iint_{K_1} (u''_{\xi\xi} H_0 + u''_{\xi\eta} G_0)\, d\xi\, d\eta.\ ^{9)}$$

Nach geläufigen potentialtheoretischen Überlegungen erkennt man aus dieser Integraldarstellung, daß $B_1(x, y; 0)$ in K_1 stetig ist. Daß $C(x, y; 0) = \lim_{h=0} C(x, y; h)$ existiert, ist selbstverständlich, da $\lim_{h=0} V_h = v'_\xi$, $\lim_{h=0} U_h = u'_\xi$ nach Voraussetzung vorhanden und stetig sind. Ebenso ist klar, daß das Randintegral $C(x, y; 0)$ eine in K_1 stetige Funktion ist. Somit existieren nach (13) $\lim_{h=0} \frac{\partial U_h}{\partial x} = u''_{xx}$, ebenso natürlich u''_{xy} und u''_{yy} überall in K_1, und sie sind in K_1 stetig, w. z. b. w.

Anmerkung. Will man den Beweis im allgemeineren Falle (1) durchführen, so sind geringe Modifikationen erforderlich. An die Stelle des Gleichungssystems (1b) ist dann etwa das System

$$\text{(1c)} \quad \begin{aligned} v'_x &= -F'_q(x, \ldots, u'_y), \qquad v'_y = F'_p(x, \ldots, u'_y) - \Phi(x, y), \\ \Phi(x, y) &= \int^x F'_u(x, \ldots, u'_y(x, y))\, dx \end{aligned}$$

zu setzen. Geht man hier zu den Differenzenquotienten (11) über, so erhält man statt des homogenen Gleichungssystems (12) ein inhomogenes, etwa mit den rechten Seiten f_h, g_h. Man erkennt leicht, daß f_h und g_h in K_0 einer Hölder-Bedingung genügen, und daß überdies ihre Hölder-Konstanten unter einer von h unabhängigen Schranke liegen. Bei der Anwendung der Integralgleichungen (6) mit (6a) bekommt man in (13) noch das entsprechende Glied $B_2(x, y; h)$. Analog wie bei B_1, jedoch einfacher, ergibt sich die Existenz von $\lim_{h=0} B_2 = B_2(x, y; 0)$. Ebenso ist diese letztere Funktion als stetig in K_1 zu erkennen. Alle übrigen Beweiseinzelheiten bleiben im wesentlichen unverändert.

---

[9]) Das Integral ist im Lebesgueschen Sinne zu verstehen.

(Eingegangen am 1. November 1928.)

Zum analytischen Character der Lösungen regulärer zweidimensionaler Variationsprobleme, *Math. Zeit.* **30 (1929) 404-413.**

Commentary

Hans Weinberger

This work is concerned with a solution $u(x,y)$ of the two-dimensional variational problem

$$\delta \int\int_B F(x,y,u,p,q)dxdy = 0, \tag{0.1}$$

F analytic in all its variables, $F_{pp}F_{qq} - F_{pq}^2 > 0$.

Hilbert had shown that $u \in C^3$ implies that u is analytic. Lichtenstein had improved this result to $u \in C^2$ implies $u \in C^3$ so that u is analytic. In this paper it is shown that $u \in C^{1.\alpha}$ implies $u \in C^2$, and hence u analytic.

Hopf obtained this result by writing the Euler equation $\frac{d}{dx}F_p + \frac{d}{dy}F_q = F_u$ as the system

$$v_x = -F_q \tag{0.2}$$

$$v_y = F_p - \int F_u dx, \tag{0.3}$$

and introducing some of the ideas of what became the theory of quasiconformal mappings.

A few years later C. Morrey carried the ideas of quasiconformal mapping sufficiently far to prove that u Lipschitz continuous implies Hopf's condition $u \in C^{1.\alpha}$. This permitted him to prove the existence of a solution of the two-dimensional Dirichlet problem for any quasilinear uniformly elliptic equation when the boundary values satisfy a 3-point condition.

Reference

Morrey, C., "On the solutions of quasi-linear elliptic partial differential equations" *AMS Transactions.* **43** (1938), 126-166.

ÜBER
EINE KLASSE SINGULÄRER INTEGRALGLEICHUNGEN

VON

N. WIENER UND **DR. E. HOPF**

PROFESSOR IN CAMBRIDGE (MASS.) IN BERLIN-DAHLEM

SONDERAUSGABE AUS DEN SITZUNGSBERICHTEN
DER PREUSSISCHEN AKADEMIE DER WISSENSCHAFTEN
PHYS.-MATH. KLASSE. 1931. XXXI

BERLIN 1931
VERLAG DER AKADEMIE DER WISSENSCHAFTEN
IN KOMMISSION BEI WALTER DE GRUYTER U. CO

(PREIS *R.M* 1.—)

(Vorgelegt von Hrn. Bieberbach).

Im folgenden wird die homogene lineare Integralgleichung

$$(1) \qquad f(x) = \int_0^\infty K(x-y) f(y)\, dy$$

aufgelöst, wo die Kernfunktion $K(x)$ für große $|x|$ wie eine Exponentialfunktion von $|x|$ klein wird. Der einfachste Spezialfall ist die Gleichung von LALESCO, $K(x) = e^{-|x|}$. Von besonderer Wichtigkeit für die Astronomie ist der Fall $K(x) = \frac{1}{2} Ei(|x|) = \frac{1}{2}\int_{|x|}^{\infty} \frac{e^{-t}}{t}\, dt$, die Gleichung von MILNE. Die Lösung ergibt hier die Temperaturverteilung in einer Sternatmosphäre im Strahlungsgleichgewicht[1]. Mit Hilfe der Theorie der LAPLACE-Transformationen und elementarfunktionentheoretischen Betrachtungen gewinnen wir die Hauptlösungen von (1), d. h. alle Lösungen $f(x)$, die für große x höchstens wie eine im Vergleich zum Kern schwächere Exponentialfunktion groß werden. Die Lösungen können durch explizite Integralformeln dargestellt werden.

Die analoge Integralgleichung

$$(2) \qquad f(x) = \int_{-\infty}^{+\infty} K(x-y) f(y)\, dy$$

ist viel leichter diskutierbar, ihre Lösungen sind im wesentlichen Exponentialfunktionen. Ist $u = u^*$ eine n-fache Nullstelle der Funktion

$$1 - \int_{-\infty}^{+\infty} K(t)\, e^{ut}\, dt,$$

so ist

$$Q(x)\, e^{-u^* x}$$

[1] Vgl. etwa E. HOPF, Mathematisches zur Strahlungsgleichgewichtstheorie der Fixsternatmosphären. Math. Zeitschr. 33 (1931). S. 109.

eine Lösung von (2), unter Q ein beliebiges Polynom von höchstens $(n-1)$-tem Grade verstanden (vorausgesetzt, daß die Integrale sinnvoll sind). Die Verwandtschaft von (1) und (2) äußert sich darin, daß die Lösungen von (1) sich für große x asymptotisch ebenso verhalten wie gewisse Lösungen von (2), eine Tatsache, die schon bei den beiden erwähnten Beispielen hervorgetreten war.

Auflösung der Integralgleichung (1). Die Kernfunktion $K(x)$ sei reell und bis auf endlich viele Stellen stetig. Die Funktion $K(x)e^{s|x|}$ sei für mindestens ein positives s im Intervalle $-\infty < x < \infty$ quadratisch integrierbar. Es bedeutet dann keine Einschränkung, wenn wir annehmen, daß

$$\int_{-\infty}^{+\infty} (K(x)e^{s|x|})^2 \, dx$$

für alle $s < 1$ konvergiert. Hieraus folgt, daß auch

$$\int_{-\infty}^{+\infty} |K(x)| e^{s|x|} dx$$

für $s < 1$ konvergiert, wie die Schwarzsche Ungleichung lehrt, wenn man den Integranden in der Form $\left(K(x)e^{\frac{s+1}{2}|x|}\right) \cdot e^{-\frac{1-s}{2}|x|}$ schreibt.

Im folgenden werden die Lösungen von (1) betrachtet, für welche

$$f(x) = O(e^{\alpha x}) \tag{3}$$

mit festem, sonst aber beliebigen $\alpha < 1$ gilt.

Um (1) unter diesen Voraussetzungen aufzulösen, schreiben wir (1) zunächst in der Form

$$g(x) = f(x) - \int_{-\infty}^{+\infty} K(x-y) f(y) \, dy \tag{4}$$

mit

$$f(x) = 0, \; x < 0; \qquad g(x) = 0, \; x > 0, \tag{5}$$

wobei $g(x)$ für $x < 0$ durch die rechte Seite in (4) definiert ist. Wir führen nun die Laplace-Transformierten

$$\phi(u) = \int_{-\infty}^{+\infty} f(x) e^{ux} dx, \quad \gamma(u) = \int_{-\infty}^{+\infty} g(x) e^{ux} dx, \quad \varkappa(u) = \int_{-\infty}^{+\infty} K(x) e^{ux} dx \tag{6}$$

von f, g, K ein, unter

$$u = s + it$$

eine komplexe Variable verstanden. $\frac{\phi(s+it)}{\sqrt{2\pi}}$, $\frac{\gamma(s+it)}{\sqrt{2\pi}}$, $\frac{\varkappa(s+it)}{\sqrt{2\pi}}$ sind als Funktionen von t also die Fourier-Transformierten von $f(x)e^{sx}$, $g(x)e^{sx}$, $K(x)e^{sx}$.

Wegen (5) ist

(6a) $$\phi(u) = \int_0^\infty f(x) e^{ux} dx, \quad \gamma(u) = \int_{-\infty}^0 g(x) e^{ux} dx.$$

Wegen (3) konvergiert das Integral (6a) für $\phi(u)$ absolut, solange $s < -\alpha$ ist. Somit gilt das

Lemma 1. $\phi(u)$ ist in der Halbebene $s < -\alpha$ regulär und in jeder Teilhalbebene beschränkt.

Wegen (4) und (5) ist für $x < 0$ mit Rücksicht auf (3)

$$|g(x)| \leqq \int_0^\infty |K(x-y)| |f(y)| dy < \text{const} \int_0^\infty |K(x-y)| e^{\alpha y} dy.$$

Für $\alpha < \lambda < 1$ ist ferner

$$\int_0^\infty |K(x-y)| e^{\alpha y} dy < \int_0^\infty |K(x-y)| e^{\lambda y} dy = e^{\lambda x} \int_{-x}^\infty |K(-y)| e^{\lambda y} dy,$$

also wegen $x < 0$

$$|g(x)| < e^{\lambda x} \int_0^\infty |K(-y)| e^{\lambda y} dy;$$

somit $g(x) = O(e^{-\lambda |x|})$ für beliebiges $\lambda < 1$. Das Laplace-Integral (6a) für $\gamma(u)$ konvergiert daher für $s > -1$ absolut, und es gilt das

Lemma 2. $\gamma(u)$ ist in der Halbebene $s > -1$ regulär und in jeder Teilhalbebene beschränkt.

Daß die Regularitäts-Halbebenen von ϕ und γ zusammen die u-Ebene ausfüllen und einen Streifen $(-1 < s < -\alpha)$ gemeinsam haben, ist der Hauptpunkt der folgenden Betrachtungen. Unter den Voraussetzungen über $K(x)$ konvergiert das Laplace-Integral von $K(x)$ absolut für $|s| < 1$, und $K(u)$ ist in diesem Streifen regulär. Wendet man nun die Laplace-Transformation auf die Gleichung (4) an, so resultiert

$$\begin{aligned}\gamma(u) &= \phi(u) - \int_{-\infty}^{+\infty} e^{ux} dx \int_{-\infty}^{+\infty} K(x-y) f(y) dy \\ &= \phi(u) - \int_{-\infty}^{+\infty} f(y) dy \int_{-\infty}^{+\infty} K(x-y) e^{ux} dx \\ &= \phi(u) - \int_{-\infty}^{+\infty} f(y) e^{uy} dy \int_{-\infty}^{+\infty} K(t) e^{ut} dt\,[1],\end{aligned}$$

also

(7) $$\gamma(u) = \phi(u)\bigl(1 - \varkappa(u)\bigr).$$

[1] Dies ist die wohlbekannte Eigenschaft der Laplace-Transformation, eine Faltung in ein Produkt von Transformierten überzuführen. Vgl. G. Doetsch. Die Integrodifferentialgleichungen vom Faltungstypus. Math. Annalen 89 (1923), S. 192.

(2*)

Die Umkehrung der Integrationsreihenfolge ist dabei $(-1 < s = \Re(u) < -\alpha)$ gewiß erlaubt, da die Integrale absolut konvergieren. Zur Diskussion von (7) benötigen wir das

Lemma 3. In jedem Streifen $|s| \leqq \beta$ $(\beta < 1)$ besitzt die Funktion $1 - \varkappa(u)$ höchstens endlich viele Nullstellen. Bezeichnet man dieselben mit $u_1, u_2, \ldots,$ u_m, so läßt sich $1 - \varkappa(u)$ im Streifen $|s| \leqq \beta$ in der Form

$$1 - \varkappa(u) = \frac{\sigma_+(u)}{\sigma_-(u)} \prod_1^m (u - u_\nu) \tag{8}$$

darstellen, wo $\sigma_+(u)$ in der Halbebene $s \geqq -\beta$, $\sigma_-(u)$ in der Halbebene $s \leqq +\beta$ regulär und nullstellenfrei ist. Die Beträge

$$\left|\sigma_+(u) u^{k+\frac{m}{2}}\right|, \quad \left|\sigma_-(u) u^{k-\frac{m}{2}}\right|$$

sind für genügend großes $|u|$ in den erwähnten Halbebenen zwischen positiven Schranken gelegen; k ist hierbei eine durch $K(x)$ wohlbestimmte ganze Zahl. Für einen symmetrischen Kern, $K = K(|x|)$, ist $\varkappa(u) = \varkappa(-u)$, m gerade, $m = 2n$, und $k = 0$.

Beweis. Es ist $\left(x = \xi + \frac{\pi}{t}\right)$

$$\varkappa(u) = \int_{-\infty}^{+\infty} K(x) e^{sx} e^{itx} dx = -\int_{-\infty}^{+\infty} K\left(\xi + \frac{\pi}{t}\right) e^{s\left(\xi + \frac{\pi}{t}\right)} e^{it\xi} d\xi,$$

somit, wenn im zweiten Integral ξ wieder durch x ersetzt wird und das erste Integral zum zweiten addiert wird,

$$2\varkappa(u) = \int_{-\infty}^{+\infty} \left\{K(x) e^{sx} - K\left(x + \frac{\pi}{t}\right) e^{s\left(x + \frac{\pi}{t}\right)}\right\} e^{itx} dx.$$

Eine einfache Zerspaltung lehrt, daß der Betrag des Integranden nicht größer als $(|s| < 1)$

$$e^{\frac{\pi}{|t|}} \left|K\left(x + \frac{\pi}{t}\right) - K(x)\right| e^{sx} + \left|e^{s\frac{\pi}{t}} - 1\right| K(x) e^{sx}$$

ist. Somit ist für $|s| \leqq s_0 < 1$

$$2|\varkappa(u)| \leqq e^{\frac{\pi}{|t|}} \int_{-\infty}^{+\infty} \left|K\left(x + \frac{\pi}{t}\right) - K(x)\right| e^{s_0|x|} dx + \left|e^{s\frac{\pi}{t}} - 1\right| \left|\int_{-\infty}^{+\infty} K(x) e^{s_0|x|} dx\right|,$$

woraus mit Rücksicht auf die Voraussetzungen über K leicht

$$\varkappa(s + it) \to 0 \quad \text{für} \quad |t| \to \infty \tag{9}$$

folgt, und zwar gleichmäßig im Streifen $|s| \leqq s_0$. Die Behauptung über die Nullstellen von $1 - \varkappa(u)$ folgt hieraus unmittelbar. Ferner ist

$$\frac{\varkappa(s+it)}{\sqrt{2\pi}}$$

als Funktion von t die FOURIER-Transformierte von $K(x)e^{sx}$, einer nach Voraussetzung quadratisch integrablen Funktion von x. Nach einem bekannten Satze von PLANCHEREL[1] ist somit $|\varkappa(s+it)|$ quadratisch summierbar in $-\infty < t < +\infty$.

Wir setzen nun

$$\tau(u) = (1 - \varkappa(u)) \frac{(u^2-1)^{\frac{m}{2}}}{\prod_1^m (u - u_\nu)} \left(\frac{u+1}{u-1}\right)^k, \tag{10}$$

wo k eine noch zu bestimmende ganze Zahl ist und unter $(u^2-1)^{\frac{m}{2}}$ derjenige in $|s|<1$ eindeutige Zweig zu verstehen ist, der sich für große $|u|$ wie u^m verhält. Es sei $\beta < \beta' < 1$, β' jedoch so gewählt, daß der etwas größere Streifen $|s| \leqq \beta'$ keine neuen Nullstellen von $1 - \varkappa(u)$ enthält. $\tau(u)$ ist dann regulär und nullstellenfrei in $|s| \leqq \beta'$, und wir haben

$$\tau(u) \to 1 \quad \text{für} \quad |t| \to \infty \tag{11}$$

gleichmäßig für $|s| \leqq \beta'$. Wir betrachten den Zuwachs von $\log \tau(s+it)$, wenn t von $-\infty$ nach $+\infty$ läuft. Er ist wegen (11) ein ganzzahliges Vielfaches von 2π und kann durch passende Wahl von k in (10) zu Null gemacht werden. Nachdem k so bestimmt ist, betrachten wir denjenigen im Streifen $|s| \leqq \beta'$ eindeutigen Zweig von $\log \tau(u)$, für welchen $\log \tau(u)$ für $t \to -\infty$ nach Null konvergiert. Dann gilt $\log \tau(u) \to 0$ auch für $t \to +\infty$, somit

$$\log \tau(u) \to 0 \quad \text{für} \quad |t| \to \infty$$

gleichzeitig im Streifen $|s| \leqq \beta'$. Fernerhin ist $\tau(u)$ für große $|t|$ von der Form $(1 - \varkappa(u))\left(1 + O\left(\frac{1}{|u|}\right)\right)$, somit ist auch $|\log \tau(u)|$ als Funktion von t quadratisch summierbar. Daher gilt die CAUCHYsche Integralformel

$$\log \tau(u) = \log \tau_+(u) - \log \tau_-(u),$$

$$\left\{ \log \tau_+(u) = -\frac{1}{2\pi i} \int_{-\beta'-i\infty}^{-\beta'+i\infty} \frac{\log \tau(v)}{v-u}\, dv, \quad \log \tau_-(u) = -\frac{1}{2\pi i} \int_{\beta'-i\infty}^{\beta'+i\infty} \frac{\log \tau(v)}{v-u}\, dv \right. \tag{12}$$

für $-\beta' < s = \Re(u) < \beta'$.

---

[1] Vgl. N. WIENER, Generalized harmonic analysis. Acta Math. 55 (1930), S. 117. insbes. S. 120—126.

Nun ist nach der Schwarzschen Ungleichung

$$|\log \tau_-(u)|^2 \leqq \frac{1}{4\pi^2}\int_{\beta'-i\infty}^{\beta'+i\infty} |\log \tau(v)|^2 |dv| \cdot \int_{\beta'-i\infty}^{\beta'+i\infty} \frac{|dv|}{|u-v|^2},$$

und $\log \tau_-(u)$ ist für $s \leqq \beta < \beta'$ regulär und beschränkt. Ebenso ist $\log \tau_+(u)$ für $s \geqq -\beta > -\beta'$ regulär und beschränkt. Setzen wir endlich

$$\sigma_+(u) = \tau_+(u)(u+1)^{-k-\frac{m}{2}}, \quad \sigma_-(u) = \tau_-(u)(u-1)^{-k+\frac{m}{2}},$$

so folgt aus (10) und (12) die Formel (8) mit allen behaupteten Eigenschaften von σ_+ und σ_-.

Für symmetrische Kerne ist offenbar $\varkappa(u) = \varkappa(-u)$ und die Anzahl der Wurzeln von $1-\varkappa(u)=0$ im Streifen $|s| \leqq \beta$ ist gerade, $m = 2n$. Wegen der Realität von $K(x)$ ist $\overline{\varkappa(u)} = \varkappa(\bar{u})$ und für imaginäre u, $\bar{u} = -u$, ist $\overline{\varkappa(u)} = \varkappa(-u) = \varkappa(u)$, d. h. $\varkappa(u)$ reell. Setzt man in (10) $k = 0$, so ist auch $\tau(u)$ reell für imaginäre u. Wegen $\tau(u) \neq 0$ ist daher der Zuwachs von $\log \tau(u)$ längs der imaginären u-Achse gleich Null. Es wird noch einmal bemerkt, daß die Funktionen σ_+ und σ_- aus der Kernfunktion $K(x)$ durch explizite Integralformeln bestimmt sind.

Die Gleichung (7) ist nun leicht lösbar. $u_1, u_2, \cdots, u_m$ seien alle im Streifen $|s| \leqq \alpha$ gelegenen Nullstellen von $1-\varkappa(u)$, wobei α das α von (3) bedeutet. Wir bestimmen β mit $\alpha < \beta < 1$ so, daß im Streifen $|s| \leqq \beta$ keine neuen Nullstellen liegen, und wenden Lemma 3 mit diesem β an. (7) und (8) ergibt dann

$$(13) \qquad \frac{\gamma(u)}{\sigma_+(u)} = \frac{\phi(u)}{\sigma_-(u)} \prod_1^m (u-u_\nu).$$

Nach Lemma 2 und 3 ist die linke Seite für $s \geqq -\beta$ regulär und für große $|u|$ gleich $O\left(|u|^{k+\frac{m}{2}}\right)$. Nach Lemma 1 und 3 ist die rechte Seite für $s \leqq -\beta$ regulär und für große $|u|$ ebenfalls gleich $O\left(|k|^{k+\frac{m}{2}}\right)$. Somit definiert (13) eine ganze Funktion, die für große $|u|$ gleich $O\left(|u|^{k+\frac{m}{2}}\right)$, somit ein Polynom vom Grade $\leqq \left[k+\frac{m}{2}\right]$ ist. Wir beschränken uns im folgenden auf die Diskussion eines symmetrischen reellen Kernes,

$$K = K(|x|).$$

Dann ist $m = 2n$, $k = 0$, und (13) definiert also ein Polynom höchstens n-ten Grades. Es kann jedoch nicht von n-tem Grade sein, da nach Lemma (3) sonst $|\phi|(u)|$ für große $|u|$ oberhalb einer positiven Schranke liegen müßte, während nach PLANCHERELS Theorem $|\phi(s+it)|$ als Funktion von t quadratisch integrierbar sein muß. Also ist

$$\phi(u) = \sigma_{-}(u) \frac{P_{n-1}(u)}{\prod\limits_{1}^{2n}(u-u_\nu)}, \quad \gamma(u) = \sigma_{+}(u) P_{n-1}(u), \tag{14}$$

unter P_{n-1} ein beliebiges Polynom von höchstens $(n-1)$-tem Grade verstanden.

Wir müssen nun noch umgekehrt zeigen, daß die durch (14) explizit gegebenen Funktionen $\phi(u)$, $\gamma(u)$ wirklichen Lösungen $f(x)$, $g(x)$ von (4), (5) entsprechen. $\phi(u)$ ist für $s \leqq \beta$ bis auf Pole in $u = u_1, \ldots, u_{2n}$ regulär und für große $|u|$ gleich $O\left(\frac{1}{|u|}\right)$. $\gamma(u)$ ist für $s \geqq -\beta$ regulär und für große $|u|$ gleich $O\left(\frac{1}{|u|}\right)$. Mit Rücksicht auf PLANCHERELS Theorem haben somit die Integrale in den MELLINschen Umkehrformeln

$$f(x) = \frac{1}{2\pi i}\int_{-\beta-i\infty}^{-\beta+i\infty} \phi(u) e^{-ux} du, \quad g(x) = \frac{1}{2\pi i}\int_{-\beta-i\infty}^{-\beta+i\infty} \gamma(u) e^{-ux} du \tag{15}$$

einen wohlbestimmten Sinn, wenn man sie als Grenzwerte en moyenne auffaßt. Im Integral (15) für $f(x)$ kann offenbar die Integrationsabszisse beliebig weit nach links verlegt werden. Somit ist

$$f(x) e^{-\lambda x} = \frac{1}{2\pi}\int_{-\infty}^{+\infty} \phi(-\lambda+it) e^{-ixt} dt, \quad \lambda \geqq \beta,$$

und $\frac{\phi(-\lambda+it)}{\sqrt{2\pi}}$ ist umgekehrt die FOURIER-Transformierte von $f(x)e^{-\lambda x}$. Wegen der Gleichheit der Absolut-Quadratintegrale von Funktion und transformierter Funktion (3) ist

$$\int_{-\infty}^{+\infty} |f(x)|^2 e^{-2\lambda x} dx = \frac{1}{2\pi}\int_{-\infty}^{+\infty} |\phi(-\lambda+it)|^2 dt$$

für beliebig große λ. Da die rechte Seite für $\lambda \geqq \beta$ beschränkt ist, muß offenbar $f(x)$ für $x < 0$ bis auf eine Nullmenge verschwinden. Analog ergibt sich $g(x) = 0$ für $x > 0$ bis auf eine Nullmenge, d. h. (5). Wir zeigen nun, daß (4) erfüllt ist, d. h. daß die Funktion

$$\delta(x) = e^{-\beta x} g(x) - e^{-\beta x} f(x) + \int_{-\infty}^{+\infty} e^{-\beta x} K(x-y) f(y) dy \tag{16}$$

verschwindet (zunächst bis auf eine Nullmenge). Nun sind wegen (15) $\frac{1}{\sqrt{2\pi}}\varphi(-\beta+it)$, $\frac{1}{\sqrt{2\pi}}\gamma(-\beta+it)$ umgekehrt die FOURIER-Transformierten von $f(x)e^{-\beta x}$, $g(x)e^{-\beta x}$. Das Integral in (16) hat die Form

$$\text{(16a)}\quad \int_{-\infty}^{+\infty} G(-y)F(y)dy\,,\quad G(y) = e^{-\beta(x+y)}K(x+y)\,,\quad F(y) = e^{-\beta y}f(y)\,,$$

wobei $|F|$ und $|G|$ gewiß quadratisch integrierbar sind. Nach der Vollständigkeitsrelation für FOURIER-Integrale (3) ist das Integral in (16a) gleich dem Integral über das Produkt der FOURIER-Transformierten von $G(y)$ und $F(y)$. Da offenbar $\frac{1}{\sqrt{2\pi}}e^{-itx}\varkappa(-\beta+it)$ als Funktion von t die FOURIER-Transformierte von $G(y)$ ist und $\frac{1}{\sqrt{2\pi}}\varphi(-\beta+it)$ die Transformierte von $F(y)$ war, folgt

$$\int_{-\infty}^{+\infty} e^{-\beta x}K(x-y)f(y)dy = \frac{1}{2\pi}\int_{-\infty}^{+\infty}\varphi(-\beta+it)\varkappa(-\beta+it)e^{-itx}dt\,.$$

Umgekehrt besagt dies, daß $\frac{1}{\sqrt{2\pi}}\varphi(-\beta+it)\varkappa(-\beta+it)$ die FOURIER-Transformierte des Integrals in (16) ist. Daher ist die FOURIER-Transformierte von $\delta(x)$ gleich $\frac{1}{\sqrt{2\pi}}(\gamma-\varphi+\varphi\varkappa)\equiv 0$, und $\delta(x)$ verschwindet bis auf eine Nullmenge. Die durch (14) und (15) definierte Funktion $f(x)$ befriedigt also die Integralgleichung (1) fast überall. Schließlich zeigt eine geläufige Schlußweise nachträglich, daß aus (1) und der quadratischen Summierbarkeit von $|f(x)|e^{-\beta x}$ die Stetigkeit und Beschränktheit von $f(x)e^{-\beta x}$ folgt; diese Funktion ist gleich dem Integral (16a), das nach der SCHWARZschen Ungleichung gewiß eine beschränkte Funktion von x ist.

Asymptotisches Verhalten der Lösungen von (1). Wir beschränken uns wieder auf symmetrische Kerne, $K = K(|x|)$. Wir haben noch das Wachstum der durch (14) und (15) definierten Lösungen $f(x)$ von (1) zu untersuchen. Im Integral (15) für $f(x)$ kann nun die Integrationsabszisse von $-\beta$ nach $+\beta$ verlegt werden; der Wert des Integrales ändert sich dabei um die negative Summe der Residuen von $\varphi(u)e^{-ux}$ in $|\Re(u)|\leqq\alpha$. Ein solches Residuum ist wegen (14) von der Form

$$-Q(x)e^{-u^* x}\,,$$

wo u^* eine in jenem Streifen gelegene Nullstelle von $1-\varkappa(u)$ und $Q(x)$ ein Polynom bedeutet, dessen Grad kleiner als die Vielfachheit der Nullstelle ist.

Somit ist

$$(17) \quad \begin{cases} f(x) = f_0(x) + r(x), \\ f_0(x) = \sum Q(x) e^{-u^* x}, \quad r(x) = \dfrac{1}{2\pi i} \displaystyle\int_{\beta - i\infty}^{\beta + i\infty} \phi(u) e^{-ux} du, \end{cases}$$

wo die Summe sich auf alle Wurzeln u^* in $|\Re(u)| \leqq \alpha$ bezieht. $f_0(x)$ ist eine Lösung der Integralgleichung (2), genügt also der Gleichung

$$(18) \qquad f_0(x) = \int_0^\infty K(x-y) f_0(y)\, dy + \int_{-\infty}^0 K(x-y) f_0(y)\, dy.$$

Nun ist gewiß $f_0(x) = O(e^{\beta|x|})$, woraus mit Rücksicht auf die Voraussetzungen über K leicht

$$(19) \qquad \int_{-\infty}^0 K(x-y) f_0(y)\, dy = O(e^{-\beta x}), \qquad x > 0$$

folgt. Nach (17) ist ferner $r(x) e^{\beta x}$ die FOURIER-Transformierte von $\dfrac{1}{\sqrt{2\pi}} \phi(\beta - it)$, somit quadratisch integrierbar. Wegen (17), (18), (19) ist

$$r(x) = \int_0^\infty K(x-y) r(y)\, dy + O(e^{-\beta x}),$$

woraus durch Anwendung der SCHWARZschen Ungleichung auf das Integral — der Integrand ist $(K(x-y) e^{-\beta y}) \cdot (r(y) e^{\beta y})$ — leicht $r(x) = O(e^{-\beta x})$ folgt. Es gilt somit unter den zugrunde gelegten Voraussetzungen über K das

Theorem. Ist $2n$ die immer endliche Anzahl der in ihrer Vielfachheit gezählten Nullstellen der Funktion

$$1 - \int_{-\infty}^{+\infty} K(x) e^{ux}\, dx \qquad K(x) = K(|x|)$$

im Streifen $|\Re(u)| \leqq \alpha < 1$, so ist die Maximalzahl der linear unabhängigen Lösungen von (1) mit $f(x) = O(e^{(\alpha+\delta)x})$ und beliebigem $\delta > 0$ genau gleich n. Die Lösungen haben die Form

$$f(x) = \sum Q(x) e^{-u^* x} + O(e^{-\beta x}).[1]$$

wobei u^* eine der obigen Nullstellen bedeutet und $Q(x)$ ein Polynom ist, dessen Grad kleiner als die Vielfachheit von u^* ist; β ist eine Zahl mit $\alpha < \beta < 1$ derart, daß die Streifen $\alpha < \Re(u) \leqq \beta$, $-\alpha > \Re(u) \geqq -\beta$ keine Nullstellen enthalten.

Als weitere Besonderheit des symmetrischen Kernes sei erwähnt, daß unter den im Hauptteil der Lösungen auftretenden Termen immer einer

[1] (3) braucht nicht erfüllt zu sein, da auf dem Rande des Streifens $|\Re(u)| \leq \alpha$ eine mehrfache Nullstelle liegen kann, und daher ein Term von der Größenordnung $x^l e^{\alpha x}$, $l < 0$, vorkommen kann.

mit $\Re(u^*) \leqq 0$ vorkommen muß, d. h. daß es keine Lösung von (1) mit $f(x) \to 0$, $x \to \infty$ gibt; denn wegen (14) und $\sigma_-(u) \neq 0$ hat $\phi(u)$ mindestens einen Pol, der nicht in der rechten Halbebene liegt.

Beispiele. Bei der *Lalescoschen Gleichung* ist $K(x) = \lambda e^{-|x|}$, soweit $\varkappa(u) = \frac{2\lambda}{1-u^2}$. Die Wurzeln von $1-\varkappa = 0$ sind $u = \pm\sqrt{1-2\lambda}$, und es ist $n = 1$. Die Darstellung (8) ergibt sich direkt,

$$\sigma_+(u) = \frac{1}{u+1}. \qquad \sigma_-(u) = u-1$$

Für $\lambda \leqq 0$ gibt es keine Lösungen mit der Eigenschaft (3) und irgendeinem $\alpha < 1$. Für $\lambda > 0$ gibt es im wesentlichen nur eine Lösung. Wegen (14) ist

$$\phi(u) = \frac{u-1}{u^2-1+2\lambda}.$$

Wegen (17) ist weiter

$$f(x) = \frac{e^{\sqrt{1-2\lambda}\,x} + e^{-\sqrt{1-2\lambda}\,x}}{2} + \frac{e^{\sqrt{1-2\lambda}\,x} - e^{-\sqrt{1-2\lambda}\,x}}{2\sqrt{1-2\lambda}} + \frac{1}{2\pi i}\int_{\beta-i\infty}^{\beta+i\infty} \phi(u)\,e^{-ux}\,du.$$

Das Integral muß hier verschwinden, da die Integrationsabszisse beliebig weit nach rechts gelegt werden kann.

Als weiteres, komplizierteres Beispiel sei die *Milnesche Gleichung* behandelt, $K(x) = \frac{1}{2}\int_{|x|}^{\infty}\frac{e^{-t}}{t}\,dt$. Hier ist

$$\begin{aligned}
\varkappa(u) &= \frac{1}{2}\int_{x=0}^{\infty}\int_{t=x}^{\infty}\frac{e^{-t}}{t}(e^{ux}+e^{-ux})\,dt\,dx \\
&= \frac{1}{2}\int_{t=0}^{\infty}\int_{x=0}^{t}\frac{e^{-t}}{t}(e^{ux}+e^{-ux})\,dx\,dt \\
&= \frac{1}{2u}\int_{0}^{\infty}\frac{e^{-(1-u)t}-e^{-(1+u)t}}{t}\,dt \\
&= \frac{1}{2u}\int_{0}^{\infty}dt\int_{1-u}^{1+u}e^{-wt}\,dw = \frac{1}{2u}\int_{1-u}^{1+u}\frac{dw}{w} = \frac{1}{2u}\log\frac{1+u}{1-u},
\end{aligned}$$

wo derjenige im Streifen $|\Re(u)| < 1$ reguläre Zweig von log genommen ist, der für $u = 0$ verschwindet. $u = 0$ ist zweifache Wurzel von $1-\varkappa = 0$, und sonst sind, wie man sich leicht überlegt, im Streifen $|\Re(u)| < 1$ keine weiteren Wurzeln vorhanden. Es gibt daher nur eine Lösung mit der Eigenschaft (3); sie verhält sich für große x wie eine lineare Funktion

$$f(x) = x + a + O(e^{-(1-\delta)x}) \tag{20}$$

mit beliebigem $\delta > 0$. Es ist uns in diesem Falle nicht gelungen, aus den allgemeinen Integraldarstellungen $f(x)$ in einfacher Gestalt zu gewinnen. Vermutlich ist sie eine neue Transzendente.

Im allgemeinen Falle ist die Voraussetzung (3) über die Lösungen von (1) durchaus natürlich, da $1-\varkappa$ unendlich viele Nullstellen im Streifen $|\Re(u)| < 1$ haben kann. Wenn jedoch $1-\varkappa$ nur endlich viele Nullstellen hat, wie in den beiden obigen Beispielen, kann man die Einseitigkeit der gewonnenen Lösungen auch unter geringeren Voraussetzungen als (3) beweisen, wie hier nicht näher ausgeführt sein möge.

Positiver Kern[1]. Es sei $K = K(|x|) > 0$ bis auf endlich viele Stellen. Ferner sei

$$\varkappa(0) \leqq 1, \quad \varkappa(1) > 1.$$

Wir zeigen, daß es dann unter den Lösungen von (1) eine für $x \geqq 0$ positive gibt. Unter diesen Voraussetzungen wächst nämlich $\varkappa(u) = \int_0^\infty K(x)(e^{ux} + e^{-ux})\,dx$ monoton, wenn u von 0 durch reelle Werte nach 1 läuft, und die Gleichung $1-\varkappa = 0$ hat im Streifen $|\Re(u)| < 1$ genau zwei reelle Wurzeln $\pm u^*$; $u^* \geqq 0$. Es gibt nun eine Lösung von (1) mit $f(x) = O(e^{u^*x})$ für $u^* > 0$, und $f(x) = O(x)$ im Falle $u^* = 0$, d. h. $\varkappa(0) = 1$. In beiden Fällen ist diese Lösung gewiß von einer Stelle ab positiv. Würde sie überhaupt Werte $\leqq 0$ annehmen, so müßte sie ihre untere Grenze irgendwo annehmen, etwa bei $x = x_0$. Wegen $K > 0$ ist daher

$$f(x_0) = \int_0^\infty K(x_0-y)\,f(y)\,dy \geqq f(x_0)\int_0^\infty K(x_0-y)\,dy,$$

oder

$$f(x_0)\left(1 - \int_0^\infty K(x_0-y)\,dy\right) \geqq 0,$$

wo das Gleichheitszeichen nur für $f \equiv f(x_0)$ gelten kann. Somit ist $f(x_0) = 0$ unmöglich. $f(x_0) < 0$ ist ebenfalls unmöglich, da das Integral in der Klammer kleiner als $\int_{-\infty}^{+\infty} K(t)\,dt = \varkappa(0) \leqq 1$ ist. Daher ist $f(x) > 0$ für $x \geqq 0$. Die Lösung (20) der MILNEschen Gleichung ist z. B. positiv.

[1] Vgl. auch E. HOPF, Über lineare Integralgleichungen mit positivem Kern. Sitzungsber. der Berliner Akad. d. Wiss. 1928. XVIII.

Ausgegeben am 28. Januar 1932.

Berlin, gedruckt in der Reichsdruckerei.

Sitzungsber. phys.-math. Kl. 1931. (3)

Über eine Klasse singularer Integralgleichungen (with Norbert Wiener), *Sitzungsberichte Preussische Akad. Wiss. Berlin* **(1931), 696-706.**

Commentary

Harold Widom

In this paper Wiener and Hopf solved integral equations of the form

$$\int_0^\infty K(x-y)\,f(y)\,dy = f(x), \qquad x > 0. \tag{0.1}$$

They were particularly interested in the Milne equation, which arises in the theory of radiative transfer, where

$$K(x) = \frac{1}{2}\int_{|x|}^\infty \frac{e^{-t}}{t}dt. \tag{0.2}$$

But they considered the general case where K was real-valued and had exponential decay at infinity, and they found all solutions f having at most a certain exponential growth (depending on the dacay of K).

If the domain in (0.1) were $(-\infty, \infty)$ rather than $(0, \infty)$ the equation could be solved immediately by taking Fourier transforms. The solutions are exponentials determined by the zeros of $1 - \kappa(u)$, where κ is the Fourier transform of K. The crucial ingredient in the solution of (0.1) was a certain representation of $1 - \kappa$, now known as the *Wiener-Hopf factorization.* This was

$$1 - \kappa(u) = \frac{\sigma_-(u)}{\sigma_+(u)} \prod_\nu (u - u_\nu), \tag{0.3}$$

valid in a closed strip about the real line, where σ_+ and σ_- extended to nonzero analytic functions in the upper and lower half-planes, respectively, and the u_ν were the zeros of $1 - \kappa$ in the strip.

If $f(x)$ is defined to be equal to 0 for $x < 0$ and one defines $g(x)$ to be the difference of the two sides of (0.1), then taking Fourier transforms leads to the equation

$$\frac{\gamma(u)}{\sigma_-(u)} = \frac{\phi(u)}{\sigma_+(u)} \prod (u - u_\nu). \tag{0.4}$$

Here γ is the Fourier transform of g and ϕ is the Fourier transform of f. The left side extends analytically to the lower half-plane and the right side to the upper, so they are analytic continuations of a single entire function. Bounds show that the function must be a polynomial of a certain degree. This determines ϕ, and so f. (Working backwards showed that all such f were actually solutions.)

In fact Wiener and Hopf used the two-sided Laplace transform rather than the Fourier transform. This is clearly unimportant but the latter tends to be used nowadays.

Variants of this beautiful solution have subsequently been used to investigate a host of problems, of which here are just a few.

(i) Equations of the second kind, where the right side of (0.1) is replaced by $f(x) + h(x)$ with h a known function.

(ii) Equations in which $K(x)$ is not necessarily real-valued or rapidly decreasing but just belongs to $L^1(-\infty, \infty)$, and the related question of invertibility of the operator $I - K$, where K is the operator with kernel $K(x - y)$. As is now known, invertibility is completely equivalent to the existence of an appropriate factorization of $1 - \kappa$. And this in turn is equivalent to $1 - \kappa$ being nonzero and having nonzero index.

(iii) Systems of equations. This requires factorization of matrix-valued functions, which is not so simple. The reason is that for scalar-valued functions the factors have explicit representations in terms of singular integrals, but this is not the case for matrix-valued functions.

(iv) The discrete analogue, where the kernel $K(x-y)$ is replaced by the *Toeplitz matrix* $(\phi_{i-j})_{i,j\geq 0}$. The subscripts denote Fourier coefficients of the function ϕ defined on the unit circle. These appeared first in Toeplitz's article (1911) studying the spectrum of the doubly-infinite matrix. The semi-infinite matrices (where $0 \leq i, j < \infty$) are the exact analogues of the Wiener-Hopf operators and there is a parallel theory. Their invertibility is connected with the Wiener-Hopf factorization of ϕ; this time the factors are analytic inside and outside the unit circle.

(v) Asymptotic solution of finite problems, in which the positive real line in (0.1) is replaced by a large interval, or the Toeplitz matrix is $n \times n$ with n large. The latter arose very early in Szegö's article (1915) where he determined the asymptotics for large n of the Toeplitz determinants. In the study of asymptotic problems the Wiener-Hopf factorizations can play a crucial role.

Good references for the above-mentioned topics, and much more, are the monographs by Böttcher and Silbermann and by Gohberg and Feldman.

The continued importance of this work of Wiener and Hopf lies in the fact that their operators and related ones appear in so many areas of mathematics and physics: boundary problems for pseudodifferential operators, probability theory, statistical mechanics, signal processing, and combinatorial theory to name a few. And new applications are constantly arising.

References

A. Böttcher and B. Silbermann, *Analysis of Toeplitz Operators,* Akademie-Verlag, Berlin 1990.

I. C. Gohberg and I. A. Feldman, "Convolution Equations and Projection methods for their Solution," *Transl. of Math. Monog. v. 41, Amer. Math. Soc.,* Providence 1974.

G. Szegö, "Ein Grenzwertsatz über die Toeplitzschen Determinanten einer reelen positiven Function," Math. Ann., **76** (1915) 490–503.

O. Toeplitz, "Zur Theorie der bescränkten Bilinearformen," *Math. Ann.*, **70** (1911) 351–376.

Über den funktionalen, insbesondere den analytischen Charakter der Lösungen elliptischer Differentialgleichungen zweiter Ordnung.

Von

Eberhard Hopf in Berlin-Dahlem.

Einleitung.

Es sei

$$\Phi\left(\frac{\partial^2 u}{\partial x_1^2}, \frac{\partial^2 u}{\partial x_1 \partial x_2}, \ldots, \frac{\partial^2 u}{\partial x_n^2}; \frac{\partial u}{\partial x_1}, \ldots, \frac{\partial u}{\partial x_n}; u; x_1, \ldots, x_n\right) = 0$$

eine partielle Differentialgleichung zweiter Ordnung in n unabhängigen reellen Veränderlichen $x_1, x_2, \ldots, x_n$. Wie allgemein bekannt, ist das Verhalten der quadratischen Form

$$\sum_{\nu \leqq \mu} \frac{\partial \Phi}{\partial \dfrac{\partial^2 u}{\partial x_\nu \partial x_\mu}} z_\nu z_\mu$$

von entscheidender Bedeutung für die Theorie der Gleichung $\Phi = 0$. Symmetrisch geschrieben lautet diese Form

$$\sum_{\nu, \mu} a_{\nu\mu} z_\nu z_\mu; \qquad a_{\nu\nu} = \frac{\partial \Phi}{\partial \dfrac{\partial^2 u}{\partial x_\nu^2}}, \qquad a_{\nu\mu} = \frac{1}{2} \frac{\partial \Phi}{\partial \dfrac{\partial^2 u}{\partial x_\nu \partial x_\mu}} \qquad (\nu \neq \mu)$$

(bei der Summation laufen ν, μ unabhängig voneinander). Die Gleichung $\Phi = 0$ heißt in bezug auf eine vorgegebene Lösung $u(x_1, \ldots, x_n)$ elliptisch an der Stelle $x_1 = \overset{0}{x_1}, \ldots, x_n = \overset{0}{x_n}$, wenn jene quadratische Form an dieser Stelle definit ist. Unbeschadet der Allgemeinheit kann sie dann als positiv definit angenommen werden.

Im elliptischen, und nur in diesem Falle beeinflußt der funktionale Charakter der Funktion Φ, wie man weiß, wesentlich den funktionalen Charakter der Lösungen $u(x_1, \ldots, x_n)$ von $\Phi = 0$. Diesen Zusammenhängen, welche im Bernsteinschen Theorem[1]) vom analytischen Charakter der dreimal stetig differenzierbaren Lösungen gipfeln, ist die vorliegende Arbeit gewidmet.

Im ersten Kapitel werden Differenzierbarkeitssätze über die Lösungen von $\Phi = 0$ bewiesen. In diesem Zusammenhange sind vor allem die zuerst von Herrn Lichtenstein[2]) für $n = 2$ bewiesenen „Ergänzungssätze" zum Analytizitätstheorem zu erwähnen, die darauf abzielen, den analytischen Charakter bei besonderer Gestalt von Φ unter möglichst geringen Differenzierbarkeits-Voraussetzungen über u zu beweisen. Sätze dieser Art werden im folgenden auf Grund einer einheitlichen Methode hergeleitet, die auf der Verwendung eines meines Wissens neuen Systems von Integralformeln für die Lösungen der linearen Gleichung,

$$\Phi = \sum_{\nu,\mu} a_{\nu\mu} \frac{\partial^2 u}{\partial x_\nu \partial x_\mu} + \sum_\nu b_\nu \frac{\partial u}{\partial x_\nu} + c u + d,$$

beruht (die Koeffizienten $a_{\nu\mu}, b_\nu, c, d$ sind Funktionen der x_i allein). Diese Formeln erhält man durch Übergang von konstanten zu variablen Koeffizienten mit Hilfe einer zu anderem Zwecke bereits von Herrn A. Korn verwandten Zerlegung[3]). Die Ausnützung jener Formeln geschieht auf Grund leichter Ausdehnungen klassischer Überlegungen der elementaren Potentialtheorie (Sätze von O. Hölder und A. Korn). Mit Hilfe einer Schrankenbestimmung[4]) für die zweiten Ableitungen der Lösungen im linearen Falle konnte die Voraussetzung der dreimaligen stetigen Derivierbarkeit im Analytizitätssatze durch die geringere der Hölder-Stetigkeit der zweiten Ableitungen ersetzt werden[5]). Ob allgemein die bloße Stetigkeit ausreicht, ist eine zur Zeit noch offene Frage[6]).

Das zweite Kapitel ist dem Analytizitätssatze selbst gewidmet. Den älteren Beweisen der Herren S. Bernstein[7]) und M. Gevrey[8]) — es sei auch auf die Arbeit von Herrn T. Radó[9]) hingewiesen — hat neuerdings Herr

---

[1]) Vgl. den Encyklopädiebericht von Herrn Lichtenstein, II C 12. Betreffs des Ideenkreises um das Analytizitätstheorem vgl. S. 1320–1324.

[2]) Vgl. L. Lichtenstein, Bulletin international de l'Académie Polonaise des Sciences et des Lettres 1913, S. 915–941.

[3]) Vgl. A. Korn, Schwarz-Festschrift, Berlin 1914, S. 215–229.

[4]) Bezüglich derartiger Dinge vgl. auch Gevrey, Annales de l'Éc. Norm. **35** (1918), insbesondere S. 145–148.

[5]) Im wesentlichen mit gleicher Methode hatte ich ein entsprechendes Resultat für die Lösungen regulärer Variationsprobleme bewiesen. Vgl. Math. Zeitschr. **30** (1929), S. 404–413.

[6]) Über die Bedeutung der entsprechenden Frage bei regulären Variationsproblemen vgl. A. Haar, Abh. des Hamburger Math. Sem. **8** (1930), S. 1–27.

[7]) Vgl. loc. cit. [1]) und die neue Arbeit in Math. Zeitschr. **28** (1928), S. 330–348.

[8]) Vgl. loc. cit. [1]) und Bull. des sciences math. **50** (1926), S. 113–128.

[9]) Math. Zeitschr. **25** (1926), S. 514–589.

13*

H. Lewy[10]) einen durchsichtigeren Beweis hinzugefügt, der auf der direkten Konstruktion der komplexen Fortsetzung der Lösung mit Hilfe der Auflösung einer hyperbolischen Anfangswertaufgabe beruht. Gegenüber diesen nur für $n = 2$ erbrachten Beweisen sei der von Herrn G. Giraud[11]) für beliebiges n geführte Beweis erwähnt. Er beruht auf eingehenden Untersuchungen über die erste Randwertaufgabe der linearen Gleichung und der analytischen Fortsetzung der Lösungen. Der Beweis des Analytizitätssatzes wird auf Grund dieser Untersuchungen durch sukzessive Approximationen unter wiederholter Auflösung erster Randwertaufgaben erbracht. Im folgenden wird der Bernsteinsche Satz für beliebige n in anderer Weise bewiesen. Die Gleichung $\Phi = 0$ wird zunächst in ein nichtlineares Integralgleichungssystem übergeführt. Mit Hilfe eines wichtigen Gedankens von Herrn E. E. Levi[12]), den in anderer Richtung auch Herr Giraud ausbeutete, lassen sich die Integrale dieses Gleichungssystems auch für komplexe Werte der Variablen definieren. Das so erhaltene komplexe Gleichungssystem, das jedenfalls von der analytischen Fortsetzung der Lösung von $\Phi = 0$ befriedigt werden muß, wird gelöst, und die Regularität der Lösung nachgewiesen. Ihrer Natur nach kann die Methode auf eine ausgedehnte Klasse nichtlinearer Integral- und Integro-Differentialgleichungen angewandt werden. Es sei noch bemerkt, daß das II. Kapitel nur die Lektüre von § 1 voraussetzt[13]).

Inhaltsübersicht.

Seite

I. Kapitel. Funktionale Eigenschaften der Lösungen elliptischer Differentialgleichungen zweiter Ordnung.

§ 1. Greensche Formeln. Die Funktion $\Gamma(a; Z)$ 197

§ 2. Hilfssätze über „Potentiale“ n-dimensionaler Belegungen . 202

§ 3. Differenzierbarkeitssätze über die Lösungen elliptischer Differentialgleichungen 206

§ 4. Verschärfung des Satzes 4 211

---

[10]) Math. Annalen **101** (1929), S. 609–619.

[11]) Annales de l'Éc. Norm. **43** (1926), S. 1–128.

[12]) Palermo Rendiconti **24** (1907), p. 275–317, insbesondere p. 297–307.

[13]) Sämtliche hier entwickelten Methoden sind durchaus auf elliptische Gleichungen höherer Ordnung, sowie auf Systeme erster Ordnung anwendbar. Der einzige springende Punkt bei dieser Ausdehnung besteht in der Konstruktion einer Grundlösung einer beliebigen linearen elliptischen Gleichung mit konstanten Koeffizienten, bei welcher nur die Ableitungen einer festen Ordnung vorkommen. Im Falle zweiter Ordnung ist dies gerade die Funktion $\Gamma(a; Z)$ von § 1. Im Falle $n = 2$ und beliebiger (natürlich gerader) Ordnung vgl. loc. cit. [12]), S. 280–284. Im allgemeinen Falle sind die wichtigen Arbeiten von Herrn Herglotz heranzuziehen. Vgl.: Abh. des Hamburger Math. Sem. **6** (1928), S. 189–197; Berichte der Sächsischen Akademie **78** (1926), S. 93–126; **78** (1926), S. 287–318; **80** (1928), S. 69–114.

Fortsetzung der Inhaltsübersicht.

Seite

II. Kapitel. Beweis des analytischen Charakters der Lösungen von $\Phi = 0$.

§ 5. Ein System nichtlinearer Integralgleichungen für die Lösungen von $\Phi = 0$ 216

§ 6. Vorbereitungen zum Übergang ins Komplexe 220

§ 7. Definition und Eigenschaften des Integrales in (6.6) für komplexe Punkte X 223

§ 8. Das komplexe Integralgleichungssystem. Existenz und Regularität des Lösungssystems 228

I. Kapitel.

Funktionale Eigenschaften der Lösungen elliptischer Differentialgleichungen zweiter Ordnung.

§ 1.

Greensche Formeln. Die Funktion $\Gamma(a; Z)$.

Mit X, Y, Z usw. seien im folgenden Punkte des n-dimensionalen kartesischen Raumes mit den Koordinaten x_ν, y_ν, z_ν usw., $\nu = 1, 2, \ldots, n$ bezeichnet. Es ist klar, was unter $aX + bY$ zu verstehen ist (a, b Konstanten). Des weiteren sei das Symbol $|X| = \sqrt{\sum_\nu x_n^2}$ benützt. Ist $\mathfrak{B}$ ein Bereich[14]) in diesem Raume und $f(X)$ eine in $\mathfrak{B}$ definierte Funktion, so sei kurz

$$\int_{\mathfrak{B}}^{(n)} \cdots \int f(X)\, dx_1\, dx_2 \ldots dx_n = \int_{\mathfrak{B}}^{(n)} f(X)\, dX$$

geschrieben. Ist $\mathfrak{B}$ ein Hyperkugel-Bereich $|X| = r \leqq R$, so ergibt sich, wenn $f(X) = \varphi(r)$ nur von r abhängt, leicht

$$\int_{r<R}^{(n)} \varphi(r)\, dX = n\, e_n \int_0^R r^{n-1} \varphi(r)\, dr, \tag{1.1}$$

unter e_n im folgenden stets den Inhalt von $|X| \leqq 1$ verstanden. Speziell ist also

$$\int_{R_1<r<R_2}^{(n)} \frac{dX}{r^\lambda} = n\, e_n \frac{R_2^{n-\lambda} - R_1^{n-\lambda}}{n-\lambda}. \tag{1.2}$$

Demnach konvergiert das Integral über eine bis auf eine Stelle $X = X_0$ in $\mathfrak{B}$ stetige Funktion $f(X)$ absolut, wenn $f(X)$ an dieser Stelle von kleinerer als n-ter Ordnung unendlich wird.

[14]) Im allgemeinen sei an üblichen Definitionen festgehalten: Gebiet = zusammenhängende offene Punktmenge, Bereich = Gebiet + Rand des Gebietes.

Der Bereich $\mathfrak{B}$ sei konvex, $\mathfrak{R}$ sei sein Rand. Bedeutet $\mathfrak{B}_i$ im Moment die Projektion von $\mathfrak{R}$ auf die Hyperebene $x_i = 0$, so lassen sich die Punkte von $\mathfrak{B}$ vollständig durch zwei Ungleichungen

$$g_i(x_1, \ldots, x_{i-1}, x_{i+1}, \ldots, x_n) \leqq x_i \leqq G_i(x_1, \ldots, x_{i-1}, x_{i+1}, \ldots, x_n)$$

charakterisieren, unter g_i, G_i zwei in $\mathfrak{B}_i$ stetige Funktion verstanden. Ist $f(X)$ auf $\mathfrak{R}$ definiert, so sei

$$\int\limits_{\mathfrak{B}_i}^{(n-1)} \{f_{x_i=G_i} - f_{x_i=g_i}\} \, dx_1 \ldots dx_{i-1} \, dx_{i+1} \ldots dx_n = \int\limits_{\mathfrak{R}}^{(n-1)} f(X) \, d_i X$$

für die Randintegrale geschrieben. Die Verwandlungsformel von Bereichintegralen in Randintegrale lautet dann

$$\int\limits_{\mathfrak{B}}^{(n)} \frac{\partial f(X)}{\partial x_i} \, dX = \int\limits_{\mathfrak{R}}^{(n-1)} f(X) \, d_i X. \tag{1.3}$$

(1.3) gilt natürlich auch, wenn $\mathfrak{B}$ ein von zwei konvexen Hyperflächen begrenzter Bereich ist; dann ist nur $\int\limits_{\mathfrak{R}} = \int\limits_{\mathfrak{R}_a} - \int\limits_{\mathfrak{R}_i}$ zu setzen, wo $\mathfrak{R}_a$ den äußeren, $\mathfrak{R}_i$ den inneren Rand von $\mathfrak{B}$ zu bedeuten hat. Ist $\mathfrak{B}$ wieder konvex, R der Radius der kleinsten $\mathfrak{B}$ umschließenden Hyperkugel, so gilt offenbar

$$\left| \int\limits_{\mathfrak{R}}^{(x-1)} f(X) \, d_i X \right| \leqq 2 e_{n-1} R^{n-1} \underset{X \text{ auf } \mathfrak{R}}{\mathrm{Max}} |f(X)|. \tag{1.4}$$

(1.3) gilt noch, wenn die Stetigkeit von f und $\frac{\partial f}{\partial x_i}$ in einem inneren Punkte X_0 von $\mathfrak{B}$ unterbrochen ist, in welchem f von kleinerer als $(n-1)$-ter, $\frac{\partial f}{\partial x_i}$ von kleinerer als n-ter Ordnung unendlich werden. In bekannter Weise sieht man dies, wenn man aus $\mathfrak{B}$ das Innere einer kleinen Hyperkugel $|X - X_0| = r$ entfernt und nach Anwendung von (1.3) und (1.4) $r \to 0$ streben läßt.

Es sei nun

$$L(u) = L_X(u) \equiv \sum_{\nu, \mu} a_{\nu\mu}(X) \frac{\partial^2 u}{\partial x_\nu \, \partial x_\mu}; \qquad a_{\nu\mu} \equiv a_{\mu\nu}$$

ein linearer partieller Differentialausdruck zweiter Ordnung,

$$M(u) = M_X(u) \equiv \sum_{\nu, \mu} \frac{\partial^2 (a_{\nu\mu} u)}{\partial x_\nu \, \partial x_\mu}$$

der adjungierte Ausdruck. Integriert man die bekannte Identität

$$v L(u) - u M(v) \equiv \sum_{\nu} \frac{\partial N_\nu}{\partial x_\nu},$$

$$N_\nu = N_\nu\left(u, v, \frac{\partial u}{\partial X}, \frac{\partial v}{\partial X}\right) = \sum_{\lambda} \left[a_{\lambda\nu} \left(v \frac{\partial u}{\partial x_\lambda} - u \frac{\partial v}{\partial x_\lambda} \right) - u v \frac{\partial a_{\lambda\nu}}{\partial x_\lambda} \right]$$

über einen konvexen Bereich $\mathfrak{B}$ — in einem $\mathfrak{B}$ enthaltenden Gebiete seien $a_{\nu\mu}(X)$, $u(X)$, $v(X)$ zweimal stetig differenzierbar —, so folgt die Greensche Formel

$$(1.5)\quad \int\limits_{\mathfrak{B}}^{(n)} \{v(Y)\, L_Y(u) - u(Y)\, M_Y(v)\}\, dY = \sum_{\nu} \int\limits_{\mathfrak{R}}^{(n-1)} N_\nu\left(u, v, \frac{\partial u}{\partial Y}, \frac{\partial v}{\partial Y}\right) d_\nu Y.$$

Sie bleibt nach dem Obigen gültig, wenn die zweiten Ableitungen $\frac{\partial^2 v}{\partial Y^2}$ in $\mathfrak{B}$ stetig sind bis auf einen inneren Punkt, in welchem sie von kleinerer als der n-ten Ordnung unendlich werden.

Nunmehr sei $L(u)$ im Bereiche $\mathfrak{B}$ vom *elliptischen* Typus, d. h. die charakteristische quadratische Form $\sum\limits_{\nu,\mu} a_{\nu\mu}(X)\, z_\nu z_\mu$ sei in jedem Punkte X von $\mathfrak{B}$ positiv definit. Die Diskriminante $\Delta = |a_{ik}(X)|$ sei gleich Eins für alle X vorausgesetzt; ist dies nicht erfüllt, so muß man vorher immer erst $L(u)$ durch $\Delta^{\frac{1}{n}}$ dividieren, um auf diesen Fall zurückzukommen[15]). Dann ist bekanntlich auch die reziproke quadratische Form

$$\Omega(a; Z) = \sum_{\nu,\mu} A_{\nu\mu} z_\nu z_\mu$$

($A_{\nu\mu}$ = algebraisches Komplement von $a_{\nu\mu}$) in $\mathfrak{B}$ positiv definit, und ihre Diskriminante in $\mathfrak{B}$ ebenfalls gleich Eins. Die Schreibweise $\Omega(a; Z)$ soll die unmittelbare Abhängigkeit von den $a = a_{\nu\mu}$ andeuten. In unserem Falle wäre genauer $\Omega(a(X); Z)$ zu schreiben. Aus Stetigkeitsgründen gibt es ein $\omega > 0$, so daß

$$(1.6)\qquad \Omega(a(X); Z) \geqq \omega\, |Z|^2$$

für alle X in $\mathfrak{B}$ gilt. Wir betrachten die Funktion

$$\Gamma(a; Z) = \frac{\Omega(a; Z)^{-\frac{n-2}{2}}}{n(n-2)\, e_n}.\ ^{16})$$

Sie hat die folgenden Eigenschaften. Aus den Formeln

$$(1.7)\quad 2n\, e_n \frac{\partial \Gamma}{\partial z_\nu} = -\Omega^{-\frac{n}{2}} \frac{\partial \Omega}{\partial z_\nu}, \qquad 2n\, e_n \frac{\partial^2 \Gamma}{\partial z_\nu\, \partial z_\mu} = \Omega^{-\frac{n+2}{2}} \left(\frac{n}{2} \frac{\partial \Omega}{\partial z_\nu} \frac{\partial \Omega}{\partial z_\mu} - \Omega \frac{\partial^2 \Omega}{\partial z_\nu\, \partial z_\mu}\right)$$

---

[15]) Bei allen Anwendungen denke man sich im folgenden vorher diese Division vorgenommen. Sie ändert nirgends etwas an den jeweiligen Voraussetzungen.

[16]) Für $n = 2$ hat man den bekannten Logarithmusausdruck. Die im folgenden gemachte Voraussetzung $n \geqq 3$ bedeutet keinen wirklichen Ausschluß des Falles $n = 2$; denn alle Sätze über die Lösungen elliptischer Differentialgleichungen in n unabhängigen Variablen enthalten die entsprechenden Sätze für weniger als n Variable, wie man sehr leicht einsieht.

erkennt man mit Rücksicht auf die bekannten Eigenschaften der reziproken Form[17]) leicht, daß Γ der Gleichung

$$(1.8)\qquad \sum_{\nu,\mu} a_{\nu\mu} \frac{\partial^2 \Gamma(a;Z)}{\partial z_\nu \partial z_\mu} = 0$$

genügt. Ferner werden nach (1.7) für $Z=0$ Γ von $(n-2)$-ter, die ersten Ableitungen $\frac{\partial \Gamma}{\partial Z}$ von $(n-1)$-ter und die zweiten Ableitungen $\frac{\partial^2 \Gamma}{\partial Z^2}$ von n-ter Ordnung unendlich.

Unter Einführung von

$$(1.9)\qquad G(X,Y) = \Gamma(a(X); Y-X)$$

zeigen wir kurz in bekannter Weise, daß jede in einem $\mathfrak{B}$ enthaltenden Gebiete zweimal stetig differenzierbare Funktion $u(X)$ für jeden inneren Punkt X von $\mathfrak{B}$ der Integralformel

$$(1.10)\quad u(X) = \int\limits_{\mathfrak{B}}^{(n)} \{u(Y) M_Y(G) - G L_Y(u)\}\, dY + \sum_\nu \int\limits_{\mathfrak{R}}^{(n-1)} N_\nu\Big(u, G, \frac{\partial u}{\partial Y}, \frac{\partial G}{\partial Y}\Big)\, d_\nu Y$$

genügt.

Man bemerkt zunächst, daß $M_Y(G)$ für $Y=X$ nur von $(n-1)$-ter Ordnung unendlich wird. Dies ist sofort ersichtlich, wenn man den wegen (1.8) und (1.9) verschwindenden Ausdruck

$$\sum_{\nu,\mu} a_{\nu\mu}(X) \frac{\partial^2 G}{\partial y_\nu \partial y_\mu}$$

abzieht. Zum Beweise von (1.10) entferne man aus $\mathfrak{B}$ das Innere eines kleinen Hyperellipsoides

$$\Omega(a(X); Y-X) = r^2$$

mit dem festgehaltenen Punkte X als Mittelpunkt und wende man auf den Restbereich $\mathfrak{B}'$ die Greensche Formel mit $v(Y) = G(X,Y)$ an,

$$\int\limits_{\mathfrak{B}'}^{(n)} \{G L_Y(u) - u M_Y(G)\}\, dY = \sum_\nu \int\limits_{\mathfrak{R}}^{(n-1)} N_\nu\Big(u, G, \frac{\partial u}{\partial Y}, \frac{\partial G}{\partial Y}\Big) - J_1 - J_2,$$

$$J_1 = -\sum_{\nu,\mu} \int\limits_{\Omega=r^2}^{(n-1)} u \frac{\partial G}{\partial y_\mu} a_{\nu\mu}\, d_\nu Y, \qquad J_2 = \sum_{\nu,\mu} \int\limits_{\Omega=r^2}^{(n-1)} \Big(G \frac{\partial u}{\partial y_\mu} a_{\nu\mu} - u G \frac{\partial a_{\nu\mu}}{\partial y_\mu}\Big)\, d_\nu Y.$$

Man sieht wegen $G = O(r^{2-n})$ auf $\Omega = r^2$ zunächst, daß für $r \to 0$ $J_2 \to 0$ strebt. Wir zeigen, daß für $r \to 0$ $J_1 \to u(X)$ konvergiert. Auf dem Hyperellipsoid ist nun wegen (1.7)

$$-\frac{\partial G}{\partial y_\mu} = \frac{r^{-n}}{2n\, e_n} \frac{\partial \Omega}{\partial y_\mu},$$

[17]) Vgl. G. Kowalewski, Determinantentheorie, Leipzig 1909, § 103.

also, wenn man in Bereichintegrale zurückverwandelt,

$$2n e_n J_1 = r^{-n} \sum_{\nu,\mu} \int\limits_{\Omega=r^2}^{(n-1)} a_{\nu\mu} u \frac{\partial \Omega}{\partial y_\mu} d_\nu Y = r^{-n} \int\limits_{\Omega<r^2}^{(n)} \left\{ \sum_{\nu,\mu} \frac{\partial^2 (u a_{\nu\mu} \Omega)}{\partial y_\nu \partial y_\mu} \right\} dY.$$

Der Integrand im rechten Integral hat für $Y = X$ den Wert

$$2u(X) \sum_{\nu,\mu} a_{\nu\mu} A_{\nu\mu} = 2n u(X),$$

somit folgt

$$J_1 = \frac{u(X)}{e_n r^n} \int\limits_{\Omega<r^2}^{(n)} dY + O(r) = \frac{u(X)}{e_n} \int\limits_{\Omega<1}^{(n)} dY + O(r).$$

Da die Diskriminante der Form Ω gleich Eins ist, muß die Determinante der linearen, das Gebiet $\Omega < 1$ in das Innere der Einheits-Hyperkugel überführenden Transformation gleich ± 1 sein. Somit ist der Inhalt des Gebietes $\Omega < 1$ gleich e_n, und man hat in der Tat $J_1 \to u(X)$ für $r \to 0$.

Die Integralformel (1.10) bleibt gültig, wenn G durch eine Funktion $G' = G + v$ ersetzt, sobald nur die zweiten Ableitungen $\frac{\partial^2 v}{\partial Y^2}$ für $Y = X$ von höchstens $(n-1)$-ter Ordnung unendlich werden; denn für u und v (bei festgehaltenem X) gilt, wie erwähnt, die Greensche Formel (1.5). Im zweiten Kapitel wird z. B. die diesen Bedingungen gewiß genügende Funktion $H(X, Y) = G(Y, X)$ verwendet.

Mit Rücksicht auf spätere Anwendungen werde die Funktion $\Gamma(a; Z)$ noch genauer untersucht. Von (1.7) ausgehend erkennt man leicht, daß die partiellen Ableitungen $\frac{\partial^i}{\partial Z^i}$ die Form

$$(1.11) \qquad \frac{\partial^i \Gamma(a; Z)}{\partial Z^i} = \frac{P_i(a; Z)}{\Omega(a; Z)^{\frac{n+2i-2}{2}}}$$

besitzen, unter P_i ein homogenes Polynom i-ten Grades in den z_ν verstanden, dessen Koeffizienten selbst Polynome in den $a = a_{\nu\mu}$ mit konstanten Koeffizienten sind. (1.11) kann man auch in der Gestalt

$$(1.12) \qquad |Z|^{n+i-2} \cdot \frac{\partial^i \Gamma(a; Z)}{\partial Z^i} = \frac{P(a; Z\,|\,Z\,|^{-1})}{\Omega(a; Z\,|\,Z\,|^{-1})^k}$$

schreiben, wo $Z|Z|^{-1}$ das Wertesystem mit den Komponenten $z_\nu |Z|^{-1}$ (ihre Quadratsumme ist gleich Eins) bedeutet und k eine positive Zahl ist; der Zähler P ist jedenfalls ein Polynom aller seiner Argumente mit konstanten Koeffizienten. Differenziert man überdies in (1.12) nach den $a_{\nu\mu}$ — dies sei kurz mit $\frac{\partial}{\partial a}$ angedeutet —, so erkennt man sofort, daß dabei die Form des Ausdrucks (1.12) erhalten bleibt,

$$(1.13)\qquad |Z|^{n+i-2}\cdot\frac{\partial^{i+k}\Gamma(a;Z)}{\partial Z^i\,\partial a^k}=\frac{P'(a;Z|Z|^{-1})}{\Omega(a;Z|Z|^{-1})^{k'}}\qquad (k'>0).$$

Man erkennt jedenfalls aus dieser Darstellung, daß sich irgendwelche Stetigkeitseigenschaften, z. B. Hölder-Stetigkeit, der Koeffizientenfunktionen $a_{\nu\mu}=a_{\nu\mu}(X)$ im obigen Bereiche $\mathfrak{B}$ automatisch auf die linke Seite in (1.13) übertragen, und zwar gleichmäßig bezüglich Z, weil der Nenner nach (1.6) $\geqq \omega^{k'}$ ist.

Mit Rücksicht auf eine Anwendung im zweiten Kapitel sei noch die ebenfalls leicht von (1.11) aus zu konstatierende Bemerkung gestattet, daß die gemischten Ableitungen genauer von der Form

$$(1.14)\qquad \frac{\partial^{i+k}\Gamma(a;Z)}{\partial Z^i\,\partial a^k}=\frac{P_{i+2k}(a;Z)}{\Omega(a;Z)^{\frac{n+2i+2k-2}{2}}}$$

sind, wo P_{i+2k} homogen vom $(i+2k)$-ten Grade in den z_ν ist.

Von der obigen Funktion $G(X,Y)$ von (1.9) gilt in leicht verständlicher symbolischer Schreibweise, $Z=Y-X$,

$$\frac{\partial^i G}{\partial Y^i}=\frac{\partial^i\Gamma}{\partial Z^i},\quad \frac{\partial^{i+1}G}{\partial Y^i\,\partial X}=\sum\frac{\partial^{i+1}\Gamma}{\partial Z^i\,\partial a}\,\frac{\partial a}{\partial X}+\frac{\partial^{i+1}\Gamma}{\partial Z^{i+1}}.$$

Es gelten also wegen (1.13) Ungleichungen der Form

$$(1.15)\qquad \left|\frac{\partial^i G}{\partial Y^i}\right|<\frac{\text{konst.}}{|Y-X|^{n+i-2}},\qquad \left|\frac{\partial^{i+1}G}{\partial Y^i\,\partial X}\right|<\frac{\text{konst.}}{|Y-X|^{n+i-1}}$$

für alle Punkte X, Y aus $\mathfrak{B}$. Die linke Ungleichung setzt nur die Stetigkeit der $a_{\nu\mu}(X)$ in $\mathfrak{B}$ voraus.

§ 2.

Hilfssätze über „Potentiale" n-dimensionaler Belegungen.

Für die weiteren Untersuchungen sind einige Zwischenbetrachtungen über Funktionen der Form

$$f(X)=\int\limits_{\mathfrak{B}}^{(n)}\varrho(Y)\,H(X,Y)\,dY$$

erforderlich. $H(X,Y)$ soll dabei eine mit Einschluß der jeweils vorkommenden partiellen Ableitungen in $\mathfrak{B}$ stetige Funktion von X und Y sein; nur für $X=Y$ darf H gewisse Singularitäten besitzen. Die „Belegung" $\varrho(X)$ soll gewisse Stetigkeitseigenschaften besitzen, etwa einer Hölder-Bedingung genügen. Die Definition ist bekanntlich die folgende. $\varrho(X)$ ist in einem Bereiche $\mathfrak{B}$ Hölder-stetig mit dem Exponenten $\alpha>0$, wenn $|\varrho(X'')-\varrho(X')|\leqq \text{konst.}\,|X''-X'|^\alpha$ für irgend zwei Punkte X'', X' aus $\mathfrak{B}$ gilt. ϱ heißt im Innern eines Bereiches $\mathfrak{B}$, oder in einem Gebiete Hölder-stetig, wenn ϱ im eben erklärten Sinne in jedem mit seinen Randpunkten ganz im Innern gelegenen Teilbereiche Hölder-stetig ist.

Hilfssatz 1. Es sei für alle Punkte X und Y aus $\mathfrak{B}$

$$|H| < \frac{\text{konst.}}{|Y-X|^{n-1}}, \quad \left|\frac{\partial H}{\partial X}\right| < \frac{\text{konst.}}{|Y-X|^{n}}.$$

Ist $\varrho(X)$ in $\mathfrak{B}$ Hölder-stetig, so ist die Funktion

$$\varphi(X) = \int_{\mathfrak{B}}^{(n)} \{\varrho(Y) - \varrho(X_0)\} H(X, Y)\, dY$$

an der Stelle $X = X_0$ (X_0 innerhalb $\mathfrak{B}$) einmal derivierbar, und es gilt an dieser Stelle

$$\frac{\partial \varphi}{\partial X} = \int_{\mathfrak{B}}^{(n)} \{\varrho(Y) - \varrho(X_0)\} \frac{\partial H}{\partial X}\, dY.$$

Den auf geläufigen potentialtheoretischen Schlußweisen beruhenden Beweis können wir uns ersparen. Eine unmittelbare Folge aus dem Hilfssatz 1 ist der

Hilfssatz 2. Sind die Voraussetzungen des Hilfssatzes 1 erfüllt, und ist die Funktion

$$\psi(X) = \int_{\mathfrak{B}}^{(n)} H(X, Y)\, dY$$

im Innern von $\mathfrak{B}$ einmal stetig differenzierbar, so besitzt die Funktion

$$f(X) = \int_{\mathfrak{B}}^{(n)} \varrho(Y) H(X, Y)\, dY$$

die gleiche Eigenschaft, und es gilt

$$\frac{\partial f(X)}{\partial X} = \varrho(X) \frac{\partial \psi(X)}{\partial X} + \int_{\mathfrak{B}}^{(n)} \{\varrho(Y) - \varrho(X)\} \frac{\partial H(X, Y)}{\partial X}\, dY.$$

Man hat nur in irgendeinem inneren Punkte X_0 von $\mathfrak{B}$ den Hilfssatz 1 auf die Funktion

$$f(X) = \varrho(X_0)\psi(X) + \int_{\mathfrak{B}}^{(n)} \{\varrho(Y) - \varrho(X_0)\} H(X, Y)\, dY$$

anzuwenden.

Hilfssatz 3. $K(X, Z)$ sei eine mit ihren partiellen Ableitungen $\frac{\partial K}{\partial Z}, \frac{\partial^2 K}{\partial Z^2}$ für alle X eines konvexen Bereichs $\mathfrak{B}$ und alle $Z \neq 0$ stetige Funktion von X, Z. Es gelte

$$|K| < \frac{\text{konst.}}{|Z|^{n-1}}, \quad \left|\frac{\partial K}{\partial Z}\right| < \frac{\text{konst.}}{|Z|^{n}}, \quad \left|\frac{\partial^2 K}{\partial Z^2}\right| < \frac{\text{konst.}}{|Z|^{n+1}}$$

für alle X aus $\mathfrak{B}$ und alle Z, und die Funktionen $|Z|^n \frac{\partial K}{\partial Z}$ seien in $\mathfrak{B}$

Hölder-stetige Funktionen von X mit dem Exponenten α $(0 < \alpha < 1)$, und zwar gleichmäßig bezüglich Z. Ist $\varrho(X)$ ebenfalls in $\mathfrak{B}$ Hölder-stetig mit dem Exponenten α, so ist auch

$$F(X) = \int\limits_{\mathfrak{B}}^{(n)} \{\varrho(Y) - \varrho(X)\} \sigma(Y) \frac{\partial K(X, Y-X)}{\partial Y} d Y$$

in $\mathfrak{B}$ Hölder-stetig, im allgemeinen jedoch mit einem Exponenten $< \alpha$.

Korollar. Ist außerdem noch $\sigma(X)$ in $\mathfrak{B}$ Hölder-stetig — auf den Exponenten kommt es dabei nicht an —, so ist $F(X)$ im Innern von $\mathfrak{B}$ Hölder-stetig mit dem Exponenten α.

Beweis. Man setze

$$F(X, Z) = \int\limits_{\mathfrak{B}}^{(n)} \{\varrho(Y) - \varrho(Z)\} \sigma(Y) \frac{\partial K(X, Y-Z)}{\partial Y} d Y.$$

Wegen $F(X) = F(X, X)$ genügt der Nachweis, daß $F(X, Z)$ als Funktion jedes Argumentpunktes (X, Z in $\mathfrak{B}$) Hölder-stetig ist, und zwar gleichmäßig bezüglich des anderen Argumentpunktes. Zunächst folgt aus den Voraussetzungen leicht

$$\begin{aligned} |F(X'', Z) - F(X', Z)| &\leqq \text{konst.} |X'' - X'|^{\alpha} \cdot \int\limits_{\mathfrak{B}}^{(n)} \frac{d Y}{|Y - Z|^{n-\alpha}} \\ &\leqq \text{konst.} |X'' - X'|^{\alpha} \end{aligned}$$

für alle X', X'', Z aus $\mathfrak{B}$, so daß $F(X, Z)$ nur noch als Funktion von Z zu untersuchen ist. Zu diesem Zweck sei für irgend zwei Punkte Z', Z'' von $\mathfrak{B}$

$$r = 2 |Z'' - Z'| \tag{2.1}$$

gesetzt und unter K_r der Durchschnitt von $\mathfrak{B}$ mit dem Hyperkugelinnern $|Y - Z'| < r$ verstanden[18]). Setzt man zur Abkürzung noch

$$H(X, Y; Z) = \{\varrho(Y) - \varrho(Z)\} \sigma(Y) \frac{\partial K(X, Y-Z)}{\partial Y}, \tag{2.2}$$

so ist

$$\begin{aligned} &F(X, Z'') - F(X, Z') = \{\varrho(Z'') - \varrho(Z')\} J + J_1 + J_2, \\ &J = \int\limits_{\mathfrak{B}-K_r}^{(n)} \sigma(Y) \frac{\partial K(X, Y-Z')}{\partial Y} d Y, \quad J_1 = \int\limits_{K_r}^{(n)} \{H(X, Y; Z'') - H(X, Y; Z')\} d Y, \\ &J_2 = \int\limits_{\mathfrak{B}-K_r}^{(n)} \{\varrho(Y) - \varrho(Z'')\} \sigma(Y) \left\{\frac{\partial K(X, Y-Z'')}{\partial Y} - \frac{\partial K(X, Y-Z')}{\partial Y}\right\} d Y. \end{aligned} \tag{2.3}$$

[18]) $\mathfrak{B}_r$ ist niemals leer, wenn man etwa r der Bedingung $r < D$ unterwirft, wo D der Durchmesser des Bereiches $\mathfrak{B}$ ist.

J_1 werde zuerst betrachtet. Wegen (2.2) ist nach den Voraussetzungen

$$|H(X, Y; Z)| < \frac{\text{konst.}}{|Y - Z|^{n-\alpha}}$$

für alle X, Y, Z aus $\mathfrak{B}$. Also hat man

$$|J_1| < \text{konst.} \int\limits_{K_r}^{(n)} \frac{dY}{|Y - Z'|^{n-\alpha}} + \text{konst.} \int\limits_{K_r}^{(n)} \frac{dY}{|Y - Z''|^{n-\alpha}}.$$

Da Z', Z'' beide in K_r liegen, vergrößert man die rechte Seite, wenn man das erste Integral über das Gebiet $|Y - Z'| < 2r$, das zweite über das Gebiet $|Y - Z''| < 2r$ erstreckt,

$$(2.4) \qquad |J_1| < \text{konst.} \int\limits_{\varrho < 2r}^{(n)} \frac{dY}{\varrho^{n-\alpha}} = \text{konst.}\, r^{\alpha}.$$

Mit $\mathfrak{B}$ ist auch K_r ein konvexer Bereich; da Z', Z'' in K_r liegen, ist nach dem Mittelwertsatze

$$(2.5) \quad \Delta = \frac{\partial K(X, Y - Z'')}{\partial Y} - \frac{\partial K(X, Y - Z')}{\partial Y} = \sum_{\nu} (z_\nu'' - z_\nu') \frac{\partial^2 K(X, Y - Z)}{\partial Y \, \partial z_\nu}\Big|_{Z = Z^*},$$

wo Z^* auf der Verbindungslinie zwischen Z' und Z'', also in K_r liegt. Nach den Voraussetzungen ist daher

$$|\Delta| \leqq \text{konst.} \frac{|Z'' - Z'|}{|Y - Z^*|^{n+1}} = \text{konst.}\, r\, |Y - Z^*|^{-n-1}.$$

Liegt nun Y in $\mathfrak{B} - K_r$, so ist $|Y - Z'| > 2|Z'' - Z'|$ und demnach $|Y - Z^*| \geqq |Y - Z'| - |Z' - Z^*| > |Y - Z'| - |Z'' - Z'| > \frac{1}{2}|Y - Z'|$, also

$$(2.6) \qquad |\Delta| < \text{konst.}\, r\, |Y - Z'|^{-n-1}, \qquad Y \text{ in } \mathfrak{B} - K_r.$$

Des weiteren ist $|Y - Z''| < |Y - Z'| + |Z'' - Z'| < \frac{3}{2}|Y - Z'|$, also

$$|\varrho(Y) - \varrho(Z'')| < \text{konst.}\, |Y - Z'|^{\alpha}, \qquad Y \text{ in } \mathfrak{B} - K_r.$$

Hieraus, sowie aus (2.5) und (2.6) folgt, wenn D den Durchmesser des Bereiches $\mathfrak{B}$ bedeutet, für $r < D$

$$|J_2| < \text{konst.}\, r \int\limits_{\mathfrak{B} - K_r}^{(n)} \frac{dY}{|X - Y'|^{n+1-\alpha}} < \text{konst.}\, r \int\limits_{r < \varrho < D}^{(n)} \frac{dY}{\varrho^{n+1-\alpha}} = \text{konst.}\, r \frac{r^{\alpha-1} - D^{\alpha-1}}{1 - \alpha},$$

also [19])

$$(2.7) \qquad |J_2| < \text{konst.}\, r^{\alpha}$$

für beliebige X, Z', Z'' aus $\mathfrak{B}$. Was endlich J betrifft, so hat man nach

---

[19]) Hierbei wird die Voraussetzung $\alpha < 1$ benützt.

Voraussetzung für $r < D$

$$|J| < \text{konst.} \int\limits_{\mathfrak{B}-K_r}^{(n)} \frac{dY}{|Y-Z'|^n} < \text{konst.} \int\limits_{r<\varrho<D}^{(n)} \frac{dY}{|Y-Z'|^n} = \text{konst.} \log \frac{D}{r}.$$

Berücksichtigt man dieses, sowie (2.4) und (2.7) in (2.3), so folgert man, daß ($r = 2\,|Z''-Z'|$) $F(X,Z)$ als Funktion von Z für beliebig vorgegebenes β mit $\beta < \alpha$ in $\mathfrak{B}$ Hölder-stetig mit dem Exponenten β ist.

Beweis des Korollars. Nach dem Obigen genügt es zu zeigen, daß $|J|$ unter einer festen Schranke gelegen ist, wenn die Punkte Z', Z'' in einem ganz innerhalb von $\mathfrak{B}$ liegenden Teilbereiche $\mathfrak{B}'$ variieren, und wenn r der Bedingung

$$r = 2\,|Z''-Z'| < d \tag{2.8}$$

genügt, unter d die Entfernung des Randes von $\mathfrak{B}'$ vom Rande $\mathfrak{R}$ von $\mathfrak{B}$ verstanden. Nun ist

$$J = \int\limits_{\mathfrak{B}-K_r}^{(n)} \{\sigma(Y)-\sigma(Z')\} \frac{\partial K(X, Y-Z')}{\partial Y}\, dY + \sigma(Z') \int\limits_{\mathfrak{B}-K_r}^{(n)} \frac{\partial K(X, Y-Z')}{\partial Y}\, dY.$$

Da $\sigma(Y)$ in $\mathfrak{B}$ Hölder-stetig sein sollte, ist das erste Integral gewiß für alle X, Z' aus $\mathfrak{B}$ absolut unter einer festen Schranke gelegen. Für das zweite Integral kann man auch, da wegen (2.8) der Y-Bereich $|Y-Z'| \leqq r$ ganz zu $\mathfrak{B}$ gehört, $\left(\frac{\partial}{\partial Y} = \frac{\partial}{\partial y_i}\right)$

$$\int\limits_{\mathfrak{R}}^{(n-1)} K(X, Y-Z')\, d_i Y - \int\limits_{|Y-Z'|=r}^{(n-1)} K(X, Y-Z')\, d_i Y$$

schreiben. Da Z' in $\mathfrak{B}'$ liegt, ist wegen $|Y-Z'| \geqq d$ das erste Integral gewiß beschränkt. Das zweite ist ebenfalls beschränkt, wie man mit Rücksicht auf die Voraussetzung über K mit Hilfe von (1.4) sofort erkennt. Daraus folgt die Behauptung über $|J|$. $F(X,Z)$ ist also als Funktion von Z in $\mathfrak{B}'$ Hölder-stetig mit dem Exponenten α, und zwar gleichmäßig bezüglich X.

§ 3.

Differenzierbarkeitssätze über die Lösungen elliptischer Differentialgleichungen.

Wir betrachten die lineare, in einem bestimmten Gebiete $\mathfrak{G}$ elliptische Differentialgleichung zweiter Ordnung

$$L(u) = \sum_{\nu,\mu} a_{\nu\mu}(X) \frac{\partial^2 u}{\partial x_\nu\, \partial x_\mu} = f(x).$$

Von den Funktionen $a_{\nu\mu}(X)$ und $f(X)$ sei jetzt nur noch die Hölder-Stetigkeit im Gebiete $\mathfrak{G}$ vorausgesetzt. Unter so geringen Annahmen wird natürlich die gewöhnliche Integralformel (1.10) hinfällig. Wir behaupten nun, daß die zweiten Ableitungen jeder in G zweimal stetig differenzierbaren Lösung u von $L(u) = f$ in irgendeinem in $\mathfrak{G}$ gelegenen Bereiche $\mathfrak{B}$ den bemerkenswerten Integralformeln[20])

$$\frac{\partial^2 u(X)}{\partial x_i \partial x_k} = J_1(X) + J_2(X) + J_{\mathfrak{R}}(X),$$

$$(3.1)\quad \begin{aligned} J_1(X) &= -\int\limits_{\mathfrak{B}}^{(n)} \{f(Y) - f(X)\} \frac{\partial^2 G}{\partial y_i \partial y_k} dY, \\ J_2(X) &= \sum_{\nu,\mu} \int\limits_{\mathfrak{B}}^{(n)} \{a_{\nu\mu}(Y) - a_{\nu\mu}(X)\} \frac{\partial^2 u}{\partial y_\nu \partial y_\mu} \frac{\partial^2 G}{\partial y_i \partial y_k} dY, \\ J_{\mathfrak{R}}(X) &= \sum_{\nu,\mu} a_{\nu\mu}(X) \int\limits_{\mathfrak{R}}^{(n-1)} \left(\frac{\partial u}{\partial y_\mu} \frac{\partial^2 G}{\partial y_i \partial y_k} - u \frac{\partial^3 G}{\partial y_i \partial y_k \partial y_\mu} \right) d_\nu Y - f(X) \int\limits_{\mathfrak{R}}^{(n-1)} \frac{\partial G}{\partial y_k} d_i Y \end{aligned}$$

genügen, wo $G(X, Y)$ die durch (1.9) eingeführte Funktion ist. Im speziellen Falle $a_{\nu\nu} = 1$, $a_{\nu\mu} = 0$ für $\nu \neq \mu$ gehen diese Formeln direkt in die O. Hölderschen Formeln für die zweiten Ableitungen der Lösungen der Poissonschen Differentialgleichung über. Die einzige im allgemeinen Falle auftretende Komplizierung besteht, wie man sieht, in dem nochmaligen Auftreten der zweiten Ableitungen in den Integralen. Diesem Umstande sind, da wir im folgenden auf die sinngemäße Verallgemeinerung der Hölderschen Sätze hinauswollen, die Überlegungen von § 2 angepaßt.

Beweis von (3.1). Die Gleichung $L(u) = f$ werde in der Form

$$L^0(u) = \sum_{\nu,\mu} a_{\nu\mu}(X_0) \frac{\partial^2 u}{\partial x_\nu \partial x_\mu} = f(X) - \sum_{\nu,\mu} \{a_{\nu\mu}(X) - a_{\nu\mu}(X_0)\} \frac{\partial^2 u}{\partial x_\nu \partial x_\mu}$$

geschrieben. Wendet man auf $L^0(u)$, einen Differentialausdruck mit *konstanten* Koeffizienten, die Integralformel (1.10) an — dabei ist $G(X, Y)$ durch $\Gamma(a(X_0); Y - X)$ zu ersetzen —, so ergibt sich, da nach (1.8) $M^0(\Gamma) = 0$ ist,

$$u(X) = -\int\limits_{\mathfrak{B}}^{(n)} \Gamma L_Y^0(u)\, dY + \sum_{\nu,\mu} a_{\nu\mu}(X_0) \int\limits_{\mathfrak{R}}^{(n-1)} \left(\Gamma \frac{\partial u}{\partial y_\mu} - u \frac{\partial \Gamma}{\partial y_\mu} \right) d_\nu Y,$$

$$\Gamma = \Gamma(a(X_0);\ Y - X).$$

Bekanntlich dürfen die ersten Ableitungen $\frac{\partial u}{\partial X}$ in jedem inneren Punkte X

[20]) Sie setzen voraus, daß die Diskriminante $|a_{ik}|$ gleich Eins ist. Vgl. jedoch die Fußnote [15]).

von $\mathfrak{B}$ durch Differentiation unter dem Integralzeichen gebildet werden[21]), da die ersten Ableitungen $\frac{\partial \Gamma}{\partial X}$ für $Y = X$ nur von $(n-1)$-ter Ordnung unendlich werden. Nach Einsetzen des obigen Wertes von $L^0(u)$ ergibt sich also mit Rücksicht auf $\frac{\partial \Gamma}{\partial X} = -\frac{\partial \Gamma}{\partial Y}$

$$(3.2) \quad \begin{aligned} \frac{\partial u}{\partial x_i} &= \int\limits_{\mathfrak{B}}^{(n)} f(Y) \frac{\partial \Gamma}{\partial y_i} dY - \sum_{\nu,\mu} \int\limits_{\mathfrak{B}}^{(n)} \{a_{\nu\mu}(Y) - a_{\nu\mu}(X_0)\} \frac{\partial^2 u}{\partial y_\nu \partial y_\mu} \frac{\partial \Gamma}{\partial y_i} dY \\ &\quad - \sum_{\nu,\mu} a_{\nu\mu}(X_0) \int\limits_{\mathfrak{R}}^{(n-1)} \left(\frac{\partial \Gamma}{\partial y_i} \frac{\partial u}{\partial y_\mu} - u \frac{\partial^2 \Gamma}{\partial y_i \partial y_\mu} \right) d_\nu Y. \end{aligned}$$

Wir bilden nunmehr $\frac{\partial^2 u}{\partial x_i \partial x_k}$. Auf das erste Integral der rechten Seite wende man den Hilfssatz 2 an. Wegen

$$\int\limits_{\mathfrak{B}}^{(n)} \frac{\partial \Gamma}{\partial y_i} dY = \int\limits_{\mathfrak{R}}^{(n-1)} \Gamma d_i Y$$

folgt $\left(\frac{\partial \Gamma}{\partial X} = -\frac{\partial \Gamma}{\partial Y}\right)$

$$\frac{\partial}{\partial x_k} \int\limits_{\mathfrak{B}}^{(n)} f(Y) \frac{\partial \Gamma}{\partial y_i} dY = -\int\limits_{\mathfrak{B}}^{(n)} \{f(Y) - f(X)\} \frac{\partial^2 \Gamma}{\partial y_i \partial y_k} dY - f(X) \int\limits_{\mathfrak{R}}^{(n-1)} \frac{\partial \Gamma}{\partial y_k} d_i Y.$$

Die Differentiation der Randintegrale unter dem Integralzeichen ist natürlich gestattet. Bei den übrigen Integralen (J_2) ist dies nach dem Hilfssatz 1 wenigstens an der Stelle $X = X_0$ möglich (X_0 = innerer Punkt von $\mathfrak{B}$). Bildet man also in dieser Weise $\frac{\partial^2 u}{\partial x_i \partial x_k}$, so erhält man, wieder unter durchgängiger Benützung von $\frac{\partial \Gamma}{\partial X} = -\frac{\partial \Gamma}{\partial Y}$, endlich die Formeln (3.1), wenn man nachträglich wieder X statt X_0 schreibt. Dabei geht $\Gamma = \Gamma(a(X_0); Y - X)$ in die Funktion $G(X, Y)$ von (1.9) über.

Wir gehen nun zu ersten Anwendungen von (3.1) über und beweisen den

Satz 1. *Sind die Funktionen $a_{\nu\mu}(X)$ und $f(X)$ im Gebiete $\mathfrak{G}$ Hölder-stetig mit dem Exponenten α $(0 < \alpha < 1)$, so sind auch die zweiten Ableitungen jeder in $\mathfrak{G}$ zweimal stetig differenzierbaren Lösung von*

$$L(u) = f$$

in $\mathfrak{G}$ Hölder-stetig mit gleichem Exponenten α.

---

[21]) Vgl. den schönen Beweis von Erhard Schmidt, Schwarz-Festschrift, S. 365—383, insbes. § 1.

Beweis. Wir zeigen zunächst gemäß (3.1), daß $J_1(X)$, $J_2(X)$, $J_{\mathfrak{R}}(X)$ im Innern irgendeines in $\mathfrak{G}$ enthaltenen konvexen Bereiches $\mathfrak{B}$ überhaupt Hölder-stetig sind. Für $J_{\mathfrak{R}}(X)$ ist das sofort klar (der Exponent ist α), da die Integranden der Randintegrale Hölder-stetige Funktionen von X mit dem Exponenten α darstellen, und zwar gleichmäßig bezüglich Y, solange Y auf $\mathfrak{R}$ und X in einem ganz innerhalb $\mathfrak{B}$ gelegenen Teilbereiche variiert.

Auf $J_1(X)$, d. h. auf die Integrale

$$(3.3) \qquad \int\limits_{\mathfrak{B}}^{(n)} \{f(Y) - f(X)\} \frac{\partial^2 G}{\partial Y^2} \, dY$$

kann man den Hilfssatz 3 nebst Korollar mit

$$K(X, Z) = \frac{\partial \Gamma(a(X); Z)}{\partial Z}, \qquad \varrho(X) = f(X), \qquad \sigma(X) = 1$$

anwenden; denn mit Rücksicht auf (1.12) und die daran angeknüpften Bemerkungen sind alle Voraussetzungen über K erfüllt. Somit ist $J_1(X)$ im Innern von $\mathfrak{B}$ Hölder-stetig mit dem Exponenten α.

Somit bleiben noch $J_2(X)$, d. h. die Integrale

$$(3.4) \qquad \int\limits_{\mathfrak{B}}^{(n)} \{a(Y) - a(X)\} \frac{\partial^2 u}{\partial Y^2} \frac{\partial^2 G}{\partial Y^2} \, dY$$

zu untersuchen. Die Anwendung des Hilfssatzes 3 mit

$$K(X, Z) = \frac{\partial \Gamma(a(X); Z)}{\partial Z}, \qquad \varrho(X) = a(X), \qquad \sigma(X) = \frac{\partial^2 u}{\partial X^2}$$

ergibt für $J_2(X)$ wegen der bloßen Stetigkeit von σ nur die Hölder-Stetigkeit schlechthin. Da $\mathfrak{B}$ in $\mathfrak{G}$ beliebig war, folgt wegen (3.1) jedenfalls überhaupt die Hölder-Stetigkeit der zweiten Ableitungen $\dfrac{\partial^2 u}{\partial X^2}$ in $\mathfrak{G}$. Die Behauptung über den Exponenten ergibt sich nunmehr nachträglich, wenn man den Hilfssatz 3 nebst Korollar noch einmal heranzieht und die eben bewiesene Hölder-Stetigkeit von $\sigma(X)$ berücksichtigt.

Satz 2. *Sind die Funktionen $a_{\nu\mu}(X)$ und $f(X)$ im Gebiete $\mathfrak{G}$ mit Hölder-stetigen Ableitungen erster Ordnung versehen, so ist jede in $\mathfrak{G}$ zweimal stetig differenzierbare Lösung von*

$$L(u) = f$$

daselbst dreimal stetig differenzierbar.

Beweis. $\mathfrak{B}$ sei irgendein konvexer Bereich in $\mathfrak{G}$. Da die $a_{\nu\mu}(X)$ in $\mathfrak{G}$ einmal stetig differenzierbar sind, gilt natürlich dasselbe von der Randintegral-Summe $J_{\mathfrak{R}}(X)$ im Innern von $\mathfrak{B}$.

Bezüglich $J_1(X)$ schreibe man

$$\{f(Y)-f(X)\}\frac{\partial^2 G}{\partial y_i\,\partial y_k}=\frac{\partial}{\partial y_k}\left[\{f(Y)-f(X)\}\frac{\partial G}{\partial y_i}\right]-\frac{\partial f(Y)}{\partial y_k}\frac{\partial G}{\partial y_i}.$$

Somit ist

$$J_1(X)=\int\limits_{\mathfrak{B}}^{(n)}\frac{\partial f(Y)}{\partial y_k}\frac{\partial G}{\partial y_i}\,dY+\text{Randintegral},$$

wobei von dem Randintegral wieder das gleiche wie von $J_{\mathfrak{R}}(X)$ gilt. Auf das n-dimensionale Integral ist der Hilfssatz 2 mit

$$H(X,Y)=\frac{\partial G(X,Y)}{\partial y_i},\qquad \varrho(X)=\frac{\partial f(X)}{\partial x_k}$$

anwendbar, da alle Voraussetzungen wegen

$$\psi(X)=\int\limits_{\mathfrak{B}}^{(n)}H(X,Y)\,dY=\int\limits_{\mathfrak{R}}^{(n-1)}G(X,Y)\,d_iY$$

erfüllt sind. Also ist $J_1(X)$ innerhalb $\mathfrak{B}$ einmal stetig differenzierbar.

Bei $J_2(X)$, d. h. den Integralen (3.4) berücksichtige man die durch Satz 1 sichergestellte Hölder-Stetigkeit der $\frac{\partial^2 u}{\partial X^2}$. Auf diese Integrale ist abermals der Hilfssatz 2 mit

$$H(X,Y)=\{a(Y)-a(X)\}\frac{\partial^2 G}{\partial Y^2},\qquad \varrho(X)=\frac{\partial^2 u}{\partial X^2}$$

anwendbar. Die Voraussetzungen des Hilfssatzes sind erfüllt, da die Funktion

$$\psi(X)=\int\limits_{\mathfrak{B}}^{(n)}H(X,Y)\,dY=\int\limits_{\mathfrak{B}}^{(n)}\{a(Y)-a(X)\}\frac{\partial^2 G}{\partial Y^2}\,dY$$

aus dem gleichen Grunde wie $J_1(X)$ — vgl. den Ausdruck (3.3) — innerhalb $\mathfrak{B}$ einmal stetig differenzierbar ist. Somit besitzt auch $J_2(X)$ diese Eigenschaft, und $u(X)$ ist also nach (3.1) innerhalb $\mathfrak{B}$, und endlich in $\mathfrak{G}$ dreimal stetig differenzierbar.

Die wiederholte Anwendung der Sätze 1 und 2 liefert leicht den

Satz 3. *Sind im Gebiete $\mathfrak{G}$ die m-ten partiellen Ableitungen der Funktionen $a_{\nu\mu}(X)$, $b_\nu(X)$, $c(X)$, $f(X)$ vorhanden und Hölder-stetig mit dem Exponenten α, so ist jede in $\mathfrak{G}$ zweimal stetig differenzierbare Lösung von*

$$\sum_{\nu,\mu}a_{\nu\mu}(X)\frac{\partial^2 u}{\partial x_\nu\,\partial x_\mu}+\sum_\nu b_\nu(X)\frac{\partial u}{\partial x_\nu}+c(X)\,u=f(X)$$

daselbst $(m+2)$-mal Hölder-stetig — mit gleichem Exponenten α — differenzierbar.

Man beweist diesen Satz mit der gleichen Überlegung wie den Satz 4. *Die partielle Differentialgleichung zweiter Ordnung*

$$\Phi\left(\frac{\partial^2 u}{\partial x_1^2}, \frac{\partial^2 u}{\partial x_1 \partial x_2}, \ldots, \frac{\partial^2 u}{\partial x_n^2}; \frac{\partial u}{\partial x_1}, \ldots, \frac{\partial u}{\partial x_n}; u; x_1, \ldots, x_n\right) = 0$$

verhalte sich bezüglich einer vorgegebenen Lösung u an der Stelle $X = X_0$ elliptisch. Ist Φ als Funktion ihrer Argumente in der Umgebung der Stelle

$$\left(\frac{\partial^2 u}{\partial x_\nu \partial x_\mu}; \frac{\partial u}{\partial x_\nu}; u; x_\nu\right)_{X=X_0}$$

mit stetigen partiellen Ableitungen aller Ordnungen versehen, so gilt das gleiche von u als Funktion von X in der Umgebung von $X = X_0$, sobald u dort dreimal stetig differenzierbar ist.

Korollar. Hängt Φ von den zweiten Ableitungen $\frac{\partial^2 u}{\partial x_\nu \partial x_\mu}$ linear ab, so genügt die Voraussetzung, daß $u(X)$ zweimal stetig differenzierbar ist.

Beweis. Man schließe durch vollständige Induktion von der Existenz und Stetigkeit der $(m+2)$-ten Ableitungen auf die Existenz und Stetigkeit der $(m+3)$-ten Ableitungen von $u(X)$ $(m \geqq 1)$. Durch m-malige Differentiation der Gleichung $\Phi = 0$ nach den x_ν erhält man offenbar lineare elliptische Differentialgleichungen

$$\sum_{\nu,\mu} a_{\nu\mu} \frac{\partial^2 u_m}{\partial x_\nu \partial x_\mu} = \Phi_m$$

für die m-ten partiellen Ableitungen $u_m = \frac{\partial^m u}{\partial X^m}$, wobei die $a_{\nu\mu}$ in der Einleitung definiert wurden, und Φ_m nur die Ableitungen bis zur $(m+1)$-ten Ordnung von $u(X)$ enthält. Da die Existenz und Stetigkeit der u_{m+2} $(m \geqq 1!)$ schon als bewiesen vorausgesetzt war, sind die $a_{\nu\mu}$ und Φ_m als Funktionen von X einmal stetig differenzierbar, also nach Satz 1 die u_{m+2} Hölder-stetig. Daher sind nunmehr $a_{\nu\mu}$ und Φ_m als Funktionen von X einmal Hölder-stetig differenzierbar, also nach Satz 2 die u_m dreimal stetig differenzierbar, d. h. $u(X)$ ist $(m+3)$-mal stetig differenzierbar.

Diese Schlußweise tritt von $m = 1$ an in Kraft. Unter der Voraussetzung des Korollars kann man offenbar schon von $m = 0$ ab schließen.

§ 4.

Verschärfung des Satzes 4.

Für eine in einem Bereiche $\mathfrak{B}$ mit dem Exponenten α Hölder-stetige Funktion $\varphi(X)$ sei die Bezeichnung

$$H_\alpha(\varphi) = \text{Ob.Gr.}\, |\varphi(X'') - \varphi(X')| \cdot |X'' - X'|^{-\alpha} \qquad (X', X'' \text{ in } \mathfrak{B})$$

14*

für die scharfe Hölder-Konstante eingeführt. Eine weitere Anwendung der Integralformeln (3.1) führt zu dem

Lemma 1. $a_{\nu\mu}(X)$ und $f(X)$ seien in einem Bereiche $\mathfrak{B}$ Hölder-stetig mit dem Exponenten α $(0 < \alpha \leqq 1)$. Für die zweiten Ableitungen jeder in $\mathfrak{B}$ zweimal stetig differenzierbaren Lösung $u(X)$ der in $\mathfrak{B}$ elliptischen linearen Differentialgleichung

$$L(u) = \sum_{\nu, \mu} a_{\nu\mu}(X) \frac{\partial^2 u}{\partial x_\nu \partial x_\mu} = f(X)$$

gelten dann Ungleichungen der Form

$$\left| \frac{\partial^2 u}{\partial X^2} \right| < \text{konst.} \left[\frac{U}{d^2} + \frac{U'}{d} + F + d^\alpha H_\alpha(f) \right],$$

unter U, U', F die oberen Grenzen der $u, \frac{\partial u}{\partial X}, f$ in $\mathfrak{B}$ verstanden, und wo d den Abstand des Punktes X vom Rande von $\mathfrak{B}$ bedeutet. Die Konstante kann ein für allemal fest gewählt werden, wenn die Funktionen $a_{\nu\mu}(X)$ hinreichend wenig abgeändert werden, jedoch so, daß dabei die Hölder-Konstanten $H_\alpha(a_{\nu\mu})$ unter einer festen Schranke bleiben[22]).

Beweis. Es bedeute K einen ganz im Innern von $\mathfrak{B}$ gelegenen Hyperkugel-Bereich, etwa mit dem Radius R. Wir operieren zunächst nur in K und verstehen unter δ den Abstand eines in K liegenden Punktes X vom Rande von K. Das Maximum

$$M = \operatorname{Max} \delta^2 \left| \frac{\partial^2 u}{\partial X^2} \right| \qquad (X \text{ in } K), \tag{4.1}$$

genommen für alle zweiten Ableitungen von u, wird gewiß in einem inneren Punkte von K, den wir als Koordinatenursprung wählen wollen, angenommen. Ist δ_0 der Abstand dieses Punktes vom Rande von K, so ist

$$\delta_0 \leqq R \tag{4.2}$$

und nach (4.1)

$$\left| \frac{\partial^2 u}{\partial X^2} \right| \leqq \frac{M}{(|X| - \delta_0)^2}, \qquad \left| \frac{\partial^2 u}{\delta X^2} \right|_{X=0} = \frac{M}{\delta_0^2} \qquad (|X| < \delta_0), \tag{4.3}$$

wobei die Gleichung so aufzufassen ist, daß das Gleichheitszeichen für mindestens eine der $\frac{\partial^2 u}{\partial x_\nu \partial x_\mu}$ zutrifft. Auf den neuen, gewiß ganz in K gelegenen Hyperkugel-Bereich $|X| < \vartheta \delta_0$ (mit noch zu bestimmendem $\vartheta < 1$) wende man nun die Formeln (3.1) an. Man beachte dabei im folgenden, daß bei hinreichend geringer Abänderung der $a_{\nu\mu}(X)$ die Kon-

[22]) Auf andere Anwendungen dieses Lemmas hoffe ich noch zurückzukommen. Von der oben angegebenen besonderen Form der Schranke wird hier kein Gebrauch gemacht.

stanten in den Ungleichungen (1.15) ein für allemal fest gewählt werden können. Dann folgt für den Punkt $X = 0$

$$(4.4)\qquad |J_1(0)| < \text{konst.}\, H_\alpha(f) \overset{(n)}{\int\limits_{|Y|<\vartheta\delta_0}} |Y|^{\alpha-n}\, dY = \text{konst.}\, H_\alpha(f)\,(\vartheta\delta_0)^\alpha.$$

Weiter erhält man, da wegen (4.3) $\left|\frac{\partial^2 u}{\partial Y^2}\right| < M\delta_0^{-2}(1-\vartheta)^{-2}$ für $|Y| < \vartheta\delta_0$ gilt,

$$(4.5)\qquad |J_2(0)| < \text{konst.}\, \frac{M}{\delta_0^2(1-\vartheta)^2} \cdot \overset{(n)}{\int\limits_{|Y|<\vartheta\delta_0}} |Y|^{\alpha-n}\, dY = \text{konst.}\, \frac{(\vartheta\delta_0)^\alpha M}{\delta_0^2(1-\vartheta)^2}.$$

Die Konstante hängt dabei nur von oberen Schranken für die $H_\alpha(a_{\nu\mu})$ ab. Für $J_{\mathfrak{R}}$, die Summe der über den Rand $|Y| = \vartheta\delta_0$ erstreckten Integrale, folgt durch leichte Anwendung von (1.4)

$$(4.6)\qquad |J_{\mathfrak{R}}(0)| < \text{konst.}\left(\frac{U}{(\vartheta\delta_0)^2} + \frac{U'}{\vartheta\delta_0} + F\right).$$

Wegen (4.3) und wegen (4.4), (4.5), (4.6) folgt dann

$$\left|\frac{\partial^2 u}{\partial X^2}\right|_{X=0} = \frac{M}{\delta_0^2} \leqq |J_1(0)| + |J_2(0)| + |J_{\mathfrak{R}}(0)|$$

$$< c\cdot\left[\frac{(\vartheta\delta_0)^\alpha}{\delta_0^2(1-\vartheta)^2} M + \frac{U}{(\vartheta\delta_0)^2} + \frac{U'}{\vartheta\delta_0} + F + (\vartheta\delta_0)^\alpha H_\alpha(f)\right],$$

wobei die Konstante c unter den Voraussetzungen des Lemmas fest gewählt werden kann. Nun ist $\delta_0 < D$, unter D den Durchmesser von $\mathfrak{B}$ verstanden, und daher

$$\frac{c(\vartheta\delta_0)^\alpha}{(1-\vartheta)^2} < cD^\alpha \cdot \frac{\vartheta^\alpha}{(1-\vartheta)^2}.$$

Wir erteilen dem $\vartheta < 1$ einen solchen festen Wert, daß hier der rechte Ausdruck $< \frac{1}{2}$ ausfällt. Dann ergibt sich wegen (4.2)

$$M < 2c\left[\frac{U}{\vartheta^2} + R\frac{U'}{\vartheta} + R^2 F + R^{2+\alpha}\vartheta^\alpha H_\alpha(f)\right].$$

Wegen (4.1) folgt also im Mittelpunkte von K $(\delta = R)$

$$\left|\frac{\partial^2 u}{\partial X^2}\right| \leqq \frac{M}{R^2} < \text{konst.}\left[\frac{U}{R^2} + \frac{U'}{R} + F + R^\alpha H_\alpha(f)\right].$$

Da nun dieser Mittelpunkt ein beliebiger innerer Punkt von $\mathfrak{B}$ sein darf, und R nur der Bedingung $R < d$ unterworfen ist, folgt endlich die gewünschte Ungleichung, wenn man $R \to d$ streben läßt.

Lemma 2. Die partielle Differentialgleichung

$$L^{(\delta)}(v) = \sum_{\nu,\mu} a_{\nu\mu}^{(\delta)}(X)\frac{\partial^2 v}{\partial x_\nu\, \partial x_\mu} = f^{(\delta)}(X)$$

sei für alle genügend kleinen $|\delta|$ im Bereiche $\mathfrak{B}$ elliptisch. Für diese $|\delta|$ seien die $a_{\nu\mu}^{(\delta)}$ und $f^{(\delta)}$ stetige Funktionen von X und δ, und die Hölder-Konstanten H_α dieser Funktionen mögen unterhalb einer von δ unabhängigen Schranke liegen. Ferner sei für jedes $\delta \neq 0$ eine Lösung v_δ vorgegeben. Konvergieren für $\delta \to 0$ v_δ und die $\frac{\partial v_\delta}{\partial X}$ gleichmäßig gegen stetige Funktionen v_0 und $\frac{\partial v_0}{\partial X}$, so ist v_0 innerhalb $\mathfrak{B}$ zweimal stetig differenzierbar.

Beweis. Nach dem Lemma 1 sind in jedem ganz innerhalb $\mathfrak{B}$ gelegenen Bereiche $\mathfrak{B}'$ die Funktionen $\left|\frac{\partial^2 v_\delta}{\partial X^2}\right|$ sicher unterhalb einer von δ unabhängigen Schranke gelegen. Man wende nun wiederum die Integralformeln (3.1) auf die Gleichung $L^{(\delta)}(v_\delta) = f^{(\delta)}$ im Bereiche $\mathfrak{B}'$ an; dabei ist $G = G_\delta = \Gamma(a_\delta(X); Y - X)$ zu setzen. Zunächst ist es klar, daß $J_{\mathfrak{R}}(X) = J_{\mathfrak{R}}^{(\delta)}(X)$ eine stetige Funktion von X und δ darstellt (mit Einschluß von $\delta = 0$). Was nun J_1 und J_2 betrifft, so bemerken wir, daß die Integranden die Form

$$K_\delta(X, Y)\,\varphi_\delta(Y)$$

besitzen, wobei $|\varphi_\delta(Y)|$ in $\mathfrak{B}'$ unterhalb einer von δ unabhängigen Schranke gelegen ist, und wo K_δ mit Rücksicht auf die Voraussetzungen des Lemmas eine für X, Y in $\mathfrak{B}$, $X \neq Y$, $|\delta| \leqq \delta_0$ bei passendem δ_0 stetige Funktion von X, Y und δ darstellt mit der Eigenschaft

$$|K_\delta(X, Y)| < \frac{\text{konst.}}{|Y - X|^{n-\alpha}};$$

die Konstante ist dabei von δ unabhängig. Auf Grund dieser Eigenschaften schließt man mit Hilfe einer wohlbekannten Stetigkeitsbetrachtung, daß das Integral

$$F_\delta(X) = \int_{\mathfrak{B}'}^{(n)} K_\delta(X, Y)\,\varphi_\delta(Y)\,dY$$

eine für alle δ mit $0 < |\delta| \leqq \delta_0$ gleichgradig stetige Funktion von X in $\mathfrak{B}'$ darstellt[23]). Da $J_1^{(\delta)}$ und $J_2^{(\delta)}$ von dieser Form sind, schließt man endlich, daß die Funktionen $\frac{\partial^2 v_\delta}{\partial X^2}$ bezüglich aller hinreichend kleinen $|\delta| > 0$ gleichgradig stetige Funktionen von X sind. Aus dieser Gleichgradigkeit folgt, wie eine leichte Überlegung lehrt, zunächst, daß die Limites

$$\lim_{\delta=0} \frac{\partial^2 v_\delta}{\partial X^2}$$

[23]) Es handelt sich um dieselbe (nur um eine geringfügige Gleichmäßigkeitsbetrachtung vermehrte) Überlegung, mit welcher man die Stetigkeit der ersten Ableitungen des Newtonschen Potentials mit beschränkter Belegungsdichte beweist. Vgl. etwa loc. cit. [21]).

existieren, und zwar mit gleichmäßiger Konvergenz im Innern von $\mathfrak{B}'$, sodann, daß diese Limites die notwendig stetigen zweiten Ableitungen von $v_0(X)$ darstellen.

Wir beweisen nun den

Satz 5. *Der Satz 4 gilt unter der geringeren Voraussetzung, daß die Lösung $u(X)$ in der Umgebung von $X = X_0$ zweimal Hölder-stetig differenzierbar ist.*

Beweis. Man setze

$$\frac{u(X+\delta) - u(X)}{\delta} = v_\delta(X), \tag{4.7}$$

unter δ den Zuwachs irgendeines x_ν verstanden. Die ersten bzw. zweiten Ableitungen von u seien kurz mit u' bzw. u'' bezeichnet. Ferner setze man

$$\begin{aligned} X_t &= X + t\delta, \\ u_t &= (1-t)\,u(X) \;\;+ t\,u(X+\delta), \\ u_t' &= (1-t)\,u'(X) \;+ t\,u'(X+\delta), \\ u_t'' &= (1-t)\,u''(X) + t\,u''(X+\delta). \end{aligned} \tag{4.8}$$

Schreibt man kurz $\Phi = \Phi(u'', u', u, X)$, so ist für $u = u(X)$ unter Anwendung einer bekannten elementaren Formel

$$\begin{aligned} 0 &= \frac{\Phi(u_1'', u_1', u_1, X_1) - \Phi(u_0'', u_0', u_0, X_0)}{\delta} \\ &= \sum \frac{u_1'' - u_0''}{\delta} \int_0^1 \frac{\partial \Phi}{\partial u''}(u_t'', u_t', u_t, X_t)\,dt \\ &\left.\begin{aligned} &+ \sum \frac{u_1' - u_0'}{\delta} \int_0^1 \frac{\partial \Phi}{\partial u'}(u_t'', u_t', u_t, X_t)\,dt \\ &+ \frac{u_1 - u_0}{\delta} \int_0^1 \frac{\partial \Phi}{\partial u}(u_t'', u_t', u_t, X_t)\,dt \\ &+ \int_0^1 \frac{\partial \Phi}{\partial X}(u_t'', u_t', u_t, X_t)\,dt \end{aligned}\right\} = -f^{(\delta)}(X). \end{aligned} \tag{4.9}$$

Die erste Summe auf der rechten Seite lautet mit Rücksicht auf die in der Einleitung definierten $a_{\nu\mu}$ und wegen

$$\frac{u_1 - u_0}{\delta} = v_\delta(X)$$

genauer

$$L^{(\delta)}(v_\delta) = \sum_{\nu,\mu} a_{\nu\mu}^{(\delta)}(X) \frac{\partial^2 v_\delta}{\partial x_\nu\, \partial x_\mu}, \qquad a_{\nu\mu}^{(\delta)}(X) = \int_0^1 a_{\nu\mu}(u_t'', u_t', u_t, X_t)\,dt.$$

Wegen der Hölder-Stetigkeit der $u''(X)$ in der Umgebung von $X = X_0$ sind die $a^{(\delta)}_{\nu\mu}(X)$, überhaupt die in (4.9) auftretenden Integrale offenbar ebenfalls Hölder-stetige Funktionen von X für alle hinreichend kleinen $|\delta|$; bei passendem $\alpha > 0$ werden sogar die auf eine passende Umgebung von $X = X_0$ bezogenen Hölder-Konstanten H_α dieser Funktionen unterhalb einer von δ unabhängigen Schranke liegen. Die Hölder-Stetigkeit der in (4.9) stehenden Faktoren

$$\frac{u_1' - u_0'}{\delta} = \frac{u'(X+\delta) - u'(X)}{\delta} = \int_0^1 u''(X + t\delta)\,dt,$$

$$\frac{u_1 - u_0}{\delta} = \frac{u(X+\delta) - u(X)}{\delta} = \int_0^1 u'(X + t\delta)\,dt$$

ist ebenfalls sofort ersichtlich, und von den Hölder-Konstanten gilt dasselbe. Somit genügen $a^{(\delta)}_{\nu\mu}(X)$ und das $f^{(\delta)}(X)$ von (4.9) allen Voraussetzungen von Lemma 2. Da wegen (4.7) für $\delta \to 0$

$$v_\delta(X) \to v_0(X) = u'(X), \qquad \frac{\partial v_\delta(X)}{\partial X} \to \frac{\partial v_0(X)}{\partial X} = u''(X)$$

gleichmäßig gilt, folgt also, daß die $u'(X)$ zweimal differenzierbar sind. Da somit $u(X)$ dreimal stetig differenzierbar ist, folgt die Behauptung aus dem Satz 4.

II. Kapitel.

Beweis des analytischen Charakters der Lösungen von $\Phi = 0$.

§ 5.

Ein System nichtlinearer Integralgleichungen für die Lösungen von $\Phi = 0$.

Die partielle Differentialgleichung

$$\Phi\left(\frac{\partial^2 u}{\partial x_1^2}, \frac{\partial^2 u}{\partial x_1 \partial x_2}, \ldots, \frac{\partial^2 u}{\partial x_n^2}; \frac{\partial u}{\partial x_1}, \ldots, \frac{\partial u}{\partial x_n}; u; x_1, \ldots, x_n\right) = 0$$

verhalte sich bezüglich der vorgegebenen Lösung $u(X)$ elliptisch und analytisch im Koordinatenursprung $X = 0$. Für den im folgenden zu erbringenden Beweis der Analytizität von $u(X)$ an der Stelle $X = 0$ setzen wir $u(X)$ als sechsmal stetig differenzierbar voraus. Nach den Resultaten des ersten Kapitels ist diese Voraussetzung von selbst erfüllt, sobald die $\frac{\partial^2 u}{\partial X^2}$ Hölder-stetig sind.

Wir schreiben zur Abkürzung $u' = \frac{\partial u}{\partial X}, \ldots, u'''' = \frac{\partial^4 u}{\partial X^4}$ für die partiellen Ableitungen von u bis zur vierten Ordnung und demgemäß $\Phi = \Phi(u'', u', u, X)$.

Es bedeutet keine Einschränkung der Allgemeinheit, wenn

$$(5.1)\qquad u(0) = u'(0) = u''(0) = u'''(0) = u''''(0) = 0$$

angenommen wird; sonst betrachte man statt u die Funktion

$$v = u - u(0) - \frac{u'(0)}{1!}X - \frac{u''(0)}{2!}X^2 - \frac{u'''(0)}{3!}X^3 - \frac{u''''(0)}{4!}X^4$$

(es ist wohl nicht mißzuverstehen, was damit angedeutet ist), welche, wie leicht zu sehen, ebenfalls einer elliptischen Differentialgleichung der obigen Art genügt. Differenziert man nun die Gleichung $\Phi = 0$ viermal nach den x_ν, so erhält man nach leichter Überlegung Differentialgleichungen der Form

$$(5.2)\qquad L(u'''') = \sum_{\nu,\mu} a_{\nu\mu} \frac{\partial^2 u''''}{\partial x_\nu \partial x_\mu} = \sum b \frac{\partial^5 u}{\partial X^5} + f \text{ [24])}$$

für jede der vierten Ableitungen u'''' mit (die $a_{\nu\mu}$ wurden bereits in der Einleitung definiert)

$$\begin{aligned} a_{\nu\mu} &= a_{\nu\mu}(u'', u', u, X), \\ b &= b(u''', u'', u', u, X), \\ f &= f(u'''', u''', u'', u', u, X), \end{aligned}$$

wobei diese Funktionen an der Stelle

$$u'''' = u''' = u'' = u' = u = X = 0$$

analytisch von ihren Argumenten abhängen. Daher läßt sich gewiß ein (im folgenden festgehaltener) Hyperkugel-Bereich

$$|X| \leqq R_0$$

so angeben, daß die Form $\sum a_{\nu\mu} z_\nu z_\mu$ für $u'' = u''(X), \ldots$ und $|X| \leqq R_0$ dauernd positiv definit ist und daß die $a_{\nu\mu}$, b, f als Funktionen von $u'''', \ldots, X$ für alle Wertesysteme $u'''' = u''''(X), \ldots$ und $|X| \leqq R_0$ analytisch sind.

Wir wenden auf (5.2) im Bereiche $|X| \leqq R_0$ die Integralformel (1.10) an, und zwar nicht mit der Funktion $G(X, Y)$ von (1.9), sondern mit der für die folgenden Zwecke geeigneteren Funktion

$$\begin{aligned} H(X, Y) &= G(Y, X) = \Gamma(a(Y); Y - X), \\ a(Y) &= a_{\nu\mu}(Y) = a_{\nu\mu}(u''(Y), u'(Y), u(Y), Y). \end{aligned}$$

[24]) D. h. die fünften Ableitungen kommen nur linear vor mit Koeffizienten, die von den u'''' nicht abhängen. Durch zweimalige Differentiation nach X erhält man zunächst

$$L(u'') = \Psi(u''', u'', u', u, X)$$

für jede der zweiten partiellen Ableitungen u''. Differenziert man noch einmal, so folgt

$$L(u''') = \sum u'''' \Psi_1(u''', u'', u', u, X) + \Psi_2(u''', u'', u', u, X),$$

woraus sich (5.2) ergibt.

Wir erhalten

$$(5.3)\quad u''''(X) = \int\limits_{|Y|<R_0}^{(n)} \{u'''' M_Y(H) - Hf\}\, dY - \sum \int\limits_{|Y|<R_0}^{(n)} Hb \frac{\partial^5 u}{\partial Y^5}\, dY$$

$$+ \sum_\nu \int\limits_{|Y|=R_0}^{(n-1)} N_\nu\left(u'''', H, \frac{\partial u''''}{\partial Y}, \frac{\partial H}{\partial Y}\right) d_\nu Y.$$

Von den Randintegralen sei gleich bemerkt, daß sie für $|X| < R_0$ analytische Funktionen von X darstellen. Für $|Y| = R_0$ sind nämlich $H, \frac{\partial H}{\partial Y}$, also auch die Ausdrücke N_ν analytische Funktionen von X im Gebiete $|X| < R_0$, und zwar gleichmäßig in Y. Im zweiten Summanden rechts integriere man partiell nach dem Schema

$$Hb \frac{\partial^5 u}{\partial Y^5} = \frac{\partial (Hbu'''')}{\partial Y} - u'''' \frac{\partial (bH)}{\partial Y},$$

wo u'''' eine passende (mit dem u'''' links in (5.3) nicht notwendig übereinstimmende) vierte Ableitung bedeutet. Dann ergibt sich endlich

$$(5.4)\quad u''''(X) = \int\limits_{|Y|<R_0}^{(n)} W\, dY + \varphi(X),$$

unter $\varphi(X)$ eine gewiß für $|X| < R_0$ analytische Funktion verstanden, mit

$$W = u''''(Y)\, M_Y(H) - Hf(u''''(Y), \ldots, Y) + \sum u''''(Y) \frac{d(bH)}{dY},$$

$$M_Y(H) = \sum_{\nu,\mu} \frac{d^2(a_{\nu\mu} H)}{dy_\nu\, dy_\mu}.$$

Wir gebrauchen im Moment das Differentiationssymbol d, wo es sich um nur mittelbare Abhängigkeit von Y handelt; z. B. ist in verständlicher symbolischer Ausdrucksweise $(Z = Y - X)$

$$\frac{dH}{dY} = \frac{\partial \Gamma}{\partial a}\frac{da}{dY} + \frac{\partial \Gamma}{\partial Z}, \quad \frac{d^2 H}{dY^2} = \frac{\partial^2 \Gamma}{\partial a^2}\left(\frac{da}{dY}\right)^2 + 2\frac{\partial^2 \Gamma}{\partial a \partial Z}\frac{da}{dY} + \frac{\partial^2 \Gamma}{\partial Z^2} + \frac{\partial \Gamma}{\partial a}\frac{d^2 a}{dY^2},$$

$$\frac{da}{dY} = \frac{\partial a}{\partial u''} u''' + \ldots + \frac{\partial a}{\partial Y}, \quad \frac{db}{dY} = \frac{\partial b}{\partial u'''} u'''' + \ldots + \frac{\partial b}{\partial Y},$$

$$\frac{d^2 a}{dY^2} = \frac{\partial a}{\partial u''} u'''' + \ldots + \frac{\partial^2 a}{\partial Y^2}.$$

Man erkennt hieraus, daß W sich als Linearkombination der Größen

$$(5.5)\quad \Gamma, \frac{\partial \Gamma}{\partial a}, \frac{\partial \Gamma}{\partial Z}, \frac{\partial^2 \Gamma}{\partial a^2}, \frac{\partial^2 \Gamma}{\partial a \partial Z} \quad (Z = X - Y)$$

— die $\frac{\partial^2 \Gamma}{\partial Z^2}$ fallen wegen (1.8) heraus — darstellt, deren Koeffizienten sich ganz und rational aus den u'''', u''', u'', u' und aus partiellen Ableitungen der $a_{\nu\mu}$, b, f nach ihren unmittelbaren Argumenten zusammensetzen. Mit Rücksicht auf die Form (1.14) der Größen (5.5) erhält man W endlich in der Gestalt

$$(5.6)\qquad W(u'''', u''', u'', u', u, Y; Z) = \frac{Q_3(u'''', \ldots, u, Y; Z)}{\Omega(a; Z)^{\frac{n+2}{2}}},$$

$$a = a_{\nu\mu} = a_{\nu\mu}(u'', u', u, Y),$$

wo Q_3 ein Polynom der z_ν mit lauter Gliedern von mindestens drittem Grade darstellt, dessen Koeffizienten analytische Funktionen der $u'''', \ldots, u, Y$ für alle Wertesysteme

$$u'''' = u''''(Y), \ldots, u = u(Y), \quad Y \text{ mit } |Y| \leqq R_0$$

derselben sind.

Fügen wir die Relation $u(X) = \sum_\nu x_\nu \int_0^1 \frac{\partial u}{\partial x_\nu}(tX)\,dt$ (man berücksichtige (5.1)) und die analogen, die u' durch die u'' usw. ausdrückenden Relationen zu (5.4) hinzu, so erhalten wir endlich das in verständlicher symbolischer Form geschriebene Integralgleichungssystem

$$(5.7)\qquad u''''(X) = \int\limits_{|Y|<R_0}^{(n)} W(u''''(X), \ldots, u(Y), Y;\ Y - X)\,dY + \varphi(X),$$

$$(5.8)\qquad \begin{cases} u'''(X) = \sum x \int_0^1 u''''(tX)\,dt, & u''(X) = \sum x \int_0^1 u'''(tX)\,dt, \\ u'(X) = \sum x \int_0^1 u''(tX)\,dt, & u(X) = \sum x \int_0^1 u'(tX)\,dt, \end{cases}$$

wobei jedenfalls die Funktionen $\varphi(X)$ gegebene analytische Funktionen in $|X| < R_0$ sind. Unser Ziel ist der Analytizitätsbeweis des vorgegebenen Lösungssystems $u(X), \ldots, u''''(X)$ von (5.7), (5.8) an der Stelle $X = 0$.

Für das Folgende sind noch ein paar Bemerkungen über die ersten Ableitungen von W nach seinen Argumenten von Nutzen. Man überzeugt sich ähnlich wie am Schluß von § 1, daß sie die Form

$$(5.9)\qquad \frac{\partial W}{\partial u^{(i)}}, \frac{\partial W}{\partial Y} = \frac{Q_5(u'''', \ldots, u, Y; Z)}{\Omega(a; Z)^{\frac{n+4}{2}}}; \quad \frac{\partial W}{\partial Z} = \frac{Q_4(u'''', \ldots, u, Y; Z)}{\Omega(a; Z)^{\frac{n+4}{2}}}$$

besitzen, wo Q_4 bzw. Q_5 Polynome der z_ν mit Gliedern vom Mindestgrad vier bzw. fünf und mit Koeffizienten derselben Natur wie oben bei Q_3 darstellen. W, $\frac{\partial W}{\partial u^{(i)}}$, $\frac{\partial W}{\partial Y}$ werden also für $Z = 0$ von $(n-1)$-ter, $\frac{\partial W}{\partial Z}$ von n-ter Ordnung unendlich.

§ 6.

Vorbereitungen zum Übergang ins Komplexe.

Aus Durchsichtigkeitsgründen soll im folgenden statt des Integralgleichungssystems (5.7), (5.8) das einfachere System

$$(6.1)\qquad \begin{aligned} u(X) &= \int\limits_{|Y|<R_0}^{(n)} W(u(Y), v(Y), Y; Y-X)\, dY + \varphi(X), \\ v(X) &= x_1 \int\limits_0^1 u(tX)\, dt \end{aligned}$$

für nur zwei Funktionen u, v betrachtet werden. $\varphi(X)$ sei wie vorhin eine für $|X| < R_0$ analytische Funktion von X, und W sei wieder von der Gestalt

$$(6.2)\qquad \begin{aligned} W(u, v, Y; Z) &= \frac{Q_3(u, v, Y; Z)}{\Omega(a, Z)^{\frac{n+2}{2}}}, \\ a_{\nu\mu} &= a_{\nu\mu}(u, v, Y). \end{aligned}$$

Die partiellen Ableitungen erster Ordnung sind dann von der Form

$$(6.3)\qquad \frac{\partial W}{\partial u}, \frac{\partial W}{\partial v}, \frac{\partial W}{\partial Y} = \frac{Q_5(u, v, Y; Z)}{\Omega(a; Z)^{\frac{n+4}{2}}}, \quad \frac{\partial W}{\partial Z} = \frac{Q_4(u, v, Y; Z)}{\Omega(a; Z)^{\frac{n+4}{2}}}.$$

Vorgegeben seien zwei im Gebiete $|X| < R_0$ beschränkte, einmal stetig differenzierbare und den Gleichungen (6.1) genügende Funktionen $u(X)$, $v(X)$ mit

$$(6.4)\qquad u(0) = v(0) = 0.$$

Die $a_{\nu\mu}$, sowie die Koeffizienten der Polynome Q_3, Q_4, Q_5 seien als Funktionen ihrer Argumente u, v, Y an allen Stellen

$$u(Y), v(Y), Y \quad \text{mit} \quad |Y| < R_0$$

und deren Häufungsstellen analytisch, und es existiere ein $\omega > 0$, so daß

$$(6.5)\qquad \begin{aligned} \Omega(a, Z) &\geqq \omega\, |Z|^2, \\ a_{\nu\mu} &= a_{\nu\mu}(u(Y), v(Y), Y), \quad |Y| < R_0 \end{aligned}$$

gilt.

Wir beweisen, daß unter diesen Voraussetzungen $u(X)$, $v(X)$ an der Stelle $X = 0$ analytisch sind. Man wird leicht bemerken, daß die folgenden auf das System (6.1) bezüglichen Überlegungen in genau derselben Weise auf das System (5.7), (5.8) anwendbar sind. Wir schreiben die Gleichungen (6.1) noch in der für das Folgende geeigneteren Form

$$(6.6)\qquad \begin{aligned} u(X) &= \int\limits_{|Y|<R}^{(n)} W(u(Y), v(Y), Y; X-Y)\, dY + \Phi(X), \\ v(X) &= x_1 \int\limits_0^1 u(tX)\, dt \end{aligned}$$

mit

$$(6.7) \qquad \Phi(X) = \int\limits_{R<|Y|<R_0}^{(n)} W(u(Y), v(Y), Y; Y-X)\,dY + \varphi(X)$$

und beschränken den Punkt X auf das kleinere Gebiet $|X| < R$ mit noch zu bestimmendem $R < R_0$.

Von jetzt ab seien auch komplexe Werte für die Koordinaten der Punkte X, Y, Z usw. zugelassen, indem

$$X = X' + iX'', \quad Y = Y' + iY'', \quad Z = Z' + iZ''$$

geschrieben werden soll, wenn $x_\nu = x'_\nu + i x''_\nu$, $y_\nu = y'_\nu + i y''_\nu$, $z_\nu = z'_\nu + i z''_\nu$ ist. Unser nächstes Ziel besteht darin, dem Gleichungssystem (6.6) auch für komplexe Punkte X einen passenden Sinn zu erteilen. Zu diesem Zweck seien komplexe Umgebungen des reellen Gebietes $|X| < R_0$ von der Form

$$(R)_\gamma\colon \quad |X'| < R, \quad |X''| < \gamma(R - |X'|)$$

$(\gamma > 0)$ eingeführt. Offenbar ist $(R)_\gamma$ konvex; denn mit X_1 und X_2 gehört auch der Punkt $X = tX_1 + (1-t)X_2$ zu $(R)_\gamma$, da

$$\begin{aligned}|X'| + \gamma|X''| &\leqq t|X'_1| + (1-t)|X'_2| + \gamma t|X''_1| + \gamma(1-t)|X''_2| \\ &\leqq \gamma t R + \gamma(1-t)R = \gamma R\end{aligned}$$

ist. Für γ sei von vornherein der durch

$$(6.8) \qquad \begin{aligned}\frac{\omega}{3\gamma^2} &= \text{Ob. Gr.} \sum_{\nu,\mu} |A_{\nu\mu}|, \\ A_{\nu\mu} &= A_{\nu\mu}(u(Y), v(Y), Y), \quad Y \text{ reell}, \quad |Y| < R_0\end{aligned}$$

gegebene feste Wert zugrunde gelegt. (Es sei noch einmal daran erinnert, daß die $A_{\nu\mu}$ die algebraischen Komplemente der $a_{\nu\mu}$ bezeichnen.)

$\Phi(X)$ für komplexe X. Wir zeigen zunächst, daß die durch (6.7) bestimmte Funktion $\Phi(X)$ bei passender Wahl von R im komplexen Gebiete $(R)_\gamma$ regulär und beschränkt ist. Für den Realteil der Form $\Omega(a; Z)$ gilt nun offenbar

$$(6.9) \qquad \begin{aligned}\Re[\Omega(a; Z)] &= \sum_{\nu,\mu} A_{\nu\mu}(z'_\nu z'_\mu - z''_\nu z''_\mu), \\ a_{\nu\mu} &= a_{\nu\mu}(u(Y), v(Y), Y), \quad Y \text{ reell}, \quad |Y| < R_0,\end{aligned}$$

also wegen $|z''_\nu| \leqq |Z''|$ und wegen (6.5) und (6.8)

$$\Re[\Omega] \geqq \omega\left(|Z'|^2 - \frac{1}{3\gamma^2}|Z''|^2\right).$$

Für $Z = Y - X$ mit irgendeinem X aus $(R)_\gamma$ und reellem Y, $|Y| > R$, ist nun

$$(6.10) \qquad |Z'| = |Y - X'|, \; |Z''| = |X''| < \gamma(R - |X'|) < \gamma|Y - X'|,$$

und daher unter der Bedingung (6.9)

$$(6.11) \qquad \Re[\Omega] \geqq \frac{2}{3}\omega|Y - X'|^2.$$

$\Omega(a; Z)$ mit (6.9) ist daher für $|Y| > R$ eine in $(R)_\gamma$ reguläre Funktion von X mit positivem Realteil, $\sqrt{\Omega}$ also — es ist die für reelle X positive Wurzel gemeint — eine in $(R)_\gamma$ eindeutige reguläre Funktion von X. Unter den Bedingungen (6.9) und für $|Y| > R$ ist also auch $W(u(Y), v(Y), Y, Y-X)$, der Integrand des Integrales in (6.7), eine in $(R)_\gamma$ eindeutige reguläre Funktion von X mit den Eigenschaften

$$(6.12)\qquad |W| < \frac{\text{konst.}}{|Y-X'|^{n-1}},\quad \left|\frac{\partial W}{\partial Z}\right| < \frac{\text{konst.}}{|Y-X'|^{n}};\qquad Z = Y-X,$$

die man leicht aus (6.2), (6.3) und (6.11) — in den Polynomen Q benütze man die aus (6.10) folgenden Ungleichungen

$$|z_\nu| \leqq |Z'| + |Z''| < (1+\gamma)\,|Y-X'|\ ^{25})$$

— ersieht. Da die Funktion $\varphi(X)$ von (6.7) im reellen Gebiete $|X| < R_0$ analytisch ist, kann man R so klein wählen, daß $\varphi(X)$ in $(R)_\gamma$ eindeutig und regulär ist. Bei passender Wahl von R wird dann nach dem Obigen die Funktion $\Phi(X)$ von (6.7) in $(R)_\gamma$ regulär und außerdem beschränkt sein, da nach (6.12)

$$|\Phi(X)| < \text{konst.} \int\limits_{R<|Y|<R_0}^{(n)} \frac{dY}{|Y-X'|^{n-1}} + \text{konst.};\quad X \text{ in } (R)_\gamma$$

ist, und das hier auftretende Integral bekanntlich beschränkt ist. Da weiterhin für reelle Y mit $|Y| > R$ und für $|X'| < R$ gewiß $|Y-X'|^{-n} < (R - |X'|)\,|Y-X'|^{-n+1}$ ist, folgt aus (6.7) und (6.12)

$$\left|\frac{\partial \Phi(X)}{\partial X}\right| < \frac{\text{konst.}}{R-|X'|} \int\limits_{R<|Y|<R_0}^{(n)} \frac{dY}{|Y-X'|^{n-1}} + \text{konst.}$$

Somit gilt bei hinreichend kleinem R

$$(6.13)\qquad |\Phi(X)| < \text{konst.},\quad \left|\frac{\partial \Phi(X)}{\partial X}\right| < \frac{\text{konst.}}{R-|X'|};\quad X \text{ in } (R)_\gamma$$

für Φ und ihre ersten Ableitungen $\dfrac{\partial \Phi}{\partial x_\nu}$.

Die Funktion W für komplexe Argumente u, v, Y, Z. Die Zahl $\varkappa > 0$ sei so gewählt, daß — man berücksichtige (6.4) — in dem komplexen Gebiete

$$(6.14)\qquad |u| < \varkappa,\quad |v| < \varkappa,\quad |Y'| < \varkappa,\quad |Y''| < \varkappa$$

die $a_{\nu\mu}$ (also auch die $A_{\nu\mu}$) sowie die Koeffizienten der Polynome Q in (6.2) und (6.3) reguläre und beschränkte Funktionen von u, v, Y sind,

---

[25]) Man hat $|Q_3| < \text{konst.}\,|Y-X'|^3$, $|Q_4| < \text{konst.}\,|Y-X'|^4$, $|Q_5| < \text{konst.}\,|Y-X'|^5$.

und daß im Gebiet (6.14)

$$(6.15)\qquad \sum_{\nu,\mu} |A_{\nu\mu} - A^0_{\nu\mu}| < \frac{\omega}{3(1+\gamma)^2},$$

$$A^0_{\nu\mu} = A_{\nu\mu}(0,0,0)$$

ausfällt. Z sei der Bedingung — ihre Bedeutung wird in § 7 klar hervortreten —

$$(6.16)\qquad |Z''| < \gamma |Z'|$$

unterworfen. Wir betrachten W in dem durch (6.14) und (6.16) bestimmten komplexen Gebiete. Zunächst ist

$$\Omega(a; Z) = \sum_{\nu,\mu} A^0_{\nu\mu} z_\nu z_\mu + \sum_{\nu,\mu} (A_{\nu\mu} - A^0_{\nu\mu}) z_\nu z_\mu$$

und daher

$$\Re[\Omega] \geqq \sum_{\nu,\mu} A^0_{\nu\mu} (z'_\nu z'_\mu - z''_\nu z''_\mu) - \sum_{\nu,\mu} |A_{\nu\mu} - A^0_{\nu\mu}| \, |z_\nu| \, |z_\mu|.$$

Wegen $|z''_\nu| \leqq |Z''| < \gamma |Z'|$ und

$$(6.17)\qquad |z_\nu| \leqq |z'_\nu| + |z''_\nu| \leqq |Z'| + |Z''| < (1+\gamma)|Z'|$$

folgt mit Rücksicht auf (6.5) $(Y=0)$

$$\Re[\Omega] > \{\omega - \gamma^2 \sum_{\nu,\mu} |A^0_{\nu\mu}| - (1+\gamma)^2 \sum_{\nu,\mu} |A_{\nu\mu} - A^0_{\nu\mu}|\} |Z'|^2,$$

und hieraus schließlich wegen (6.8) und (6.15)

$$(6.18)\qquad \Re[\Omega] > \frac{\omega}{3} |Z'|^2$$

im Gebiete (6.14), (6.16). Wie oben folgt hieraus, daß die für reelle Z positiv bestimmte Wurzel $\sqrt{\Omega}$ in diesem Gebiete eine eindeutige reguläre und von Null verschiedene Funktion von u, v, Y, Z darstellt. In diesem Gebiete ist also auch W eine reguläre Funktion ihrer Argumente u, v, Y, Z mit den aus (6.2), (6.3) und (6.17), (6.18) folgenden Eigenschaften

$$(6.19)\qquad |W|, \left|\frac{\partial W}{\partial u}\right|, \left|\frac{\partial W}{\partial v}\right|, \left|\frac{\partial W}{\partial Y}\right| < \frac{M}{|Z'|^{n-1}}, \quad \left|\frac{\partial W}{\partial Z}\right| < \frac{M^*}{|Z'|^n};$$

die Zahlen M und M^* hängen hier zwar noch von einer oberen Schranke für $|Z'|$ ab (die Polynome Q sind nicht homogen), können indessen von vornherein festgelegt werden, da im folgenden stets $|Z'| < 2R < 2R_0$ sein wird.

§ 7.

Definition und Eigenschaften des Integrales in (6.6) für komplexe Punkte X.

Dem in der Fundamentalgleichung (6.6) auftretenden Integral

$$F(X) = \int_{|Y|<R}^{(n)} W(u(Y), v(Y), Y; Y - X)\, dY$$

soll nunmehr mit Hilfe einer schönen Idee von E. E. Levi auch für komplexe X ein Sinn erteilt werden. Zu diesem Zweck denke man sich die für reelle Y gegebenen Funktionen $u(Y)$, $v(Y)$ irgendwie stetig in das komplexe Gebiet $(R)_\gamma$ hinein fortgesetzt. Wir definieren dann ein ganz bestimmtes „Funktional", welches diesen stetigen Fortsetzungen u, v eine in $(R)_\gamma$ erklärte, für reelle X sich auf das obige (gegebene) $F(X)$ reduzierende Funktion von X in $(R)_\gamma$ zuordnet. Zur Motivierung der in § 8 anzustellenden Überlegungen sei bemerkt, daß dieses Funktional eine in $(R)_\gamma$ reguläre Funktion von X liefert, wenn u, v in $(R)_\gamma$ regulär sind. Infolge dieser Eigenschaft müßten $u(X)$ und $v(X)$, falls sie wirklich in $(R)_\gamma$ regulär sind, das „komplexe" Integralgleichungssystem (6.6), wo das Integral durch jenes Funktional ersetzt ist, wirklich befriedigen. Wir kehren den Gedankengang um, indem wir in § 8 das komplexe Integralgleichungssystem auflösen und die Regularität des Lösungssystems $u(X)$, $v(X)$ in $(R)_\gamma$ beweisen.

Ein wesentlicher Punkt bei der Fortsetzung des obigen Integrales für komplexe X besteht in der gleichzeitigen Verlegung des Integrationsgebietes in den komplexen Y-Raum. Ist X irgendein Punkt aus $(R)_\gamma$, so wählen wir als Integrationsgebiet die Mannigfaltigkeit $[X, R]$ aller Punkte Y, welche auf den offenen geradlinigen Verbindungsstrecken des Punktes X mit den Punkten P der reellen Hyperkugel $|Y| = R$ liegen. Wegen der Konvexität von $(R)_\gamma$ liegen mit X auch alle Punkte der Mannigfaltigkeit $[X, R]$ in $(R)_\gamma$. Eine Parameterdarstellung von $[X, R]$ ist sofort durch die Formeln

$$(7.1)\qquad \begin{aligned} Y &= X + t(P - X); \qquad 0 < t < 1, \\ P &= P(\varphi_1, \varphi_2, \ldots, \varphi_{n-1}) \end{aligned}$$

gegeben, wobei $\varphi_1, \varphi_2, \ldots, \varphi_{n-1}$ Parameter auf der Hyperkugel

$$|P| = \sqrt{\Sigma p_\nu^2} = R$$

bedeuten. $\varphi_1, \ldots, \varphi_{n-1}, t$ sind dann die Parameter der n-dimensionalen Mannigfaltigkeit $[X, R]$. Das über die letztere erstreckte Integral über eine Funktion $f(Y)$ wird durch die Formel

$$(7.2)\qquad \int\limits_{[X,R]}^{(n)} f(Y)\,dY = \int\limits_{t=0}^{1}\int\limits_{(\varphi)}^{(n-1)} f(Y)\frac{\partial(y_1, \ldots, y_{n-1}, y_n)}{\partial(\varphi_1, \ldots, \varphi_{n-1}, t)} \prod_\nu d\varphi_\nu\, dt$$

definiert; (φ) bedeutet eine gewisse Punktmenge im $(n-1)$-dimensionalen Parameterraum, deren Punkten die Punkte der Hyperkugel umkehrbar eindeutig entsprechen. Die Funktionaldeterminante ergibt sich leicht in der Form

$$(7.3)\qquad \frac{\partial(y_1, \ldots, y_{n-1}, y_n)}{\partial(\varphi_1, \ldots, \varphi_{n-1}, t)} = t^{n-1}\sum_1^n g_\nu(P)(p_\nu - x_\nu) = t^{n-1} E(P, P - X),$$

wobei die $g_\nu(P)$ die $(n-1)$-reihigen Determinanten der Funktionalmatrix $\left\| \frac{\partial p_k}{\partial \varphi_i} \right\|$ darstellen. Hieraus resultiert sogleich die geometrische Bedeutung des Linearausdrucks E; bis auf den Faktor

$$G(P) = \sqrt{\sum g_\nu^2(P)}$$

ist nämlich $E(P, P-X)$ bei reellem X gleich der Entfernung des Punktes X von derjenigen Hyperebene, welche die Hyperkugel $|P| = R$ im Punkte P berührt[26]). Hieraus folgen unmittelbar die Ungleichungen

$$(7.4) \quad G(P)(R - |X|) \leqq E(P, P-X) \leqq G(P)|P-X|; \quad X \text{ reell}, \ |X| \leqq R.$$

Allgemein ist

$$E(P, T) = \sum_\nu g_\nu(P) t_\nu \leqq G(P)|T|.$$

Für komplexe $X = X' + iX''$ aus $(R)_\gamma$ ist also

$$|E(P, P-X)| \leqq E(P, P-X') + |E(P, X'')|,$$

$$|E(P, X'')| \leqq G(P)|X''| < \gamma G(P)(R - |X'|) \leqq \gamma E(P, P-X'),$$

und daher

$$|E(P, P-X)| < (1+\gamma) E(P, P-X'); \qquad X \text{ in } (R)_\gamma.$$

Eine einfache Folge hieraus und aus (7.2), (7.3) ist die Ungleichung

$$(7.5) \qquad \left| \int\limits_{[X,R]}^{(n)} f(Y)\,dY \right| \leqq (1+\gamma) \int\limits_{|Y'|<R}^{(n)} |f(Y)|\,dY',$$

wobei im rechten Integral Y den durch seine reelle Projektion Y' eindeutig auf $[X, R]$ bestimmten Punkt bedeutet.

Wir wenden uns nun der genaueren Betrachtung des Integrales

$$(7.6) \qquad \begin{aligned} F(X) &= \int\limits_{[X,R]}^{(n)} W(u(Y), v(Y), Y; Y-X)\,dY \\ &= \int\limits_{t=0}^{1} \int\limits_{(\varphi)}^{(n-1)} W(u(Y), v(Y), Y; Y-X)\, t^{n-1} E(P, P-X) \prod_\nu d\varphi_\nu\, dt, \\ &\qquad Y = X + t(P-X) \end{aligned}$$

zu und bemerken zunächst, daß sämtliche Argumentwerte u, v, Y, Z mit

$$(7.7) \qquad \begin{aligned} &|u| < \varkappa, \quad |v| < \varkappa, \quad |Y'| < \varkappa, \quad |Y''| < \varkappa; \\ &Z = Y - X, \quad X \text{ in } (R)_\gamma, \quad Y \text{ auf } [X, R] \end{aligned}$$

in das Regularitätsgebiet (6.14), (6.16) der Funktion $W(u, v, Y; Z)$ hineinfallen; denn es ist wegen (7.1)

$$|Z''| = t|X''| < t\gamma(R - |X'|) \leqq t\gamma|P - X'| = \gamma|Z'|.$$

---

[26]) Man kann die φ_ν immer so wählen, daß $E(P, P-X') \geqq 0$ für $|X'| \leqq R$ ausfällt.

Wegen (6.19) — man berücksichtige die Bemerkung am Schluß von § 6 — gilt ferner, sobald (7.7) erfüllt ist,

$$(7.8)\quad |W|,\ \left|\frac{\partial W}{\partial u}\right|,\ \left|\frac{\partial W}{\partial v}\right|,\ \left|\frac{\partial W}{\partial Y}\right| < \frac{M}{|Y'-X'|^{n-1}},\qquad \left|\frac{\partial W}{\partial Z}\right| < \frac{M^*}{|Y'-X'|^{n}}.$$

Damit das Integral (7.6) überhaupt einen Sinn hat, verlangen wir, daß die in $(R)_\gamma$ hinein stetig fortgesetzten $u(X)$, $v(X)$ dort ständig den Bedingungen $|u| < \varkappa$, $|v| < \varkappa$ genügen; ferner sei von vornherein

$$(7.9)\qquad R < \frac{\varkappa}{1+\gamma}$$

gewählt; denn dann ist sicher für alle Punkte Y aus $(R)_\gamma$ $|Y'| < R < \varkappa$, $|Y''| < \gamma R < \varkappa$.

Stetigkeit von $F(X)$ in $(R)_\gamma$. Denkt man sich $u(X)$, $v(X)$ unter Einhaltung der obigen Bedingungen in $(R)_\gamma$ hinein stetig fortgesetzt, so wird die durch (7.6) dargestellte Funktion $F(X)$ ebenfalls in $(R)_\gamma$ stetig sein. Denn der Integrand im rechten Integral $(Y = X + t(P - X))$ ist zunächst eine für

$$0 < t < 1,\qquad X \text{ in } (R)_\gamma$$

stetige Funktion von X, P, t, und es gilt wegen (7.8)

$$(7.10)\qquad t^{n-1}|W| < \frac{M}{|P-X'|^{n-1}} \leqq \frac{M}{(R-|X'|)^{n-1}}.$$

Bezeichnet man daher mit $F_\varepsilon(X)$ das aus (7.6) entstehende Integral, wenn man nur über $\varepsilon < t < 1 - \varepsilon$ integriert, so ist $F_\varepsilon(X)$ gewiß in $(R)_\gamma$ stetig. Somit ist auch $F = F_0$ in $(R)_\gamma$ stetig, da wegen (7.10) auf jedem abgeschlossenen Teil von $(R)_\gamma$ gleichmäßig $F_\varepsilon \to F_0$ konvergiert.

Differenzierbarkeit von $F(X)$ in $(R)_\gamma$. Die ins Komplexe fortgesetzten Funktionen $u(X)$, $v(X)$, $X = X' + iX''$ seien nun als einmal stetig differenzierbar nach X', X'' in $(R)_\gamma$ vorausgesetzt. Eine solche Fortsetzung ist jedenfalls möglich, da die für reelle X gegebenen Funktionen $u(X)$, $v(X)$ als einmal stetig differenzierbar vorausgesetzt waren[27]) (z. B. setze man $u(X) = u(X')$, $v(X) = v(X')$). Genügen nun die Ableitungen Bedingungen der Form

$$(7.11)\quad \left|\frac{\partial u}{\partial X'}\right|,\ \left|\frac{\partial u}{\partial X''}\right|,\ \left|\frac{\partial v}{\partial X'}\right|,\ \left|\frac{\partial v}{\partial X''}\right| < \frac{\text{konst.}}{R-|X'|},\qquad X \text{ in } (R)_\gamma,$$

so ist $F(X)$ in $(R)_\gamma$ einmal stetig nach X', X'' differenzierbar, und es gelten Ungleichungen derselben Form für das Integral F

$$(7.12)\qquad \left|\frac{\partial F}{\partial X'}\right|,\ \left|\frac{\partial F}{\partial X''}\right| < \frac{\text{konst.}}{R-|X'|},\qquad X \text{ in } (R)_\gamma.$$

[27]) Im Original-Gleichungssystem (57), (58) war die Existenz und Stetigkeit der sechsten Ableitungen von u vorausgesetzt worden.

Zum Beweise überlegt man sich leicht, daß im rechten Integral in (7.6) unter dem Integralzeichen differenziert werden darf. Rein formal ergibt sich zunächst mit Rücksicht auf den analytischen Charakter von W als Funktion von u, v, Y, Z, d. h. wegen

$$\frac{\partial W}{\partial y_\lambda'} = -i\frac{\partial W}{\partial y_\lambda''} = \frac{\partial W}{\partial y_\lambda}, \qquad \frac{\partial W}{\partial z_\lambda'} = -i\frac{\partial W}{\partial z_\lambda''} = \frac{\partial W}{\partial z_\lambda} \qquad (i=\sqrt{-1}),$$

$$\frac{\partial F(X)}{\partial x_\lambda'} = J_1 + J_2 + J_3,$$

$$(7.13)\quad \begin{aligned} J_1 &= \int\limits_{t=0}^{1}\int\limits_{(\varphi)}^{(n-1)} \left\{\frac{\partial W}{\partial u}\frac{\partial u}{\partial y_\lambda'} + \frac{\partial W}{\partial v}\frac{\partial v}{\partial y_\lambda'} + \frac{\partial W}{\partial y_\lambda}\right\}(1-t)\,t^{n-1}E(P,P-X)\prod_\nu d\varphi_\nu\,dt,\\ J_2 &= -\int\limits_{t=0}^{1}\int\limits_{(\varphi)}^{(n-1)} \frac{\partial W}{\partial z_\lambda}\,t^n E(P,P-X)\prod_\nu d\varphi_\nu\,dt,\\ J_3 &= -\int\limits_{t=0}^{1}\int\limits_{(\varphi)}^{(n-1)} W t^{n-1} g_\lambda(P)\prod_\nu d\varphi_\nu\,dt, \end{aligned}$$

wobei unter den Integralzeichen $Y = X + t(P-X)$, $Z = Y - X$ eingesetzt zu denken ist. Statt dessen kann man die Integrale auch in der Form

$$(7.14)\quad \begin{aligned} J_1 &= \int\limits_{[X,R]}^{(n)} \left\{\frac{\partial W}{\partial u}\frac{\partial u}{\partial y_\lambda'} + \frac{\partial W}{\partial v}\frac{\partial v}{\partial y_\lambda'} + \frac{\partial W}{\partial y_\lambda}\right\}(1-t)\,dY,\\ J_2 &= -\int\limits_{[R,X]}^{(n)} \frac{\partial W}{\partial z_\lambda}\,t\,dY, \qquad J_3 = -\int\limits_{[X,R]}^{(n)} W\frac{g_2(P)}{E(P,P-X)}\,dY \end{aligned}$$

schreiben. Analoge Formeln erhält man natürlich für $\frac{\partial}{\partial x_\lambda''}$; bei diesen sind jedoch die Größen $\frac{\partial W}{\partial y_\lambda}$, $\frac{\partial W}{\partial z_\lambda}$ und $g_\lambda(P)$ in den Integranden von J_1, J_2 und J_3 mit dem Faktor $i=\sqrt{-1}$ zu multiplizieren. Hieraus folgt die wichtige Formel

$$(7.15)\quad \begin{aligned} &\frac{\partial F(X)}{\partial x_\lambda'} + i\frac{\partial F(X)}{\partial x_\lambda''}\\ &= \int\limits_{[X,R]}^{(n)} \left\{\frac{\partial W}{\partial u}\left(\frac{\partial u}{\partial y_\lambda'} + i\frac{\partial u}{\partial y_\lambda''}\right) + \frac{\partial W}{\partial v}\left(\frac{\partial v}{\partial y_\lambda'} + i\frac{\partial v}{\partial y_\lambda''}\right)\right\}(1-t)\,dY, \end{aligned}$$

in welcher der analytische Charakter von W wesentlich benützt wird.

Die oben vorgenommene Differentiation unter dem Integralzeichen kann leicht gerechtfertigt werden; denn die Integranden in (7.13) sind für $0 < t < 1$, X in $(R)_\gamma$ wieder stetige Funktionen von X, P, t, welche außerdem — das geht aus sogleich anzugebenden Ungleichungen hervor — für alle X eines abgeschlossenen Teiles von $(R)_\gamma$, sowie für alle t und P beschränkt sind. Es führt dann eine der obigen analoge ε-Überlegung zum Ziel.

15*

Aus (7.8) und (7.11) folgt nun leicht

$$(7.16)\quad \left|\frac{\partial W}{\partial u}\frac{\partial u}{\partial y_\lambda'}+\frac{\partial W}{\partial v}\frac{\partial v}{\partial y_\lambda'}+\frac{\partial W}{\partial y_\lambda}\right| < \frac{\text{konst.}}{R-|Y'|}\cdot\frac{1}{|Y'-X'|^{n-1}}=\frac{\text{konst.}}{R-|Y'|}\frac{t^{-n+1}}{|P-X'|^{n-1}},$$

$Y=X+t(P-X)$, $Z=Y-X$. Da weiter $Y'-(1-t)X'=tP$, und somit $|Y'|-(1-t)|X'|\leqq t|P|=tR=R-(1-t)R$ ist, folgt

$$(7.17)\qquad 1-t\leqq\frac{R-|Y'|}{R-|X'|}.$$

Ferner ist nach (7.8)

$$(7.18)\quad |W|<\frac{M}{|Y'-X'|^{n-1}}=\frac{M t^{-n+1}}{|P-X'|^{n-1}},\qquad \left|\frac{\partial W}{\partial z_\lambda}\right|<\frac{M^*}{|Y'-X'|^n}=\frac{M^* t^{-n}}{|P-X'|^n}$$

und

$$(7.19)\qquad t=\frac{|Y'-X'|}{|P-X'|}\leqq\frac{|Y'-X'|}{R-|X'|}.$$

Außerdem ist für alle X aus $(R)_\gamma$

$$|E(P,P-X)|\geqq\Re[E]=E(P,P-X')\geqq G(P)(R-|X'|)$$

(vgl. (7.4)) und daher wegen $|g_\lambda|\leqq G$

$$(7.20)\qquad \left|\frac{g_\lambda(P)}{E(P,P-X)}\right|\leqq\frac{1}{R-|X'|}.$$

Für die absoluten Beträge der Integranden in (7.14) ergibt sich damit, und zwar für J_1 aus (7.16) und (7.17), für J_2 aus (7.18) und (7.19), für J_3 aus (7.18) und (7.20) eine gemeinsame obere Schranke der Form

$$\frac{\text{konst.}}{R-|X'|}\,\frac{1}{|Y'-X'|^{n-1}}.$$

Die Anwendung der Ungleichung (7.5) auf (7.14) ergibt also

$$\left|\frac{\partial F}{\partial x_\lambda'}\right|\leqq|J_1|+|J_2|+|J_3|<\frac{\text{konst.}}{R-|X'|}\cdot\int\limits_{|Y'|<R}^{(n)}\frac{dY}{|Y'-X'|^{n-1}}$$

und Entsprechendes für $\frac{\partial F}{\partial x_\lambda''}$, woraus schließlich wegen der Beschränktheit des hier auftretenden Integrales die Behauptung folgt.

§ 8.

Das komplexe Integralgleichungssystem. Existenz und Regularität der Lösung.

Die für reelle X gegebenen, den Gleichungen (6.6) genügenden Funktionen $u(X)$, $v(X)$ sollen nun so in das komplexe Gebiet $(R)_\gamma$ fortgesetzt werden, daß sie für alle X aus $(R)_\gamma$ die beiden Integralgleichungen

$$(8.1)\qquad \begin{aligned} u(X) &= \int\limits_{[X,R]}^{(n)} W(u(Y),\, v(Y),\, Y;\; Y-X)\,dY + \Phi(X), \\ v(X) &= x_1 \int\limits_0^1 u(tX)\,dt \end{aligned}$$

befriedigen. Diese Fortsetzung kann in einfachster Weise durch sukzessive Approximationen konstruiert werden.

Wir definieren durch die Gleichungen

$$(8.2)\qquad \begin{aligned} u_{\nu+1}(X) &= \int\limits_{[X,R]}^{(n)} W_\nu\,dY + \Phi(X), \qquad v_{\nu+1}(X) = x_1 \int\limits_0^1 u_\nu(tX)\,dt, \\ W_\nu &= W(u_\nu(Y),\, v_\nu(Y),\, Y;\, Y-X), \end{aligned}$$

$$(8.3)\qquad u_0(X) = u(X'), \qquad v_0(X) = v(X')$$

zwei Folgen von Funktionen $u_\nu(X)$, $v_\nu(X)$ in $(R)_\gamma$ und setzen

$$(8.4)\qquad \varphi_\nu(X) = u_{\nu+1}(X) - u_\nu(X), \qquad \psi_\nu(X) = v_{\nu+1}(X) - v_\nu(X),$$

$$(8.5)\qquad \sigma_\nu = \text{Ob. Gr.}\,(|\varphi_\nu(X)|,\, |\psi_\nu(X)|), \quad X \text{ in } (R)_\gamma.$$

Da sich $u_0(X)$, $v_0(X)$ für reelle X auf die gegebenen Funktionen $u(X)$, $v(X)$ reduzieren, wird dies auch von allen $u_\nu(X)$, $v_\nu(X)$ gelten. Wir müssen zunächst durch passende Wahl von R dafür sorgen, daß überhaupt alle Approximationen u_ν, v_ν gebildet werden können. Natürlich muß R von vornherein so klein gewählt werden, daß $\Phi(X)$ in $(R)_\gamma$ regulär ist und die Eigenschaften (6.13) besitzt, und daß alle $u_\nu(Y)$, $v_\nu(Y)$, Y in $(R)_\gamma$, die Grundbedingungen (7.7) erfüllen (vgl. (7.9)). Zu diesem Zweck sei zunächst R so klein angesetzt, daß — man berücksichtige (6.4) und (8.3) —

$$(8.6)\qquad |u_0(X)|, \quad |v_0(X)| < \frac{\varkappa}{3}, \qquad X \text{ in } (R)_\gamma$$

ausfällt. Da weiter wegen (8.2), (8.3), (8.4)

$$\varphi_0(X) = \{u_1(X) - \Phi(X)\} - \{u(X') - \Phi(X)\} = \int\limits_{[X,R]}^{(n)} W_0\,dY - \int\limits_{|Y'|<R}^{(n)} W_0\,dY'$$

ist, wird für genügend klein gewähltes R, wie man leicht einsieht,

$$(8.7)\qquad \sigma_0 < \frac{\varkappa}{3}$$

ausfallen. Schließlich sei R so klein gewählt, daß außerdem noch für $|X'| < R$

$$(8.8)\qquad 2(1+\gamma)\,M \int\limits_{|Y'|<R}^{(n)} \frac{dY}{|Y'-X'|^{n-1}} < \frac{1}{2}; \qquad (1+\gamma)\,R < \frac{1}{2}$$

ausfällt. Wir legen ein allen diesen Bedingungen genügendes R zugrunde und zeigen, daß dann die Approximationen u_ν, v_ν ständig in $(R)_\gamma$ den Bedingungen $|u|, |v| < \varkappa$ genügen und gleichmäßig in $(R)_\gamma$ gegen stetige Grenzfunktionen u, v konvergieren. Aus (8.2) und (8.4) folgt zunächst

$$(8.9)\quad \varphi_{\nu+1}(X) = \int\limits_{[X,R]}^{(n)} (W_{\nu+1} - W_\nu)\, dY, \qquad \psi_{\nu+1}(X) = x_1 \int\limits_0^1 \varphi_\nu(tX)\, dt.$$

Nehmen wir an, wir hätten schon bewiesen, daß $u_0, v_0, \ldots, u_{\nu+1}, v_{\nu+1}$ in $(R)_\gamma$ den Bedingungen $|u|, |v| < \varkappa$ genügen[28]). Wir schließen dann, daß dies auch für $u_{\nu+2}, v_{\nu+2}$ gilt. Zunächst ist nach einer bekannten Formel

$$W_{\nu+1} - W_\nu = \varphi_\nu(Y) \int\limits_0^1 \frac{\partial W}{\partial u} dt + \psi_\nu(Y) \int\limits_0^1 \frac{\partial W}{\partial v} dt,$$

wobei in den Integralen für die Argumente u, v

$$u = t\, u_{\nu+1}(Y) + (1-t)\, u_\nu(Y), \qquad v = t\, v_{\nu+1}(Y) + (1-t)\, v_\nu(Y)$$

eingesetzt zu denken ist. Diese Argumentwerte genügen sicherlich den Bedingungen $|u|, |v| < \varkappa$, und man hat wegen (7.8) und (8.5)

$$|W_{\nu+1} - W_\nu| \leqq \frac{2M\sigma_\nu}{|Y' - X'|^{n-1}},$$

somit unter Benutzung der Ungleichung (7.5) und wegen (8.8)

$$|\varphi_{\nu+1}(X)| \leqq 2(1+\gamma) M \sigma_\nu \int\limits_{|Y'|<R}^{(n)} \frac{dY}{|Y' - X'|^{n-1}} < \frac{1}{2}\sigma_\nu.$$

Ferner ist wegen (8.9) und $|x_1| \leqq |X'| + |X''| < R + \gamma R$ mit Rücksicht auf (8.8)

$$|\psi_{\nu+1}(X)| \leqq (1+\gamma) R\, \sigma_\nu \leqq \frac{1}{2}\sigma_\nu,$$

also endlich $\sigma_{\nu+1} \leqq \frac{1}{2}\sigma_\nu$ und damit $\sigma_\mu \leqq 2^{-\mu}\sigma_0$ für $\mu \leqq \nu + 1$. Dann folgt

$$\begin{aligned} |u_{\nu+2}(X)| = |u_0(X) + \varphi_0(X) + \ldots + \varphi_{\nu+1}(X)| &\leqq |u_0(X)| + \sigma_0 + \ldots + \sigma_{\nu+1} \\ &\leqq |u_0(X)| + 2\sigma_0 \end{aligned}$$

und Analoges für $v_{\nu+2}$, also wegen (8.6) und (8.7)

$$|u_{\nu+2}(X)|,\ |v_{\nu+2}(X)| < \frac{\varkappa}{3} + 2\frac{\varkappa}{3} = \varkappa.$$

Daher erfüllen sämtliche Approximationen für alle X in $(R)_\gamma$ die Bedingungen $|u|, |v| < \varkappa$, und gleichzeitig konvergieren wegen $\sigma_\nu \leqq 2^{-\nu}\sigma_0$ die

[28]) u_0, v_0, u_1, v_1 erfüllen wegen (8.6), (8.7), (8.4) und (8.5) gewiß diese Bedingungen.

$u_\nu(X)$, $v_\nu(X)$ in $(R)_\gamma$ gleichmäßig gegen stetige Grenzfunktionen $u(X)$, $v(X)$, welche natürlich die Gleichungen (8.1) befriedigen und für reelle X mit den gegebenen Funktionen übereinstimmen müssen.

Regularität von $u(X)$, $v(X)$ *in* $(R)_\gamma$. Wir beweisen dieselbe, indem wir noch einmal auf die sukzessiven Approximationen zurückgreifen und zeigen, daß $u_\nu(X)$, $v_\nu(X)$ für $\nu \to \infty$ sozusagen „immer regulärer" werden. Zunächst ist leicht zu beweisen, daß für alle ν stetige Ableitungen

$$\frac{\partial u_\nu}{\partial X'}, \quad \frac{\partial u_\nu}{\partial X''}, \quad \frac{\partial v_\nu}{\partial X'}, \quad \frac{\partial v_\nu}{\partial X''}$$

in $(R)_\gamma$ existieren und dem Betrage nach unter Schranken der Form

$$\frac{\text{konst.}}{R - |X'|}$$

gelegen sein müssen. Wegen (8.3) sind zunächst $u_0(X)$, $v_0(X)$ gewiß einmal stetig differenzierbar in $(R)_\gamma$, da $u(X)$ und $v(X)$ für reelle X als einmal stetig differenzierbar vorausgesetzt waren. Man hat daher nach (8.2)

$$\frac{\partial u_1(X)}{\partial X'} = \frac{\partial}{\partial X'} \int\limits_{[X,R]}^{(n)} W_1 \, dY + \frac{\partial \Phi}{\partial X}$$

und Analoges für $\dfrac{\partial u_1}{\partial X''}$, wobei das Integral nach dem Differenzierbarkeitssatz von § 7 in $(R)_\gamma$ gewiß einmal stetig nach X', X'' derivierbar ist. Wegen der dort angegebenen Betragsschranken für die Ableitungen und wegen (6.13) folgt daher die Richtigkeit der Behauptung für $\nu = 1$. In der gleichen Weise kann man sukzessive für $\nu = 2, 3, \ldots$ weiterschließen.

Wir führen nun die Operatoren

$$\nabla_1 = \frac{\partial}{\partial x_1'} + i\frac{\partial}{\partial x_1''}, \quad \nabla_2 = \frac{\partial}{\partial x_2'} + i\frac{\partial}{\partial x_2''}, \ldots, \nabla_n = \frac{\partial}{\partial x_n'} + i\frac{\partial}{\partial x_n''}$$

und das abkürzende Symbol

$$\nabla = \frac{\partial}{\partial X'} + i\frac{\partial}{\partial X''}$$

ein. Diese Operation darf auf alle $u_\nu(X)$, $v_\nu(X)$ in $(R)_\gamma$ angewendet werden, und die obere Grenze

(8.10) $d_\nu = \text{Ob. Gr.}\{(R - |X'|)|\nabla u_\nu(X)|, (R - |X'|)|\nabla v_\nu(X)|\}$,

genommen für alle X aus $(R)_\gamma$ und alle $\nabla = \nabla_1, \ldots, \nabla_n$, ist nach dem Obigen für jedes ν endlich. Da $\Phi(X)$ in $(R)_\gamma$ regulär ist, hat man $\nabla\Phi = 0$, und die Anwendung der Formel (7.15) auf die Gleichungen (8.3) ergibt in verständlicher Weise

$$(8.11)\qquad \nabla u_{\nu+1}(X) = \int\limits_{[X,R]}^{(n)} \left\{ \frac{\partial W}{\partial u} \nabla u_\nu(Y) + \frac{\partial W}{\partial v} \nabla v_\nu(Y) \right\} (1-t)\, dY,$$

$$(8.12)\qquad \nabla v_{\nu+1}(X) = x_1 \int\limits_0^1 t \nabla u_\nu(tX)\, dt,$$

wobei $\frac{\partial W}{\partial u}, \frac{\partial W}{\partial v}$ an den Stellen $u = u_\nu(Y)$, $v = v_\nu(Y)$ genommen sind. Für das $1-t$ in (8.11) gilt nach (7.17)

$$0 < 1 - t \leqq \frac{R - |Y'|}{R - |X'|}.$$

Mit Rücksicht auf (7.8) und durch Anwendung von (7.5) folgt also wegen (8.10)

$$|\nabla u_{\nu+1}(X)| \leqq \frac{2(1+\gamma) M d_\nu}{R - |X'|} \int\limits_{|Y'|<R}^{(n)} \frac{dY}{|Y' - X'|^{n-1}}.$$

Ferner ist, da in $(R)_\gamma$ $|x_1| < (1+\gamma) R$ gilt,

$$|\nabla u_{\nu+1}(X)| \leqq (1+\gamma) R\, d_\nu \cdot \int\limits_0^1 \frac{dt}{R - t|X'|} \leqq \frac{(1+\gamma) R d_\nu}{R - |X'|}.$$

Wegen (8.8) ergibt sich daher $d_{\nu+1} \leqq \frac{1}{2} d_\nu$, somit $d_\nu \leqq 2^{-\nu} d_0$ und nach (8.10) für alle ν

$$(8.13)\qquad |\nabla u_\nu(X)|,\ |\nabla v_\nu(X)| \leqq \frac{d_0}{2^\nu (R - |X'|)},\quad X \text{ in } (R)_\gamma$$

für $\nabla = \nabla_1, \ldots, \nabla_n$. Hieraus läßt sich nun die Regularität der Funktionen $u(X) = \lim u_\nu(X)$, $v(X) = \lim v_\nu(X)$ in $(R)_\gamma$, und damit die Analytizität der für reelle X gegebenen Funktionen $u(X)$, $v(X)$ leicht folgendermaßen erschließen. Halten wir im Moment $x_2, \ldots, x_n$ fest und betrachten wir die $u_\nu(X)$, $v_\nu(X)$ als Funktionen von x_1 in einem Bereiche der komplexen x_1-Ebene. Es bedeute Δ irgendein nebst seinem Inneren in diesem Bereiche gelegenes Dreieck. Für das längs Δ über u_ν erstreckte Integral gilt dann

$$\left| \int\limits_\Delta u_\nu\, dx_1 \right| \leqq U^2 \cdot \text{Ob. Gr.} |\nabla_1 u_\nu|,$$ [29])

---

[29]) Ist $f(x) = f(x' + ix'')$ eine in einem einfach zusammenhängenden Bereiche einmal stetig nach x', x'' differenzierbare Funktion, so gilt allgemein für jedes in diesem Bereiche gelegene Dreieck Δ ($U =$ Umfang von Δ) die Ungleichung

$$\left| \int\limits_\Delta f(x)\, dx \right| \leqq U^2 \{\text{Ob. Or.}\, |\nabla f(x)|;\ x \text{ in } \Delta\}.$$

Zum Beweise kann man dieselbe Überlegung heranziehen, die man zum heute üblichen Beweise des Cauchyschen Integralsatzes anstellt (vgl. etwa Bieberbach, Funktionentheorie I (1921), § 6). Man berücksichtige dabei die leicht einzusehende Tatsache, daß $|\nabla f(x)|$ den Maximalbetrag der Schwankung der im Punkte x nach allen möglichen Richtungen $\overrightarrow{dx}$ gebildeten Ableitungen $\frac{df}{dx}$ darstellt.

unter U den Umfang von Δ verstanden; die obere Grenze ist für alle x_1 aus Δ zu nehmen. Wegen (8.13) konvergiert dieselbe für $\nu \to \infty$ gegen Null, und aus der Gleichmäßigkeit der Konvergenz in $u_\nu \to u$, $v_\nu \to v$ folgt

$$\int_\Delta u\, d x_1 = \int_\Delta v\, d x_1 = 0.$$

Bekanntlich folgt daraus das Verschwinden der Kurvenintegrale über jeden geschlossenen, nebst seinem Inneren in jenem x_1-Bereiche enthaltenen Weg der x_1-Ebene, nach dem Moreraschen Satze also die Regularität von u, v als Funktionen von x_1. Ebenso folgt natürlich die Regularität in $x_2, \ldots, x_n$ und damit die Regularität von $u(X), v(X)$ im Gebiete $(R)_\gamma$, w. z. b. w.

(Eingegangen am 9. Mai 1930.)

Über den funktionalen, insbesonders den analytischen Charakter der Lösungen elliptischer Differentialgleichungen zweiter Ordnung, *Math. Zeit.* **34 (1932) 194-233.**

Commentary

Hans Weinberger

This paper shows that if a solution of the fully nonlinear elliptic equation $\Phi(D^2u, Du, u, x) = 0$ in n dimensions has Hölder continuous second derivatives, and if Φ is analytic in all its variables, then u is analytic. This improves an older result of S. Bernstein, who obtained this conclusion under the stronger hypothesis that u has continuous third derivatives. $C^{2+\alpha}$ turns out to be the "correct" hypothesis.

The proof is done in two parts. The first part deals with the linear elliptic equation

$$a_{\mu\nu}(x)u_{x_\mu x_\nu} = f.$$

Potential theory and a trick of A. Korn are used to show that if f and the $a_{\mu\nu}$ are Hölder continuous with exponent α, then every solution u has second partial derivatives which are Hölder continuous with exponent α. It is also shown that if f and the $a_{\mu\nu}$ have Hölder continuous first derivatives, then u has continuous third derivatives.

The analyticity of a solution of $\Phi(D^2u, Du, u, x) = 0$ which has Hölder continuous second derivatives follows from the latter result and and the Bernstein result. However, Hopf gives an independent and more natural proof of analyticity in the second part of the paper.

For the case of minimizers of first order variational problems, the sufficient condition for analyticity $u \in C^{2,\alpha}$ given by Hopf was successively weakened to $u \in C^{1,\alpha}$ by G. Stampacchia (1952), to $u \in C^1$ by C. Morrey (1954), and finally to $u \in W^{1,2}$ by E. De Giorgi (1957), and J. Nash (1958).

While Hopf's first result can be considered as a qualitative precursor of the quantitative Schauder estimates , J. P. Schauder (1934) seems to have obtained his estimates independently.

References

De Giorgi, E., "Sulla differenziabilità e l'analiticità delle estremali degli integrali multipli," Mem. dell'Accad. delle Scienze di Torino, *Cl. Sci. Fis., Mat. e Nat.* (3) **3** (1957), 25-43.

Gilbarg, D. and N. Trudinger, "Partial Differential Equations of Second Order," *Grundlehren der mathematischen Wissenschaften* 224, Springer, (1977), p. 133.

MORREY, C., "Second order elliptic systems of differential equations," *C. Annals of Math. Studies.* **33**, Princeton Univ. Press, 1954.

NASH, J., "Continuity of solutions of parabolic and elliptic equations," *American Journal of Mathematics.* **80** (1958), 931-954.

SCHAUDER, J. P., "Über lineare elliptische Differentialgleichungen zweiter Ordnung," *Math. Zeit.* **38** (1934), pp. 257-282.

STAMPACCHIA, G., "Sistemí di equazione de tipo ellittico a derivate parziali dell primo ordina e proprietà delle estremali degli integrali multipli," *Ricerche di Mat.* **1** (1952), 200-226.

ABDRUCK
AUS DEN BERICHTEN DER MATHEMATISCH-PHYSISCHEN KLASSE DER SÄCHSISCHEN AKADEMIE DER WISSENSCHAFTEN ZU LEIPZIG
XCIV. BAND
SITZUNG VOM 19. JANUAR 1942

Abzweigung einer periodischen Lösung von einer stationären Lösung eines Differentialsystems.

PAUL KOEBE
zum sechzigsten Geburtstag gewidmet.

Von
Eberhard Hopf.

1. Einleitung.

Es sei

$$\dot{x}_i = F_i\,(x_1, \ldots, x_n, \mu) \quad (i = 1, \ldots, n)$$

oder in Vektorschreibweise

$$\dot{\mathfrak{x}} = \mathfrak{F}\,(\mathfrak{x}, \mu) \tag{1. 1}$$

ein reelles Differentialsystem mit reellem Parameter μ. $\mathfrak{F}$ sei analytisch in $\mathfrak{x}$ und μ, wenn $\mathfrak{x}$ in einem Gebiete G liegt und $|\mu| < c$ ist. (1. 1) soll eine für $|\mu| < c$ analytische Schar stationärer Lösungen $\mathfrak{x} = \tilde{\mathfrak{x}}\,(\mu)$ in G besitzen,

$$\mathfrak{F}\,(\tilde{\mathfrak{x}}\,(\mu), \mu) = 0.$$

Die charakteristischen Exponenten der stationären Lösung sind bekanntlich die Eigenwerte der Eigenwertaufgabe

$$\lambda\,\mathfrak{a} = \mathfrak{L}_\mu\,(\mathfrak{a}),$$

wo $\mathfrak{L}_\mu$ den nur von μ abhängigen linearen Operator bedeutet, welcher durch Weglassen der nicht linearen Glieder in der Reihenentwicklung von $\mathfrak{F}$ um $\mathfrak{x} = \tilde{\mathfrak{x}}$ entsteht. Die Exponenten sind entweder reell oder paarweise konjugiert komplex und hängen von μ ab.

Setzt man lediglich voraus, daß für den speziellen Wert $\mu = 0$ eine stationäre Lösung $\mathfrak{x}_0$ in G vorhanden ist und daß kein char. Exp. Null ist, so gibt es bekanntlich von selbst für jedes μ genügend kleinen Betrages in einer passenden Umgebung von $\mathfrak{x} = \mathfrak{x}_0$ genau eine stationäre Lösung $\tilde{\mathfrak{x}}\,(\mu)$ mit $\tilde{\mathfrak{x}}\,(0) = \mathfrak{x}_0$; und $\tilde{\mathfrak{x}}\,(\mu)$ ist analytisch bei $\mu = 0$.

Beim Durchgang durch $\mu = 0$ soll nun kein char. Exp. verschwinden, aber es soll einer (und damit der konjugierte) die imaginäre Achse überschreiten.

Diese Situation tritt bei nichtkonservativen mechanischen Systemen, z. B. in der Hydrodynamik häufig auf. Unter jener Voraussetzung gibt es, wie folgender Satz aussagt, in der Umgebung der Werte $\mathfrak{x} = \mathfrak{x}_0$, $\mu = 0$ stets periodische Lösungen von (1. 1).

Satz. Für $\mu = 0$ seien genau zwei charakteristische Exponenten rein imaginär. Ihre stetigen Fortsetzungen $\alpha(\mu)$, $\bar{\alpha}(\mu)$ mögen den Bedingungen

$$(1.2) \qquad \alpha(0) = -\bar{\alpha}(0) \neq 0, \qquad \Re(\alpha'(0)) \neq 0$$

genügen. Dann existiert eine Schar reeller periodischer Lösungen $\mathfrak{x} = \mathfrak{x}(t, \varepsilon)$, $\mu = \mu(\varepsilon)$ mit den Eigenschaften $\mu(0) = 0$ und $\mathfrak{x}(t, 0) = \tilde{\mathfrak{x}}(0)$, aber $\mathfrak{x}(t, \varepsilon) \neq \tilde{\mathfrak{x}}(\mu(\varepsilon))$ für alle hinreichend kleinen $\varepsilon \neq 0$. $\mu(\varepsilon)$ und $\mathfrak{x}(t, \varepsilon)$ sind an der Stelle $\varepsilon = 0$ bzw. an jeder Stelle $(t, 0)$ analytisch. Dasselbe gilt von der Periode $T(\varepsilon)$, und es ist

$$T(0) = \frac{2\pi}{|\alpha(0)|}.$$

Zu beliebig großem L gibt es zwei positive Zahlen a und b derart, daß für $|\mu| < b$ außer der stationären Lösung und den Lösungen der Scharhälfte $\varepsilon > 0$ keine periodischen Lösungen existieren, deren Periode kleiner als L ist, und die ganz in $|\mathfrak{x} - \tilde{\mathfrak{x}}(\mu)| < a$ liegen.[1] *Die periodischen Lösungen existieren bei hinreichend kleinem μ entweder nur für $\mu > 0$ oder nur für $\mu < 0$ (Allgemeiner Fall), oder aber nur für $\mu = 0$.*

Die charakteristischen Exponenten der periodischen Lösung $\mathfrak{x}(t, \varepsilon)$ sind bekanntlich die Eigenwerte der Eigenwertaufgabe

$$(1.3) \qquad \dot{\mathfrak{y}} + \lambda \mathfrak{y} = \mathfrak{L}_{t, \varepsilon}(\mathfrak{y}),$$

wo $\mathfrak{y}(t)$ dieselbe Periode $T = T(\varepsilon)$ wie die Lösung haben soll. $\mathfrak{L}$ ist der längs der periodischen Lösung gebildete lineare Operator; er hängt von t periodisch mit der Periode T ab und ist in t, ε bei $\varepsilon = 0$ analytisch. Die char. Exp. sind nur mod $(2\pi i/T)$ bestimmt und hängen stetig von ε ab. Einer von ihnen ist bekanntlich Null; denn $\mathfrak{F}$ hängt nicht explizit von t ab, also ist

$$\lambda = 0, \quad \mathfrak{y} = \dot{\mathfrak{x}}(t, \varepsilon)$$

Lösung der Eigenwertaufgabe. Die Exponenten gehen mod $(2\pi i/T_0)$ für $\varepsilon \to 0$ stetig in diejenigen der stationären Lösung $\tilde{\mathfrak{x}}(0)$ von (1. 1), $\mu = 0$, über. Nach Voraussetzung streben dabei genau zwei Exponenten gegen die imaginäre Achse. Einer von ihnen ist der identisch verschwindende. Der andere $\beta = \beta(\varepsilon)$ muß reell und bei $\varepsilon = 0$ analytisch sein, $\beta(0) = 0$. Von den Koeffizienten in

---

1) Die andere Scharhälfte muß demnach dieselben Lösungskurven darstellen.

$$\mu = \mu_1 \varepsilon + \mu_2 \varepsilon^2 + \ldots,$$
$$\beta = \beta_1 \varepsilon + \beta_2 \varepsilon^2 + \ldots.$$

folgt direkt aus dem obigen Satz $\mu_1 = \beta_1 = 0$. Darüber hinaus wird im folgenden die einfache Beziehung

$$\beta_2 = -2\mu_2 \Re\left(\alpha'(0)\right) \tag{1.4}$$

bewiesen, die mir vorher nicht begegnet ist.

Sie gibt im allgemeinen Falle $\mu_2 \neq 0$ über die Stabilitätsverhältnisse Auskunft. Haben z. B. für $\mu < 0$ alle char. Exp. der stationären Lösung $\mathfrak{x} = \tilde{\mathfrak{x}}(\mu)$ negativen Realteil (Stabilität, eine kleine Umgebung von $\tilde{\mathfrak{x}}$ schrumpft für $t \to \infty$ auf $\tilde{\mathfrak{x}}$ zusammen), so gilt folgende Alternative. Entweder zweigen die period. Lös. nach Unstabilwerden der stat. Lös. von dieser ab ($\mu > 0$); dann haben alle char. Exp. der period. Lös. negativen Realteil (Stabilität, ein dünner Schlauch um die period. Lös. schrumpft für $t \to \infty$ auf diese zusammen). Oder die Schar existiert vorher ($\mu < 0$); dann sind die period. Lös. unstabil.[1])

Da man in der Natur bei genügend langer Beobachtungszeit nur stabile Lösungen beobachten kann, ist die Abzweigung einer period. Lös. von einer stat. Lös. nur nach dem Unstabilwerden der letzteren beobachtbar. Solche Beobachtungen sind in der Hydromechanik wohlbekannt. Z. B. bei Umströmung eines festen Körpers, wo die Strömung bei kleiner Anströmungsgeschwindigkeit stationär ist, nach genügender Steigerung derselben jedoch periodisch wird (periodische Wirbelablösung). Es handelt sich hier um Beispiele nicht konservativer Systeme (Zähigkeit der Flüssigkeit).[2]) Bei konservativen Systemen ist die Voraussetzung (1. 2) bekanntlich nie erfüllt; mit λ ist stets auch $-\lambda$ char. Exponent.

Obwohl mir die Behandlung der Abzweigungsaufgabe auf Grund der Voraussetzung (1. 2) in der Literatur nicht begegnet ist, glaube ich kaum, daß an dem obigen Satz etwas wesentlich Neues ist; die Methoden sind von Poincaré vor etwa 50 Jahren entwickelt worden[3]) und gehören heute zum

---

1) In $n = 2$ Dimensionen sofort in die Augen springend.

2) Ein hydrodynamisches Beispiel, wo man bei vorsichtigstem Experimentieren (sehr langsame Parameteränderung) immer an derselben Stelle plötzliches Abbrechen der stationären Bewegung beobachtet, wo man also auf Eintreten des zweiten Falles der Alternative schließen kann, ist mir nicht bekannt.

3) Les méthodes nouvelles de la mécanique céleste. Die obigen periodischen Lösungen stellen den einfachsten Grenzfall von Poincarés periodischen Lösungen zweiter Gattung (genre) dar. Vgl. Tome III, Kap. 28, 30, 31. Poincaré hat diese Lösungen im Hinblick auf himmelsmechanische Anwendungen nur bei kanonischen Differential-

klassischen Gedankengut der Theorie der periodischen Lösungen im Kleinen. Da aber der Satz von Interesse in der nichtkonservativen Mechanik ist, schien mir eine ausführliche Darstellung nicht unnütz zu sein. Um die Durchführung bei Systemen mit unendlich vielen Freiheitsgraden, z. B. bei den Grundgleichungen der Bewegung einer zähen Flüssigkeit, zu erleichtern, habe ich allgemeinere Methoden der linearen Algebra speziellen Hilfsmitteln (z. B. Wahl eines speziellen Koordinatensystems) nach Möglichkeit vorgezogen.

Natürlich kann es ebensogut vorkommen, daß bei $\mu = 0$ ein reeller char. Exp. $\alpha(\mu)$ der stat. Lös. $\bar{\mathfrak{x}}(\mu)$ die imaginäre Achse überschreitet,

$$\alpha(0) = 0, \quad \alpha'(0) \neq 0,$$

während die anderen dieser Achse fernbleiben. In diesem Falle zweigen nicht period., sondern wieder stat. Lös. ab.[1]) Wir begnügen uns mit der Anführung der Sätze in diesem einfacheren Fall. Es gibt eine von $\bar{\mathfrak{x}}$ verschiedene analytische Schar $\mathfrak{x} = \mathfrak{x}^*(\varepsilon)$, $\mu = \mu(\varepsilon)$, stationärer Lös. mit $\mu(0) = 0$, $\mathfrak{x}^*(0) = \bar{\mathfrak{x}}(0)$. Ist $\mu_1 \neq 0$ (allgemeiner Fall) so existieren die Lös. für $\mu \gtrless 0$. Von dem durch Null gehenden char. Exp. $\beta(\varepsilon)$ gilt die Analogie zu (1. 4)

$$\beta_1 = -\mu_1 \alpha'(0).$$

Ist $\bar{\mathfrak{x}}$ stabil für $\mu < 0$, unstabil für $\mu > 0$, so folgt daraus das Umgekehrte von $\mathfrak{x}^*$ (Beobachtet man $\bar{\mathfrak{x}}$ für $\mu < 0$, so beobachtet man $\mathfrak{x}^*$ für $\mu > 0$). Im Ausnahmefalle $\mu_1 = 0$ ist die Sachlage anders. Ist $\mu_2 \neq 0$, so existieren die neuen Lösungen entweder nur für $\mu > 0$ oder nur für $\mu < 0$. Es gibt jeweilig zwei Lösungen für festes μ (eine mit $\varepsilon > 0$, eine mit $\varepsilon < 0$). Hier gilt

$$\beta_2 = -2\mu_2 \alpha'(0),$$

woraus sich eine analoge Alternative wie oben über die Stabilität ergibt. Dabei sind entweder beide Lösungen $\mathfrak{x}^*$ stabil oder beide unstabil.

---

systemen ausführlich studiert (mit Hilfe der Integralinvarianten dieser Systeme), wo die Dinge schwieriger sind als oben. Poincaré benützt den Hilfsparameter ε in Kap. 30 bei der Koeffizientenberechnung (die Rechnung in unserem § 4 ist im wesentlichen dieselbe) aber nicht beim Existenzbeweis, der dadurch einfacher wird.

An eine kurze Bemerkung in Tome I, S. 156, knüpft Painlevé an: Les petits mouvements périodiques des systèmes. Comptes Rendus Paris XXIV (1897), S. 1222. Der dort ausgesprochene allgemeine Satz bezieht sich auf den Fall $\mu = 0$ in unserem System (1. 1), kann aber nicht allgemein richtig sein. Für die Gültigkeit dieser Aussage muß $\mathfrak{F}$ speziellen Bedingungen genügen.

1) Ein Beispiel aus der Hydrodynamik ist die Flüssigkeitsbewegung zwischen zwei konzentrisch rotierenden Zylindern (G. I. Taylor).

2. Die Existenz der periodischen Lösungen.

Man kann ohne Beschränkung der Allgemeinheit annehmen, daß die stationäre Lösung in den Nullpunkt fällt,

$$\mathfrak{F}(0, \mu) = 0.$$

Die Entwicklung von $\mathfrak{F}$ nach Potenzen der x_i sei

(2. 1) $$\mathfrak{F}(\mathfrak{x}, \mu) = \mathfrak{L}_\mu(\mathfrak{x}) + \mathfrak{Q}_\mu(\mathfrak{x}, \mathfrak{x}) + \mathfrak{K}_\mu(\mathfrak{x}, \mathfrak{x}, \mathfrak{x}) + \ldots,$$

wo die Vektorfunktionen

$$\mathfrak{L}_\mu(\mathfrak{x}), \quad \mathfrak{Q}_\mu(\mathfrak{x}, \mathfrak{y}), \quad \mathfrak{K}_\mu(\mathfrak{x}, \mathfrak{y}, \mathfrak{z}), \ldots$$

von jedem Argumentvektor linear abhängen und symmetrische Funktionen dieser Vektoren sind.

Die Substitution

(2. 2) $$\mathfrak{x} = \varepsilon \mathfrak{y}$$

führt (1. 1) in

(2. 3) $$\dot{\mathfrak{y}} = \mathfrak{L}_\mu(\mathfrak{y}) + \varepsilon \mathfrak{Q}_\mu(\mathfrak{y}, \mathfrak{y}) + \varepsilon^2 \mathfrak{K}_\mu(\mathfrak{y}, \mathfrak{y}, \mathfrak{y}) + \ldots.$$

über. Die rechte Seite ist an der Stelle $\varepsilon = \mu = 0$, $\mathfrak{y} = \mathfrak{y}^0$ ($\mathfrak{y}^0$ beliebig) analytisch von ε, μ, $\mathfrak{y}^0$ abhängig.

Wir betrachten den Fall $\varepsilon = 0$ in (2. 3) der für die Existenzfrage entscheidende Bedeutung hat, d. h. die homogene lineare Differentialgleichung

(2. 4) $$\dot{\mathfrak{z}} = \mathfrak{L}_\mu(\mathfrak{z}).$$

Die konjugiert komplexen charakteristischen Exponenten $\alpha(\mu)$, $\bar{\alpha}(\mu)$, von denen in der Voraussetzung die Rede war, sind für alle kleinen $|\mu|$ einfach. In den zugehörigen Lösungen

(2. 5) $$e^{\alpha t}\mathfrak{a}, \quad e^{\bar{\alpha} t}\bar{\mathfrak{a}}$$

von (2. 4) ist daher der komplexe Vektor $\mathfrak{a}$ bis auf einen komplexen Skalarfaktor eindeutig bestimmt; $\bar{\mathfrak{a}}$ ist der konjugierte Vektor. Ferner gibt es keine Lösungen der Form

(2. 6) $$e^{\alpha t}(t\mathfrak{b} + \mathfrak{c}), \quad \mathfrak{b} \neq 0.$$

$\alpha(\mu)$ ist bei $\mu = 0$ analytisch. Man kann einen festen reellen Vektor $\mathfrak{e} \neq 0$ so wählen, daß für alle kleinen $|\mu|$ $\mathfrak{a} \cdot \mathfrak{e} \neq 0$ ausfällt, sobald $\mathfrak{a} \neq 0$ ist. $\mathfrak{a} = \mathfrak{a}(\mu)$ wird dann durch die Forderung

(2. 7) $$\mathfrak{a}(\mu) \cdot \mathfrak{e} = \frac{1}{\alpha(\mu) - \bar{\alpha}(\mu)} \qquad (\bar{\mathfrak{e}} = \mathfrak{e} \neq 0)$$

eindeutig festgelegt. Nach Voraussetzung ist

(2. 8) $$\bar{\alpha}(0) = -\alpha(0) \neq 0.$$

$\mathfrak{a}(\mu)$ ist bei $\mu = 0$ analytisch.

Die reellen Lösungen von (2. 4), die lineare Kombination von (2. 5) sind, haben die Form

(2. 9) $$\mathfrak{z} = c e^{\alpha t} \mathfrak{a} + \bar{c} e^{\bar{\alpha} t} \bar{\mathfrak{a}}$$

mit komplexem Skalar c. Sie bilden eine von zwei reellen Parametern abhängige Schar; der eine Parameter ist ein Proportionalitätsfaktor, während der andere eine additive Konstante in t darstellt (die Lösungen bilden eine nur einparametrige Kurvenschar). Wegen $\bar{\mathfrak{e}} = \mathfrak{e}$ ist

$$\left.\begin{aligned} \mathfrak{z} \cdot \mathfrak{e} &= c\, \mathfrak{a} \cdot \mathfrak{e} + \bar{c}\, \overline{\mathfrak{a} \cdot \mathfrak{e}} \\ \dot{\mathfrak{z}} \cdot \mathfrak{e} &= c \alpha\, \mathfrak{a} \cdot \mathfrak{e} + \bar{c} \bar{\alpha}\, \overline{\mathfrak{a} \cdot \mathfrak{e}} \end{aligned}\right\} t = 0.$$

Für $c = 1$ erfüllt also (2. 9),

(2. 10) $$\mathfrak{z} = e^{\alpha t} \mathfrak{a} + e^{\bar{\alpha} t} \bar{\mathfrak{a}} = \mathfrak{z}(t, \mu),$$

wegen (2. 7) die Bedingungen

(2. 11) $$t = 0: \quad \mathfrak{z} \cdot \mathfrak{e} = 0, \quad \frac{d}{dt} (\mathfrak{z} \cdot \mathfrak{e}) = 1.$$

Durch diese Bedingungen ist auch die Lösung (2. 9) eindeutig bestimmt; denn aus

$$t = 0: \quad \mathfrak{z} \cdot \mathfrak{e} = \dot{\mathfrak{z}} \cdot \mathfrak{e} = 0$$

und aus (2. 9) folgt wegen (2. 7) und (2. 8) $c = 0$; also $\mathfrak{z} = 0$.

Da nach Voraussetzung α, $\bar{\alpha}$ für $\mu = 0$ die einzigen rein imaginären unter den charakteristischen Exponenten sind, stellen (2. 9) für $\mu = 0$ sämtliche reellen und periodischen Lösungen von (2. 4) dar. Ihre Periode ist

(2. 12) $$T_0 = \frac{2\pi}{|\alpha(0)|}.$$

Insbesondere ist (2. 10) für $\mu = 0$ die einzige reelle und periodische Lösung mit den Eigenschaften (2. 11).

Für späteren Gebrauch sei noch bemerkt, daß (2. 4) für $\mu = 0$ keine Lösungen der Form

$$t\, \mathfrak{p}(t) + \mathfrak{q}(t)$$

haben kann, wo $\mathfrak{p}$ und $\mathfrak{q}$ eine gemeinsame Periode haben und $\mathfrak{p}$ nicht identisch Null ist. Im entgegengesetzten Falle würde sich (2. 4) in die beiden Gleichungen

$$\dot{\mathfrak{p}} = \mathfrak{L}_0(\mathfrak{p}), \qquad \mathfrak{p} + \dot{\mathfrak{q}} = \mathfrak{L}_0(\mathfrak{q})$$

aufspalten, und $\mathfrak{p}$ wäre eine nichttriviale Linearkombination der Lösungen (2. 5). Die Fourierentwicklung von $\mathfrak{q}(t)$ würde dann zu einer Lösung der Form (2. 6) führen.

Durch Differenzieren von (2. 4) nach μ für $\mu = 0$ ergibt sich die inhomogene Differentialgleichung

$$\dot{\mathfrak{z}}' = \mathfrak{L}_0(\mathfrak{z}') + \mathfrak{L}_0'(\mathfrak{z}); \quad \mathfrak{L}_0' = \frac{d}{d\mu}\mathfrak{L}_\mu, \quad \mu = 0, \tag{2. 13}$$

für die μ-Ableitung von (2. 10)

$$\mathfrak{z}' = t\,(\alpha' e^{\alpha t}\mathfrak{a} + \bar{\alpha}' e^{\bar{\alpha} t}\bar{\mathfrak{a}}) + (e^{\alpha t}\mathfrak{a}' + e^{\bar{\alpha} t}\bar{\mathfrak{a}}') \qquad (\mu = 0).$$

Der Faktor von t ist Lösung von (2. 4). Drückt man ihn linear durch die Lösung (2. 10) und durch $\dot{\mathfrak{z}}$ aus, so folgt mit Rücksicht auf (2. 8)

$$\mathfrak{z}' = t\left(\Re(\alpha')\,\mathfrak{z} + \frac{\Im(\alpha')}{\alpha}\dot{\mathfrak{z}}\right) + \mathfrak{h}(t) \tag{2. 14}$$

mit

$$\mathfrak{h}(t + T_0) = \mathfrak{h}(t). \tag{2. 15}$$

Nun sei

$$\mathfrak{y} = \mathfrak{y}(t, \mu, \varepsilon, \mathfrak{y}^0)$$

die Lösung von (2. 3), welche die Anfangsbedingung $\mathfrak{y} = \mathfrak{y}^0$ für $t = 0$ erfüllt. Sie hängt nach bekannten Sätzen an jeder Stelle $(t, 0, 0, \mathfrak{y}^0)$ analytisch von allen Argumenten ab. Sie ist genau dann periodisch mit der Periode T, wenn die Gleichung

$$\mathfrak{y}(T, \mu, \varepsilon, \mathfrak{y}^0) - \mathfrak{y}^0 = 0 \tag{2. 16}$$

besteht. Bezeichnet man mit $\mathfrak{z}^0$ den festen Anfangswert der festen Lösung (2. 10) von (2. 4), $\mu = 0$, so ist nach Obigem (2. 16) durch die Werte

$$T = T_0, \quad \mu = \varepsilon = 0, \quad \mathfrak{y}^0 = \mathfrak{z}^0 \tag{2. 17}$$

erfüllt. Die zu lösende Aufgabe ist, (2. 16) bei vorgegebenem ε nach $T, \mu, \mathfrak{y}^0$ aufzulösen. Es sind dies n Gleichungen mit $n + 2$ Unbekannten. Um die Lösung eindeutig zu machen, fügen wir die zwei Gleichungen

$$\mathfrak{y}^0 \cdot \mathfrak{e} = 0, \quad \dot{\mathfrak{y}}^0 \cdot \mathfrak{e} = 1 \tag{2. 18}$$

hinzu, wo $\mathfrak{e}$ der oben eingeführte reelle Vektor ist, und wo $\dot{\mathfrak{y}}^0 = \dot{\mathfrak{y}}$ für $t = 0$ ist. Ihre Hinzufügung bedeutet keine Beschränkung der Lösungsgesamtheit im kleinen, wie im nächsten Abschnitt mitgezeigt wird. Für die Ausgangswerte $\mu = \varepsilon = 0$, $\mathfrak{y}^0 = \mathfrak{z}^0$ erfüllt nach (2. 11) die Lösung diese Gleichungen.

(2. 16) und (2. 18) haben nun bei beliebigem, hinreichend kleinem $|\varepsilon|$ genau eine Lösung

$$T = T(\varepsilon), \quad \mu = \mu(\varepsilon), \quad \mathfrak{y}^0 = \mathfrak{y}^0(\varepsilon) \tag{2. 19}$$

in einer passenden Umgebung des Wertesystems

$$T = T_0, \quad \mu = 0, \quad \mathfrak{y}^0 = \mathfrak{z}^0, \tag{2. 20}$$

wenn folgendes der Fall ist: die durch Differentialbildung (an der Stelle [2. 17]) nach den Variablen T, μ, ε, $\mathfrak{y}^0$ gebildeten linearen Gleichungen sind für gegebenes $d\varepsilon$ eindeutig lösbar. Oder: wenn diese linearen Gleichungen für $d\varepsilon = 0$ nur die Nullösung $dT = d\mu = d\mathfrak{y}^0 = 0$ haben. Dies ist der Fall, wie nun gezeigt werden soll.

Es ist

(2. 21) $$\dot{\mathfrak{y}} = \mathfrak{L}_\mu(\mathfrak{y}), \qquad \mathfrak{y} = \mathfrak{y}(t, \mu, 0, \mathfrak{y}^0).$$

Speziell ist

(2. 22) $$\mathfrak{y}(t, \mu, 0, \mathfrak{z}^0) = \mathfrak{z}(t, \mu)$$

die Lösung (2. 10). Das Differential $d\mathfrak{y}(t, \mu, 0, \mathfrak{y}^0)$ setzt sich additiv aus den Differentialen nach den einzelnen Argumenten bei festbleibenden übrigen zusammen. Führt man für die Differentiale

$$dt, \quad d\mu, \quad d\mathfrak{y}^0$$

als unabhängige Konstante bzw. Vektoren die Bezeichnungen

$$\varrho, \quad \sigma, \quad \mathfrak{u}^0$$

ein, so wird das besagte Differential gleich

$$\varrho\dot{\mathfrak{y}} + \sigma\mathfrak{y}' + \mathfrak{u},$$

wo $\dot{\mathfrak{y}}$ und $\mathfrak{y}' = \partial\mathfrak{y}/\partial\mu$ für $T = T_0$, $\mu = 0$, $\mathfrak{y}^0 = \mathfrak{z}^0$ genommen sind und wo $\mathfrak{u}$ die Lösung von

$$\dot{\mathfrak{u}} = \mathfrak{L}_0(\mathfrak{u})$$

mit dem Anfangswert $\mathfrak{u}^0$ für $t = 0$ bedeutet. Wegen (2. 22) ist $\dot{\mathfrak{y}} = \dot{\mathfrak{z}}(t, 0)$. Setzt man $\mathfrak{y}' = \mathfrak{v}$, so ist $\mathfrak{v}(t)$ die Lösung von

(2. 23) $$\dot{\mathfrak{v}} = \mathfrak{L}_0(\mathfrak{v}) + \mathfrak{L}_0'(\mathfrak{z}), \qquad \mathfrak{v}(0) = 0.$$

Die aus (2. 16) entstehende lineare Vektorgleichung ist nach Obigem

(2. 24) $$\varrho\dot{\mathfrak{z}}(T_0) + \sigma\mathfrak{v}(T_0) + \mathfrak{u}(T_0) - \mathfrak{u}(0) = 0,$$

wo $\mathfrak{z}(t)$ die Lösung (2. 10) von

(2. 25) $$\dot{\mathfrak{z}} = \mathfrak{L}_0(\mathfrak{z}),$$

$\mathfrak{u}(t)$ irgendeine Lösung dieser homogenen linearen Differentialgleichung mit konstantem $\mathfrak{L}_0$ und $\mathfrak{v}(t)$ die Lösung von (2. 23) bedeutet. Wir beweisen nun, daß (2. 24) nur für $\varrho = \sigma = 0$ und $\mathfrak{u}(t) = 0$ möglich ist.

Es ist nun für alle t

(2. 26) $$\varrho\dot{\mathfrak{z}}(t) + \sigma[\mathfrak{v}(t + T_0) - \mathfrak{v}(t)] + \mathfrak{u}(t + T_0) - \mathfrak{u}(t) = 0.$$

Da nämlich $\mathfrak{z}(t)$ die Periode T_0 hat, ist die eckige Klammer wegen (2. 23) Lösung von (2. 25). Und da auch $\dot{\mathfrak{z}}$ Lösung von (2. 23) ist, ist die ganze linke Seite von (2. 26) Lösung von (2. 25); ihr Anfangswert ist wegen (2. 24) und wegen $\mathfrak{v}(0) = 0$ Null, also ist sie identisch Null. Aus (2. 13) und (2. 23) folgt nun

$$\mathfrak{v}(t) = \mathfrak{z}'(t) + \mathfrak{g}(t), \qquad \dot{\mathfrak{g}} = \mathfrak{L}_0(\mathfrak{g}).$$

Die eckige Klammer in (2. 26) hat also nach (2. 14) und (2. 15) den Wert

$$T_0 \left[\Re(\alpha')\,\mathfrak{z}(t) + \frac{\Im(\alpha')}{\alpha}\,\dot{\mathfrak{z}}(t)\right] + [\mathfrak{g}(t + T_0) - \mathfrak{g}(t)].$$

Setzt man $\mathfrak{u} + \sigma\mathfrak{g} = \mathfrak{w}$ und

(2. 27) $$\sigma T_0 \Re(\alpha')\,\mathfrak{z}(t) + \left[\varrho + \sigma T_0 \frac{\Im(\alpha')}{\alpha}\right] \dot{\mathfrak{z}}(t) = \tilde{\mathfrak{z}}(t),$$

so folgt

$$\tilde{\mathfrak{z}}(t) + \mathfrak{w}(t + T_0) - \mathfrak{w}(t) = 0,$$

wo $\mathfrak{w}(t)$ Lösung und $\tilde{\mathfrak{z}}(t)$ eine periodische Lösung von (2. 25) ist. Dies bedeutet aber, daß

$$\mathfrak{w}(t) = -\frac{t}{T_0}\,\tilde{\mathfrak{z}}(t) + \mathfrak{q}(t)$$

mit periodischen $\mathfrak{q}$ ist. Solche Lösungen kann es aber, wie erwähnt, nicht geben, wenn nicht $\tilde{\mathfrak{z}} = 0$ ist. Da $\mathfrak{z}, \dot{\mathfrak{z}}$ linear unabhängig sind, folgt aus (2. 27) und aus der Voraussetzung (1. 2) $\sigma = 0$ und $\varrho = 0$. Wegen (2. 24) hat also $\mathfrak{u}(t)$ die Periode T_0.

Schließlich folgt aus den Gleichungen (2. 18) wegen $d\mathfrak{y}^0 = \mathfrak{u}^0$, und da $d\dot{\mathfrak{y}}^0 = \dot{\mathfrak{u}}$, $t = 0$, ist,

$$\mathfrak{u} \cdot \mathfrak{e} = \dot{\mathfrak{u}} \cdot \mathfrak{e} = 0, \qquad t = 0.$$

Eine periodische Lösung von $\dot{\mathfrak{u}} = \mathfrak{L}_0(\mathfrak{u})$ mit diesen Eigenschaften muß, wie oben hervorgehoben, verschwinden. Damit ist der Beweis für die Existenz einer periodischen Schar beendet.

Die Lösungen (2. 19) sind bei $\varepsilon = 0$ analytisch,

(2. 28) $$\begin{aligned} T &= T_0(1 + \tau_1\varepsilon + \tau_2\varepsilon^2 + \ldots), \\ \mu &= \mu_1\varepsilon + \mu_2\varepsilon^2 + \ldots. \end{aligned}$$

Die periodische Lösung $\mathfrak{y}(t, \varepsilon)$ von (2. 3), und damit die periodische Lösungsschar

(2. 29) $$\mathfrak{x}(t, \varepsilon) = \varepsilon\,\mathfrak{y}(t, \varepsilon)$$

von (1), ist an jeder Stelle $(t, 0)$ analytisch.

Übrigens bekommt man genau dieselben periodischen Lösungen, wenn man von einem Vielfachen $m T_0$ der Periode statt von T_0 ausgeht, d. h. wenn man in einer Umgebung des Wertesystems

(2. 30) $$T = m T_0, \quad \mu = 0, \quad \mathfrak{y}^0 = \mathfrak{z}^0$$

statt (2. 20) operiert. Am Beweise ändert sich nichts Wesentliches.

3. Beendigung des Beweises für den Satz.

Zu beliebig großem $L > T_0$ gibt es zwei positive Zahlen a und b mit folgender Eigenschaft. Jede periodische Lösung $\mathfrak{x}(t) \neq 0$ von (1), deren Periode kleiner als L ist, die zu einem μ mit $|\mu| < b$ gehört und welche in $|\mathfrak{x}| < a$ liegt, kommt bei passender Wahl des t-Nullpunktes in der Schar (2. 29), (2. 28), $\varepsilon > 0$, vor.

Wäre das nicht der Fall, so gäbe es eine Folge von periodischen Lösungen $\mathfrak{x}(t) \neq 0$ beschränkter Periode $T < L$ und von entsprechenden μ-Werten mit

$$\varkappa = \operatorname*{Max}_t |\mathfrak{x}(t)| \to 0, \quad \mu \to 0 \tag{3. 1}$$

und derart, daß kein Paar $\mathfrak{x}(t), \mu$ der obigen Schar angehört. Von den Lösungen

$$\mathfrak{y}(t) = \frac{1}{\varkappa}\mathfrak{x}(t)$$

von (2. 3), mit $\varkappa$ statt ε, gilt dann

$$\operatorname*{Max}_t |\mathfrak{y}(t)| = 1.$$

Man betrachte zunächst eine Teilfolge, für welche die Anfangswerte konvergieren, $\mathfrak{y}^0 \to \mathfrak{z}^0$. Dann gilt für $|t| < L$ gleichmäßig $\mathfrak{y}(t) \to \mathfrak{z}(t)$, wo $\dot{\mathfrak{z}} = \mathfrak{L}_0(\mathfrak{z})$ mit $\mathfrak{z}(0) = \mathfrak{z}^0$ ist. Wegen $\operatorname{Max}|\mathfrak{z}| = 1$ ist $\mathfrak{z}$ nicht identisch Null. $\mathfrak{z}$ ist von der Form (2. 8), $c \neq 0$, und hat die Grundperiode T_0. Man verlege in $\mathfrak{z}(t)$ den t-Nullpunkt an diejenige Stelle, wo $\mathfrak{z} \cdot \mathfrak{e} = 0$ wird. Dann ist dort $\dot{\mathfrak{z}} \cdot \mathfrak{e} \neq 0$. Diese Größe kann > 0 angenommen werden, denn sonst könnte man es wegen

$$\mathfrak{z}(t + \tfrac{1}{2} T_0) = -\mathfrak{z}(t)$$

durch Verlegung von $t = 0$ um $\frac{1}{2} T_0$ erreichen. Es ist also

$$\mathfrak{z}^0 \cdot \mathfrak{e} = 0, \quad \dot{\mathfrak{z}}^0 \cdot \mathfrak{e} > 0.$$

Hieraus folgt, daß in der Nähe von $\mathfrak{z}^0$ und für kleine $\varkappa$ und $|\mu|$ alle Lösungen der Differentialgleichung (2. 3) ($\varkappa$ statt ε) einmal die Hyperebene $\mathfrak{y} \cdot \mathfrak{e} = 0$ schneiden. In diesen Schnittpunkt werde $t = 0$ gelegt. Von der betrachteten Folge $\mathfrak{y}(t), \varkappa, \mu$ gilt dann nach dieser Wahl ebenfalls $\mathfrak{y}^0 \to \mathfrak{z}^0$ und

$$\mathfrak{y}^0 \cdot \mathfrak{e} = 0, \quad \varrho = \dot{\mathfrak{y}}^0 \cdot \mathfrak{e} \to \dot{\mathfrak{z}}^0 \cdot \mathfrak{e} = \varrho > 0$$

sowie $\varkappa \to 0$, $\mu \to 0$. Setzt man nun

$$\tilde{\mathfrak{y}}(t) = \frac{1}{\varrho}\mathfrak{y}(t) = \frac{1}{\varrho\varkappa}\mathfrak{x}(t), \quad \varrho\varkappa = \varepsilon,$$

so ist $\tilde{\mathfrak{y}}$ Lösung von (3) für die Parameterwerte $\varepsilon > 0$ und μ. Für sie gilt (2. 18),

$$\tilde{\mathfrak{y}} \cdot \mathfrak{e} = 0, \quad \dot{\tilde{\mathfrak{y}}} \cdot \mathfrak{e} = 1, \quad t = 0.$$

Die Perioden in der Lösungsfolge konvergieren notwendig gegen ein Vielfaches von T_0, $m T_0$. Ferner gilt $\varepsilon \to 0$. Hieraus ergibt sich aber, daß man in der Folge von einer geeigneten Stelle ab in die erwähnte Umgebung von (2. 20) bzw. von (2. 30) kommt, in welcher es für alle genügend kleinen ε nur eine Lösung des betrachteten Gleichungssystems gibt. Die Lösungen unserer Folge müßten dann doch der obigen Schar, und zwar mit $\varepsilon > 0$, angehören, im Widerspruch zur Annahme. Damit ist die Behauptung bewiesen.

Aus der eben bewiesenen Tatsache folgt nun: Wenn nicht $\mu(\varepsilon) \equiv 0$ ist, so ist der erste von Null verschiedene Koeffizient in $\mu = \mu_1 \varepsilon + \mu_2 \varepsilon^2 + \ldots$ von gerader Ordnung; dasselbe gilt von der Entwicklung $T = T_0 (1 + \tau_1 \varepsilon + \tau_2 \varepsilon^2 + \ldots)$. Denn die Lösungen der Schar für $\varepsilon < 0$ und die zugehörigen μ- und T-Werte müssen unter denen für $\varepsilon > 0$ bereits vorkommen.[1])

Insbesondere ist

(3. 2) $$\mu_1 = \tau_1 = 0.$$

Die periodischen Lösungen existieren, wenn man sich auf hinreichend kleine $|\mu|$ und $|\mathfrak{x}|$ beschränkt, entweder nur für $\mu > 0$ oder nur für $\mu < 0$ oder nur für $\mu = 0$.

4. Bestimmung der Koeffizienten.

Wir benötigen folgendes Kriterium über die Auflösung der inhomogenen Differentialgleichung

(4. 1) $$\dot{\mathfrak{w}} = \mathfrak{L}(\mathfrak{w}) + \mathfrak{q}, \qquad (\mathfrak{L} = \mathfrak{L}_0)$$

wo $\mathfrak{q}(t)$ die Periode T_0 hat. Es sei

(4. 2) $$\dot{\mathfrak{z}}^* = -\mathfrak{L}^*(\mathfrak{z}^*)$$

die zur homogenen adjungierte Differentialgleichung; $\mathfrak{L}^*$ ist der zu $\mathfrak{L}$ adjungierte Operator (transponierte Matrix), definiert durch

$$\mathfrak{L}(\mathfrak{u}) \cdot \mathfrak{v} \equiv \mathfrak{u} \cdot \mathfrak{L}^*(\mathfrak{v}).$$[2])

(4. 1), wo $\mathfrak{q}(t)$ die Periode T_0 hat, hat dann und nur dann eine periodische Lösung $\mathfrak{w}$ mit der Periode T_0, wenn

(4. 3) $$\int_0^{T_0} \mathfrak{q} \cdot \mathfrak{z}^* dt = 0$$

für alle Lösungen von (4. 2), die die Periode T_0 haben, gilt.

---

1) Und zwar mit einer Verschiebung des t-Nullpunktes um nahezu $T_0/2$.

2) Das innere Produkt zweier komplexer Vektoren $\mathfrak{a}$, $\mathfrak{b}$ wird im folgenden durch $\sum a_i b_i$ definiert.

Dieses Kriterium folgt aus dem bekannten Kriterium für die Lösbarkeit eines gewöhnlichen linearen Gleichungssystems. Die Notwendigkeit folgt direkt aus (4. 1) und (4. 2). Daß die Bedingung hinreicht, folgt so. Die adjungierte Gleichung hat die gleichen charakteristischen Exponenten und daher ebenfalls zwei Lösungen der Form

$$e^{\alpha t}\mathfrak{a}^*, \quad e^{-\alpha t}\bar{\mathfrak{a}}^*, \quad \alpha = \alpha(0) = -\bar{\alpha}(0), \tag{4. 4}$$

aus denen sich alle periodischen Lösungen linear kombinieren lassen. Ferner lehrt die Entwicklung von $\mathfrak{q}(t)$ in eine Fourierreihe, daß man sich auf den Fall

$$\mathfrak{q} = e^{-\alpha t}\mathfrak{b}$$

und den analogen Fall mit α statt $-\alpha$ beschränken kann. In (4. 1) werde

$$\mathfrak{w} = e^{-\alpha t}\mathfrak{c}$$

angesetzt. (4. 1) geht dann in

$$(\alpha\mathfrak{E} + \mathfrak{L})\,\mathfrak{c} = \mathfrak{b}$$

über. (4. 4) und (4. 2) sind mit

$$(\alpha\mathfrak{E} + \mathfrak{L})^*\,\mathfrak{a}^* = 0$$

gleichbedeutend, während (4. 3) $\mathfrak{b}\cdot\mathfrak{a}^* = 0$ lautet. Daraus folgt alles mit Hilfe des besagten Satzes.

Zweitens benötigen wir folgende Tatsache. Zu irgendeiner Lösung $\mathfrak{z} \neq 0$ von $\dot{\mathfrak{z}} = \mathfrak{L}(\mathfrak{z})$ mit der Periode T_0 gibt es stets eine Lösung $\mathfrak{z}^*$ der adjungierten Gleichung mit gleicher Periode derart, daß

$$\int_0^{T_0} \mathfrak{z}\cdot\mathfrak{z}^*\,dt \neq 0\ ^{1)}$$

ist.

Im entgegengesetzten Falle hätte nämlich $\dot{\mathfrak{w}} = \mathfrak{L}(\mathfrak{w}) + \mathfrak{z}$ eine Lösung $\mathfrak{w}$ mit der Periode T_0 und $\mathfrak{w} + t\mathfrak{z}$ wäre Lösung der homogenen Differentialgleichung, im Widerspruch zur Einfachheit des charakteristischen Exponenten α.

$\mathfrak{z}_1^*$ und $\mathfrak{z}_2^*$ seien zwei linear unabhängige Lösungen von (4. 2) mit der Periode T_0. Macht man von den Symbolen

$$[\mathfrak{q}]_i = \int_0^{T_0} \mathfrak{q}\cdot\mathfrak{z}_i^*\,dt \quad (i = 1, 2)$$

1) Übrigens ist der Integrand immer konstant.

Gebrauch, so lautet das Kriterium für die Lösbarkeit von (4. 1) unter den angegebenen Bedingungen

$$[\mathfrak{q}]_1 = [\mathfrak{q}]_2 = 0. \tag{4.5}$$

Im übrigen kann man $\mathfrak{z}_1^*$, $\mathfrak{z}_2^*$ so wählen, daß

$$[\mathfrak{z}]_1 = [\dot{\mathfrak{z}}]_2 = 1, \quad [\mathfrak{z}]_2 = [\dot{\mathfrak{z}}]_1 = 0 \tag{4.6}$$

wird, wo $\mathfrak{z}$ die Lösung (2. 10) von (2. 4), $\mu = 0$, ist (Biorthogonalisierung).

Die Aufgabe der Koeffizientenbestimmung für die Potenzreihen-Darstellung der periodischen Schar kann nun übersichtlich gelöst werden. Führt man durch

$$t = s\,(1 + \tau_2 \varepsilon^2 + \tau_3 \varepsilon^3 + \ldots) \tag{4.7}$$

die neue unabhängige Variable s ein, so wird wegen (2. 28) die Periode in der Lösungsschar $\mathfrak{y} = \mathfrak{y}\,(s, \varepsilon)$ konstant gleich T_0. $\mathfrak{y}$ ist als Funktion von s, ε ebenfalls an jeder Stelle $(s, 0)$ analytisch. Man hat

$$\mathfrak{y} = \mathfrak{y}_0\,(s) + \varepsilon\,\mathfrak{y}_1\,(s) + \varepsilon^2\,\mathfrak{y}_2\,(s) + \ldots, \tag{4.8}$$

wo alle $\mathfrak{y}_i$ die Periode T_0 haben. Die Ableitung nach s sei wieder durch einen Punkt gekennzeichnet. Schreibt man einfacher

$$\mathfrak{L}_0 = \mathfrak{L}, \quad \mathfrak{L}_0' = \mathfrak{L}', \quad \mathfrak{Q}_0 = \mathfrak{Q}, \quad \mathfrak{K}_0 = \mathfrak{K}, \quad \ldots,$$

so erhält man durch Einsetzen von (4. 7) und (4. 8) in (2. 3) unter Berücksichtigung von (3. 2) die rekursiven Gleichungen

$$\dot{\mathfrak{y}}_0 = \mathfrak{L}\,(\mathfrak{y}_0) \qquad (\mathfrak{y}_0 = \mathfrak{z}) \tag{4.9}$$

$$\dot{\mathfrak{y}}_1 = \mathfrak{L}\,(\mathfrak{y}_1) + \mathfrak{Q}\,(\mathfrak{y}_0, \mathfrak{y}_0) \tag{4.10}$$

$$-\tau_2\,\dot{\mathfrak{y}}_0 + \dot{\mathfrak{y}}_2 = \mathfrak{L}\,(\mathfrak{y}_2) + \mu_2\,\mathfrak{L}'\,(\mathfrak{y}_0) + 2\,\mathfrak{Q}\,(\mathfrak{y}_0, \mathfrak{y}_1) + \mathfrak{K}\,(\mathfrak{y}_0, \mathfrak{y}_0, \mathfrak{y}_0), \tag{4.11}$$

. . . .

aus denen die $\mathfrak{y}_i$, μ_i, τ_i zu bestimmen sind. Dazu kommen noch die aus (2. 18) folgenden Bedingungen

$$\mathfrak{y}_k \cdot \mathfrak{e} = \dot{\mathfrak{y}}_k \cdot \mathfrak{e} = 0, \quad s = 0, \tag{4.12}$$

für $k = 1, 2, \ldots$. In den Gleichungen kann wieder t statt s geschrieben werden. Durch (4. 10) und (4. 12) ist $\mathfrak{y}_1$ als periodische Funktion mit der Periode T_0 eindeutig bestimmt. Aus (4. 11) muß erst $\mathfrak{L}'$ mit Hilfe von (2. 13) eliminiert werden. Da die Klammer im ersten Summanden von (2. 14) Lösung von $\dot{\mathfrak{z}} = \mathfrak{L}\,(\mathfrak{z})$ ist, kann (2. 13)

$$\mathfrak{R}\,(\alpha')\,\mathfrak{z} + \frac{\mathfrak{J}\,(\alpha')}{\alpha}\,\dot{\mathfrak{z}} + \dot{\mathfrak{h}} = \mathfrak{L}\,(\mathfrak{h}) + \mathfrak{L}'\,(\mathfrak{z}) \tag{4.13}$$

geschrieben werden. Setzt man

(4. 14) $$\mathfrak{y}_2 - \mu_2 \mathfrak{h} = \mathfrak{v},$$

was wegen (2. 15) die Periode T_0 hat, so folgt wegen $\mathfrak{z} = \mathfrak{y}_0$

(4. 15) $$\begin{aligned} &-\mu_2 \Re(\alpha')\, \mathfrak{y}_0 - \left(\tau_2 + \mu_2 \frac{\Im(\alpha')}{\alpha}\right) \dot{\mathfrak{y}}_0 \\ &+ \dot{\mathfrak{v}} = \mathfrak{L}(\mathfrak{v}) + 2\mathfrak{Q}(\mathfrak{y}_0, \mathfrak{y}_1) + \mathfrak{K}(\mathfrak{y}_0, \mathfrak{y}_0, \mathfrak{y}_0). \end{aligned}$$

Wegen (4. 6) ist daher

(4. 16) $$\begin{aligned} \mu_2 \Re(\alpha') &= -[2\mathfrak{Q}(\mathfrak{y}_0, \mathfrak{y}_1) + \mathfrak{K}(\mathfrak{y}_0, \mathfrak{y}_0, \mathfrak{y}_0)]_1, \\ \tau_2 + \mu_2 \frac{\Im(\alpha')}{\alpha} &= -[2\mathfrak{Q}(\mathfrak{y}_0, \mathfrak{y}_1) + \mathfrak{K}(\mathfrak{y}_0, \mathfrak{y}_0, \mathfrak{y}_0)]_2. \end{aligned}$$

Hierdurch sind nach Voraussetzung (1. 2) μ_2 und τ_2 bestimmt. Man löst dann (4. 15) nach $\mathfrak{v}$ auf und gewinnt aus (4. 14) und (4. 12), $k = 2$, $\mathfrak{y}_2$ eindeutig.

In analoger Weise bestimmen sich die weiteren Koeffizienten aus den späteren Rekursionsformeln. Im allgemeinen ist $\mu_2 \neq 0$. Ist $\mu_2 > 0$, so existieren die periodischen Lösungen nur für $\mu > 0$; entsprechendes gilt für $\mu_2 < 0$.

5. Die charakteristischen Exponenten der periodischen Lösungen.

Im folgenden wird teilweise von Determinanten Gebrauch gemacht, was aber vermeidbar ist. In der längs den periodischen Lösungen von (2. 3) gebildeten Variationsgleichung

(5. 1) $$\dot{\mathfrak{u}} = \mathfrak{L}_{t,\varepsilon}(\mathfrak{u})$$

ist wegen (2. 3)

(5. 2) $$\mathfrak{L}_{t,\varepsilon}(\mathfrak{u}) = \mathfrak{L}_\mu(\mathfrak{u}) + 2\varepsilon \mathfrak{Q}_\mu(\mathfrak{y}, \mathfrak{u}) + 3\varepsilon^2 \mathfrak{K}_\mu(\mathfrak{y}, \mathfrak{y}, \mathfrak{u}) + \dots .$$

Ein mit festen Anfangswerten gebildetes Fundamentalsystem $\mathfrak{u}_i(t, \varepsilon)$ hängt analytisch von (t, ε) ab. Die Koeffizienten in $\mathfrak{u}_i(T, \varepsilon) = \sum a_{i\nu}(\varepsilon)\, \mathfrak{u}_\nu(0)$ sind bei $\varepsilon = 0$ analytisch. Die Determinantengleichung

(5. 3) $$\| a_{ik}(\varepsilon) - z\,\delta_{ik} \| = 0, \qquad z = e^{\lambda T(\varepsilon)},$$

bestimmt die charakt. Exp. λ_k und die Lösungen $\mathfrak{v}$,

$$\mathfrak{u} = e^{\lambda t}\, \mathfrak{v},$$

von (1. 3). Da (5. 1) durch $\mathfrak{u} = \dot{\mathfrak{y}}$ gelöst wird, ist $z = 1$ Wurzel von (5. 3).

Der Exponent β, von dem in der Einleitung die Rede war, entspricht einer einfachen Wurzel der durch $z-1$ dividierten Gleichung. $\beta(\varepsilon)$ ist also reell und bei $\varepsilon=0$ analytisch, $\beta=\beta_2\varepsilon^2+\ldots$ (β_1 ist aus demselben Grunde wie μ_1 und τ_1 Null). Ist nun nicht $\beta\equiv 0$, so wird in der Determinante (5. 3) (mit entsprechendem z) nicht alle Minoren $(u-1)$-ter Ordnung $\equiv 0$. Hieraus folgt, daß (1. 3), $\lambda=\beta$, eine bei $\varepsilon=0$ analytische Lösung $\mathfrak{v}\not\equiv 0$ hat. Ist jedoch $\beta\equiv 0$, so sind nicht alle Minoren $(n-2)$-ter Ordnung Null. Dann gibt es bekanntlich eine bei $\varepsilon=0$ analytische Lösung von (5. 1) der Form $\mathfrak{u}=t\mathfrak{v}+\mathfrak{w}$ mit periodischen $\mathfrak{v},\mathfrak{w}$, wo entweder $\mathfrak{v}\not\equiv 0$ ist, oder $\mathfrak{v}=0$ ist und $\mathfrak{w}$ von der Lösung $\mathfrak{u}=\dot{\mathfrak{y}}$ linear unabhängig ist.[1]) Daß $t\mathfrak{v}+\mathfrak{w}$ Lösung ist, drückt sich auch durch

(5. 4) $$\dot{\mathfrak{v}}=\mathfrak{L}_{t,\varepsilon}(\mathfrak{v}),\qquad \mathfrak{v}+\dot{\mathfrak{w}}=\mathfrak{L}_{t,\varepsilon}(\mathfrak{w})$$

aus.

Nach diesen Vorbemerkungen berechnen wir β_2. Wir setzen dabei $\mu_2\neq 0$ voraus. $\beta\equiv 0$ ist dann unmöglich, wie nachträglich bewiesen wird. Führt man in (1. 3) vermöge (4. 7) s als neues t ein, so wird

$$(1-\tau_2\varepsilon^2+\ldots)\dot{\mathfrak{v}}+\beta\mathfrak{v}=\mathfrak{L}_{t,\varepsilon}(\mathfrak{v})$$

und es ist (mit dem neuen t)

$$\mathfrak{v}=\mathfrak{v}_0(t)+\varepsilon\mathfrak{v}_1(t)+\varepsilon^2\mathfrak{v}_2(t)+\ldots,$$

wo alle $\mathfrak{v}_i$ dieselbe Periode T_0 haben. Führt man die Potenzreihen für μ, β, $\mathfrak{v}$, $\mathfrak{y}$ ein, so folgt (der Index Null in den Operatoren wird wie vorher weggelassen)

(5. 5) $$\dot{\mathfrak{v}}_0=\mathfrak{L}(\mathfrak{v}_0),$$

(5. 6) $$\dot{\mathfrak{v}}_1=\mathfrak{L}(\mathfrak{v}_1)+2\mathfrak{Q}(\mathfrak{y}_0,\mathfrak{v}_0),$$

(5. 7) $$\begin{aligned}\beta_2\mathfrak{v}_0-\tau_2\dot{\mathfrak{v}}_0+\dot{\mathfrak{v}}_2&=\mathfrak{L}(\mathfrak{v}_2)+\mu_2\mathfrak{L}'(\mathfrak{v}_0)\\&\quad+2\mathfrak{Q}(\mathfrak{y}_1,\mathfrak{v}_0)+2\mathfrak{Q}(\mathfrak{y}_0,\mathfrak{v}_1)+3\mathfrak{K}(\mathfrak{y}_0,\mathfrak{y}_0,\mathfrak{v}_0),\end{aligned}$$

Diese Gleichungen haben die triviale Lösung

(5. 8) $$\beta_i=0,\quad \mathfrak{v}_i=\dot{\mathfrak{y}}_i\quad (i=0,1,\ldots).$$

Man hat also

(5. 9) $$\ddot{\mathfrak{y}}_1=\mathfrak{L}(\dot{\mathfrak{y}}_1)+2\mathfrak{Q}(\mathfrak{y}_0,\dot{\mathfrak{y}}_0),$$

(5. 10) $$\begin{aligned}-\tau_2\ddot{\mathfrak{y}}_0+\ddot{\mathfrak{y}}_2&=\mathfrak{L}(\dot{\mathfrak{y}}_2)+\mu_2\mathfrak{L}'(\dot{\mathfrak{y}}_0)\\&\quad+2\mathfrak{Q}(\mathfrak{y}_1,\dot{\mathfrak{y}}_0)+2\mathfrak{Q}(\mathfrak{y}_0,\dot{\mathfrak{y}}_1)+3\mathfrak{K}(\mathfrak{y}_0,\mathfrak{y}_0,\dot{\mathfrak{y}}_0).\end{aligned}$$

---

1) Vgl. z. B. F. R. Moulton, Periodic Orbits, Washington 1920, S. 26.

Da $\mathfrak{v}_0 \neq 0$ vorausgesetzt werden darf, ist

$$\mathfrak{v}_0 = \varrho \mathfrak{y}_0 + \sigma \dot{\mathfrak{y}}_0, \tag{5.11}$$

wo nicht $\varrho = \sigma = 0$ ist. Setzt man

$$\mathfrak{v}_1 - 2\varrho \mathfrak{y}_1 - \sigma \dot{\mathfrak{y}}_1 = \mathfrak{w}, \tag{5.12}$$

so folgt aus (4. 10), (5. 6) und (5. 9) $\dot{\mathfrak{w}} = \mathfrak{L}(\mathfrak{w})$, also

$$\mathfrak{w} = \varrho' \mathfrak{y}_0 + \sigma' \dot{\mathfrak{y}}_0. \tag{5.13}$$

Bildet man die Kombination

$$(5.7) - \varrho\,(4.11) - \sigma\,(5.10),$$

in welcher $\mathfrak{L}'$ herausfällt, und setzt man

$$\mathfrak{v}_2 - \varrho \mathfrak{y}_2 - \sigma \dot{\mathfrak{y}}_2 = \mathfrak{u},$$

so ergibt sich mit Berücksichtigung von (5. 11) und (5. 12)

$$\beta_2 \mathfrak{v}_0 + \dot{\mathfrak{u}} = \mathfrak{L}(\mathfrak{u}) + 2\varrho\left(2\mathfrak{Q}(\mathfrak{y}_0, \mathfrak{y}_1) + \mathfrak{K}(\mathfrak{y}_0, \mathfrak{y}_0, \mathfrak{y}_0)\right) + \mathfrak{R} \tag{5.14}$$

mit

$$\mathfrak{R} = 2\mathfrak{Q}(\mathfrak{y}_0, \mathfrak{w}).$$

Wendet man nun auf (4. 10) und (5. 9) das Klammerkriterium des vorigen Abschnitts an, so folgt nach (5. 13)

$$[\mathfrak{R}]_1 = [\mathfrak{R}]_2 = 0.$$

Wendet man es auf (5. 14) an — $\mathfrak{u}$ hat die Periode T_0 —, so folgt wegen (4. 6), $\mathfrak{z} = \mathfrak{y}_0$,

$$\varrho \beta_2 = 2\varrho\,[2\mathfrak{Q}(\mathfrak{y}_0, \mathfrak{y}_1) + \mathfrak{K}(\mathfrak{y}_0, \mathfrak{y}_0, \mathfrak{y}_0]_1,$$

also wegen (4.16)

$$\varrho \beta_2 = -2\varrho \mu_2 \Re(\alpha').$$

Ebenso folgt

$$\sigma \beta_2 = -2\varrho\left(\tau_2 + \mu_2 \frac{\Im(\alpha')}{\alpha}\right).$$

Demnach ist entweder β_2 durch (1. 4) gegeben — dann ist β_2 wegen $\mu_2 \neq 0$ nicht Null —, oder es ist $\beta_2 = 0$. In jedem Falle ist $\varrho : \sigma$ vollständig bestimmt (Im zweiten Fall ist $\varrho = 0$).

Zum Nachweis, daß der erste Fall wirklich eintritt, mußten wir eine längere Betrachtung anstellen. Man denke sich das Verfahren folgendermaßen schematisiert. Die Gleichung für β und $\mathfrak{v}$ (sie ist die auf (5. 4) folgende Gleichung) dividiere man durch den Klammerfaktor. Sie ist dann wieder von der Form

$$\dot{\mathfrak{v}} + \beta\mathfrak{v} = \mathfrak{L}_{t,\varepsilon}(\mathfrak{v})$$

mit

$$\mathfrak{L}_{t,\varepsilon} = \mathfrak{L}_0 + \varepsilon\mathfrak{L}_1 + \varepsilon^2\mathfrak{L}_2 + \dots,$$

wo $\mathfrak{L}_0$ ein konstanter Operator ist, während $\mathfrak{L}_i$, $i > 0$, von t mit der Periode T_0 abhängt. Die Koeffizienten von $1, \varepsilon$ werden durch obige Division nicht geändert. Einsetzen der Potenzreihen liefert

$$\begin{aligned} \dot{\mathfrak{v}}_0 &= \mathfrak{L}_0(\mathfrak{v}_0), \\ \dot{\mathfrak{v}}_1 &= \mathfrak{L}_0(\mathfrak{v}_1) + \mathfrak{L}_1(\mathfrak{v}_0),^{1)} \\ \beta_2\mathfrak{v}_0 + \dot{\mathfrak{v}}_2 &= \mathfrak{L}_0(\mathfrak{v}_2) + \mathfrak{L}_1(\mathfrak{v}_1) + \mathfrak{L}_2(\mathfrak{v}_0), \\ \beta_3\mathfrak{v}_0 + \beta_2\mathfrak{v}_1 + \dot{\mathfrak{v}}_3 &= \mathfrak{L}_0(\mathfrak{v}_3) + \mathfrak{L}_1(\mathfrak{v}_2) + \mathfrak{L}_2(\mathfrak{v}_1) + \mathfrak{L}_3(\mathfrak{v}_0) \end{aligned} \tag{5.15}$$

usw. Die Situation ist folgende. Für $\varepsilon = 0$ gibt es zwei Lösungen $\mathfrak{z}, \dot{\mathfrak{z}}$ der Periode T_0. Es ist ferner

$$\mathfrak{v}_0 = \varrho\mathfrak{z} + \sigma\dot{\mathfrak{z}} \tag{5.16}$$

und

$$[\mathfrak{L}_1(\mathfrak{z})] = [\mathfrak{L}_1(\dot{\mathfrak{z}})] = 0 \tag{5.17}$$

für beide Klammerindizes. Es ergibt sich

$$\mathfrak{v}_1 = \varrho\mathfrak{g} + \sigma\mathfrak{h} + \varrho'\mathfrak{z} + \sigma'\dot{\mathfrak{z}} \tag{5.18}$$

mit festen periodischen $\mathfrak{g}$, $\mathfrak{h}$. Das Klammerkriterium ergibt für die dritte Gleichung (5. 15)

$$\begin{aligned} \beta_2\varrho &= A_1\varrho + B_1\sigma \\ \beta_2\sigma &= A_2\varrho + B_2\sigma \end{aligned} \tag{5.19}$$

mit

$$A_i = [\mathfrak{L}_1(\mathfrak{g}) + \mathfrak{L}_2(\mathfrak{z})]_i, \qquad B_i = [\mathfrak{L}_1(\mathfrak{h}) + \mathfrak{L}_2(\dot{\mathfrak{z}})]_i, \tag{5.20}$$

während ϱ', σ' wegen (5. 17) herausfallen. Die Situation ist nun die, daß die Gleichungen (5. 18) mit den Unbekannten β_2, ϱ, σ zwei verschiedene reelle Lösungen β_2 haben.[2] Zu ihnen gehören zwei linear unabhängige Paare (ϱ, σ). Jedes der beiden Lösungssysteme führt nun zu einer eindeutigen Bestimmung der β_i, $\mathfrak{v}_i$ durch die Rekursionsformeln, wenn man $\mathfrak{v}$ geeignet

1) Man hätte $\beta_1 = 0$ nicht vorauszusetzen brauchen. Nach dem Klammerkriterium ist es eine Folge von (5. 17).

2) Im allgemeinen Falle, d. h. wenn die speziellen Bedingungen (5. 17) nicht erfüllt sind, tritt die Verzweigung in zwei Fälle bereits bei der zweiten Gleichung (5. 15) ein. Die Lösung der Aufgabe in diesem Falle findet sich in F. R. Moulton, Periodic Orbits. Vgl. Kap. I, insbesondere S. 34 und S. 40.

normiert. Hierzu wähle man einen konstanten Vektor $\mathfrak{a} \neq 0$ derart, daß $\mathfrak{v}_0 \cdot \mathfrak{a} = 1$ $(t = 0)$ für beide Paare (ϱ, σ) in (5. 16) ausfällt. Man fordere dann

$$\mathfrak{v} \cdot \mathfrak{a} = 1, \quad t = 0,$$

d. h. $\mathfrak{v}_i \cdot \mathfrak{a} = 0$ $(t = 0)$ für $i > 0$. Setzt man

$$\mathfrak{z} \cdot \mathfrak{a} = C, \quad \dot{\mathfrak{z}} \cdot \mathfrak{a} = D \quad (t = 0),$$

so folgt, daß das Gleichungssystem

$$\begin{aligned} (A_1 - \beta_2)\,\varrho + B_1 \sigma &= 0 \\ A_2 \varrho + (B_2 - \beta_2)\,\sigma &= 0 \\ C\varrho + D\sigma &= 1 \end{aligned} \tag{5. 21}$$

für jeden der beiden Werte von β_2 die Unbekannten ϱ, σ eindeutig festlegt. Bis jetzt sind β_2, ϱ, σ, $\mathfrak{v}_0$ bestimmt. Aus der dritten Gleichung (5. 15) erhält man nach Einsetzen von (5. 18) mit Berücksichtigung der Definition von $\mathfrak{g}$, $\mathfrak{h}$

$$\mathfrak{v}_2 = \varrho' \mathfrak{g} + \sigma' \mathfrak{h} + \varrho'' \mathfrak{z} + \sigma'' \dot{\mathfrak{z}} + \dots, \tag{5. 22}$$

wo die weggelassenen Glieder bereits bekannt sind. Setzt man (5. 18) und (5. 22) in die vierte Gleichung (5. 15) ein, so erhält man nach Anwendung des Klammerkriteriums mit Rücksicht auf (5. 20) die Gleichungen

$$\begin{aligned} \varrho\beta_3 - (A_1 - \beta_2)\,\varrho' - B_1 \sigma' &= \dots, \\ \sigma\beta_3 - A_2 \varrho' - (B_2 - \beta_2)\,\sigma' &= \dots. \end{aligned}$$

Dazu tritt noch wegen $\mathfrak{v}_1 \cdot \mathfrak{a} = 0$ $(t = 0)$ die Gleichung

$$C\varrho' + D\sigma' = \dots.$$

Durch die drei Gleichungen sind nun die drei Größen β_3, ϱ', σ' eindeutig bestimmt. Für die Determinante errechnet man mit Hilfe von (5. 21) den Wert

$$A_1 + B_2 - 2\beta_2.$$

Er ist $\neq 0$ Null, da nach Voraussetzung (5. 19) zwei verschiedene Lösungen β_2 hat. Damit sind β_3, ϱ', σ' und $\mathfrak{v}_1$ bestimmt.

Man überzeugt sich nun leicht, daß im nächsten Schritt β_4, ϱ'', σ'' durch Gleichungen mit genau denselben linken Seiten bestimmt sind, und daß durch die weiteren analogen Schritte alles festgelegt wird.

Wir kehren noch einmal zu der uns interessierenden speziellen Aufgabe zurück und stellen fest, daß es bei geeigneter Normierung zwei verschiedene formale Potenzreihenpaare $(\beta, \mathfrak{v})$ gibt, die die Gleichung

$$(1 - \tau_2 \varepsilon^2 + \dots)\,\dot{\mathfrak{v}} + \beta \mathfrak{v} = \mathfrak{L}_{t,\varepsilon}(\mathfrak{v})$$

lösen. Andererseits wurde vorher festgestellt, daß es unter der Voraussetzung $\beta \not\equiv 0$ zwei wirkliche Lösungen gibt, von denen eine bekannt ist, nämlich (5. 8). Unter dieser Voraussetzung ist also die zweite (normierte) Lösung durch die betreffenden Potenzreihen darstellbar und es gilt wirklich die Formel (1. 4) für β_2. Zur vollständigen Erledigung muß noch gezeigt werden, daß nicht $\beta \equiv 0$ sein kann, wenn $\mu_2 \neq 0$ ist. Wir zeigen das wieder an Hand der schematisierten Aufgabe. Da (5. 19) die Lösung $\beta_2 = \varrho = 0$ hat und das zweite $\beta_2 \neq 0$ ist, ist

(5. 23) $$B_1 = B_2 = 0, \quad A_1 \neq 0.$$

Wäre $\beta \equiv 0$, so hätte (5. 4) eine Lösung mit den dort angegebenen Eigenschaften.

Einsetzen der Potenzreihen für $\mathfrak{v}$, $\mathfrak{w}$ ergibt

(5. 24) $$\begin{aligned} \mathfrak{v}_0 + \dot{\mathfrak{w}}_0 &= \mathfrak{L}_0(\mathfrak{w}_0), \\ \mathfrak{v}_1 + \dot{\mathfrak{w}}_1 &= \mathfrak{L}_0(\mathfrak{w}_1) + \mathfrak{L}_1(\mathfrak{w}_0), \\ \mathfrak{v}_2 + \dot{\mathfrak{w}}_2 &= \mathfrak{L}_0(\mathfrak{w}_2) + \mathfrak{L}_1(\mathfrak{w}_1) + \mathfrak{L}_2(\mathfrak{w}_0). \end{aligned}$$

Es ist

(5. 25) $$\mathfrak{w}_0 = \varrho \mathfrak{z} + \sigma \dot{\mathfrak{z}}.$$

Da $\mathfrak{v}_0$ auch von dieser Form ist, muß nach dem Klammerkriterium $\mathfrak{v}_0 = 0$ sein. Wegen (5. 17) folgt analog $\mathfrak{v}_1 = 0$. Ähnlich wie in (5. 18) wird

$$\mathfrak{w}_1 = \varrho \mathfrak{g} + \sigma \mathfrak{h} + \varrho' \mathfrak{z} + \sigma' \dot{\mathfrak{z}}.$$

Oben war bewiesen worden, daß $\dot{\mathfrak{v}} = \mathfrak{L}_{t,\varepsilon}(\mathfrak{v})$ eine bis auf einen Faktor eindeutige Lösung der Periode T_0 hat ($\dot{\mathfrak{y}}$). Es ist also sicher

$$\mathfrak{v}_2 = \lambda \dot{\mathfrak{z}}.$$

Anwendung der Klammerregel auf (5. 24) ergibt ähnlich wie oben wegen (5. 20) die Gleichungen (in denen ϱ', σ' wieder herausfallen)

$$\begin{aligned} 0 &= A_1 \varrho + B_1 \sigma, \\ \lambda &= A_2 \varrho + B_2 \sigma. \end{aligned}$$

Wegen (5. 23) folgt also $\varrho = \lambda = 0$, und daher $\mathfrak{v}_2 = 0$. Nach (5. 25) ist $\mathfrak{w}_0 = \sigma \dot{\mathfrak{z}}$. Zieht man nun von der zweiten Gleichung (5. 4) die Lösung $\sigma \dot{\mathfrak{y}}$ von $\dot{\mathfrak{w}} = \mathfrak{L}(\mathfrak{w})$ ab und dividiert man durch ε, so kann man das ganze Verfahren wiederholen, und es folgt schrittweise $\mathfrak{v}_i = 0$, also $\mathfrak{v} = 0$. Damit ist bewiesen, daß nicht $\beta = 0$ sein kann.

Die Rechtfertigung der Formel (1. 4) ist damit unter der Voraussetzung $\mu_2 \neq 0$ vollständig. Diese Voraussetzung könnte durch $\mu \not\equiv 0$ ersetzt werden.

Die Betrachtung würde sich allein darin ändern, daß in der Koeffizientenberechnung der Verzweigungsfall später auftritt.

Die Umständlichkeit der Überlegung könnte folgendermaßen vermieden werden. Man berechnet zuerst rein formal wie oben die Koeffizienten der Potenzreihen für β und $\mathfrak{v}$ und beweist dann die Konvergenz direkt durch geeignete Anwendung der Majorantenmethode. Dies entspräche unserer Absicht, die Anwendung auf partielle Differentialsysteme zu erleichtern. Man kann aber auch die Diskussion des Verzweigungsfalles und den Beweis von (1. 4) ausschließlich mit Determinanten führen.

Abzweigung einer periodischen Lösung von einer stationaeren Lösung eines Differentialsystems, *Akad. Wiss.* **(Leipzig) 94 (1942), 3-22.**

Commentary

Martin Golubitsky and Paul H. Rabinowitz

Introduction

The Hopf Bifurcation Theorem provides the simplest criterion for a family of periodic solutions to bifurcate from a known family of equilibrium solutions of an evolution equation. A second theorem gives information about the stability or instability of the bifurcating branch of solutions. We do not know exactly what motivated Hopf to study these questions. When asked about it many years later, he could not recall how he come to them. However in his paper, he mentions that the qualitative phenomena they describe are well known in hydrodynamics, for example, periodic vortex shedding in flow past an obstacle when the velocity is large enough. Thus we might speculate that such problems were the origin of his interest in bifurcation.

The paper appeared in 1942 and in the introduction Hopf writes about the first theorem: "I scarcely think that there is anything new in the above theorem. The methods have been developed by Poincaré perhaps 50 years ago and belong today to the classical conceptual structure of the theory of periodic solutions in the small". In fact using such methods, Andronov obtained the bifurcation and stability results for two dimensional systems as can be seen already in his book, Andronov-Vitt-Khaikin (1937). Even earlier work of Andronov which was not available to us was cited by Arnold (1983). These points not withstanding, the Hopf Bifurcation Theorem has become a paradigm of a useful and elementary result that has been extremely influential. New proofs have been given and extensions have been made in many directions. There are now degenerate and equivariant and Hamiltonian and global and infinite dimensional versions of the theorem. Unexpected connections have been found to the much older Liapunov Center Theorem. Several numerical codes have been written to implement the theorem. And of course there are many physical applications. In what follows, we will discuss the two theorems, illustrating them in a simple setting and giving a sketch of a modern proof of the first. There will also be a brief discussion of the extensions and some applications.

1. The Hopf Theorem

Hopf considered the system of ordinary differential equations:

$$\dot{x} = F(x, \lambda), \quad x \in \mathbb{R}^n \tag{1.1.1}$$

where λ belongs to an interval $I \subset \mathbb{R}$ and (1.1.1) possesses a known family of equilibrium solutions $(x(\lambda), \lambda), \lambda \in I$. Making a change of variables, it can be assumed that $x(\lambda) \equiv 0$ and $0 \in I$. Let

$$J(\lambda) = d_x F|_{(0,\lambda)},$$

the Jacobian matrix or Fréchet derivative of F evaluated at $(0, \lambda)$. Suppose that

$$(H_1) \quad J(\lambda) \text{ has a pair of simple complex conjugate eigenvalues } a(\lambda), \overline{a}(\lambda) \text{ with } a(0) \text{ purely imaginary and } a'(0) \neq 0.$$

The condition $a'(0) \neq 0$ is called the *eigenvalue crossing condition.* Assume also that

$$(H_2) \quad \pm a(0) \text{ are the only eigenvalues of } J(0) \text{ on the imaginary axis.}$$

Under assumptions (H1) and (H2), the first Hopf theorem asserts the existence of a one parameter family of solutions of (1.1.1) that are periodic in t (with period near $2\pi/|a'(0)|$) and bifurcate from the equilibrium solution $(0,0)$. The theorem also contains some information about the form of the bifurcating solutions. In particular, roughly speaking, the solution is parametrized by its amplitude.

The simplest example of the result occurs in the planar linear system

$$\begin{aligned} \dot{x}_1 &= \lambda x_1 - x_2 \\ \dot{x}_2 &= x_1 + \lambda x_2. \end{aligned} \tag{1.1.2}$$

It is easy to check that the origin is a spiral sink when $\lambda < 0$ and a spiral source when $\lambda > 0$. When $\lambda = 0$, the origin is a center, and there is a continuous family of periodic solutions surrounding this center. See Figure 1.

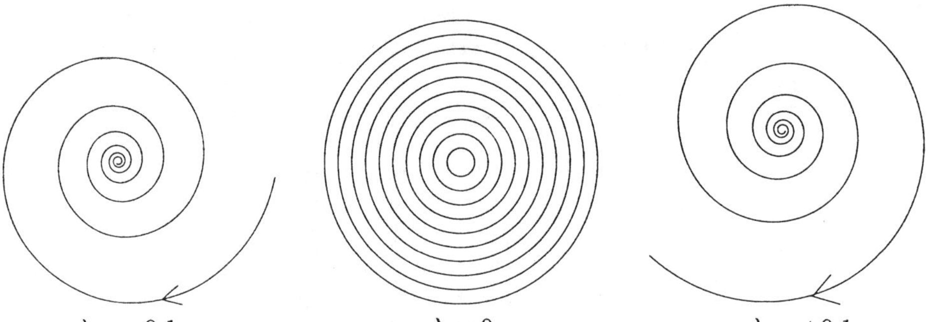

FIGURE 1. Phase planes for (1.1.2).

The second Hopf theorem which is sometimes referred to as the *exchange of stability theorem* provides information about the stability of the bifurcating branch. To simplify the discussion, suppose that the equilibria $(0, \lambda)$ with $\lambda < 0$, are asymptotically stable. Then the theorem states that if a certain number μ_2 (which can be computed explicitly from the linear, quadratic, and cubic terms of F) is nonzero, the bifurcating solutions are either supercritical (occur when $\lambda > 0$) and asymptotically stable or are subcritical (occur when $\lambda < 0$) and unstable.

A simple example of exchange of stability is given by adding a nonlinear term to the linear system (1.1.2).

$$\begin{aligned} \dot{x}_1 &= \lambda x_1 - x_2 - (x_1^2 + x_2^2)x_1 \\ \dot{x}_2 &= x_1 + \lambda x_2 - (x_1^2 + x_2^2)x_2. \end{aligned} \tag{1.1.3}$$

Note that when $\lambda < 0$ the origin is still a spiral sink and that the nonlinear terms $(x_1^2 + x_2^2)(x_1, x_2)^t$ point toward the origin, so that it is not surprising that globally solutions spiral toward the origin. However, when $\lambda > 0$, the origin is a spiral source but the nonlinear terms still point toward the origin. This interaction between the linear and nonlinear effects is resolved by the existence of a stable limit cycle.

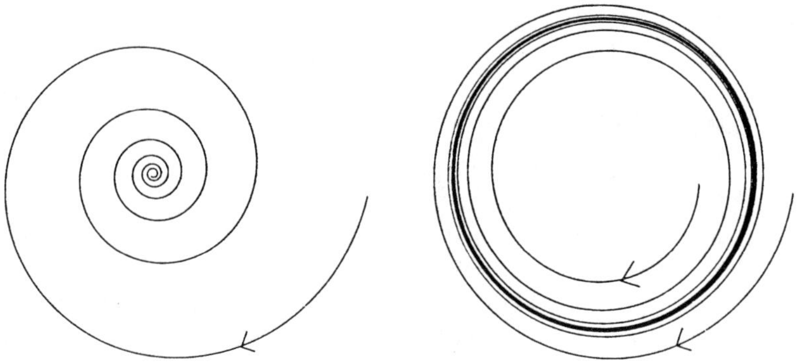

$\lambda = -0.1$ $\lambda = +0.1$

FIGURE 2. Phase planes for (1.1.3).

Both Hopf theorems can be proved by two very different methods: (i) Liapunov-Schmidt reduction, or (ii) center manifold reduction coupled with Poincaré-Birkhoff normal form theory. It is noteworthy that the existence proof using Liapunov-Schmidt reduction is relatively straightforward, whereas the stability result is straightforward when the center manifold approach is invoked. In this review we sketch the existence proof by Liapunov-Schmidt reduction. This approach is due to Cesari and Hale (1969).

The Liapunov-Schmidt proof of the existence of periodic solutions proceeds in three steps. First, the ordinary differential equations in n dimensions are posed as an operator on infinite-dimensional loop space whose zeros are the desired periodic solutions. Second, the implicit function theorem is used to reduce the space from infinite dimensions back to two dimensions (the real eigenspace corresponding to the complex conjugate eigenvalues of $J(0)$). Finally, phase-shift $\mathbf{S}^1$ symmetry is used to simplify the search for zeros in two dimensions. Hopf's approach to the problem is related to the Liapunov-Schmidt reduction but, in line with the techniques of his day, uses a Poincaré map and the implicit function theorem in $\mathbf{R}^n$.

Step 1: Loop Space. To simplify the discussion, assume that the complex conjugate eigenvalues of $J(0)$ are $\pm i$. (Rescaling time in (1.1.1) will accomplish this task.) The linearized system

$$\dot{x} = J(0)x \tag{1.1.4}$$

then has periodic solutions of period 2π. The important observation that follows is that the bifurcating periodic solutions in the nonlinear system are parametrized by the periodic solutions for the linear system and that the period of the new periodic solutions of the nonlinear system are approximately 2π. Rescaling time allows us to search only for periodic solutions that have period exactly 2π. This clever trick is what makes the approach work.

Introduce the loop space $\mathcal{C}^0_{2\pi}$ consisting of $\mathcal{C}^0$ maps $\mathbf{S}^1 \to \mathbb{R}^n$ and let $\mathcal{C}^1_{2\pi}$ be the corresponding subspace of $\mathcal{C}^1$ maps. Observe that zeros of the operator equation

$$\mathcal{F} : \mathcal{C}^1_{2\pi} \times \mathbb{R} \times \mathbb{R} \to \mathcal{C}^0_{2\pi}$$

where

$$\mathcal{F}(u, \lambda, \tau) = (1 + \tau)\frac{du}{ds} - f(u, \lambda),$$

correspond to $2\pi/(1+\tau)$ periodic solutions to the original system (1.1.1); so we think of τ as the perturbed period parameter.

Step 2: Reduction to Two Dimensions. The linearization of $\mathcal{F}$ at the origin is just the linear system of differential equations

$$\mathcal{L}(u) = \frac{du}{ds} - J(0)u.$$

The eigenvalue assumptions on $J(0)$ imply that $\mathcal{K} = \ker \mathcal{L}$ is the two-dimensional space of 2π periodic solutions to the linear differential equation (1.1.4). Let $\mathcal{R}$ denote the range of $\mathcal{L}$ and let $P : \mathcal{C}^0_{2\pi} \to \mathcal{R}$ be a projection. The Fredholm alternative can be used to show that $\ker P$ is also two-dimensional.

It follows that solving the nonlinear operator equation $\mathcal{F} = 0$ can be divided into two parts

$$\begin{aligned} P\mathcal{F} &= 0 \\ (I - P)\mathcal{F} &= 0 \end{aligned}$$

The first equation can be solved near the origin using the implicit function theorem. Let $\mathcal{W}$ be a complement to the kernel $\mathcal{K}$ in the domain space $\mathcal{C}^1_{2\pi}$ and observe that $\mathcal{L}|_{\mathcal{W}} : \mathcal{W} \to \mathcal{R}$ is invertible. Thus, there exists an implicit function $W : \mathcal{K} \times \mathbb{R} \times \mathbb{R} \to \mathcal{W}$ such that

$$P\mathcal{F}(k, W(k, \lambda, \tau), \lambda, \tau) \equiv 0.$$

It now follows that periodic solutions to (1.1.1) near the origin and with period near 2π are parametrized by zeros of the equation

$$G(k, \lambda, \tau) = (I - P)\mathcal{F}(k, W(k, \lambda, \tau), \lambda, \tau)$$

where $G : \mathcal{K} \times \mathbb{R} \times \mathbb{R} \to \ker P$. We can identify the two-dimensional subspaces $\mathcal{K}$ and $\ker P$ with $\mathbf{C}$ and the proof then reduces to finding zeros of a map

$$g : \mathbf{C} \times \mathbb{R} \times \mathbb{R} \to \mathbf{C}$$

near the origin. Since g is defined only implicitly and since simultaneously solving two nonlinear equations in two variables that depend on two parameters is not straightforward, what remains is still a difficult problem. However, phase-shift symmetry comes to the rescue.

Step 3: The Use of $\mathbf{S}^1$ *Phase Shift Symmetry.* The circle group $\mathbf{S}^1$ acts naturally on a periodic function $u(t)$ by

$$\theta u(t) = u(t - \theta).$$

It is easy to check that the operator $\mathcal{F}$ commutes with this action of $\mathbf{S}^1$. It is also possible to set up the implicit function theorem (correct choices for $\mathcal{W}$ and $\ker P$) so that the $\mathbf{S}^1$ action survives reduction. That is, we may assume that

$$g(e^{i\theta}z, \lambda, \tau) = e^{i\theta}g(z, \lambda, \tau).$$

It follows from invariant theory that g has the form

$$g(z, \lambda, \tau) = p(|z|^2, \lambda, \tau)z + q(|z|^2, \lambda, \tau)iz$$

where p and q are real-valued functions.

It is now possible to solve for the zeros of g. Since we are looking only for trajectories of periodic solutions, we can apply $\mathbf{S}^1$ symmetry to assume that $z \in \mathbb{R}$. Observe that solving $g = 0$ is equivalent to solving $p = q = 0$. Finally some calculations (based on implicit differentiation) are needed. These calculations do require substantial work to complete. In particular $p_\lambda(0,0,0) = a'(0)$, $p_\tau(0,0,0) = 0$, and $q_\tau(0,0,0) = -1$ where $a'(0)$ is the speed with which the critical eigenvalues of $J(\lambda)$ cross the imaginary axis.

We can now apply the implicit function theorem to obtain a function $\tau(|z|^2, \lambda)$ such that

$$q(|z|^2, \lambda, \tau(|z|^2, \lambda)) \equiv 0.$$

So our desired periodic solutions are obtained by solving

$$A(|z|^2, \lambda) \equiv p(|z|^2, \lambda, \tau(|z|^2, \lambda)) = 0.$$

Second $A_\lambda(0) = a'(0)$, so we can now apply the implicit function theorem (for the third time) to obtain a branch of solutions to $A = 0$ parametrized by $|z|^2$. These calculations complete the proof of the first Hopf theorem.

Finally, turning to the second theorem, let $\mu_2 = A_{|z|^2}(0)$. Recall that μ_2 was the number alluded to above that determined whether the branch of new periodic solutions were supercritical or subcritical. To lowest order

$$A(|z|^2, \lambda) = \mu_2 |z|^2 + a'(0)\lambda + \cdots .$$

When $\mu_2 \neq 0$ we see that the branch has the form

$$\lambda = -\frac{\mu_2}{a'(0)}|z|^2 + \cdots .$$

which decides super- or subcriticality. This is roughly the computation Hopf made.

The discussion of stability of solutions requires Floquet theory and careful control of Floquet exponents in the reduction process. To be a bit more precise, returning to $\mathcal{F}(u, \lambda, \tau) = 0$ or equivalently

$$\frac{du}{ds} = \frac{1}{1+\tau} f(u, \lambda), \tag{1.1.5}$$

the stability of the solution $u(s, z)$ depends on the Floquet exponents of the linearization of (1.1.5) about $u(s, z)$. In a physical problem where $(x(\lambda), \lambda)$ is actually observed e.g. for $\lambda < 0$, it is implicit that this equilibrium solution is stable. Hence at $(x(0), 0) = (0, 0)$, we expect that $n - 2$ Floquet exponents of

$$\frac{dv}{ds} = f_u(0,0)v \tag{1.1.6}$$

have negative real parts and 0 is a Floquet exponent of multiplicity 2. The Floquet exponents of (1.1.6) are the values $-\kappa$ for which

$$\frac{dw}{ds} - f_u(0,0)w = \kappa w, \quad w(0) = w(2\pi) \tag{1.1.7}$$

has a nontrivial solution. Differentiating (1.1.5) with respect to s shows that $w = \frac{du}{ds}$ satisfies (1.1.6) with $\kappa = 0$, i.e. 0 continues as a Floquet exponent along the bifurcating branch, $\mathcal{B}$, of solutions. Since the $n-2$ Floquet exponents with negative real parts will also continue to be negative on $\mathcal{B}$ near the bifurcation point, the stability of the solutions on $\mathcal{B}$ is governed by how the second zero exponent of (1.1.7) continues along $\mathcal{B}$. It turns out that $\kappa(z)$ and $\lambda'(z)$ have the same zeroes and whenever $\lambda'(z) \neq 0$, $\kappa(z)$ and — $a'(0)z\lambda'(z)$ have the same sign. In particular, if $a'(0) < 0$ and $z\lambda'(z) > 0$ if $z \neq 0$ (i.e. bifurcation is supercritical), then the solutions on $\mathcal{B}$ near the bifurcation point are stable. This generalizes the remarks made about μ_2 above. Such results can be found in Joseph-Nield (1975), Weinberger (1977), and Crandall-Rabinowitz (1977).

2. Some Extensions

The Hopf Bifurcation Theorem is a local theorem; it describes the structure of the branch of periodic solutions of (1.1.1) near the bifurcation point $(x, \lambda) = (0, 0)$. Using topological methods, Alexander and Yorke (1978) have given a global version of the theorem. While they pose their result for an n-manifold, in the setting of the Hopf theorem, they allow more general spectral conditions than $(H_1) - (H_2)$. In particular (H_2) is dropped and (H_1) is replaced by a milder condition that will not be made explicit here. To describe the conclusions, let $G(\lambda, t, x_0)$ denote the solution of (1.1.1) with $G(\lambda, 0, x_0) = x_0$. Thus equilibrium solutions have $G(\lambda, t, x_0) = x_0$ for all $t \in \mathbb{R}$. If $x(t) = G(\lambda, t, x_0)$ is not an equilibrium solution and there is a $T > 0$ such that $x(0) = x(T)$, then $x(t)$ is a nontrivial T-periodic solution of (1.1.1). Set

$$\begin{aligned}\mathcal{N} &= \{(\lambda, T, x_0) \in \mathbb{R} \times (0, \infty) \times \mathbb{R}^n | G(\lambda, T, x_0) = x_0 \\ &\qquad \text{and } x(t) \text{ is a nontrivial T-periodic solution of (1.1.1) } \},\end{aligned}$$

i.e. $\mathcal{N}$ corresponds to the set of nonequilibrium periodic solutions of (1.1.1). The main result of Alexander and Yorke (1978) is that $\mathcal{N} \cup \{(0, \frac{2\pi i}{a(0)}, 0)\}$ contains a connected subset $\mathcal{N}_0$ which is either unbounded in $\mathbb{R} \times [0, \infty) \times \mathbb{R}^n$ or meets $(\overline{\lambda}, \overline{T}, \overline{x}) \in \overline{\mathcal{N}}_0 \backslash \mathcal{N}_0$ with $G(\overline{\lambda}, \overline{T}, \overline{x})$ an equilibrium solution of (1.1.1). Stated more informally, there is a global branch of periodic solutions of (1.1.1) which is either unbounded in the triple (λ, T, x_0) or meets an equilibrium solution other than $(0, \frac{2\pi i}{a(0)}, 0)$. Thus bifurcation here, like bifurcation from equilibrium to equilibrium solutions as described by the so-called Global Bifurcation Theorem (Rabinowitz (1971)) is not a local but a global phenomenon.

The Liapunov Center Theorem (Liapunov (1907)) is an early bifurcation theorem that predates the Hopf Theorem. It considers the n dimensional autonomous system:

$$\dot{x} = Ax + g(x) \tag{2.2.1}$$

where A is an n matrix having eigenvalues $\pm i\beta, \lambda_3, \ldots, \lambda_n$, with $\beta \neq 0$ and $g(x) = o(|x|)$ as $x \to 0$. Assume (2.2.1) has an integral I, (that is, $I(z(t)) \equiv$ constant for any solution $z(t)$ of (2.2.1)) with

$$I(x) = \frac{1}{2} x \cdot Sx + o(|x|^2)$$

as $x \to 0$, S being symmetric and nonsingular. Suppose further that $\lambda_j / i\beta \notin \mathbf{Z}$ for $j = 3, \ldots, n$. Then the Liapunov Theorem states that (2.2.1) possesses a 1-parameter family of solutions $x(t, s)$ of period $T(s)$ for s near 0 with $x(t, 0) = 0$ and $T(0) = 2\pi/\beta$.

Being autonomous, (2.2.1) looks rather different from (1.1.1), but as was observed by Schmidt (1976) – see also Alexander-Yorke (1978) – by a nice trick one can prove the Liapunov Theorem by a simple application of the Hopf theorem. To see how, consider

$$\dot{x} = F(\lambda, x) \equiv Ax + g(x) + \lambda \text{ grad } I(x) \tag{2.2.2}$$

For this choice of F, it is not difficult to verify that $J(\lambda) = A + \lambda S$ and satisfies $(H_1) - (H_2)$ so by (a small generalization of) the Hopf theorem, there is a branch of solutions of (2.2.2) $(x(t,s), \lambda(s))$, with $x(\cdot, s)$ periodic in t, bifurcating from $(0,0)$ in $\mathbb{R} \times \mathbb{R}^n$. Therefore

$$\frac{dI(x)}{dt} = \text{grad } I(x) \cdot (Ax + g(x) + \lambda \text{ grad } I(x)) \tag{2.2.3}$$

with $x = x(t,s)$. Since I is an integral for (2.2.1), for all $z \in \mathbb{R}^n$,

$$\text{grad } I(z) \cdot (Az + g(z)) = 0. \tag{2.2.4}$$

Hence

$$\frac{dI(x)}{dt} = \lambda |\text{grad } I(x)|^2 \tag{2.2.5}$$

Since $I(x(t,s))$ is periodic in t for $s \neq 0$, the right hand side of (2.2.5) must equal 0 for all t. The form of I shows grad $I(x(t,s)) \not\equiv 0$ so $\lambda(s) \equiv 0$, i.e. $x(t,s)$ is a solution of (2.2.1).

There have been many infinite dimensional versions of the Hopf theorem. The first that we know of appeared in the 1970's motivated by attempts to establish the bifurcation of periodic solutions of the Navier-Stokes equations. See Iudovich (1971), Sattinger (1971), Iooss (1972), Joseph and Sattinger (1972), Crandall-Rabinowitz (1977), ... Subsequently there have been applications in many other directions such as reaction-diffusion problems (Henry (1981)), vortex shedding (Provansal, Mathis, and Boyer (1987)), convection in binary fluids (Knobloch (1986)) and in double-diffusion systems (Knobloch and Proctor (1981)), panel flutter (Holmes-Marsden (1978)), predator prey problems, ..., and much more. Considerably different technical settings and tools are required to treat these problems. Often, however, after a nontrivial Liapunov-Schmidt or center manifold reduction, they play back to the two basic approaches to the Hopf setting.

The finite-dimensional Hopf bifurcation theorems can be generalized to include degenerate cases ($a'(0) = 0$ or $\mu_2 = 0$ or both). These degeneracies appear in systems with several parameters and can lead to the existence of multiple periodic solutions for a given λ. See Kielhofer (1979) and Golubitsky-Langford (1981). Symmetry, which is often present in fluid mechanics problems, can force the critical eigenvalues to be multiple and can lead to multiple branches of periodic solutions with intriguing spatio-temporal symmetry. See Chossat-Iooss (1985) and Golubitsky-Stewart (1985).

References

ALEXANDER, J. C. AND J. A. YORKE, (1978) "Global bifurcations of periodic orbits." *Amer. J. Math.* **100**, no. 2, 263–292.

ANDRONOV, A. A. , A. A. VITT AND S. E. KHAIKIN, (1937) *Theory of Oscillations*, Moscow.

ARNOLD, V. , (1983) "Geometrical Methods in the Theory of Ordinary Differential Equations," *Fundamental Principles of Math. Sc* **250**, Springer, New York.

CHOSSAT, P. AND G. IOOSS, (1985) "Primary and secondary bifurcations in the Couette-Taylor problem." *Japan J. Appl. Math.* **2**, no. 1, 37–68.

CRANDALL, M. G. AND P. H. RABINOWITZ, (1977) "The Hopf bifurcation theorem in infinite dimensions." *Arch. Rational Mech. Anal.* **67**, no. 1, 53–72.

GOLUBITSKY, M. AND W. F. LANGFORD, (1981) "Classification and unfoldings of degenerate Hopf bifurcation." *J. Diff. Eqns.* **41** (1981) 375–415.

GOLUBITSKY, M. AND I. N. STEWART, (1985) "Hopf bifurcation in the presence of symmetry." *Arch. Rational Mech. Anal.* **87**, no. 2, 107–165.

HALE, J. , (1969) "Ordinary Differential Equations." *Pure and Applied Mathematics* **XXI**. Wiley-Interscience, New York

HENRY, D. , (1981) "Geometric Theory of Semilinear Parabolic Equations." *Lecture Notes in Mathematics* **840**, Springer-Verlag, New York.

HOLMES, P. AND J. E. MARSDEN (1978) "Bifurcation to divergence and flutter in flow-induced oscillations: an infinite dimensional analysis." *Automatica — J. IFAC* **14**, no. 4, 367–384.

IOOSS, G. , (1972) "Existence et stabilité de la solution périodique secondaire intervenant dans les problèmes d'evolution du type Navier-Stokes." *Arch. Rational Mech. Anal.* **47**, 301–329.

IUDOVICH, V. I. , (1971) "The onset of auto-oscillations in a fluid," *J. Applied Math. Mech* **35**, 587–603.

JOSEPH, D. D. AND D. A. NIELD, (1975) "Stability of bifurcating time-periodic and steady solutions of arbitrary amplitude." *Arch. Rational Mech. Anal.* **58**, no. 4, 369–380.

KIELHOFER, H. , (1979) "Hopf bifurcation at multiple eigenvalues." *Arch. Rational Mech. Anal.* **69** 53–83.

KNOBLOCH, E. , (1986) "Oscillatory convection in binary mixtures." *Phs. Rev. A* **34** No. 2, 1538–1549.

KNOBLOCH, E. AND M. R. E. PROCTOR (1981) "Nonlinear periodic convection in double-diffusive systems." *J. Fluid Mech.* **108** 291–316.

LIAPUNOV, A., (1907) "Probléme générale de la stabilité du mouvement." *Ann. Fac. Sci. Toulouse* **2**, 203–474.

PROVANSAL, M., C. MATHIS, AND L. BOYER, (1987) "Bénard - von Karman instability: transient and forced regimes." *J. Fluid Mech.* **182** No. 1, 1–22.

RABINOWITZ, P. H., (1971) "A global theorem for nonlinear eigenvalue problems and applications." *Contributions to Nonlinear Functional Analysis*, Academic Press, New York, 11–36.

SATTINGER, D. H., (1971) "Bifurcation of periodic solutions of the Navier-Stokes equations." *Arch. Rational Mech. Anal.* **41** 66–80.

D. S. SCHMIDT, D. S., (1976) "Hopf's Bifurcation Theorem and the Center Theorem of Liapunov," appearing in *The Hopf Bifurcation and its Applications*, edited by J. E. Marsden and M. McCracken, Applied Math. Sc. **19**, Springer, New York, 95–104.

H. F. WEINBERGER, H. F., (1977) "The stability of solutions bifurcating from steady or periodic solutions." *Dynamical Systems*, Academic Press, New York, 349–366.

REPEATED BRANCHING THROUGH LOSS OF STABILITY AN EXAMPLE

BY

Eberhard Hopf

Indiana University

Consider an incompressible fluid of density one and of constant viscosity $\mu > 0$. Suppose that the fluid is contained between given fixed material walls which do not move in space. Consider the initial value problem, given an initial velocity field $u(x)$ in the part of x-space occupied by the fluid, determine the flow $u(x, t)$, $t > 0$, such that $u(x, 0) = u(x)$. In order to render this problem a perfectly determined one certain side conditions may have to be imposed in addition to the condition that $u = 0$ on the boundary. These additional conditions too are supposed to be given once for all and to be independent of t. Two typical examples are the flow around a fixed obstacle with velocity at infinity prescribed independent of t, and the flow through an infinite cylindrical pipe with average pressure drop prescribed independent of t. In such a definitely given and temporally constant arrangement the solution $u(x, t)$ of the initial value problem, $u(x, 0)$ given, depends on the value of the viscosity μ, $u = u(x, t; \mu)$.

There is the question how the solutions $u(x, t; \mu)$ behave as $t \to \infty$ and how this behavior changes with the value of μ. Here is what the experimental evidence seems to indicate about this behavior. There is a time-independent (the laminar) solution $u = U(x; \mu)$ which, above a certain critical value of $\mu, \mu > \mu_1$, attracts all the other solutions $u(x, t; \mu)$ as $t \to \infty$ regardless of what $u(x, 0)$ is. The mathematical indications are that, for $\mu > \mu_1$, $U(x; \mu)$ still exists and varies continuously with $\mu, \mu > 0$. The fact that it is no longer observed for $\mu < \mu_1$ indicates its instability at those values of μ. Instead, a stable periodic solution $U(x, t; \mu)$ is observed which branches away continuously from the laminar $U(x; \mu)$ at $\mu = \mu_1$. However, below a second critical value, $\mu < \mu_2$ $(< \mu_1)$, the periodic flow is no longer observed and has given way to a more complicated type of motion. As $\mu \to 0$ the complexity of the motion increases (highly turbulent flow) and we are led to surmise that an infinite number of critical values $\mu_n, \mu_n \downarrow 0$, exist such that at each crossing the previous type of motion loses its stability and that another stable motion filling densely a manifold with one more dimension branches away from it. Several years ago the author constructed a simpler example of a system of space-time equations the solutions of which can be proved to show this behavior [1].

In this paper we present and treat a still simpler example. As in the first example, the x-space is taken to be one-dimensional and x is an angle variable (mod 2π). All functions of x are taken to be periodic in x with period 2π. Instead of the former three we now have two dependent real variables $u(x,t)$, $v(x,t)$. Everything will be written in complex form, $w=u+iv$, $w^*=u-iv$. The space-time system is like the first one of integro-differential form,

$$w_t' = \mathfrak{T}(w) + \mathfrak{L}(w) + \mu w_{xx}'' \tag{1}$$

where

$$\mathfrak{L}(w) = \frac{1}{2\pi}\int_0^{2\pi} w(x+y)F^*(y)\,dy \tag{2}$$

and

$$\mathfrak{T}(w) = -\frac{1}{4\pi^2}\int_0^{2\pi}\int_0^{2\pi} w(y)w(y')w^*(y+y'-x)\,dy\,dy'. \tag{3}$$

$F(y)$ is to be a given periodic function of y (integrable). $\mathfrak{L}$ and $\mathfrak{T}$ are functional operators of w of first and third degree, respectively. They are of integral type in contrast to the differential operators in the hydrodynamical case. Also $\mathfrak{T}$ is of third, and not of second degree. However, it has a similar effect in limiting the solutions as $t\to\infty$. Only the term $\mu w_{xx}''$ is the perfect analogue of the frictional terms.

A solution of (1) for $0<t<T$ is understood to be a function $w(x,t)$ such that w, w_x', w_{xx}'' are continuous in x,t for $0<t<T$ and such that (1) is satisfied for all x,t, $0<t<T$. The initial value problem is to find a solution w, $0<t<T$, such that

$$\lim_{t\to 0}\int_0^{2\pi} |w(x,t)-w^0(x)|^2\,dx = 0 \tag{4}$$

where $w^0(x)$ is a given square-integrable function.

The discussion right below furnishes the proof that the initial value problem has, for every square integrable $w^0(x)$, a unique solution for $0<t<\infty$. We use the formal Fourier-expansions for w and F

$$w(x,t)=\sum w_n(t)e^{inx}, \qquad F(x)=\sum F_n e^{inx}. \tag{5}$$

On multiplying both sides of (1) by e^{-inx} and on averaging through a period we find that, for $t > 0$, $n = 0, \pm 1, \cdots$,

$$\frac{dw_n}{dt} = -w_n{}^2 w_n{}^* + (F_n{}^* - \mu n^2) w_n, \qquad F_n = a_n - ib_n. \tag{6}$$

Interchanging x-integrations among each other and with t-differentiation is, by virtue of the definition of a solution, permitted. The splitting of the differential system for the Fourier-coefficients into completely independent ones for each n is a consequence of the fact that our integral operators are convolutions. If

$$w_0(x) = \sum w_n{}^0 e^{inx} \tag{7}$$

is the given initial function and if $w(x,t)$ is a solution of (1), $t > 0$, satisfying (4) then it follows from (4) that $w_n(t) \to w_n{}^0$ as $t \to 0$ and, hence, that $w_n(t)$ must satisfy (6) also at $t = 0$. Consequently, for every solution of (1) which satisfies (4) the Fourier-coefficients must satisfy (6) and $w_n(0) = w_n{}^0$. This implies the uniqueness of the solution of (1), $t > 0$, and (4).

We now discuss the solutions of (6). From this discussion, the existence of the solution of (1), (4) for all $t > 0$ is proved. The equations (6) are all of the same form

$$\frac{dw}{dt} = -w^2 w^* + (F^* - \nu) w, \qquad F = a - ib, \tag{8}$$

where ν is a non-negative parameter and where F is to be some fixed (complex) constant. The symbols w, F used in connection with (8) are, of course, to be kept apart from the same symbols used in (1), (2), (3). In (8) they are Fourier-coefficients of the functions w, F in (1). On using polar coordinates

$$w = re^{i\phi}$$

one finds that (8) splits into

$$\frac{dr}{dt} = (a - \nu - r^2) r \tag{9}$$

and

$$\frac{d\phi}{dt} = b. \tag{10}$$

The connection betwen (9), (10) and (6) is that

$$\begin{gathered} \nu = \mu n^2, \qquad a = a_n = \operatorname{Re} F_n, \qquad b = b_n = -\operatorname{Im} F_n, \\ r = r_n, \qquad \phi = \phi_n, \qquad w_n = r_n e^{i\phi_n}. \end{gathered} \tag{11}$$

The behavior of the solutions of (9) as $t \to \infty$ is evidently this:

$$(12)\qquad \begin{aligned} &\nu \geqq a:\ \lim_{t\to\infty} r = 0,\\ &\nu < a:\ r \equiv 0 \text{ or } r > 0 \text{ and} \lim_{t\to\infty} r = +\sqrt{a-\nu}. \end{aligned}$$

$r = \sqrt{a-\nu}$ is a stationary solution in the second case. Every solution which starts at $t=0$ exists for all $t > 0$.

We first show that (1) has a solution w for all $t > 0$ which satisfies (4). We consider, for a fixed $\mu > 0$, the solutions of (6), $w_n(0) = w_n{}^0$. If we integrate the linear differential inequality obtained from (9) on omitting the term r^2, we find that

$$(13)\qquad | w_n(t) | \leqq | w_n{}^0 | \exp[(a_n - \mu n^2)t], \qquad t > 0.$$

The numbers $a_n = \operatorname{Re} F_n$ and $w_n{}^0$ are certainly bounded by virtue of the assumptions made about $F(x)$ and $w^0(x)$. If we now define

$$w(x,t) = \sum w_n(t) e^{inx}$$

we infer from well-known theorems, on using (13), that x-differentiations are possible term by term and that $w, w_x', w_{xx}'', \cdots$ are continuous functions of x, t as $t > 0$. As the right hand sides of (6) satisfy similar inequalities as in (13) it follows that

$$\sum \frac{dw_n(t)}{dt} e^{inx}$$

is a continuous function of x, t as $t > 0$. Consequently, it equals $\partial w/\partial t$, $t > 0$. Finally we substitute w into (1). It was found before that the n-th Fourier-coefficients of both sides of (1) equal the two sides of (6). So, by virtue of (6), both sides of (1), $t > 0$, have the same Fourier-expansion and are therefore identical.

To prove (4) observe that the integral equals

$$\sum | w_n(t) - w_n{}^0 |^2.$$

Every term $\to 0$ as $t \to 0$, and the tail can be made $< \epsilon$ for all $t < 1$ as follows from the assumed convergence of

$$\sum | w_n{}^0 |^2$$

and from (13).

The problem of the behavior of the solutions $w(x,t)$ of (1) as $t \to \infty$, for any fixed value $\mu > 0$, is completely solved by the following

THEOREM. *Let $F(x) = \sum F_n e^{inx}$ be a fixed integrable function of period*

2π and let $a_n = \operatorname{Re} F_n$, $b_n = -\operatorname{Im} F_n$. For any fixed $\mu > 0$, the functions $w = W$,

$$W(x,t) = \sum_{n^2\mu < a_n} \sqrt{a_n - n^2\mu}\, \exp[i(nx + b_n t + \lambda_n)]$$

are solutions of (1) for any real constants λ_n. So is any function which is obtained from this sum by omission of any terms in it. Let $\mu > 0$ be fixed. To every solution $w(x,t)$ of (1) whose Fourier-coefficients are all $\neq 0$ at $t = 0$ there belongs a solution $W(x,t)$ such that

$$\lim_{t\to\infty} |w - W| = \lim_{t\to\infty} |w_x' - W_x'| = 0$$

holds uniformly with respect to x. This remains true for any solution w provided that in the sum for W the terms are omitted for precisely those n for which the Fourier-coefficient of w is zero at $t = 0$.

We prove the theorem before we discuss its meaning. From (9)-(12) we infer the truth of the first and second statement concerning the special solutions. The sum for W extends over those integers n for which $a_n > n^2\mu$. $n = 0$ belongs to them if and only if $a_0 > 0$. For given $\mu > 0$, only finitely many n can satisfy this inequality. If this is the case for no n at all then we are to let $W = 0$. Let, for fixed $\mu > 0$,

$$w(x,t) = \sum w_n(t) e^{inx}$$

be any solution of (1) with initial values

$$w_n^0 = r_n^0 e^{i\lambda_n}.$$

From (10), (11) $\phi_n = b_n t + \lambda_n$. From (12), (11) we have, for any fixed n, complete information about $w_n(t)$ as $t \to \infty$. If $n^2\mu \geqq a_n$, then $w_n \to 0$. If, however, $n^2\mu < a_n$ then we have $w_n \equiv 0$ if $w_n(0) = 0$, and $w_n \neq 0$ and

$$w_n - e^{i(b_n t + \lambda_n)}\sqrt{a_n - n^2\mu} = e^{i(b_n t + \lambda_n)}(r_n - \sqrt{a_n - n^2\mu}) \to 0$$

if $w_n(0) \neq 0$. This makes the statements concerning w and the corresponding solution W perfectly obvious if we understand convergence in the formal sense, i.e., for each Fourier-coefficient. But we must remember that the corresponding non-zero terms of W are only finite in number. For all n sufficiently large in absolute value we have $n^2\mu > a_n$ and hence, $w_n \to 0$ as $t \to \infty$. Moreover, if we let

$$A = \sup_n a_n$$

then, by virtue of (13),

$$| w_n(t) | < \text{const.} \exp[(A - \mu n^2) t]$$

holds for all t. If we now split the series for w into a sum over all n such that $| n | \leqq A/\mu$ and a sum over all the other integers n then the first sum as a function of x, t has already the property stated in the theorem for the full w, uniformity of convergence with respect to x. The second sum is in absolute value less than

$$\text{const.} \sum_n \exp[(A - \mu n^2) t]$$

extended over all n with $n^2 > A/\mu$, regardless of x, and this latter sum is known to tend to zero as $t \to \infty$. This proves the statement concerning w. The statement about w_x' follows similarly, and the theorem is thereby completely proved.

In order to obtain a qualitative analogue to the hydrodynamical situation outlined in the beginning we choose the fixed function

$$F(x) = \sum F_n e^{inx}, \qquad F_n = a_n - ib_n$$

so as to have the following properties:

$$a_n = b_n = 0, \quad n \leqq 0; \qquad a_n/n^2 \downarrow 0, \quad n > 0.$$

We also suppose that any finite set of the numbers $b_1, b_2, \cdots$ is linearly independent.

Define

$$\mu_n = a_n/n^2, \quad n > 0.$$

The condition in the summation defining the W of the theorem means that $n \geqq 1$ and $\mu_n > \mu$. By the way, no error in W is committed if we include the equality in the inequality as the term in question vanishes then.

It is convenient to consider the phase space of the system (1)—called w-space—whose elements or points are the different functions $w(x)$ admitted, periodic in x of period 2π. This space may be metrized by the norm

$$\| w \| = \sup_x | w(x) | + \sup_x | w_x'(x) |.$$

We denote the trajectory in w-space defined by a solution $w(x, t)$ of (1) simply by $w(t)$. In the case that

$$\mu \geqq \mu_1$$

the sum defining the W of the theorem is void and the result is that every trajectory $w(t)$ tends in this case to the point $w = 0$ as $t \to \infty$. $w = 0$ is obviously a stationary solution of (1) for any value of μ. In the next case,

$$\mu_1 > \mu \geqq \mu_2$$

we have

$$W = \sqrt{\mu_1 - \mu}\, e^{i\psi} e^{ix}, \qquad \psi = b_1 t + \lambda_1$$

which is a periodic solution in t or closed trajectory $W(\psi)$ in w-space. It exists only for $\mu \leqq \mu_1$ and it branches from the point $w = 0$ continuously at $\mu = \mu_1$. The theorem says that the "majority" of the trajectories $w(t)$ tend toward this closed curve as $t \to \infty$. The point solution $w = 0$ still exists but has yielded its stability to the new closed trajectory. In the case

$$\mu_2 > \mu \geqq \mu_3$$

we have

$$W = \sqrt{\mu_1 - \mu}\, e^{i\psi_1} e^{ix} + 2\sqrt{\mu_2 - \mu}\, e^{i\psi_2} e^{2ix} \tag{14}$$

and

$$\psi_\alpha = b_\alpha t + \lambda_\alpha. \tag{15}$$

The functions $W(x;\psi_1,\psi_2;\mu)$ of x or w-points $W(\psi_1,\psi_2;\mu)$ with the angle-variables ψ_α form a manifold of the type of a 2-torus in w-space which exists for $\mu \leqq \mu_2$ and which branches away from the former closed curve continuously at $\mu = \mu_2$. The trajectories (15) lie on this torus and every one of them is, by virtue of the assumed linear independence of b_1 and b_2, dense on the torus. Also, these trajectories $W(t)$ are almost periodic functions in the earlier sense of Bohl. The theorem says again that now the "majority" of all possible trajectories $w(t)$ tend toward this torus as $t \to \infty$, more exactly, toward certain special trajectories (15) on this torus. The former closed trajectory (or 1-torus) still exists but it has lost its stability. It is clear how this goes on. For

$$\mu_n > \mu \geqq \mu_{n+1}$$

any trajectory $w(t)$ out of a majority of all possible trajectories tends as $t \to \infty$ toward a special trajectory

$$\psi_\alpha = b_\alpha t + \lambda_\alpha, \qquad \alpha = 1, \cdots, n,$$

lying on the n-dimensional torus

$$W = \sum_{\alpha=1}^{n} \alpha \sqrt{\mu_\alpha - \mu}\, e^{i\psi_\alpha} e^{i\alpha x} \tag{16}$$

in w-space. Again, each of these special trajectories is dense on the torus. The next $(n+1)$-dimensional torus branches away from the n-dimensional one continuously at $\mu = \mu_{n+1}$.

Suppose, now, that $\mathfrak{F}(w)$ is a number-valued functional in w-space continuous in the sense of the metric induced by the norm. Follow its values along a trajectory $w(t)$ and denote the asymptotic trajectory belonging to $w(t)$ by $W(t)$. We clearly have that

$$|\mathfrak{F}(w(t))-\mathfrak{F}(W(t))|\to 0, \qquad t\to\infty,$$

and, hence, that

$$\bar{\mathfrak{F}}=\lim_{T\to\infty}\frac{1}{T}\int_0^T\mathfrak{F}(w(t))\,dt=\lim_{T\to\infty}\frac{1}{T}\int_0^T\mathfrak{F}(W(t))\,dt$$

if the second limit exists. The second limit exists, however, in consequence of Weyl's equidistribution theorem. In the case where

$$\mu_n>\mu\geqq\mu_{n+1}$$

the time average $\bar{\mathfrak{F}}$ equals, for the majority of possible solutions $w(t)$, the space average of $\mathfrak{F}(W)$ on the torus (16) based upon the volume element $d\psi=d\psi_1\cdots d\psi_n$ on this torus. This makes it possible to study "statistical mechanics" of the system (1) and to study the limit form of its laws as $\mu\to 0$. A detailed examination furnishes the result that a normal distribution is obtained in the limit $\mu\to 0$ if the function $F(x)$ belongs to a certain restricted class of functions (see the author's paper [1] where everything is formulated and proved for the former example).

REFERENCE

[1] E. Hopf, "A mathematical example displaying features of turbulence," Comm. Pure Appl. Math., vol. 1 (1948), pp. 303-322.

A Mathematical Example Displaying Features of Turbulence

By EBERHARD HOPF

Introduction

Before entering upon the study of the example in question we want to make some introductory remarks about the actual hydrodynamic problems, in particular, about what is known and what is conjectured concerning the future behavior of the solutions. Consider an incompressible and homogeneous viscous fluid within given material boundaries under given exterior forces. The boundary conditions and the outside forces are assumed to be stationary, i.e. independent of time. For that, it is not necessary that the walls be at rest themselves. Parts of the material walls may move in a stationary movement provided that the geometrical boundary as a whole stays at rest. An instance is a fluid between two concentric cylinders rotating with prescribed constant velocities or a fluid between two parallel planes which are translated within themselves with given constant velocities. As to the stationarity of the exterior forces we may cite the case of a flow through an infinitely long pipe with a pressure drop (regarded as an outside force). In this case the pressure drop is required to be a given constant independent of time.

Each motion of the fluid that is theoretically possible under these conditions satisfies the Navier-Stokes equations ($\rho = 1$)

$$\frac{\partial u_i}{\partial t} = - \sum_\nu u_\nu \frac{\partial u_i}{\partial x_\nu} - \frac{\partial p}{\partial x_i} + \mu \Delta u_i \tag{0.1}$$

together with the incompressibility condition

$$\sum \frac{\partial u_\nu}{\partial x_\nu} = 0 \tag{0.2}$$

and the given stationary boundary conditions. To an arbitrarily prescribed initial velocity field $u(x, 0)$ satisfying (0.2) and the boundary conditions there is expected to belong a unique solution $u(x, t; \mu)$ ($t \geq 0$) of (0.1) and (0.2) that fulfills these boundary conditions. The pressure $p(x, t; \mu)$ may be considered as an auxiliary variable which, at every moment t is (up to an additive constant) perfectly well determined by the instantaneous velocity field $u(x, t)$ (solution of a Neumann problem of potential theory). If p is eliminated in this manner the Navier-Stokes equations appear in the form of an integrodifferential space-time system for the u_i alone where the right hand sides consist of first and second degree terms in the u_ν .

It is convenient to visualize the solutions in the phase space Ω of the problem. A phase or state of the fluid is a vector field $u(x)$ in the fluid space that satisfies (0.2) and the boundary conditions. The totality Ω of these phases is therefore a functional space with infinitely many dimensions. A flow of the fluid represents a point motion in Ω and the totality of these phase motions forms a stationary flow in the phase space Ω, which, of course, is to be distinguished from the fluid flow itself. What is the asymptotic future behavior of the solutions, how does the phase flow behave for $t \to \infty$? And how does this behavior change as μ decreases more and more? How do the solutions which represent the observed turbulent motions fit into the phase picture? The great mathematical difficulties of these important problems are well known and at present the way to a successful attack on them seems hopelessly barred. There is no doubt, however, that many characteristic features of the hydrodynamical phase flow occur in a much larger class of similar problems governed by nonlinear space-time systems. In order to gain insight into the nature of hydrodynamical phase flows we are, at present, forced to find and to treat simplified examples within that class. The study of such models has been originated by J. M. Burgers in a well known memoir.[1] His principal example is essentially

$$\frac{\partial v}{\partial t} = -v\frac{\partial v}{\partial x} + w\frac{\partial w}{\partial x} + v - w + \mu\frac{\partial^2 v}{\partial x^2}$$

$$\frac{\partial w}{\partial t} = w\frac{\partial v}{\partial x} + v\frac{\partial w}{\partial x} + v + w + \mu\frac{\partial^2 w}{\partial x^2}$$

where $0 \le x \le 1$ and where the boundary conditions are $v = w = 0$ at $x = 0$ and $x = 1$. Though simpler in form than the hydrodynamic equations this example presents essentially the same difficulties and the future behavior of the solutions for small values of μ still is an unsolved problem.

In this paper another nonlinear example is presented and studied that differs from Burgers' model in that the future behavior of its solutions can be completely determined. In this respect our example seems to us to be the first of its kind. The detailed study of this space-time system reveals geometrical features of the phase flow which come close to the qualitative picture we believe to prevail in the hydrodynamic cases. It must, however, be said that, for reasons to appear later in the paper, the analogy does not extend to the quantitative relations found to hold in turbulent fluid flow.

The observational facts about hydrodynamic flow reduced to the case of fixed side conditions and with μ as the only variable parameter are essentially these: For μ sufficiently large, $\mu > \mu_0$, the only flow observed in the long run is a stationary one (laminar flow). This flow is stable against arbitrary initial

[1]J. M. Burgers, *Mathematical examples illustrating relations occurring in the theory of turbulent fluid motion.* Akademie van Wetenschappen, Amsterdam, Eerste Sectie, Deel XVII, No. 2, pp. 1–53, 1939.

disturbances. Theoretically, the corresponding exact solution is known to exist for every value of $\mu > 0$ and its stability in the large can be rigorously proved, though only for sufficiently large values of μ. The corresponding phase flow in phase space Ω thus possesses an extremely simple structure. The laminar solution represents a single point in Ω invariant under the phase flow. For $\mu > \mu_0$, every phase motion tends, as $t \rightarrow \infty$, toward this laminar point. For sufficiently small values of μ, however, the laminar solution is never observed. The turbulent flow observed instead displays a complicated pattern of apparently irregularly moving "eddies" of varying sizes. The view widely held at present is that, for $\mu > 0$ having a fixed value, there is a "smallest size" of eddies present in the fluid depending on μ and tending to zero as $\mu \rightarrow 0$. Thus, macroscopically, the flow has the appearance of an intricate chance movement whereas, if observed with sufficient magnifying power, the regularity of the flow would never be doubted.

The qualitative mathematical picture which the author conjectures to correspond to the known facts about hydrodynamic flow is this: To the flows observed in the long run after the influence of the initial conditions has died down there correspond certain solutions of the Navier-Stokes equations. These solutions constitute a certain manifold $\mathfrak{M} = \mathfrak{M}(\mu)$ in phase space invariant under the phase flow. Presumably owing to viscosity $\mathfrak{M}$ has a finite number $N = N(\mu)$ of dimensions. This effect of viscosity is most evident in the simplest case of μ sufficiently large. In this case $\mathfrak{M}$ is simply a single point, $N = 0$. Also the complete stability of $\mathfrak{M}$ is in this simplest case obviously due to viscosity. On the other hand, for smaller and smaller values of μ, the increasing chance character of the observed flow suggests that $N(\mu) \rightarrow \infty$ monotonically as $\mu \rightarrow 0$. This can happen only if at certain "critical" values

$$\mu_0 > \mu_1 > \mu_2 > \cdots \rightarrow 0$$

the number $N(\mu)$ jumps. The manifold $\mathfrak{M}(\mu)$ itself presumably changes analytically as long as no critical value is passed. Now we believe that when μ decreases through such a value μ_k a continuous branching phenomenon occurs. The manifold $\mathfrak{M}(\mu)$ of motions observed in the long run (more precisely its analytical continuation for $\mu < \mu_k$) loses its stability. The notion of stability here refers to the whole manifold and not to the single motions contained in it. The loss of stability implies that the motions on the analytically continued $\mathfrak{M}$ are no longer observed. What we observe after passing μ_k is not the analytical continuation of the previous $\mathfrak{M}$ but a new manifold $\mathfrak{M}(\mu)$ continuously branching away from $\mathfrak{M}(\mu_k)$ and slightly swelling in a new dimension. This new $\mathfrak{M}(\mu)$ takes over stability from the old one. Stability here means that the "majority" of phase motions tends for $t \rightarrow \infty$ toward $\mathfrak{M}(\mu)$. We must expect that there is a "minority" of exceptional motions that do not converge toward $\mathfrak{M}$ (for instance the motions on the analytical continuation of the old $\mathfrak{M}$ and of all the other manifolds left over from all the previous branchings). The simplest case of such a bifurcation with corresponding change of stability is the branching of a periodic motion from a stationary one. This case is clearly observed in the flow around an obstacle (transition from the laminar flow to a periodic one with periodic discharges of eddies from the boundary). The next simplest case is the branching of a one-parameter family of almost periodic solutions from a periodic one. The new solutions are expressed by functions

$$u(\phi_1, \phi_2; \mu)$$

periodic in each ϕ with period 2π where

$$\phi_i = a_i t + c_i, \qquad a_i = a_i(\mu),$$

and where the c_i are arbitrary constants (we can without loss of generality assume $c_1 = 0$). The functions f with ϕ_i arbitrary describe the manifold $\mathfrak{M}(\mu)$ which, in our case, is of the type of a torus. If $\mathfrak{M}$, quite generally, continuously develops out of the laminar point there is a reasonable expectation that $\mathfrak{M}$ is a multidimensional torus-manifold described by functions

$$u(\phi_1, \cdots, \phi_N; \mu)$$

with period 2π in each of the ϕ and that the turbulent solutions are given by linear functions $\phi_i = a_i(\mu)t + c_i$ as before. This is what happens in our example which precisely exhibits this phenomenon of continuous growth of almost periodic solutions out of the laminar one with an infinite succession of branchings of the type described above.

The geometrical picture of the phase flow is, however, not the most important problem of the theory of turbulence. Of greater importance is the determination of the probability distributions associated with the phase flow, particularly of their asymptotic limiting forms for small μ. In the case of our example these distributions have limiting forms (normal distribution). Recent investigations, however, suggest that there are essential deviations from normality in the hydrodynamic case. It seems that the influence of the second degree terms is in this case essentially different and much more complicated than in the case of our over-simplified model.

Another observation on our model case is this: If we proceed to the limit $\mu \to 0$ within the "observed," i.e. the turbulent solutions the turbulent fluctuations are found to disappear and we obtain a special stationary solution in the "ideal case" (equations with $\mu = 0$). This shows, by way of analogy, how important a role viscosity plays in turbulence.

Formulation of the Problem

The space of our model is a one-dimensional circular line and our space variable is an angular variable x mod 2π. All space functions are thus periodic functions of x with period 2π. For two arbitrary space functions f, g we denote by $$f \circ g = \frac{1}{2\pi}\int_0^{2\pi} f(x+y)g(y)\,dy$$

their convolution product which is again a space function. $f \circ 1$ is a constant, the mean value of $f(x)$ over a period. Throughout this paper z^* denotes the conjugate of the complex number z. Our integrodifferential system written in complex form is

$$\begin{aligned} \partial u/\partial t &= -z \circ z^* - u \circ 1 + \mu\,(\partial^2 u/\partial x^2), \\ \partial z/\partial t &= z \circ u^* + z \circ F^* + \mu\,(\partial^2 z/\partial x^2), \end{aligned} \tag{1.0}$$

where $\mu > 0$ is a parameter and where

$$F(x) = a(x) + ib(x) \tag{1.1}$$

is an arbitrarily given complex-valued space function. $F(x)$ is supposed to be an even and absolutely integrable function of x,

$$F(-x) = F(x). \tag{1.2}$$

Further conditions upon F will be stated when they are needed. The unknowns are the two complex-valued functions $u(x, t)$ and $z(x, t)$. The real equations into which (1.0) splits up are four in number.

In what follows we confine ourselves to those solutions of (1.0) for which u, z are even functions of x and for which u is real. It will be proved, by using (1.2), that any solution u, z which is even for $t = 0$ must be even for all $t > 0$ and that, for such a solution, u is always real if it is real for $t = 0$. If we confine ourselves to the even solutions with u real, (1.0) splits upon setting

$$z = v + iw$$

into three real equations for u, v, w

$$\begin{aligned} \partial u/\partial t &= -v \circ v - w \circ w - u \circ 1 + \mu\,(\partial^2 u/\partial x^2), \\ \partial v/\partial t &= v \circ u + v \circ a + w \circ b + \mu\,(\partial^2 v/\partial x^2), \\ \partial w/\partial t &= w \circ u - v \circ b + w \circ a + \mu\,(\partial^2 w/\partial x^2), \end{aligned} \tag{2.0}$$

where $F(x) = a(x) + ib(x)$. Our problem is to study the real solutions of (2.0) which are even functions of x with period 2π.

Another equivalent but in some respects more straightforward formulation of our problem is obtained if we confine ourselves to the interval

$$0 \leq x \leq \pi.$$

We look for the real solutions of (2.0), where

$$f \circ g = \frac{1}{2\pi}\int_0^{\pi} f(|\,x - y\,|)g(y)\,dy + \frac{1}{2\pi}\int_0^{\pi} f(\pi - |\,x - y\,|)g(\pi - y)\,dy,$$

satisfying the boundary conditions

(3.0) $\partial u/\partial x = \partial v/\partial x = \partial w/\partial x = 0$ at $x = 0$ and $x = \pi$.

The equivalence of this formulation is a consequence of the following facts. An even function of period 2π may be arbitrarily prescribed in the interval $(0, \pi)$. If the first derivative exists for all x, (3.0) must be satisfied. On the other hand, if a function in $[0, \pi]$ has a second derivative in this closed interval and if (3.0) is satisfied the corresponding even and periodic function has a second derivative at every x (in particular, the first derivative is continuous throughout). That the convolution of two even space functions reduces, for $0 \leq x \leq \pi$, to the expression mentioned above follows from a simple calculation.

In what follows we use the handier complex form (1.0) of the problem with restriction to the even solutions whether u is real or not. Our second degree terms share an important property with those in the hydrodynamic case. When the time derivative of the kinetic energy

$$\frac{1}{2}\int_0^{2\pi} (uu^* + zz^*)\,dx, \qquad \frac{1}{2}\int_0^{2\pi} (u^2 + v^2 + w^2)\,dx$$

is computed from the equations the third degree terms obtained on the right hand side are found to cancel. In our case this follows from the identity

$$\int_0^{2\pi} (f \circ g)h dx = \int_0^{2\pi} (f \circ h) g dx.$$

Our second degree terms, however, strikingly differ in nature from the hydrodynamic ones in that they are pure integrals. The fact that they are convolutions enables one to calculate the solutions by spatial Fourier analysis.

Properties of the Even Solutions of (1.0) and (2.0)

There is an infinite number of critical values $\mu_1 > \mu_2 > \mu_3 > \cdots \to 0$ for μ with the following properties. For $\mu > \mu_1$ there is a stationary "laminar solution" which is stable in the large for $t \to +\infty$, i.e. which will, as $t \to +\infty$, be approached by any other solution for the same value of μ. For $\mu_n > \mu > \mu_{n+1}$ there is an n-dimensional manifold of "turbulent solutions" essentially stable in the large as $t \to +\infty$, i.e. "almost" all other solutions will approach some of the turbulent solutions with the same value of μ. These turbulent solutions are represented by almost periodic functions of t of very simple type. These solutions persist for $\mu < \mu_{n+1}$ but they are no longer stable. For any given value of μ there is a definite statistical distribution within the totality of the "velocity fields," i.e. in the function space of the sets of three arbitrary functions $u(x)$, $v(x)$, $w(x)$ satisfying the given boundary conditions. These statistics are simply defined by time averages:. The statistical average $\overline{\mathfrak{F}}$ of an arbitrary functional $\mathfrak{F}[u(x), v(x), w(x)]$ of the three functions is

$$\overline{\mathfrak{F}} = \lim_{T \to +\infty} \frac{1}{T} \int_0^T \mathfrak{F}[u(x, t), v(x, t), w(x, t)]\, dt,$$

where $u(x, t)$, $\cdots$ is the solution of our equations with $u(x)$, $\cdots$ as initial values for $t = 0$. If the function $F(x)$ given in our model satisfies certain requirements (which will be fulfilled "in general") the average $\overline{\mathfrak{F}}$ turns out to be essentially independent of the initial values $u(x)$, $v(x)$, $w(x)$ (property of ergodicity). Of course it will depend on μ. The probability that the point $[u(x), \cdots]$ of our function space falls into a given subset of this space is defined by the average of the "characteristic functional" of this set (1 inside, 0 outside). The statistics defined in this natural manner varies with μ. For $\mu > \mu_1$ it is a trivial one while, for μ decreasing, it will be more complex. The following fact must, however, be noted. If the real part $a(x)$ of our given function $F(x)$ is not too smooth a function our probability distribution becomes, in the limit $\mu \to 0$, more and more a Gaussian or normal one. The values u, v, w of the solutions at arbitrarily given fixed points $x_1, x_2, \cdots, x_i$ may be regarded as chance variables. If these chance variables are denoted by $u_1, u_2, \cdots, u_k$ respectively then the probability that

$$u_1 > a_1, u_2 > a_2, \cdots, u_k > a_k$$

differs, as $\mu \to 0$, less and less (uniformly with respect to the a's) from

$$K \int_{a_1}^{\infty} \cdots\cdots \int_{a_k}^{\infty} \exp\left\{-\frac{1}{2} \sum A_{\nu\mu}(\xi_\nu - m_\nu)(\xi_\mu - m_\mu)\right\} d\xi_1 \cdots d\xi_k,$$

where K is a constant chosen in such a way that the expression equals 1 if each $a_\nu = -\infty$. We have $m_\nu = \overline{u}_\nu$ and the matrix (A_{ik}) is the inverse of the correlation matrix $\overline{(u_i - m_i)(u_k - m_k)}$. For very small values of μ the distribution is therefore nearly determined by its moments of first and second degree. In our case these moments, i.e. the mean values $\overline{u(x)}$, $\overline{v(x)}$, $\overline{w(x)}$ and the correlation functions $\overline{u(x)u(x')}$ etc., are easily evaluated. Their asymptotic forms for $\mu \to 0$ will be investigated. Appreciable statistical interdependence is found only in points x, x' sufficiently close to each other. Approximately the correlations depend only on the mutual differences $|x' - x|$. This is analogous to the tendency of turbulence toward spatial isotropy and homogeneity in certain hydrodynamic cases.

The asymptotic evaluation of these moments is the cardinal problem in the theory of turbulent flow in hydrodynamics. Its great mathematical difficulties apparently arise from the fact that the spatial Fourier components of the motion are interrelated with each other, in contrast to our simple model where there is no interaction between the different frequencies of the spatial Fourier pattern. The mathematical nature and the formulation of this problem will be the subject of a later paper on the foundations of statistical hydrodynamics. Still, the continued study of models of our particular kind seems not without interest to the author. There might perhaps be a starting point in devising and discussing simple models with slight interaction.

Reduction to a Four-Dimensional Problem

We use Fourier series

$$u = \sum u_n e^{inx}, \qquad z = \sum z_n e^{inx}, \qquad F = \sum F_n e^{inx}$$

and corresponding notation for the Fourier coefficients of v, w, a, b. The summation extends from minus to plus infinity. Making use of the identity

$$f \circ g^* = \sum f_n g_n^* e^{inx}$$

and inserting into (1.0) the differential equations for the complex Fourier coefficients we obtain

$$\dot{u}_0 = -z_0 z_0^* - u_0$$

(4.0)

$$\dot{z}_0 = z_0 u_0^* + F_0^* z_0$$

and for $n \gtrless 0$

$$\dot{u}_n = -z_n z_n^* \qquad - n^2 \mu u_n ,$$

(5.0) $\qquad (F_n = a_n + ib_n)$

$$\dot{z}_n = z_n u_n^* + F_n^* z_n - n^2 \mu z_n ,$$

which shows that the Fourier components belonging to different frequencies behave completely independently of each other. The equations (5.0) are all of the same type

$$\dot{u} = -zz^* \qquad - \nu u,$$

(6.0) $\qquad (F = a + ib)$

$$\dot{z} = zu^* + F^* z - \nu z,$$

where $u(t)$, $z(t)$ are the complex-valued unknowns and where $\nu > 0$ is a parameter. $F = a + ib$ is a given complex constant.

(1.2) means that

(7.0) $$F_{-n} = F_n$$

for all n. Let us simultaneously consider the system (5.0) with some n and the same system with the index $-n$. If the initial values at $t = 0$ of two respective solutions coincide, i.e. if $u_{-n} = u_n$, $z_{-n} = z_n$ at $t = 0$, they must, according to the uniqueness theorem for ordinary differential equations and according to (7.0), coincide for all t. If, furthermore, such an even solution satisfies $u_{-n} = u_n^*$, i.e. $u_n = u_n^*$, at $t = 0$ then it obviously follows from the first equation (5.0) that this relation must hold for all t. According to the uniqueness of the Fourier expansion, this proves a remark made in the introduction: If a solution of (1.0) at $t = 0$ is even in x it must be even for all t; if for such an even solution u is real at $t = 0$ then u is real for all t.

The Solutions of (4.0) and (6.0)

Relations (4.0) and (6.0) have trivial stationary solutions,

(4.1) $$u_0 = z_0 = 0$$

and

(6.1) $$u = z = 0 \text{ (for all } \nu)$$

respectively. They correspond to the trivial solution $u = z = 0$ of (1.0). If $a > 0$ in $F = a + ib$ the system (6.0) has, in the ν-interval $0 < \nu < a$, a periodic solution besides,

(6.2) $$u = -a + \nu, \qquad z = (\nu(a - \nu))^{1/2} \exp\{-ib(t + \alpha)\} \qquad \text{for } 0 < \nu < a$$

where α is an arbitrary real constant. It obviously branches off the stationary solution (6.1) at $\nu = a$. If, however, $a \leq 0$ there is no such a periodic solution for $\nu > 0$. The behavior, as $t \to +\infty$, of all other solutions of the fundamental systems (4.0) and (6.0) is described by the following

LEMMA. *If, in* (4.0),

$$a_0 < 0$$

every solution u_0, z_0 *of* (4.0) *converges, as* $t \to \infty$, *toward the solution* $u_0 = z_0 = 0$.

If, in (6.0),

$$a \leq 0$$

every solution of (6.0), *no matter what the value of* $\nu > 0$, *converges toward* $u = z = 0$.

Suppose that

$$a > 0$$

in (6.0). *If* $\nu > a$ *every solution of* (6.0) *converges toward* $u = z = 0$. *If, however,* $0 < \nu \leq a$, *every solution of* (6.0), *with the exception of those where* $z = 0$ *at* $t = 0$, *converges toward the periodic solution* (6.2), *i.e. for every such solution there can be found an* α *in* (6.2) *such that the difference of the two solutions tends to zero. The exceptional solutions tend to* $u = z = 0$.

In this section we confine ourselves to the proof of the lemma. The first two assertions and the first part of the third one are obvious consequences of the energy equations

$$\frac{1}{2}\frac{d}{dt}(u_0u_0^* + z_0z_0^*) = -(u_0u_0^* - a_0z_0z_0^*), \qquad a_0 < 0,$$

and

(8.0) $$\frac{1}{2}\frac{d}{dt}(uu^* + zz^*) = -(\nu uu^* + (\nu - a)zz^*).$$

The right side of (8.0) is, for $\nu > a$ and $a > 0$, not greater than

$$-(\nu - a)(uu^* + zz^*).$$

By integration we therefore find that every solution of (6.0) satisfies ($t \geq 0$)

$$uu^* + zz^* \leqq (uu^* + zz^*)_{t=0} \exp \{-2(\nu - a)t\} \qquad (\nu > a > 0).^2 \tag{9.0}$$

The case $0 < \nu \leq a$ in (6.0) requires more elaborate considerations. On introducing new real variables

$$u = q + ip, \qquad z = re^{i\varphi} \qquad (r \geq 0) \tag{10.0}$$

(6.0) is transformed into

$$\begin{aligned} \dot{q} &= -r^2 - \nu q \\ \dot{p} &= - \nu p \\ \dot{r} &= (q + a - \nu)r \\ \dot{\varphi} &= -b - p. \end{aligned}$$

From the second and fourth equation,

$$p = \beta e^{-\nu t}, \qquad \varphi = -(\beta/\nu)\, e^{-\nu t} - b(t + \alpha), \tag{11.0}$$

where α, β are constants of integration. It remains to study the equations

$$\begin{aligned} \dot{q} &= -r^2 - \nu q \\ \dot{r} &= (q + a - \nu)r \end{aligned} \qquad (r \geq 0). \tag{12.0}$$

To (6.2) there corresponds the stationary solution

$$q = -a + \nu, \qquad r = +(\nu(a - \nu))^{1/2} \tag{13.0}$$

of (12.0). All we have to prove is that, for $0 < \nu \leq a$, every solution of (12.0), with $r > 0$ at $t = 0$, must tend to the point (13.0) in the q-r-plane. Since, for such a solution, there is always $r > 0$ we may write

$$r^2 = e^Q.$$

(12.0) transforms into

$$\dot{q} = -e^Q - \nu q, \qquad \dot{Q} = 2(q + a - \nu). \tag{14.0}$$

We have to prove that, for $0 < \nu \leq a$, every solution of (14.0) tends as $t \to +\infty$ toward the point (q_0, Q_0) where

$$q_0 = -a + \nu, \qquad e^{Q_0} = \nu(a - \nu).$$

[2]In the case in which $a \leq 0$ this inequality stays true if a is dropped in the exponential factor.

On eliminating q from (14.0) we obtain the second order equation

$$\ddot{Q} + \nu \dot{Q} = H'(Q) \tag{15.0}$$

where

$$H'(Q) = 2\nu(a - \nu) - 2e^{Q} = 2(e^{Q_0} - e^{Q}). \tag{16.0}$$

It remains to be shown that, for $0 < \nu \leq a$, every solution of (15.0) has the property

$$Q(t) \to Q_0 , \qquad \dot{Q}(t) \to 0 \qquad \text{for} \qquad t \to +\infty. \tag{17.0}$$

Now, the function

$$H(Q) = \int_{Q_0}^{Q} H'(x)\, dx$$

has the properties

$$H(Q_0) = H'(Q_0) = 0, \qquad H''(Q) < 0, \qquad H(Q) \to -\infty \text{ as } |Q| \to \infty \tag{18.0}$$

for $\nu < a$, i.e. for Q_0 finite. In the case where $\nu = a$, i.e. where $Q_0 = -\infty$, the latter limit relation holds for $Q \to +\infty$ only. Furthermore, in any neighborhood of $Q = Q_0$ the functions $H'(Q)$, $H''(Q)$ remain bounded. By neighborhood we mean any finite Q interval around Q_0 if Q_0 is finite and any semi-infinite interval reaching to $-\infty$ if $Q_0 = -\infty$.

On multiplying (15.0) by $\dot{Q}$ and integrating we obtain

$$\frac{1}{2}\dot{Q}^2 + \nu \int_0^t \dot{Q}^2\, dt = H(Q) + C_1 . \tag{19.0}$$

As $H \leq 0$ we infer from this relation first that $\dot{Q}$ remains bounded and that $\int_0^\infty \dot{Q}^2\, dt$ is finite, and second—on using (18.0)—that Q stays in a neighborhood of Q_0. (19.0) can now be written

$$\frac{1}{2}\dot{Q}^2 = H(Q) + \delta(t) + C_2 , \qquad \delta(t) \to 0 \qquad \text{as} \qquad t \to +\infty. \tag{20.0}$$

On multiplying (15.0) by $\ddot{Q}$ we obtain by integration

$$\int_0^t \ddot{Q}^2\, dt + \frac{\nu}{2}\dot{Q}^2 = \int_0^t \ddot{Q}\, H'(Q)\, dt + C_3$$

$$= \dot{Q}\, H'(Q) - \int_0^t \dot{Q}^2\, H''(Q)\, dt + C_4 .$$

Since $Q(t)$ stays in a neighborhood of Q_0 , $H'(Q)$ and $H''(Q)$ remain bounded. It was stated already that $\dot{Q}$ stays bounded. Hence one infers that $\int_0^\infty \ddot{Q}^2\, dt$ must be finite. As $\int_0^\infty (\dot{Q}^2 + \ddot{Q}^2)\, dt$ is finite there must exist a sequence $t_n \to +\infty$ such that $\dot{Q}(t_n) \to 0$ and $\ddot{Q}(t_n) \to 0$. If Q_∞ denotes any value of accumulation of the $Q(t_n)$ we must, on account of (15.0), have $H'(Q_\infty) = 0$ which, according to (18.0), is compatible only with $Q_\infty = Q_0$. On inserting $t = t_n$ in (20.0) we find, taking account of $H(Q_0) = 0$, that $C_2 = 0$ and that

$$\frac{1}{2}\dot{Q}^2 - H(Q) = \delta(t), \qquad \lim_{t=+\infty} \delta(t) = 0.$$

As $H(Q_0) = 0$ and $H < 0$ elsewhere this relation implies what we had to prove,

$$\dot{Q} \to 0, \qquad H(Q) \to 0, \qquad \text{i.e. } Q \to Q_0 \quad \text{as} \quad t \to +\infty.$$

Behavior of the Solutions of (1.0) as $t \to +\infty$

The following conditions will now be imposed upon the given function

$$F(x) = a(x) + ib(x),$$

$$a(x) = a_0 + 2\sum_1^\infty a_n \cos nx, \tag{21.0}$$

$$b(x) = 2\sum_1^\infty b_n \cos nx.$$

We demand that

$$a_0 < 0, \qquad a_n > 0 \text{ for infinitely many } n \tag{22.0}$$

and that any of the b_n in finite number be linearly independent with respect to integer coefficients (for instance $b_n = \kappa^n$ where κ is a transcendental number with $0 < \kappa < 1$).

Our lemma, now, furnishes complete information about the behavior of the solutions u, z of (1.0) (μ fixed) as $t \to \infty$. The Fourier coefficients of order n are solutions of (6.0) where $\nu = n^2\mu$ and $F = F_n = a_n + ib_n$. If n satisfies $n^2\mu > a_n$ the corresponding coefficient tends to zero. For $n^2\mu = a_n$ the same is true since in this case the periodic solution is $= 0$. For $n^2\mu < a_n$ (which can happen only if $a_n > 0$), however, the n-th coefficients $u_n(t)$, $z_n(t)$ will, unless $z_n(0) = 0$, tend toward

$$-a_n + n^2\mu, \qquad (n^2\mu(a_n - n^2\mu))^{1/2} \exp\{-ib_n(t + \alpha_n)\}$$

where α_n is a suitable constant (convergence in the sense expressed in the lemma). For $\mu > 0$ fixed, the latter case can only occur for finitely many values of the index n (the a_n are the Fourier coefficients of an absolutely integrable function and must, therefore, tend to zero). The coefficients which fall under the first case, $n^2\mu > a_n$, satisfy according to (9) and footnote 3 below the inequality

$$u_n u_n^* + z_n z_n^* \le (u_n u_n^* + z_n z_n^*)_{t=0} \exp\{-2(n^2\mu - a_n)t\}$$

[3] a_n is to be omitted if $a_n \le 0$.

for all $t > 0$. This makes it obvious that for $t \to \infty$ not only the corresponding terms of the Fourier expansion but also their sum (which is a function of x and t) tends to zero uniformly with respect to x. We have hereby proved the following

THEOREM.

Every solution $u(x, t)$, $z(x, t)$ of (1.0) ($\mu > 0$), *except the solutions described right afterward, tends for $t \to \infty$ to the special solution*

$$(23.0)\qquad \begin{aligned} u &= \sum_{n^2\mu < a_n} (-a_n + n^2\mu)e^{inx}, \\ z &= \sum_{n^2\mu < a_n} (n^2\mu(a_n - n^2\mu))^{1/2} e^{inx} \exp\{-ib_n(t + \alpha_n)\} \end{aligned}$$

with suitable real values of the α_n, i.e. the difference between the two solutions tends to zero uniformly with respect to x. The exceptional solutions are precisely those for which some Fourier coefficient of z with an index satisfying $n^2\mu < a_n$ vanishes.[4]

If only those solutions of (1.0) *are considered (and we will do so in the sequel) for which u, z are even functions of x then the limit solution* (23.0) *is*[5]

$$(23.1)\qquad \begin{aligned} u &= 2 \sum_{n^2\mu < a_n} (-a_n + n^2\mu) \cos nx, \\ z &= 2 \sum_{n^2\mu < a_n} (n^2\mu(a_n - n^2\mu))^{1/2} \exp\{-ib_n(t + \alpha_n)\} \cos nx \end{aligned}$$

where the indices are confined to positive integers.[6]

From now on we restrict ourselves to the solutions of (1.0) with u, z even.

Let us describe the situation brought to light by the theorem in more detail. For $\mu \geq \mu_1 = \max (a_n/n^2)$ every solution of (1.0) tends to $u = z = 0$ (the "laminar solution"). This solution exists for any μ but it is unstable for $\mu < \mu_1$. The number $N = N(\mu)$ of terms in each sum (23.1) will, according to the hypothesis (22.0), increase beyond limit as $\mu \to 0$. Let us visualize the general case in which the positive among the a_n/n^2 are different from each other and let us arrange these numbers in a decreasing sequence

$$\mu_1 > \mu_2 > \mu_3 > \cdots \to 0.$$

Every time μ decreases through such a critical value $N(\mu)$ increases by one.

The limit solutions (23.1) constitute an $N(\mu)$-dimensional torus-like manifold $\mathfrak{M} = \mathfrak{M}(\mu)$ in our functional phase space.

---

[4] These solutions also tend to limit solutions which are obtained from (23.0) simply by dropping the terms with that index n.

[5] According to (7), $a_{-n} = a_n$, $b_{-n} = b_n$. The index $n = 0$ does not occur in the sums (23.0).

[6] A drastic difference between our model and what is conjectured in the hydrodynamic cases is that the time periods $2\pi/b_n$ of the partial waves become longer ($b_n \to 0$) instead of shorter.

$$u = 2 \sum_{n^2\mu < a_n} (-a_n + n^2\mu) \cos nx,$$

(24.0)

$$z = 2 \sum_{n^2\mu < a_n} (n^2\mu(a_n - n^2\mu))^{1/2} e^{i\varphi_n} \cos nx$$

where the N angular variables φ_n are the parameters. The limit solutions are

$$\varphi_n = -b_n(t + \alpha_n), \tag{25.0}$$

α_n being arbitrary real constants. Since the b_n were supposed to be linearly independent numbers each of those solutions will be everywhere dense on the manifold $\mathfrak{M}$.

The manifold $\mathfrak{M}(\mu)$ varies continuously with μ though not analytically in the phase space of our problem (the function space of the $u(x)$, $z(x)$). For $\mu > \mu_1$ it is simply the point $u = z = 0$. As μ decreases and passes μ_1, $\mathfrak{M}$ branches off this point as a small and gradually enlarging closed curve. On passing μ_2 $\mathfrak{M}$ branches again off this curve and forms a thin and gradually swelling tire. The curve continues to exist but it has yielded its stability to the tire. As μ passes through μ_3 another branching occurs and so forth. The number $N(\mu)$ of dimensions of $\mathfrak{M}(\mu)$ (the "number of degrees of freedom of turbulence") increases beyond limit.

Limit of the Solutions of (1.0) for $\mu \to 0$

If the initial values $u(x)$, $z(x)$ from which a solution of our model, with μ given, starts are chosen at random then the solution actually observed in the long run will be precisely a solution (23.1). Passing to the limit $\mu \to 0$ in the observed solution therefore means letting $\mu \to 0$ in (23.1).

We suppose (only in this section) that $F(x)$ satisfies the following condition (which could, however, be replaced by a less restrictive one)

$$\sum | a_n | < \infty.$$

Under this condition we can easily prove:
As $\mu \to 0$ the functions (23.1) converge towards the time independent functions

$$u = -2 \sum_{a_n > 0} a_n \cos nx, \qquad z = 0 \tag{26.0}$$

uniformly with respect to x and t.

The turbulent fluctuations which, in our model, occur only in the z-component disappear in the limit $\mu \to 0$. Incidentally (26.0) is a special stationary solution of the equations (1.0) with $\mu = 0$. These equations have infinitely many stationary solutions: every pair $u = u(x)$, $z = 0$ where u has a vanishing mean value is such a solution for $\mu = 0$.

The proof of the limit relation is simply carried out by splitting each sum in (23.1) into two sums

$$\sum = \sum_{n<k} + \sum_{n\geq k}$$

with k fixed. In the sums of the first type $\mu \to 0$ can be effected without difficulty. The absolute value of the sums of second type is less than $\sum_k^\infty |a_n|$ which can, by choosing k sufficiently large, be made arbitrarily small[7].

The Statistics in Phase Space

We consider the non-trivial case $\mu < \mu_1$ in our problem. Let $\mathfrak{F}[u(x), z(x)]$ be an arbitrary functional. $\mathfrak{F}$ is supposed to vary continuously if the argument functions (and their first derivatives) change uniformly continuously.[8] If $[u(x), z(x)]$ does not belong to the exceptional set as stated in our main theorem the difference between $f(t) = \mathfrak{F}[u(x, t), z(x, t)]$, where $u(x, 0) = u(x)$, $z(x, 0) = z(x)$, and the same function, where a suitable solution (23.1) is substituted, tends to zero as $t \to \infty$. Therefore,

$$\overline{\mathfrak{F}} \equiv \lim_{T=\infty} \frac{1}{T} \int_0^T \mathfrak{F}[u(x, t), z(x, t)]\, dt$$

(27.0)

$$= \lim_{T=\infty} \frac{1}{T} \int_0^T \mathfrak{F}_{u,z=(23.1)}\, dt$$

if the latter limit exists. The statistics is reduced to a statistics on the restricted finite-dimensional manifold (24.0) of solutions. On $\mathfrak{M}$ the functional $\mathfrak{F}$ is simply a continuous function $f(\varphi_1, \cdots, \varphi_N)$ of the parameters φ_n with period 2π in each of them. According to (25.0) where the b_n are linearly independent and according to the equidistribution theorem of H. Weyl the right hand limit in (27.0) exists and is independent of the initial phases. The phase flow on $\mathfrak{M}$ is ergodic. It follows that the time average of any continuous functional $\mathfrak{F}[u(x), z(x)]$ has, for the majority of initial phases, the constant value

$$\overline{\mathfrak{F}} = \frac{1}{(2\pi)^N} \int_0^{2\pi} \cdots\cdots \int_0^{2\pi} \mathfrak{F}_{u,z=(24)} \prod_{n^2\mu<a_n} d\varphi_n\,. \tag{28.0}$$

For instance, the averages of polynomial functionals of degree one and two are determined by ($z = v + iw$)

$$\overline{v(x)} = \overline{w(x)} = 0 \tag{29.0}$$

and[9]

$$\overline{v(x)w(x')} = 0, \qquad \overline{v(x)v(x')} = \overline{w(x)w(x')} = h_\mu(x + x') + h_\mu(x - x') \tag{30.0}$$

---

[7] In the second sum of (23.1) use the inequality $\nu(a - \nu) \leq \frac{1}{4}a$.

[8] Instances of such functionals are $z(x_1)$, $z(x_1)z^*(x_2)$, where x_1, x_2 are fixed points, and

$$\int_0^\pi z(x)z^*(x)\, dx, \qquad \int_0^\pi \frac{dz(x)}{dx}\frac{dz^*(x)}{dx}\, dx.$$

[9] u can be left out of consideration since, on $\mathfrak{M}$, u is independent of the time, $u = \bar{u}$.

where

$$h_\mu(y) = \sum_{n^2\mu < a_n} n^2\mu(a_n - n^2\mu) \cos ny. \tag{31.0}$$

In (29.0) and (30.0) x, x' are arbitrary fixed points in $(0, \pi)$. The functionals considered in (30.0) (products of values of u, v, w at fixed points) are the simplest of degree two. Every other quadratic functional can be written as a linear functional of the functions (30.0) of x, x'.[10]

Asymptotic Form of the Correlations for $\mu \to 0$

In order to avoid too lengthy considerations we will, from now on, restrict the given function $a(x)$ to the class of functions $(a_0 < 0)$

$$a(x) = a_0 + \sum_1^\infty n^{-s} \cos nx \qquad (s > 0). \tag{32.0}$$

All these functions are absolutely integrable. We shall first show that if

$$\sum n^2 a_n \text{ diverges, i.e. } s \leq 3, \tag{33.0}$$

the correlations (30.0) have in the interval $(0, \pi)$ a tendency toward homogeneity as $\mu \to 0$. The main fact is that

$$\lim_{\mu=0} \frac{h_\mu(y)}{h_\mu(0)} = 0, \qquad 0 < y < 2\pi \tag{34.0}$$

holds uniformly in any closed subinterval of $(0, 2\pi)$. This would obviously no longer be true if $(s > 3)$ $\sum n^2 a_n$ converges, i.e. if the function $a(x)$ is too smooth. The limit behavior, for $\mu \to 0$, of the correlation quantities is now evident from (30.0). The dispersions about the mean values $\bar{v} = \bar{w} = 0$

$$\overline{v^2(x)} = \overline{w^2(x)} = h_\mu(0) + h_\mu(2x)$$

are proportionally nearly constant in the fundamental interval $0 < x < \pi$. At its endpoints there is a sharp rise to the double value. The coefficient of correlation at two points x, x' in $(0, \pi)$

$$\frac{\overline{v(x)v(x')}}{(\overline{v^2(x)})^{1/2}(\overline{v^2(x')})^{1/2}} = \frac{\overline{w(x)w(x')}}{(\overline{w^2(x)})^{1/2}(\overline{w^2(x')})^{1/2}} \sim \frac{h_\mu(x' - x)}{h_\mu(0)}$$

has appreciable values only for sufficiently small distances $|x' - x|$. The statistical distributions at two not too close points are nearly independent of each other.

[10]In the hydrodynamic case of the flow through a channel with a given net flow through a cross section, for instance, the force exerted by the fluid on the wall is a second degree functional of the instantaneous velocity field. It can be computed, since the correlations (of second degree) between the velocities at different points are known functions of these points.

Relation (34.0) will be proved further below. An asymptotic expression for

$$(35.0) \qquad h_\mu(y) = \mu \sum_1^N n^2(n^{-s} - n^2\mu) \cos ny$$

where N is the largest integer less than $\mu^{-1/(2+s)}$ is easily found by setting $y = \eta/N$ where η is kept fixed and by inserting $\mu = (N + \vartheta)^{-2-s}$ where $0 < \vartheta < 1$. The sum (35.0) can then be written as a Riemann approximating sum for a definite integral. The asymptotic expression, for μ small, is

$$(36.0) \qquad h_\mu(y) \sim N^{2s-1} \int_0^1 \alpha^2(\alpha^{-s} - \alpha^2) \cos Ny\alpha \, d\alpha, \qquad N \sim \mu^{-1/(2+s)}.$$

This shows that appreciable correlation occurs only at distances of the order of $N^{-1} \sim \mu^{1/(2+s)}$. The case $s = 3$ where the asymptotic behavior is different has been excluded here. (36.0) implies that

$$h_\mu(0) \sim \frac{2+s}{5(3-s)} \mu^{(2s-1)/(s+2)}.$$

This shows, in extension of the result of a previous section, that the turbulent fluctuations die down for $\mu \to 0$ if $s > 1/2$. For $s < 1/2$, however, they are seen to increase beyond limit.

Relation (34.0) is proved by application of Abel's partial summation to the sum in the denumerator (35.0) of (34.0). The transformed sum is a sum of products where one factor is a partial sum of $\sum \cos ny$ and where the other factor is the difference of two successive coefficients in the Fourier sum (35.0). Those partial sums are known to be uniformly bounded in any closed subinterval of $(0, 2\pi)$. On the other hand, the sum of the absolute values of those differences is easily found to be bounded if $s \geq 2$ and to be of the order of N^{2-s} if $s < 2$. The denominator sum in (34.0), however, is of the order of N^{3-s} for $s < 3$ and of the order of $\log N$ for $s = 3$. This proves (34.0).

There is no reason to investigate this matter any further because we have, now, arrived at a point at which the analogy to the hydrodynamic case breaks down. The limiting form of the correlation curves in our model case is quite different from the one in hydrodynamics. Recent investigations by Kolmogoroff, von Weizsäcker, and Heisenberg indicate that the principal correlation coefficient in hydrodynamics has approximately the universal form

$$1 - \text{const.} \mid x' - x \mid^{2/3}$$

if the distance is small compared to the dimensions of the boundaries and large as compared to a length depending on the viscosity μ and tending to zero with μ. Their reasoning essentially uses the fact that there is interaction between the different frequencies of the spatial Fourier pattern (which is entirely absent in our case). The arguments are based on the Prandtl mixing length or similar highly intuitive concepts. These notions in turn rest on a semiempirical picture of turbulent fluid motion. The ultimate goal, however, must be a rational theory of statistical hydrodynamics where those important results and other properties of turbulent flow can be mathematically deduced from the fundamental equations of hydromechanics.

Normality of the Distribution

In the hydrodynamic case we are to expect that the values of the velocities in an arbitrarily given finite set of fixed points at rest become more and more normally distributed as $\mu \to 0$. In our model case this can be actually proven.

The values $w(x_1), \cdots, w(x_j)$ at j fixed points $x_1, \cdots, x_j$ in the open interval $(0, \pi)$ may be regarded as a set of chance variables the distribution law of which has been determined before. We use the parametric representation (24.0),

$$v(x) = \sum_{n^2\mu<a_n} v^{(n)}(x), \qquad w(x) = \sum_{n^2\mu<a_n} w^{(n)}(x) \tag{37.0}$$

where

$$\left\{\begin{matrix} v^{(n)} \\ \\ w^{(n)} \end{matrix}(x)\right\} = 2(n^2\mu(a_n - n^2\mu))^{1/2} \cos nx \left\{\begin{matrix} \cos \\ \\ \sin \end{matrix} \varphi_n\right\} \tag{38.0}$$

and where the φ_n are the fundamental chance variables with the probability differential element

$$\prod (d\varphi_n/2\pi).$$

If we regard the set of values

$$(v(x_1), w(x_1), \cdots, w(x_j)) \tag{39.0}$$

as a chance vector we see that this vector is the sum of certain chance vectors

$$(v^{(n)}(x_1), w^{(n)}(x_1), \cdots, w^{(n)}(x_j)) \tag{40.0}$$

where the summation index runs through all values that satisfy $n^2\mu < a_n$. Now, these different chance vectors are obviously statistically independent of each other which means that any set of components where no two of them belong to the same vector constitutes a set of independent chance variables. The components of one and the same vector (40.0) may, however, be statistically correlated among each other. As the number of terms $N(\mu)$ in our sum tends $\to \infty$ for $\mu \to 0$ we should expect the probability distribution of the variables (39.0) (with points $x_1, \cdots$ arbitrarily fixed in advance) to deviate less and less from a multidimensional normal distribution as formulated above in the introduction. The normal distribution will, in this case, be the one with the same first and second degree moments (mean values and correlations) as the actual distribution. The central limit theorem of probability theory affirms the truth of this limit relation provided that the following two conditions are satisfied. The first one roughly says that none of the vectors (40.0) plays too dominant a statistical role in their sums (39.0). The second condition means that the correlation hyperellipsoid of the variables (39.0) does not degenerate as $\mu \to 0$. The first condition (this is the well known Lindeberg condition) splits up into a set of conditions referring to the single components of (39.0) i.e. to the chance variable $v(x)$ or to $w(x)$ in (37.0), (38.0) with a fixed point x in $(0, \pi)$. As all the terms in a sum (38.0), apart from the values of their first and second degree moments, have the same distribution the Lindeberg condition on the sum $v(x)$ or $w(x)$ simply reduces to the condition that, for $\mu \to 0$,

$$\frac{\overline{(v^{(n)}(x))^2}}{\overline{(v(x))^2}} = \frac{\overline{(w^{(n)}(x))^2}}{\overline{(w(x))^2}} = \frac{n^2(a_n - n^2\mu)\cos^2 nx}{\sum_{a_i > i^2\mu} i^2(a_i - i^2\mu)\cos^2 ix} \qquad (a_n > n^2\mu)$$

converge to zero uniformly with respect to n.

As in the preceding section we now restrict ourselves to the case where the given function $a(x)$ is of the special form (32.0). We shall verify that the Lindeberg condition and the dimensionality condition are fulfilled if

$$\text{(41.0)} \qquad \sum n^2 a_n \text{ diverges}, \qquad s \leq 3,$$

i.e. under the same hypothesis as in the preceding section. The dispersion ratios in question are not greater than

$$\text{(42.0)} \qquad \frac{n^2(n^{-s} - n^2\mu)}{\sum_1^N i^2(i^{-s} - i^2\mu)\cos^2 ix} \qquad (n \leq N)$$

respectively where $N(\mu)$ is the largest integer $<\mu^{-1/(2+s)}$. The denominator is $\frac{1}{2}h_\mu(0) + \frac{1}{2}h_\mu(2x)$. According to (34.0) for the purpose of investigating the order of magnitude the cos-factors in the denominator may be dropped. Now, in the preceding section, we have studied the ratio where the denominator was the same but where the denumerator was the sum of the absolute values of the differences of successive denumerators in (42.0). That ratio was found to tend to zero for $\mu \to 0$ if (41.0) is fulfilled. For $n = 1$ (42.0) obviously tends $\to 0$ under condition (41.0). Hence we infer that (42.0) a fortiori tends to zero uniformly with respect to n.

As to the dimensionality condition we need only observe that the correlation matrix of the chance variables (39.0), $x_1, \cdots, x_j$ being fixed points, becomes, for $\mu \to 0$, more and more the unit matrix times a scalar factor, i.e. that the correlation hyperellipsoid becomes a hypersphere if the condition (41.0) is satisfied. It was manifestly this fact which resulted from the preceding section.

The distribution of the values at arbitrarily preassigned fixed points $x_1, \cdots, x_j$ in $(0, \pi)$ becomes, for $\mu \to 0$, a spherical normal distribution.

The result can, of course, be generalized by dropping the condition that the points x_ν be kept fixed while $\mu \to 0$. If the points $x_1, \cdots, x_j$ (j fixed) are allowed to vary while $\mu \to 0$ the only (sufficient) condition to be imposed on them will be that the distances from 0 and π and between any two of them stay greater than const./$N(\mu)$ while $\mu \to 0$. This was found to be the order of distance with appreciable correlation. In this general case, of course, the correlations will enter in the limit distribution.

The fundamental reason why we obtained a normal distribution in the limit was the fact that in our model the components of the spatial Fourier pattern behave statistically independently. In the hydrodynamic case where we have interaction the dependence will probably be only slight in the sense that it will not extend far over the spatial spectrum. With such an approximate independence the central limit theorem of probability theory might still hold. The distribution of the simultaneous velocities at a fixed finite set of space points might be expected to become normal. It is, however, not certain if the space derivatives $\partial u/\partial x, \cdots$ also behave like this. It is quite possible that their distribution does not approach a normal one.

Repeated branching through loss of stability, an example. *Proceedings of the Conference on Differential Equations (dedicated to A. Weinstein)*, **pp. 49–56. University of Maryland Book Store, College Park, Md., 1956, and**

A mathematical example displaying features of turbulence. *Communications on Pure and Appl. Math.* **1, (1948). 303–322.**

Commentaries

Roger Temam

In these two articles, E. Hopf studies the mechanism of the route to turbulence. It is interesting to put these two articles in perspective in the evolution of ideas on turbulence and in fluid mechanics.

The mathematical theory of fluid mechanics began in the 1930's. For the Navier Stokes equations (NSE) there was the pioneering work of the French mathematician Jean Leray who set the basis of the theory of these equations in three articles published in 1932 and 1933; in particular J. Leray introduced the concept of weak solutions for the three dimensional incompressible NSE and proved existence for all time for flows in $\mathbb{R}^3$. In the mid to late 1930's J. Leray established with J.P. Schauder the fixed point theorem which bears their name and used it to prove the existence of stationary solutions. The mathematical theory of the Euler equations developed independently in the 1930s with the works of N. Gunther, L. Lichtenstein and W. Wolibner (which was nearly lost during World War II, but rediscovered about thirty years later).

The mathematical theory of fluid mechanics entered then in a long ten year lull and E. Hopf was one of the first to bring it back to life in the late 1940s and early 1950s.

During this lull, fluid mechanics was not dormant, and in fact many important results were derived based on more physical grounds. One can mention in particular the works of L. Prandtl and T. von Karman on boundary layer theory and other aspects of fluid mechanics; in 1941 N. Kolmogorov published his famous papers on the statistical approach to turbulence; also J. von Neuman, after designing the first computers, used them in the 1940s for the first numerical simulations in fluid mechanics and meteorology.

When he introduced the concept of weak solutions for the NSE, the motivation of Jean Leray was turbulence: he conjectured that vortices were associated with points or lines where the curl vector becomes infinite, and he looked for a class of non smooth solutions to the NSE which can display such singularities; this is, by the way, one of the first utilizations of Sobolev spaces in nonlinear partial differential equations, the natural framework before that being spaces of continuously differentiable functions. The hypothesis of J. Leray has not yet been proven nor disproven, and, in fact, the answer to his question is now the object of one of the famous Clay Prizes.

J. Leray abandoned fluid mechanics during World War II (for reasons recalled in his collected work [Leray (1998)]), and E. Hopf continued J. Leray's work on NSE in the late 1940s. His major mathematical work in this direction is the proof of existence of weak solutions for all time for the three dimensional NSE in a bounded (limited) domain (see Hopf bibliography #57 and p. XXX of these selection). E. Hopf was also interested in turbulence and he developed ideas of his own in these two articles.

The route to turbulence that he proposed is known in the turbulence literature as the Landau route to chaos; namely in the presence of a specified time independent forcing, the flow will perform bifurcations into more and more complex flows, from stationary to stationary solutions, from stationary to time periodic (what is now called Hopf's bifurcation, (see Hopf bibliography #43 and page XXX of these selection), and from periodic into tori of higher and higher dimensions up to full turbulence: as the viscosity μ decreases, at bifurcation points, a given solution loses its stability and then another stable motion appears filling densely a manifold with one more dimension which branches away from the previous one. This route to full turbulence was also proposed, independently it appears by L. Landau (see Landau and Lifschitz (1991), Ch.III, §27, but this may have been mentioned in an earlier article of Landau).

Hence E. Hopf formulated this same hypothesis. Unable to establish such a result for the NSE, E. Hopf constructed in these two articles, equations for which this infinite sequence of bifurcations actually occurs. This route to turbulence does not prevail anymore; laboratory experiments do not support this long sequence of bifurcations and, on the contrary, a more favored view is the idea of Ruelle and Takens (1971) who conjectured that, generically, turbulence occurs after a small number of bifurcations. Despite their lack of (current) physical relevance, these two articles are a technical tour de force, although, by now, we know a simpler system producing this picture (Minea (1997)).

In the first article the system considered reads (with slightly different notations),

$$\frac{\partial \mathbf{u}}{\partial t} - \mu \frac{\partial^2 \mathbf{u}}{\partial x^2} + B(\mathbf{u}, \mathbf{u}) + E(\mathbf{u}) = 0. \tag{1}$$

Here $\mathbf{u} = (u, v, w)$ is a real function of t and $x \in (0, 2\pi), 2\pi$-periodic in x and even; B is a quadratic functional involving convolution products and E is a linear operator; $\mu > 0$ is the viscosity. There exist coefficients a_n, and, for μ fixed, as $t \to +\infty$, all but some exceptional solutions of (1) tend to $\tilde{\mathbf{u}} = (\tilde{u}, \tilde{v}, \tilde{w})$, with

$$\begin{aligned} \tilde{u} &= 2 \sum_{n^2\mu < a_n} (-a_n + n^2\mu) \cos nx, \\ \tilde{w} &= 2 \sum_{n^{\mu} < a_n} (n^2\mu(a_n - n^2\mu))^{1/2} \exp(-ib_n(t + \alpha_n)) \cos nx, \end{aligned} \tag{2}$$

(see Theorem p.315). Hopf furthermore studied the statistics of these solutions.

In the second article, Hopf considered a simpler model with only two scalar functions u, v, and he introduces the complex variable $w = u + iv(z = v + iw$ in the case of (1)). For w the structure of the equation is similar to (1), with $\mathbf{u}$ replaced by w, B is now a cubic convolution term and E is a linear convolution term involving a forcing $F = F(x) = \sum f_n e^{inx}$. The analysis is much simpler, the conclusions are similar, involving now the forcing F (Theorem p.52); the statistics of the solutions is also briefly discussed.

References

HOPF, EBERHARD, "Abzweigung einer periodischen Lösung von einer stationäre eines Differential systems." (German) Ber. Verh. Sächs. Akad. Wiss. Leipzig. *Math.-Nat. Kl.* **95**, (1943) no. 1, 3–22.

HOPF, EBERHARD, "Über die Anfangswertaufgabe für die hydrodynamischen Grundgleichungen." *(German) Math. Nachr.* **4**, (1951) 213–231

LANDAU, L. D. AND LIFSCHITZ, E. M., *Hydrodynamics*, Translated from the Russia by Wolfgang Weller and Adolf Khnel. Fifth edition. Akademie-Verlag, Berlin, (1991) xvi+683 pp.

LERAY, JEAN, *Selected papers. Œuvres scientifiques.* (French) Edited by Paul Malliavin. Springer-Verlag, Berlin; Société Mathématique de France, Paris, (1998) x+507 pp.

MINEA, GHEORGHE, "Remarque sur les équations d'Euler dans un domaine possédant une symétrie de révolution." *C. R. Acad. Sci. Paris Sr. A-B* **284**, no. 9, (1997) pp. A477–A479.

RUELLE, D. AND TAKENS, F., "On the nature of turbulence." *Comm. Math. Phys.* **20**, (1971) pp. 167-192.

ON S. BERNSTEIN'S THEOREM ON SURFACES $z(x, y)$ OF NONPOSITIVE CURVATURE

EBERHARD HOPF

The following classical theorem is due to S. Bernstein: If $z=z(x, y)$ is of class C'' in the whole x-y-plane and if

$$(1) \qquad z_{xx}z_{yy} - z_{xy}^2 \leqq 0, \quad z_{xx}z_{yy} - z_{xy}^2 \not\equiv 0,$$

then $z(x, y)$ cannot be bounded.[1] The original proof was found to contain a gap of topological nature. It is the purpose of this note to bridge this gap and to prove a somewhat more general theorem.

THEOREM. *If $z(x, y)$ belongs to C'' in the whole x-y-plane and satisfies* (1) *then $z(x, y)$ cannot be $o(r)$ where r is the distance of (x, y) from an arbitrarily chosen fixed point.*

That this estimate of the order of magnitude at infinity cannot be essentially improved is shown by examples of the form $z=f(x)-g(y)$, $f''>0$, $g''>0$, where f and g can be chosen such that the order is just $O(r)$. A still open question is whether $z(x, y)$ can or cannot be $o(r)$ along a special sequence of radii $r_i \to \infty$.

In proving the theorem we shall, essentially, follow Bernstein's original arguments. For the sake of completeness the arguments will be repeated.

LEMMA 1 OF BERNSTEIN. *Let $z(x, y)$ be of class C'' in a bounded open set R and let $z_{xx}z_{yy}-z_{xy}^2 \leqq 0$ in R. If z is continuous on the boundary B of R and if $z \leqq 0$ on B, then $z \leqq 0$ in the whole of R.*

PROOF (according to M. Shiffman). Let C be a circle in the plane $z=0$ whose interior contains $R+B$. Consider the parts $z \geqq 0$ of all possible spheres which intersect the plane $z=0$ in C. They form a

Received by the editors October 1, 1948.

[1] S. Bernstein, *Ueber ein geometrisches Theorem und seine Anwendung auf die partiellen Differentialgleichungen vom elliptischen Typus*, Math. Zeit. vol. 26 (1927) pp. 151–158. This is the translation of a paper written in French (*Sur une théorème de géométrie et son application aux équations aux dérivées partielles du type elliptique*) and published in 1914 by the Mathematical Society of Charkow.

Note added in proof. It had escaped the author's attention that Bernstein himself had already published stronger results, which include the generalized theorem. See his paper, *Renforcement de mon théorème sur les surfaces à courbure negative*, Bull. Acad. Sci. URSS. Sér. Math. (1942). His proof, however, contains the same gap as the original one. The argument fails if all four domains Ω spiral infinitely often around the origin.

monotonic family of surfaces $z=Z(x, y; \lambda)$ above $R+B$ with $\lambda=\max Z \geqq 0$ taken on the sphere. We have $Z=0$ for $\lambda=0$, $Z>0$ in $R+B$ for $\lambda>0$, and $Z \to \infty$ uniformly in $R+B$ for $\lambda \to \infty$. Let λ^* be the greatest lower bound of all λ for which everywhere

$$Z(x, y; \lambda) \geqq z(x, y) \qquad \text{in } R+B. \tag{2}$$

It is to be proved that $\lambda^*=0$. Suppose that $\lambda^*>0$. For reasons of continuity (2) must hold also for $\lambda=\lambda^*$ and there must be some joint x^*, y^* in $R+B$ where $=$ holds in (2), $\lambda=\lambda^*$. This point must lie in R because, according to hypothesis, $z \leqq 0$ and $Z>0$ $(\lambda>0)$ on B. In the space point $x^*, y^*, z^*=z(x^*, y^*)$ the surface must, according to (2), $\lambda=\lambda^*$, be internally tangent to a sphere. Evidently the point of tangency would be a point of positive Gaussian curvature for $z(x, y)$, in contradiction to the hypothesis.

LEMMA 2 (*essentially due to Bernstein*). *Let $z(x, y)$ be of class C'' and let $z_{xx}z_{yy}-z_{xy}^2 \leqq 0$ in a connected open set R. Let $z>0$ in R and let z be continuous and equal to 0 on the boundary B of R. Suppose that R can be placed in an angle less than π. Consider the chord of this angle perpendicular to the line bisecting the angle and at distance u from the vertex. Then the maximum $M(u)$ of $z(x, y)$ on the $(R+B)$-part of this chord is defined in an infinite u-interval and nowhere convex in this interval,*

$$\begin{gathered}(u_3-u_1)M(u_2) \leqq (u_3-u_2)M(u_1)+(u_2-u_1)M(u_3), \\ u_1<u_2<u_3.\end{gathered} \tag{3}$$

COROLLARY. *Under the hypothesis of Lemma 2, there exists a positive constant c such that $M(u)>cu$ for all u sufficiently large.*

PROOF. According to Lemma 1, R must be unbounded. Since R is connected, $M(u)$ must be defined for all $u \geqq u_0$, and $M(u) \geqq 0$, $M(u_0)=0$. Suppose, now, that (3) were false for three fixed values $u_1<u_2<u_3$,

$$(u_3-u_1)M(u_2)>(u_3-u_2)M(u_1)+(u_2-u_1)M(u_3). \tag{4}$$

As the linear function of x, y

$$l(x, y)=au+b \qquad (a, b \text{ constants})$$

always satisfies the equality in (3), the chord maxima $M^*(u)=M(u)-(au+b)$ of the function

$$z^*(x, y)=z(x, y)-l(x, y) \tag{5}$$

again satisfy (4) for those fixed u_i,

(6) $(u_3 - u_1)M^*(u_2) > (u_3 - u_2)M^*(u_1) + (u_2 - u_1)M^*(u_3).$

If a, b are chosen such that

(7) $$l(u_i) = M(u_i) \geqq 0, \qquad i = 1, 3,$$

we have $M^*(u_i) = 0$ for $i = 1, 3$ and, therefore,

(8) $$z^* \leqq 0 \qquad \text{on } u = u_1 \text{ and } u = u_3.$$

(6) now implies that $M^*(u_2) > 0$ and that

(9) $$z^* > 0 \qquad \text{somewhere on } u = u_2.$$

From (7) we infer that $l \geqq 0$ for $u_1 < u < u_3$. This shows, in conjunction with the hypothesis that $z = 0$ on B, that

(10) $$z^* \leqq 0 \qquad \text{on } B \text{ for } u_1 \leqq u \leqq u_3.$$

If R' denotes the set common to R and to $u_1 < u < u_3$, (8) and (10) mean that $z^* \leqq 0$ on the entire boundary of the bounded open set R'. Since

$$\overset{*}{z}_{xx}\overset{*}{z}_{xy} - (\overset{*}{z}_{xy})^2 \leqq 0$$

in R', Lemma 1 shows that $z^* \leqq 0$ in the whole of R'. This contradiction to (9) proves Lemma 2.

The corollary is an obvious consequence of the properties of nowhere convex functions $M(u) \geqq 0$, $M(u_0) = 0$.

LEMMA 3. *Suppose that the hypotheses of Lemma 2 are fulfilled. Let* $N(r)$ *be the maximum of* $z(x, y)$ *on the* $(R+B)$*-part of the circle of radius* r *about some fixed point. Then there exists a positive constant* c' *such that* $N(r) > c'r$ *for all* r *sufficiently large.*

PROOF. For all r sufficiently large the circle about the fixed point intersects each of the two rays of the angle just once. To each such value r_1 of r belongs a number $u = u_1$ defined as the maximum value of u at which the chord lies in the area $r \leqq r_1$. It is geometrically obvious that

(11) $$u_1/r_1 \geqq c'' > 0 \qquad (c'' \text{ constant})$$

for all sufficiently large values of r_1. We further restrict r_1 to a range $r_1 \geqq a$ where a is chosen such that both the chord u_1 and the circle r_1 always intersect $R+B$. Consider the region R_1 common to R and to $r < r_1$. On the part of the boundary of R_1 with $r < r_1$ we have, according to hypothesis, $z = 0$. On $r = r_1$, $z \leqq N(r_1)$ and $N(r_1) \geqq 0$. The function $z^* = z - N(r_1)$ has, therefore, values not greater than 0 on the whole boundary of the bounded open set R_1. From Lemma 1 one infers that $z^* \leqq 0$ everywhere in R_1 and, in particular, on the chord $u = u_1$ which we know to lie in $r \leqq r_1$. Therefore,

(12) $$M(u_1) \leqq N(r_1)$$

holds for all r_1 sufficiently large and the lemma evidently follows from (11), (12), and from the corollary of Lemma 2.

PROOF OF THE THEOREM. The points x, y where $z = z(x, y)$ has negative curvature form an open set. Among these points is surely one where $z_x = z_y = 0$ does not hold. We may suppose that

$$z = z_x = 0, \qquad z_y = q_0 > 0,$$
$$z_{xx}z_{yy} - z_{xy}^2 < 0 \qquad \text{at } x = y = 0.$$

The function $\xi = z - q_0 y$ satisfies $\xi_{xx}\xi_{yy} - \xi_{xy}^2 \leqq 0$ and

(13) $$\xi = \xi_x = \xi_y = 0, \qquad \xi_{xx}\xi_{yy} - \xi_{xy}^2 < 0 \qquad \text{at } x = y = 0.$$

Suppose, now, the theorem were false. There would exist a function $\epsilon(r)$, $\epsilon \to \infty$, such that

(14) $$|z(x, y)| < r\epsilon(r),$$

where r is the distance of x, y from some fixed point. The statement is easily seen to be independent of the location of this initial point. We may suppose that $r^2 = x^2 + y^2$. We can also assume that ϵ is continuous and decreasing for $r \geqq r_0$ and that

$$\epsilon(r_0) = q_0 \qquad (r_0 > 0).$$

The region S defined by the inequalities $r < r_0$ and, for $r \geqq r_0$,

(15) $$|y| < \frac{r}{q_0}\epsilon(r)$$

is bounded by two continuous curves L^+, L^-, defined respectively by

(16) $$y = \pm \frac{r}{q_0}\epsilon(r), \qquad r \geqq r_0,$$

or, in polar coordinates, by $q_0 \sin \phi = \pm\epsilon(r)$. Each of these curves reaches towards $x = \pm\infty$. We also mention that $|y/r| \to 0$ along both curves as $|x| \to \infty$. It follows from (14), (15), and (16) that $\xi = z - q_0 y$ satisfies

(17) $$\xi < 0 \text{ on } L^+, \qquad \xi > 0 \text{ on } L^-,$$

and

(18) $$|\xi| < 2r\epsilon(r) \text{ in } S, \qquad r > r_0.$$

These properties of $\xi(x, y)$ together with the ones mentioned before are contradictory. We proceed to prove this.

(13) implies the existence of two straight line segments crossing each other in (0, 0) such that, except at (0, 0), $\xi>0$ on one and $\xi<0$ on the other segment. Two points on the same segment but on opposite sides of (0, 0) can never be joined by a Jordan arc on which ξ does not change sign. Otherwise a bounded region would be enclosed in which $\xi\neq 0$ and on whose boundary $\xi=0$, which contradicts Lemma 1. The open set where $\xi\neq 0$, therefore, contains exactly four components that contain, respectively, the four partial segments obtained by removing the point (0, 0),

$$\xi > 0 \text{ in } \overset{+}{\Omega}_i, \qquad \xi < 0 \text{ in } \overset{-}{\Omega}_i, \qquad i = 1, 2. \tag{19}$$

$\xi=0$ on the boundary of each of these four regions. Choose a $p>0$ such that each of the four connected sets Ω contains a point in which

$$|\xi| > p.$$

The set where $|\xi|>p$ must contain four different components Ω' such that each of them contains one of those four points,

$$\overset{+}{\Omega}{}'_i \subset \overset{+}{\Omega}_i, \qquad \overset{-}{\Omega}{}'_i \subset \overset{-}{\Omega}_i. \tag{20}$$

Continuity of ξ implies that these inequalities remain true if the left-hand sets are replaced by their closures. On using these auxiliary regions Ω' we shall be able to bridge the gap mentioned in the beginning.

We prove that each of the regions Ω contains a Jordan curve $x(t)$, $y(t)$ that lies in the strip S and for which $x\to\infty$ as $t\to\infty$ and $x\to-\infty$ as $t\to-\infty$. This is trivial if Ω contains a point of one of the lines L because, according to (17) and to the definition of Ω, Ω would then contain all points of L. Suppose, now, that Ω contains no point of L^++L^-. Since (0, 0) is an interior point of S and a boundary point of Ω the connected open set must be entirely contained in the strip S,

$$\Omega \subset S. \tag{21}$$

Consider the component Ω' of the set $|\xi|>p$ whose closure is, according to (20), contained in Ω. The notion of closure was hitherto understood in the sense of adding all finite points of accumulation. We now add the two infinite points $x=\pm\infty$ to the boundary of the strip S. In the new sense of closure thus involved the situation is this. Ω' is a connected open subset of Ω. The boundaries (in the extended sense) of Ω and Ω' cannot have common points except $x=-\infty$

and $x=+\infty$. We now show that they must be common boundary points of Ω and Ω'. It is sufficient to show that $x=+\infty$ is a boundary point of Ω'. If this were false, Ω' would lie in some half-plane $x<x_0$. The common part of S and of this half-plane could be placed within an angle less than π that is bisected by the x-axis (even within an angle of arbitrarily small opening). The definition of Ω' implies that one of the functions $-\xi-p$, $\xi-p$ is greater than 0 in Ω' and equal to 0 on its boundary. Lemma 3 can, therefore, be applied to this function in Ω' (which region must in view of Lemma 1 be unbounded). The statement of this lemma is, however, incompatible with the property (18) of ξ, $\epsilon \to 0$ for $r \to \infty$. $x=\infty$ is, therefore, a boundary point of Ω' (and $x=-\infty$ as well). It now follows from the topological theorem proved in the preceding paper[2] (by mapping S on a bounded set it is seen that the theorem applies to our case) that each of the two boundary points $x=\pm\infty$ is accessible from Ω, which is precisely what we wanted to prove.[3]

The statement just proved, namely that each of the four regions Ω contains a Jordan curve in S that joins the two infinite boundary points $|x|=\infty$, is, however, contradictory. Let α_i be a Jordan curve that joins (0, 0) (which is a boundary point of each Ω) through $S\Omega_i^+$ to $x=+\infty$. The "closed" curve $\alpha_1+\alpha_2$ must enclose one of the regions Ω^-. $S\Omega^-$ clearly could not contain a Jordan curve running towards $x=-\infty$, which is in contradiction to the statement proved above. This finishes the proof of the theorem.

NEW YORK UNIVERSITY

[2] E. Hopf, *A theorem on the accessibility of boundary parts of an open point set.*

[3] The error in the original proof of the theorem was, essentially, to infer the accessibility of $x=\infty$ from the mere fact that it is a boundary point of Ω.

On S. Bernstein's theorem on surfaces $z(x, y)$ of nonpositive curvature
***Proc. Amer. Math. Soc.* 1, (1950), pp. 80–85.**

Commentary

Louis Nirenberg

The famous theorem of S. Bernstein on minimal surfaces says that if such a surface is a graph of a function z on all of $\mathbb{R}^2$, then the surface is a plane. His first proof was based on a geometric theorem published in 1915. It says that if the graph of a function z defined on all of $\mathbb{R}^2$ has nonpositive curvature, i.e.,

$$z_{xx}z_{yy} - z_{xy}^2 \leq 0,$$

and if $|z|$ is bounded, then

$$z_{xx}z_{yy} - z_{xy}^2 \equiv 0.$$

A German translation of this paper appeared in 1927.

In the late 1940's when I was giving a seminar talk on the paper, Eberhard Hopf, who was in the audience, questioned an argument. Indeed the proof was not correct and no one had noticed it before. The proof involves a study of four regions where $\zeta = z-$ a certain affine function, changes sign. The difficulty, which had gone unobserved before, is that these regions may spiral infinitely often as one goes to infinity.

Hopf constructed a correct proof under a weaker condition than $|z|$ bounded, namely

$$z(x, y) = o(x^2 + y^2)^{1/2} \quad \text{near infinity.}$$

The proof is rather tricky. (For someone wishing to study it, just above (14), $\epsilon \to \infty$ should read $\epsilon \to 0$.)

Some years later, Peter Ungar made a slight simplification of Hopf's proof (private communication).

Remarks about Bernstein's theorem on minimal surfaces: There exist many proofs; perhaps the simplest is by C. C. Nitsche in *Annals of Math.* **66** (1957) 543–544. The theorem holds in dimensions n up to 7 but fails for $n = 8$ (see E. Bombieri, E. De Giorgi, E. Giusti, *Invent. Math* **7** (1969) 243–269).

References

Bernstein, S. "Über ein geometrisches Theorem und seine Anwendung auf die partiellen Differentialgleichungen vom elliptischen Typus," *Math. Zeit.* vol. 26 (1927) pp. 151-158.

The Partial Differential Equation $u_t + uu_x = \mu_{xx}$*

By EBERHARD HOPF
Department of Mathematics, Indiana University

1. *Introduction*

In the last decades, mathematicians have become increasingly interested in problems connected with the behavior of the solutions of partial differential space-time systems in which the highest order terms occur linearly with small coefficients. These problems have originated from physical applications, mainly from modern fluid dynamics (compressible fluids of small viscosity μ and of small heat conductivity λ). Research in these fields has led to some general mathematical conjectures, such as the following two: The solution of the initial value problem (the solution is prescribed at $t = 0$) for the general equations of fluid flow tends in general, i.e. for "most" values of the space-time-coordinates, towards a limit function as $\mu \to 0$ and $\lambda \to 0$. The limit function is, in general, discontinuous and pieced together by solutions of the system in which those highest order coefficients have the value zero (ideal fluid with contact- and shock-discontinuities). These conjectures are probably valid in a much wider range of partial differential systems. The second one is restricted to non-linear systems, but it seems to point out a typical occurrence in this general case. Exact formulation and rigorous proof of these conjectures are still tasks for the future. These problems are closely tied up with the present or future theory of functional spaces. Continued study of special problems is still a commendable way towards greater insight into this matter.

Among the partial differential systems studied in these directions, we have never met one in which the totality of its solutions has been rigorously determined and in which those limit problems can thus be studied in all detail. In this paper we present such a complete solution for the case of the equation

$$(1) \qquad u_t + uu_x = \mu u_{xx}, \qquad \mu > 0.$$

It was first introduced by J. M. Burgers[1] as the simplest model for the differ-

*Prepared under Navy Contract N6onr-180, Task Order No. 5, with Indiana University.

[1](a) Application of a model system to illustrate some points of the statistical theory of free turbulence, *Proc. Acad. Sci. Amsterdam*, Volume 43 (1940), pp. 2–12. (b) A mathematical model illustrating the theory of turbulence; *Advances in Applied Mechanics*, edited by R. v. Mises and T. v. Kármán, Volume 1, 1948, pp. 171–199, in particular pp. 182–184.

Burgers treats in (b) the equation $v_t + 2vv_y = \nu v_{yy}$ which goes over into equation (1) by

ential equations of fluid flow. There is a close analogy between the left of (1) and the terms $u_{i,t} + u_\nu u_{i,\nu}$, and between the right of (1) and the viscosity terms $\mu u_{i,\nu\nu}$ of those equations. However, no additional dependent variables such as pressure, density or temperature appear in (1). Nonetheless Burgers observed analogies between certain solutions of (1) and certain one-dimensional flows of a compressible fluid. He had an intuitive picture of the limit case $\mu \to 0$ in the solutions of (1) and determined the origin and the law of propagation of a discontinuity. Like Burgers we study the boundary-free initial value problem: u given for all x at $t = 0$, u wanted for all x and all $t > 0$. The solution is achieved by an exact integration of (1).[2] Both problems, the behavior of the solutions as $t \to \infty$ while μ stays constant, and their behavior as $\mu \to 0$ while the initial values are kept fixed, are solved. The second problem turns out to be connected with the support lines of a plane (in general non-convex) curve. A fact brought out by the discussion is that the reversal of order in the successive limit passages $t \to \infty$, $\mu \to 0$ leads to different results. From a general point of view this is not surprising, but it is a reminder that, for the turbulence problem of hydrodynamics (behavior of the flow of a slightly viscous fluid as $t \to \infty$), the relevant order is the one stated. The limit function obtained if $\mu \to 0$ is in general discontinuous. Under the wide assumptions we have made about the initial values, the points x, t of discontinuity could be everywhere dense; but we show that they are always lined up along curves $x(t)$ that have certain differentiability properties (lines of discontinuity) and that the points of continuity on the limit surface $u = u(x, t)$ are lined up along characteristics of (1), $\mu = 0$. A line of discontinuity can only originate but never terminate as t increases. That the irreversibility is retained in the limit case $\mu \to 0$ of the solutions was already duly emphasized by Burgers (dissipation of energy in discontinuities). This is analogous to the apparent irreversibility of the general flow of an ideal fluid with its shock-discontinuities.

A question of general interest (considered in Section 8) is that of the natural functional equations which determine the limit functions of the solutions ($\mu \to 0$, $\lambda \to 0$ in fluid flow, $\mu \to 0$ in the solutions of (1)) directly and independently of that limit passage in the solutions of the higher order system. The answer to this question is a prerequisite to a mathematical theory of the general initial value problem of ideal fluid flow. We use the case of (1) to illustrate the answer. We put (1) into the following form (the double integrals are extended over the semiplane $t > 0$): The equation

---

the substitution $u = 2v$, $x = y$, $\mu = \nu$. We doubt that Burgers' equation fully illustrates the statistics of free turbulence. Kolmogoroff's idea about the probability distribution of the turbulent fluctuations in the small is essentially concerned with the velocity differences, not the velocities themselves. Equation (1) is too simple a model to display chance fluctuations of these differences.

[2]The reduction of (1) to the heat equation was known to me since the end of 1946. However, it was not until 1949 that I became sufficiently acquainted with the recent development of fluid dynamics to be convinced that a theory of (1) could serve as an instructive introduction into some of the mathematical problems involved.

$$\iint \left[uf_t + \frac{u^2}{2} f_x + \mu u f_{xx} \right] dx\, dt = 0$$

is to be satisfied by each function $f(x, t)$ of class C'' in $t > 0$ that vanishes outside some circle contained in $t > 0$. It can be shown that this problem has essentially the same solutions as (1) provided that $\mu > 0$. Nothing is thus gained by this transformation of the differential problem. In Section 8 we rigorously prove, however, that the limit functions u of the solutions of (1) obtained as $\mu \to 0$ satisfy the relation

$$\iint \left[uf_t + \frac{u^2}{2} f_x \right] dx\, dt = 0$$

(for each f as specified) which is simply the case $\mu = 0$ of the preceding relation. While the case $\mu = 0$ of the differential equation (1) cannot completely determine the limit functions, the indicated integral formulation of the problem grasps them. It is probable (though not proved by us) that, under very general assumptions about u, those limit functions are the only solutions of the problem in $t > 0$.

Analogous considerations must hold in a much wider class of such limit problems and certainly in the case of fluid dynamics. This concept of generalized solutions of differential systems is not new in itself. It has been successfully applied to other differential problems. But the modern concept of ideal fluid flow with all its discontinuities is perhaps the most striking example in which the classical differential description is insufficient and which demands the integral form of the fundamental equations for its complete and general description.

2. *The explicit solution for* $\mu > 0$

We introduce a new dependent variable $\varphi = \varphi(x, t)$ into Burgers' equation

$$u_t = \left(\mu u_x - \frac{u^2}{2} \right)_x \tag{2}$$

by means of the substitution

$$\varphi = \exp\left\{ -\frac{1}{2\mu} \int u\, \partial x \right\} \tag{3}$$

whose inverse is

$$u = -2\mu(\log \varphi)_x = -2\mu\, (\varphi_x/\varphi). \tag{4}$$

Then, (1) becomes

$$-2\mu(\log \varphi)_{xt} = -2\mu(\log \varphi)_{tx} = -2\mu\left(\frac{\varphi_t}{\varphi}\right)_x = -\left(2\mu^2 \frac{\varphi_{xx}}{\varphi}\right)_x,$$

or upon integration with respect to x

$$\varphi_t = \mu\varphi_{xx} + C(t)\varphi,$$

where C is a suitable function of t only. If

$$\varphi \cdot \exp\left\{-\int C\, dt\right\}$$

is introduced as a new dependent variable φ one simply obtains the heat equation

$$\varphi_t = \mu\varphi_{xx} . \tag{5}$$

Precisely stated: If u solves (1) in an open rectangle R of the x, t-plane and if u, u_x, u_{xx} are continuous in R then there exists a positive function φ of the form (3) that solves the heat equation in R and for which φ, φ_x, φ_{xx}, φ_{xxx} are continuous in R. One easily shows that, conversely, every positive solution φ of (5) with the mentioned properties goes, by means of (4), over into a solution of (1) of the described general type. Let us call a function u that solves (1) in an x, t-domain D *a regular solution in* D if u, u_x, u_{xx} (and consequently u_t) are continuous in D.

Theorem 1. *Suppose that $u_0(x)$ is integrable in every finite x-interval and that*

$$\int_0^x u_0(\xi)\, d\xi = o(x^2) \tag{6}$$

for $|x|$ large. Then

$$u(x,t) = \frac{\displaystyle\int_{-\infty}^{\infty} \frac{x-y}{t} \exp\left\{-\frac{1}{2\mu} F(x, y, t)\right\} dy}{\displaystyle\int_{-\infty}^{\infty} \exp\left\{-\frac{1}{2\mu} F(x, y, t)\right\} dy}, \tag{7}$$

where

$$F(x, y, t) = \frac{(x-y)^2}{2t} + \int_0^y u_0(\eta)\, d\eta, \tag{8}$$

is a regular solution of (1) *in the half plane $t > 0$ that satisfies the initial condition*

$$\int_0^x u(\xi, t)\, d\xi \to \int_0^a u_0(\xi)\, d\xi \quad \text{as} \quad x \to a,\ t \to 0, \tag{9}$$

for every a. If, in addition, $u_0(x)$ is continuous at $x = a$ then

$$u(x, t) \to u_0(a) \quad \text{as} \quad x \to a,\ t \to 0. \tag{10}$$

A solution of (1) *which is regular in some strip $0 < t < T$ and which satisfies* (9) *for each value of the number a necessarily coincides with* (7) *in the strip.*

The condition (6) on the initial values merely insures the existence of the solution for all $t > 0$. The weaker condition

$$\int_0^x u_0(\xi)\, d\xi = O(x^2)$$

entails the existence and regularity of the solution only in some finite strip $0 < t < T$. The example of the solution $u = x/(t - T)$ of (1) shows that this restriction of the conclusion is natural. The exact hydrodynamic equations for a homogeneous and incompressible fluid have, by the way, analogous solutions $u_i = a_{i\nu}x_\nu/(t - T)$, $a_{ik} = a_{ki}$, $a_{\nu\nu} = 0$. Note the generality and naturalness of the uniqueness statement. No restriction is imposed upon the behavior of $u(x, t)$ for $|x|$ large except, of course, in the initial case $t = 0$.

Proof of Theorem 1. The continuous function

$$\varphi_0(y) = \exp\left\{-\frac{1}{2\mu}\int_0^y u_0(\eta)\, d\eta\right\} \tag{11}$$

is, in virtue of (6), of the order

$$\varphi_0(y) = e^{o(x^2)}$$

for large $|x|$. The integral

$$\varphi(x, t) = \frac{1}{\sqrt{4\pi\mu t}}\int_{-\infty}^{\infty} \varphi_0(y) \exp\left\{-\frac{(x - y)^2}{4\mu t}\right\} dy \tag{12}$$

therefore converges for all x and all $t > 0$. The formal derivatives of any order with respect to x, t are again absolutely convergent integrals for $t > 0$ which represent continuous functions of x, t. Therefore φ has, in $t > 0$, continuous partial derivatives of any order which are obtained by differentiation under the integral sign. It is well known that, for $t > 0$, $\varphi(x, t)$ solves (5) and satisfies the initial condition

$$\lim_{\substack{x=a\\ t=0}} \varphi(x, t) = \varphi_0(a)$$

for every a. For $t > 0$, φ is positive. The function (7)

$$u(x, t) = -2\mu \frac{\varphi_x(x, t)}{\varphi(x, t)}$$

solves, therefore, (1) in the half plane $t > 0$ and has continuous partial derivatives of all orders. It evidently satisfies (9) for each value of a.

Suppose, now, that in addition to (6) $u_0(x)$ is continuous at $x = a$. The formula

$$\varphi_x(x, t) = \frac{1}{\sqrt{4\pi\mu t}}\int_{-\infty}^{\infty} \varphi_0'(y) \exp\left\{-\frac{(x - y)^2}{4\mu t}\right\} dy \tag{13}$$

is valid for all x and all $t > 0$. It can, for instance, be obtained from (12) by differentiating under the integral sign and integrating by parts the differentiated integral. Integration by parts is legitimate under the assumption that $u_0(x)$ is integrable in every x-interval and that it satisfies (6). If, at $x = a$, $u_0(x)$ is continuous then $\varphi_0'(x)$ is continuous also. Hence

$$\lim_{\substack{x=a \\ t=a}} \varphi_x(x, t) = \varphi_0'(a)$$

and therefore

$$u(x, t) = -2\mu \frac{\varphi_x(x, t)}{\varphi(x, t)} \longrightarrow -2\mu \frac{\varphi_0'(a)}{\varphi_0(a)} = u_0(a)$$

as $x \longrightarrow a$, $t \longrightarrow 0$.

It remains to prove the uniqueness statement. Suppose that $u(x, t)$ is a regular solution of (1) in $0 < t < T$ and that it satisfies (9) for each value of a. We know that

$$u = -2\mu(\varphi_x/\varphi) \tag{14}$$

where $\varphi = \varphi(x, t)$,

$$\varphi(x, t) = \varphi(0, t)\psi(x, t), \qquad \psi(x, t) = \exp\left\{-\frac{1}{2\mu}\int_0^x u(\xi, t)\, d\xi\right\}, \tag{15}$$

is a positive and regular solution of (5) in this strip. $\varphi(0, t)$ is continuous and positive for $t > 0$. That it has a positive limit as $t \longrightarrow 0$ is not obvious. This we prove in the following way. From (9) we infer that

$$\lim_{t=0} \psi(y, t) = \psi_0(y) = \exp\left\{-\frac{1}{2\mu}\int_0^y u_0(\xi)\, d\xi\right\} \tag{16}$$

holds uniformly in every finite y-interval. We apply a theorem on non-negative solutions of the heat equation, which was proved by D. V. Widder[3] in 1944. A solution of (5) which is positive and regular in a strip $\alpha < t < \beta$ and continuous on $t = \alpha$ is uniquely determined by the values on $t = \alpha$ and is, in the strip, represented by the classical integral formula. Applying this theorem to $\varphi(x, t + \epsilon)$ in the strip $0 < t < T - \epsilon$, $\epsilon > 0$, we find that

$$\varphi(0, t + \epsilon)\psi(x, t + \epsilon) = \frac{1}{\sqrt{4\pi\mu t}}\int_{-\infty}^{\infty} \varphi(0, \epsilon)\psi(y, \epsilon) \exp\left\{-\frac{(x - y)^2}{4\mu t}\right\} dy \tag{17}$$

holds for $0 < t < T$. Let

$$\varphi_0 = \limsup_{\epsilon=0} \varphi(0, \epsilon), \qquad 0 \leqq \varphi_0 \leqq \infty,$$

---

[3]D. V. Widder, Positive temperatures on an infinite rod, *Trans. Am. Math. Soc.*, Volume 55 (1944), pp. 85–95.

and let ϵ_ν be a sequence of positive numbers satisfying

$$\epsilon_\nu \to 0, \qquad \varphi(0, \epsilon_\nu) \to \varphi_0 . \tag{18}$$

Let now $\epsilon = \epsilon_\nu$ and $\nu \to \infty$ in (17) while x, t are kept fixed. Since

$$\int_{-\infty}^{\infty} \geqq \int_{-A}^{A}$$

we infer from the uniformity of (16) in the interval $(-A, A)$, which we keep fixed, and from (18) that

$$\varphi(0, t)\psi(x, t) \geqq \frac{\varphi_0}{\sqrt{4\pi\mu t}} \int_{-A}^{A} \psi_0(y) \exp\left\{-\frac{(x-y)^2}{4\mu t}\right\} dy. \tag{19}$$

This shows that $0 \leqq \varphi_0 < \infty$. Inequality (19) holds for arbitrary A and the integral converges, in view of (6), also for $A = \infty$. Hence (19) is true for $A = \infty$. As $\psi_0(y)$ is continuous, the right hand side converges, for a fixed value of x, towards $\varphi_0\psi_0(x)$ as $t \to 0$. Hence, applying (16) to the left hand side of (19), we conclude that

$$[\liminf_{t=0} \varphi(0, t)]\psi_0(x) \geqq \varphi_0\psi_0(x).$$

Remembering the definition of φ_0, we see that

$$\lim_{t=0} \varphi(0, t) = \varphi_0$$

exists. According to (15) the non-negative solution $\varphi(x, t)$ of (5) is therefore continuous for $0 \leqq t < T$. The application of Widder's theorem to the full strip $0 < t < T$ is now permitted. The result is that $\varphi(x, t)$ is uniquely determined by the initial values $\varphi_0\psi_0(x) = \text{const.}\ \psi_0(x)$ where $\psi_0(x)$ is a given function. $\varphi(x, t) > 0$ now implies that $\varphi_0 > 0$. $\varphi(x, t)$ is, up to a constant factor, uniquely determined. $u(x, t)$ is, therefore, completely unique.

3. *Behavior of the solutions as* $t \to +\infty$, $\mu > 0$

In this section we consider only initial data $u_0(x)$ for which the integral

$$M = \int_{-\infty}^{\infty} u_0(x)\, dx$$

exists as a sum of two improper integrals: $\int_0^\infty + \int_{-\infty}^0$. Burgers calls it the moment (at time $t = 0$) of the velocity distribution. The existence of M means for the function $\varphi_0(x)$, defined by (4), that the limits $\varphi_0(-\infty)$ and $\varphi_0(\infty)$ both exist (they are > 0) and that

$$2\mu \log \frac{\varphi_0(-\infty)}{\varphi_0(\infty)} = M. \tag{20}$$

The solution $\varphi(x, t)$ of (5) with the initial values $\varphi_0(x)$ can be expressed by the equivalent formula

$$\tag{21} \begin{aligned} \varphi(\bar{x}\sqrt{2\mu t}, t) &= \frac{1}{\sqrt{2\pi}} \int_{-\infty}^{\infty} \varphi_0(\sqrt{2\mu t}\, y) \exp\left\{-\frac{(\bar{x}-y)^2}{2}\right\} dy \\ &= \frac{1}{\sqrt{2\pi}} \int_{-\infty}^{\infty} \varphi_0(\sqrt{2\mu t}\,(\bar{x}-y)) \exp\left\{-\frac{y^2}{2}\right\} dy. \end{aligned}$$

If we let $\bar{x} \to \infty$ while t is kept fixed, the extreme right hand side evidently converges toward

$$\frac{\varphi_0(\infty)}{\sqrt{2\pi}} \int_{-\infty}^{\infty} \exp\left\{-\frac{y^2}{2}\right\} dy = \varphi_0(\infty).$$

An analogous result is found if $\bar{x} \to -\infty$. The limits $\varphi(-\infty, t)$, $\varphi(\infty, t)$ therefore exist and their values are independent of t. By virtue of (4) it follows: *The solution $u(x, t)$ of* (1) *with the initial values $u_0(x)$ in the sense of theorem* 1 *possesses a moment*

$$\int_{-\infty}^{\infty} u(x, t)\, dx$$

for every $t > 0$ if it possesses one for $t = 0$ (i.e. for $u = u_0$). The value of the moment is that assumed at $t = 0$. The moment is an "integral" of the differential equation (1).

It is easily seen what happens in the limit $t \to \infty$ as $\bar{x}$ and μ stay fixed. If the last integral in (21) is split up into

$$\int_{-\infty}^{\infty} = \int_{-\infty}^{\bar{x}-\epsilon} + \int_{\bar{x}+\epsilon}^{\infty} + \int_{\bar{x}-\epsilon}^{\bar{x}+\epsilon}, \qquad \epsilon > 0,$$

the φ_0-factor of the integrand behaves in the following way: In the first integral it tends uniformly toward $\varphi_0(\infty)$, in the second it tends uniformly to $\varphi_0(-\infty)$, while in the third term it is bounded by an upper bound for $\varphi_0(x)$. From this it easily follows that

$$\tag{22} \begin{aligned} &\sqrt{2\pi} \lim_{t=\infty} \varphi(\bar{x}\sqrt{2\mu t}, t) \\ &= \varphi_0(\infty) \int_{-\infty}^{\bar{x}} \exp\left\{-\frac{y^2}{2}\right\} dy + \varphi_0(-\infty) \int_{\bar{x}}^{\infty} \exp\left\{-\frac{y^2}{2}\right\} dy \end{aligned}$$

holds uniformly in $\bar{x}$. The limit relation differentiated with respect to $\bar{x}$,

$$\tag{23} \lim_{t=\infty} \sqrt{4\pi\mu t}\, \varphi_x(\bar{x}\sqrt{2\mu t}, t) = [\varphi_0(\infty) - \varphi_0(-\infty)] \exp\left\{-\frac{\bar{x}^2}{2}\right\},$$

also is readily shown to hold uniformly for all $\bar{x}$. One has only to differentiate the first integral in (21) and then to transform it by the substitution $y \to \bar{x} - y$.

The resulting integral is handled again like the second integral of (21). Since (22) is positively bounded from below, the limit

$$\lim_{t\to\infty} \sqrt{2\mu t}\, \frac{\varphi_x(\bar{x}\sqrt{2\mu t},\, t)}{\varphi(\bar{x}\sqrt{2\mu t},\, t)}$$

exists uniformly for all $\bar{x}$. This limit relation can again be differentiated any number of times, and the limit relations obtained hold uniformly with respect to $\bar{x}$ as is shown by perfectly analogous reasoning. According to (4) we have hereby proved the following:

Theorem 2. *Let $u(x, t)$ be a regular solution of* (1), $t > 0$, *of finite moment* $M = 2\mu K$. *Let*

$$\bar{x} = \frac{x}{\sqrt{2\mu t}}, \qquad \bar{u} = \sqrt{\frac{t}{2\mu}}\, u, \qquad \bar{u} = \bar{u}(\bar{x}\,, t), \tag{24}$$

and

$$G(\xi) = e^{-\frac{K}{2}} \int_{-\infty}^{\xi} e^{-\frac{y^2}{2}}\, dy + e^{\frac{K}{2}} \int_{\xi}^{\infty} e^{-\frac{y^2}{2}}\, dy. \tag{25}$$

Then the limit relation

$$\lim_{t\to\infty} \bar{u}(\bar{x}, t) = -\frac{G'(\bar{x})}{G(\bar{x})} \tag{26}$$

holds uniformly with respect to $\bar{x}$. This relation may be differentiated any number of times with respect to $\bar{x}$ and the differentiated relation again holds uniformly with respect to $\bar{x}$.

The transformation (24), if coupled with

$$\bar{t} = \log t, \tag{24'}$$

carries Burgers' equation over into

$$\frac{\partial \bar{u}}{\partial \bar{t}} + \bar{u}\frac{\partial \bar{u}}{\partial \bar{x}} = \frac{1}{2}\frac{\partial^2 \bar{u}}{\partial \bar{x}^2} + X \tag{27}$$

where

$$X = \frac{1}{2}\left(\bar{u} + \bar{x}\frac{\partial \bar{u}}{\partial \bar{x}}\right). \tag{27'}$$

The moment of a solution becomes

$$\int_{-\infty}^{\infty} \bar{u}(\bar{x}, t)\, d\bar{x} = \frac{M}{2\mu} = K.$$

Theorem 2 simply asserts that any solution of (27), (27′) of finite moment K tends toward the time-independent solution of this equation with the same moment K. These stationary solutions form a family with exactly one member for every value of K.

Precisely the same transformation (24), (24′) of the coordinates and velocities carriers the Navier-Stokes equations of a homogeneous ($\rho = 1$) and incompressible fluid occupying the entire space

$$\frac{\partial u_i}{\partial t} + u_\nu \frac{\partial u_i}{\partial x_\nu} = -\frac{\partial p}{\partial x_i} + \mu \frac{\partial^2 u_i}{\partial x_\nu \, \partial x_\nu}, \qquad \frac{\partial u_\nu}{\partial x_\nu} = 0$$

over into

$$\frac{\partial \bar{u}_i}{\partial \bar{t}} + \bar{u}_\nu \frac{\partial \bar{u}_i}{\partial \bar{x}_\nu} = -\frac{\partial \bar{p}}{\partial \bar{x}_i} + \frac{1}{2}\frac{\partial^2 \bar{u}_i}{\partial \bar{x}_\nu \, \partial \bar{x}_\nu} + X_i, \qquad \frac{\partial \bar{u}_\nu}{\partial \bar{x}_\nu} = 0$$

where

$$X_i = \frac{1}{2}\left(\bar{u}_i + \bar{x}_\nu \frac{\partial \bar{u}_i}{\partial \bar{x}_\nu}\right).$$

The transformation is of a similar importance for the spreading of a turbulent disturbance. However, the hydrodynamic case presents additional complications with no analogue in Burgers' oversimplified case.

It is of interest to uncover the mathematical reason for the significance of the transformation. The transformed equation $\partial \bar{u}/\partial \bar{t} = \cdots$ has the property that the right hand side does not explicitly contain the time variable in common with the original equation $\partial u/\partial t = \cdots$. In general this is not the case if the equation is transformed by an arbitrary transformation

$$\bar{x}(x, t, u), \qquad \bar{t}(x, t, u), \qquad \bar{u}(x, t, u).$$

Why it is the case for the special transformation (24), (24′) is made plain by the following observation which applies to the hydrodynamical case as well. The statement that the right hand side of $\partial \bar{u}/\partial \bar{t} = \cdots$ does not contain $\bar{t}$ explicitly is synonymous with the statement that the equation is invariant under the substitutions (indicated by an arrow)

$$\bar{t} \to \bar{t} + a. \tag{28}$$

Now (1) is carried into itself by the well-known group of substitutions

$$\begin{aligned} x &\to cx \\ t &\to c^2 t \\ u &\to c^{-1} u. \end{aligned} \tag{29}$$

If we introduce variables $\bar{x}$, $\bar{t}$, $\bar{u}$, then the transformed equation (1) must be invariant under the group (29) expressed in these new variables. The transformed equation will not contain $\bar{t}$ explicitly, if we choose $\bar{x}$, $\bar{t}$, $\bar{u}$ such that (29) has, in these new variables, the form of a translation in $\bar{t}$

$$\bar{x} \to \bar{x}$$

$$\bar{t} \to \bar{t} + a$$

$$\bar{u} \to \bar{u}.$$

This is obviously accomplished by choosing

$$\bar{x} = x/\sqrt{t}, \qquad \bar{u} = u\sqrt{t}, \qquad \bar{t} = \log t \tag{30}$$

which, essentially, is the transformation (24), (24′).

Using transformation (30) (which does not contain μ) rather than (24) our limit relation becomes

$$\lim_{t=\infty} \sqrt{t}\, u(\bar{x}\sqrt{t},\, t) = -\sqrt{2\mu}\,\frac{G'(\bar{x}/\sqrt{2\mu})}{G(\bar{x}/\sqrt{2\mu})} = -2\mu \frac{d}{d\bar{x}} \log G(\bar{x}/\sqrt{2\mu}), \tag{31}$$

where $K = M/2\mu$ and where M is the moment of the solution $u(x, t)$ of (1). The right hand side of (31), as a time-independent solution of equation (1) transformed by (30), had been studied by Burgers.[4] He found that its limit as $\mu \to 0$, while M stays fixed, is the discontinuous function $\alpha f(\bar{x}/\alpha)$ where $\alpha \mid \alpha \mid = 2M$ and

$$f(\xi) = \begin{cases} 0, & \xi < 0, \\ \xi, & 0 < \xi < 1, \\ 0, & \xi > 1. \end{cases}$$

4. *Passage to the limit* $\mu \to 0$

We proceed to treat rigorously the behavior of the solutions $u(x, t; \mu)$ of (1) as $\mu \to 0$ while the initial values $u_0(x)$ are kept fixed. It will be shown that the limit function $u(x, t; 0)$ exists, but is in general discontinuous. Away from the discontinuities it fulfills essentially differential equation (1), $\mu = 0$,

$$u_t + uu_x = 0. \tag{32}$$

More precisely, the characteristic equations of (32)

$$\frac{dx}{dt} = u, \qquad \frac{du}{dt} = 0 \tag{33}$$

(which, under somewhat stricter assumptions as to differentiability, are equivalent to (32)) are strictly fulfilled. A simple formula is derived that expresses $u(x, t; 0)$ in terms of $u_0(x)$ and that comprises, in particular, the results which Burgers obtained by intuitive reasoning.

[4]See Burgers, footnote 1(b), p. 183.

The formula makes use of certain properties of the function $F(x, y, t)$, $t > 0$, of (8) which we have to derive first. The properties refer to F as a function of y (which is the variable of integration in (7)) with x, t being fixed. F is continuous in y and has, according to (6), the property

$$\frac{F}{y^2} \to \frac{1}{2t} > 0, \qquad |y| \to \infty. \tag{34}$$

Hence F attains its smallest value for one or several values of y, the smallest and the largest of which are denoted by y_* and y^*, respectively,

$$y_*(x, t) \leqq y^*(x, t).$$

Lemma 1. *The functions y_* and y^* have the properties*

$$\text{(a)} \quad y^*(x, t) \leqq y_*(x', t) \qquad \text{if} \qquad x < x',$$

$$\text{(b)} \quad y_*(x - 0, t) = y_*(x, t), \qquad y^*(x + 0, t) = y^*(x, t),$$

$$\text{(c)} \quad y_*(+\infty, t) = +\infty, \qquad y^*(-\infty, t) = -\infty.$$

As a function of y,

$$G(x, y, t) = F(x, y, t) - x^2/2t \tag{35}$$

attains its absolute minimum at the same values of y as F. Since

$$G(x, y, t) = G(0, y, t) - xy/t \tag{36}$$

the lemma expresses properties of the points in the y, z-plane which the curve $z = G(0, y, t)$, with $t > 0$ fixed, has in common with the line of support $z = xy/t + c$ of a given slope x/t. As we saw, such a line that supports the curve from below exists for every value of the slope. $y_*(x, t)$ and $y^*(x, t)$ are the smallest and largest y-coordinates of the points which the line of support has in common with the curve.

Proof of the lemma. According to the definition of y^*

$$G(x, y, t) - G(x, y^*, t) \begin{cases} \geqq 0, & y < y^*, \\ > 0, & y > y^*, \end{cases} \tag{37}$$

where

$$y^* = y^*(x, t). \tag{38}$$

From (36) one obtains

$$G(x + a, y, t) - G(x + a, y^*, t) = G(x, y, t) - G(x, y^*, t) - \frac{a}{t}(y - y^*). \tag{39}$$

We identify y^* with (38) and let

$$a > 0.$$

Now the upper inequality (37) shows that the right hand side of (39) and, therefore, the left hand side, is positive for all $y < y^*$. As it vanishes at $y = y^*$ we infer that $G(x + a, y, t)$ as a function of y can attain its smallest value only for $y \geqq y^*$. In virtue of (35) this establishes property (a) if we let $x' = x + a$.

According to the lower inequality (37),

$$\frac{G(x, y, t) - G(x, y^*, t)}{y - y^*}$$

is a positive and continuous function of y if $y > y^*$. By (34) and (35) this expression tends to $+\infty$ as $y \to +\infty$. Hence, if $\epsilon > 0$ is chosen arbitrarily, this function has a positive minimum for $y \geqq y^* + \epsilon$. $a > 0$ can, therefore, be chosen such that the right hand side of (39) is positive whenever $y > y^* + \epsilon$. With this fixed value of a, the function of y on the left of (39) is thus positive outside the interval $y^* \leqq y \leqq y^* + \epsilon$ and vanishes at $y = y^*$. Hence $G(x + a, y, t)$ as a function of y can reach its minimum only in this interval, in other words $y^*(x + a, t)$ is not greater than $y^*(x, t) + \epsilon$ if $a > 0$ is sufficiently small. This proves the second of the properties (b). The first property (b) follows immediately by applying the second one to the function $F(-x, -y, t)$.

Only the first property (c) needs to be proved. Let m denote the smallest value of $G(0, y, t)$ for a fixed $t > 0$. The function of y

$$G(0, y, t) - m - \frac{x}{t}(y - A), \qquad \frac{x}{t} = G(0, A + 1, t) - m$$

is $\geqq 0$ for $y < A$ and $=0$ for $y = A + 1$. Hence it attains its smallest value at some $y \geqq A$. According to (36), the function $G(x, y, t)$ where x has the indicated value must have the same property, in other words $y^*(x, t) \geqq A$ must hold for this value of x. Since A is arbitrary, it follows that $y^*(x, t)$ as a function of x takes on arbitrarily large values. The desired property then follows from the fact that $y^*(x, t)$ is a nowhere decreasing function of x and from property (a). This concludes the proof of lemma 1.

(a) and (b) signify that $y_*(x, t)$ and $y^*(x, t)$ are two nowhere decreasing functions of x with the same points x of discontinuity and with the same limit value on the left of x and the same limit value on the right of x. These limit values are $y_*(x, t)$ and $y^*(x, t)$ respectively. From the fact that a monotonic function has only a denumerable set of discontinuities, we infer: at any given moment $t > 0$, $y_*(x, t) = y^*(x, t)$ holds for all x with the possible exception of a denumerable set of values of x where $y_* < y^*$.

Lemma 2. *The minimum of $F(x, y, t)$ if x, t are fixed,*

$$m(x, t) = F(x, y_*(x, t), t) = F(x, y^*(x, t), t)$$

is a continuous function of x, t in the semiplane $t > 0$.

Proof. Let x', t' be an arbitrary fixed point in $t > 0$ and put

$$y_{\prime} = y_*(x', t'), \qquad y' = y^*(x', t'), \qquad m' = F(x', y_{\prime}, t') = F(x', y', t').$$

(34) and the continuity of F in y imply the existence of a positive number p such that

$$F(x', y, t') > m' + p \quad \text{whenever} \quad y < y_{\prime} - 1 \quad \text{or} \quad y > y' + 1.$$

F is continuous in all three variables and

$$F(x, y, t) \to +\infty \quad \text{as} \quad |y| \to \infty \tag{40}$$

holds uniformly with respect to x, t in the neighborhood of x', t'. Hence there exists a positive number q, $q < t'$, such that

$$F(x, y, t) > m' + (p/2)$$

holds whenever

$$y < y_{\prime} - 1 \text{ or } y > y' + 1, \qquad \text{and} \qquad |x - x'| + |t - t'| \leqq q.$$

On the other hand, there certainly exists an $r > 0$, $r < q$, such that

$$F(x, y', t) < m' + \frac{p}{2} \qquad \text{whenever } |x - x'| + |t - t'| \leqq r.$$

Hence $F(x, y, t) - m'$, as a function of y, attains its minimum in the interval

$$y_{\prime} - 1 \leqq y \leqq y' + 1 \qquad \text{whenever} \qquad |x - x'| + |t - t'| \leqq r.$$

We therefore may confine ourselves to the closed set

$$y_{\prime} - 1 \leqq y \leqq y' + 1, \qquad |x - x'| + |t - t'| \leqq r$$

in the space of x, y, t and to min F, as x, t are kept fixed, within this set. The fact that this function min F is continuous at $x = x'$, $t = t'$ is readily proved.

Lemma 3. *As functions of x and t, $y_*(x, t)$ and $y^*(x, t)$ are lower- and upper-semicontinuous, respectively, in the semiplane $t > 0$. At a point where $y_* = y^*$ both functions are continuous.*

Proof. Let x', t' be an arbitrary fixed point in $t > 0$. Let the point (x, t) run through a sequence of points that converge toward (x', t'). Since (40) holds uniformly with respect to x, t in a neighborhood of (x', t'), the values of y_* (and of y^* as well) along this sequence must be bounded. Denote by $y_{\prime}$ the limes inferior of the values y_* along the sequence. If in the inequality

$$F(x, y, t) \geqq F(x, y_*(x, t), t)$$

y is kept fixed, and if (x, t) runs toward (x', t') along a suitable subsequence, one infers that

$$F(x', y, t') \geqq F(x', y_{\prime}, t')$$

must hold for every y. From the definition of y_* and y^* it therefore follows that $y_*(x', t') \leqq y_{\prime}$. Hence $y_*(x, t)$ is lower-semicontinuous at x', t'. The statement concerning y^* is proved in the same way. This concludes the proof of lemma 3.

Theorem 3. *Let $u(x, t; \mu)$, $t > 0$, be the solution of* (1) *having arbitrarly given initial values $u_0(x)$ which satisfy* (6). *Let $y_*(x, t)$ and $y^*(x, t)$ be determined from these initial values as stated above. Then, for every x and $t > 0$,*

$$\frac{x - y^*(x, t)}{t} \leqq \liminf_{\substack{\mu=0\\ \xi=x\\ \tau=t}} u(\xi, \tau; \mu) \leqq \limsup_{\substack{\mu=0\\ \xi=x\\ \tau=t}} u(\xi, \tau; \mu) \leqq \frac{x - y_*(x, t)}{t}.$$

In particular,

$$\lim_{\substack{\mu=0\\ \xi=x\\ \tau=t}} u(\xi, \tau; \mu) = \frac{x - y_*(x, t)}{t} = \frac{x - y^*(x, t)}{t}$$

holds at every point, $t > 0$, in which $y_* = y^*$.

Proof. Obviously, we may write[5]

$$u(\xi, \tau; \mu) = \frac{\displaystyle\int_{-\infty}^{\infty} \frac{\xi - y}{\tau} \exp\left\{- \frac{P(\xi, y, \tau)}{\mu}\right\} dy}{\displaystyle\int_{-\infty}^{\infty} \exp\left\{- \frac{P(\xi, y, \tau)}{\mu}\right\} dy} \tag{41}$$

where

$$P(\xi, y, \tau) = \tfrac{1}{2}[F(\xi, y, \tau) - m(\xi, \tau)]$$

and where

$$m(\xi, \tau) = F(\xi, y_*(\xi, \tau), \tau) = F(\xi, y^*(\xi, \tau), \tau)$$

is the smallest value of $F(\xi, y, \tau)$ with ξ, τ being fixed. P has the properties

$$\begin{aligned} &P(\xi, y, \tau) > 0 \quad \text{as} \quad y < y_*(\xi, \tau) \quad \text{and as} \quad y > y^*(\xi, \tau), \\ &P(\xi, y, \tau) = 0 \quad \text{as} \quad y = y_*(\xi, \tau) \quad \text{and as} \quad y = y^*(\xi, \tau). \end{aligned} \tag{42}$$

P is, in virtue of lemma 2, continuous in ξ, y, τ if $\tau > 0$ and

$$\lim_{|y|=\infty} y^{-2} P(\xi, y, \tau) = \frac{1}{4\tau} \tag{43}$$

holds uniformly with respect to ξ, τ in every closed set in $\tau > 0$.

We choose arbitrarily a point x, t in $t > 0$ and put in this fixed point

$$Y_* = y_*(x, t), \qquad Y^* = y^*(x, t).$$

We also choose arbitrarily a number $\epsilon > 0$. We take the positive numbers a, b, $a < t$, so small that

---

[5]We have avoided the application of the asymptotic Laplace-formula to numerator and denominator of this expression because that would have required an assumption which does not lie in the nature of things.

The form of the expression suggests a probability interpretation as Max A. Zorn remarked to the author. We presume that this interpretation will be brought out by the study of the stochastic process back of (1).

$$(44)\qquad l = \frac{x - Y^*}{t} - \epsilon < \frac{\xi - y}{\tau} < \frac{x - Y_*}{t} + \epsilon = L$$

holds whenever ξ, y, τ satisfy

$$(45)\qquad |\xi - x| + |\tau - t| < a \qquad \text{and} \qquad Y_* - 2b < y < Y^* + 2b.$$

By lemma 3, the number a can at the same time be chosen such that

$$(46)\qquad Y_* - b < y_*(\xi, \tau) \leqq y^*(\xi, \tau) < Y^* + b$$

holds whenever ξ, τ satisfy

$$(47)\qquad |\xi - x| + |\tau - t| < a.$$

We keep the numbers a, b fixed. The numerator on the right of (41) is, by (44) and (45), greater than

$$l \int_{-\infty}^{\infty} \exp\{\}\, dy + \int_{-\infty}^{Y_*-2b} \left(\frac{\xi - y}{\tau} - l\right) \exp\{\}\, dy + \int_{Y^*+2b}^{\infty} \left(\frac{\xi - y}{\tau} - l\right)$$

and less than

$$L \int_{-\infty}^{\infty} \exp\{\}\, dy + \int_{-\infty}^{Y_*-2b} \left(\frac{\xi - y}{\tau} - L\right) \exp\{\}\, dy$$

$$+ \int_{Y^*+2b}^{\infty} \left(\frac{\xi - y}{\tau} - L\right) \exp\{\}\, dy$$

whenever ξ, τ satisfy (47). The absolute values of the brackets (. . .) in the second and third terms can be majorised by

$$\omega(Y_* - y) \qquad \text{and} \qquad \omega(y - Y^*),$$

respectively, where ω is independent of ξ, τ and μ as long as (47) is satisfied. Clearly, the main part of theorem 3 will be proved if we can show that the two expressions

$$(48)\qquad \frac{\displaystyle\int_{\infty}^{Y_*-2b} (Y_* - y) \exp\{\}\, dy}{\displaystyle\int_{-\infty}^{y_*(\xi,\tau)} \exp\{\}\, dy}, \qquad \frac{\displaystyle\int_{Y^*+2b}^{\infty} (y - Y^*) \exp\{\}\, dy}{\displaystyle\int_{y^*(\xi,\tau)}^{\infty} \exp\{\}\, dy}$$

tend to zero as $\mu \to 0$ uniformly with respect to ξ, τ provided that (47) is fulfilled. Since both expressions can be treated in the same way, we may confine ourselves to the second one. With the exponential expression

$$\exp\left\{-\frac{P(\xi, y, \tau)}{\mu}\right\}$$

we deal in the following way. By virtue of (42), (43) and (46)

$$\frac{P(\xi, y, \tau)}{(y - Y^*)^2}$$

is positively bounded from below if ξ, y, τ satisfy $y > Y^* + 2b$ and (47) (remember that (46) is then fulfilled). If $A/2$ is a positive lower bound, we infer that the numerator of the second ratio (48) is less than

$$\int_{Y^*+2b}^{\infty} (y - Y^*) \exp\left\{-\frac{A}{2\mu}(y - Y^*)^2\right\} dy = \frac{\mu}{A} \exp\left\{-\frac{2Ab^2}{\mu}\right\}$$

if ξ, τ satisfy (47). On the other hand, the uniform continuity of P and (42) imply the existence of a fixed $\delta > 0$ such that $P < 2Ab^2$ holds whenever $y^*(\xi, \tau) < y < y^*(\xi, \tau) + \delta$ and $| \xi - x | + | \tau - t | < a$. Hence the denominator of the second ratio (48) is greater than the integral extended from y^* to $y^* + \delta$ which, in turn, is greater than

$$\delta \exp\left\{-\frac{2Ab^2}{\mu}\right\}$$

whenever (47) is satisfied. The second ratio is, therefore, less than $\mu/A\delta$ whenever ξ, τ satisfy (47). Hence we infer the truth of the statement concerning (48). Theorem 3 is thus proved.

We now define the function

$$u(x, t) = \lim_{\mu=0} u(x, t; \mu)$$

in every point x, t, $t > 0$, in which this limit exists. By virtue of theorem 3 this is the case in every point where $y_* = y^*$. These points constitute the normal case insofar as, for any given $t > 0$, the exceptional points form a denumerable set. Theorem 3 moreover implies that at any normal point, $t > 0$, $u(x, t)$ is not only defined but also continuous in both variables. From theorem 3 and from property (b) of lemma 1 one infers that generally

$$u(x - 0, t) = \frac{x - y_*(x, t)}{t}, \qquad u(x + 0, t) = \frac{x - y^*(x, t)}{t} \tag{49}$$

holds in each point of the semiplane $t > 0$. In particular,

$$u(x - 0, t) \geqq u(x + 0, t)$$

holds everywhere in $t > 0$. At a point in which these two limits coincide, $u(x, t)$ is defined and continuous in both variables.

Theorem 4. *At each point of $t > 0$ the inequalities*

$$\limsup_{y=y_*-0} \frac{\int_{y_*}^{y} u_0(\eta)\, d\eta}{y - y_*} \leqq u(x - 0, t) \leqq \liminf_{y=y_*+0} \frac{\int_{y_*}^{y} u_0(\eta)\, d\eta}{y - y_*}$$

hold if $y_* = y_*(x, t)$. These inequalities stay valid if y_* is replaced by y^* and $u(x - 0, t)$ by $u(x + 0, t)$.

Proof. This is an immediate consequence of theorem 3, of the fact that (8) as a function of y attains its minimum at $y = y_*$ and at $y = y^*$, and of (49).

If $u_0(y)$ has left and right limits at $y = y_*$ as well as at $y = y^*$ the inequalities are

$$u_0(y_* - 0) \leqq u(x - 0, t) \leqq u_0(y_* + 0)$$

and

$$u_0(y^* - 0) \leqq u(x + 0, t) \leqq u_0(y^* + 0).$$

In particular, we have $u_0(y_*) = u(x, t)$ if $y_* = y^*$ and if u_0 is continuous at $y = y_*$.

5. *Continued study of the limit function u(x,t)*

We now examine the relations between the limit function $u(x, t)$ and equation (32) which is the case $\mu = 0$ of Burgers' equation (1).

The characteristics of (32) are the solutions of (33),

$$x = a + bt, \qquad u = b.$$

They are the straight lines in the space of x, t, u which are parallel to the x, t-plane and whose x, t-projections have a slope dx/dt equal to the distance from that plane. We shall speak of a (generalized) solution $\tilde{u}(x, t)$ of (32) in an open set R at the x, t-plane, if $\tilde{u}(x, t)$ is single-valued in R and if, above R, each segment of a characteristic that has a point in common with the surface $u = \tilde{u}$ belongs entirely to it. If this is the case, the x, t-projections of the characteristic segments obviously form a continuous field in R. Conversely, a set of characteristic segments above R forms a single-valued function $\tilde{u}$ (and hence a generalized solution) in R if that field condition is satisfied by the x, t-projections in R. If $\tilde{u}$ is a generalized solution of class C' in R it actually solves (32) in R. However, the differentiability question which offers no difficulty in our case, will be left aside.

Returning to our particular function $u(x, t)$ we consider an arbitrary point x_1, t_1 in the semiplane $t > 0$ and let $y_1 = y_*(x_1, t_1)$, $y^1 = y^*(x_1, t_1)$. To this point we attach the two characteristics

$$(50_,) \qquad x = x_1 + \frac{x_1 - y_1}{t_1}(t - t_1), \qquad u = \frac{x_1 - y_1}{t_1}$$

and

$$(50') \qquad x = x_1 + \frac{x_1 - y^1}{t_1}(t - t_1), \qquad u = \frac{x_1 - y^1}{t_1}.$$

(50,) and (50′) are characterized by either one of the two following properties: They pass through the two limit-points x_1, t_1, u_1 and x_1, t_1, u^1 of the surface, respectively, where $u_1 = u(x_1 - 0, t_1)$ and $u^1 = u(x_1 + 0, t_1)$ (see (49)). Their x, t-projections pass through x_1, t_1 and meet the line $t = 0$ at $x = y_1$ and $x = y^1$, respectively. We call their open segments $0 < t < t_1$ the characteristic segments and characteristic x, t-segments, respectively, belonging to x_1, t_1 (and to the surface). In the normal case where $y_1 = y^1$ (50,) and (50′) coincide and simply represent the segment $0 < t < t_1$ of the characteristic through the point $x_1, t_1, u(x_1, t_1)$ of the surface.

Theorem 5. *At every point x, t of a characteristic x, t-segment $u(x, t)$ is defined and continuous. It has the constant value $u = dx/dt$ along the segment, that is, the surface $u = u(x, y)$ contains the full characteristic segment. Two characteristic x, t-segments (no matter whether they belong to the same point or to different points) have either no point in common or one is part of the other (remember that the characteristic segments are defined as open segments $0 < t < t_1$).*

Proof. Let the segment belong to the point $x_1, t_1, t_1 > 0$. Suppose it is the segment formed with y_1 (the other one with y^1 is treated in exactly the same way). Let

$$u_1 = \frac{x_1 - y_1}{t_1}.$$

According to the definition of y_* by means of F of (8),

$$F(x_1, y, t_1) - F(x_1, y_1, t_1) = \int_{y_1}^{y} \left[u_0(\eta) + \frac{\eta - x_1}{t_1} \right] d\eta \geqq 0$$

for all y. Now, for a point x, t on the segment,

$$x = x_1 - (t_1 - t)u_1 = y_1 + tu_1 = \frac{tx_1 + (t_1 - t)y_1}{t_1}, \qquad 0 < t < t_1, \tag{51}$$

and hence

$$\frac{\eta - x}{t} = \frac{\eta - x_1}{t_1} + \frac{t_1 - t}{t_1 t}(\eta - y_1).$$

Therefore,

$$F(x, y, t) - F(x, y_1, t) = \int_{y_1}^{y} \left[u_0(\eta) + \frac{\eta - x}{t} \right] d\eta$$

$$= F(x_1, y, t_1) - F(x_1, y_1, t_1) + \frac{t_1 - t}{2t_1 t}(y - y_1)^2.$$

This proves that $F(x, y, t)$ as a function of y reaches its minimum only at $y = y_1$ if x, t is an arbitrary point on the characteristic x, t-segment belonging to x_1, t_1, that is, that $y_*(x, t) = y^*(x, t) = y_1$ holds along this segment. By virtue of the remarks made at the end of the preceding paragraph and of theorem 3, $u(x, t)$ is continuous at all points of the segment and it has the value

$$u(x, t) = (x - y_1)/t$$

along it. In consequence of (51), this value is constant and equal to $u_1 = dx/dt$.

The last statement of the theorem remains to be proved. Consider two characteristic x, t-segments. They may belong to two different points x_1, t_1 and either segment may be of either kind (50_*) or (50^*). On each such straight line segment $u(x, t)$ is defined and equal to the slope dx/dt of the segment.

Hence, if the two segments have a point in common, they must have the same slope and consequently one must be contained in the other. This concludes the proof of theorem 5.

Lemma 4. *If $u(x, t_1)$, $t_1 > 0$ fixed, is continuous for $a \leqq x \leqq b$ the characteristic x, t-segments belonging to the points $a \leqq x \leqq b$, $t = t_1$ form a continuous field.*

Proof. In virtue of theorem 5 two such segments belonging to different points of $a \leqq x \leqq b$ can have no point in common. The slope of the segment which is $u(x, t_1)$ varies continuously with x. Hence the segments form a continuous field, q.e.d.

The characteristic x, t-segments of lemma 4 completely and simply cover the region contained between $t = 0$ and $t = t_1$ and between the two extreme segments belonging to the points (a, t_1) and (b, t_1). As mentioned before, the lower end points of the segments are the points $y_*(x, t_1) = y^*(x, t_1)$ on the line $t = 0$. As x increases this lower endpoint moves continuously and never to the left. $y_*(x, t_1)$ may well have an interval of constancy which means that the corresponding x, t-segments have a common lower endpoint, say y_0. Suppose that $u_0(y_0 \pm 0)$ both exist. (49) and theorem 4 then show that the initial values $u_0(y)$ must make an upward jump at $y = y_0$. It is seen that such a discontinuity of the initial values is, for $t > 0$, immediately dissolved into a linear change of $u(x, t)$, $u = (x - y_0)/t$.

Theorem 6. *The complementary set S in the semiplane $t > 0$ of the closure of the set of points x, t of discontinuity for $u(x, t)$ is either empty or open. In the latter case the part of the surface $u = u(x, t)$ above S is formed by segments $0 < t < t$ (in general t_1 varies with the segment) of characteristics. Their x, t-projections form a continuous field in S.*

This is an easy consequence of theorem 5 and lemma 4. If the initial values are merely continuous, it can presumably happen that S is empty. We leave the question of the structure of S aside and quite generally construct a set of continuous curves which fill the semiplane $t > 0$ and which contain the characteristic x, t-segments of the surface $u = u(x, t)$ as well as the "lines of discontinuity for $u(x, t)$".

To every point x, t in $t > 0$ there belongs a characteristic triangle formed by $t = 0$ and the two characteristic x, t-segments (50). The endpoints of its basis on $t = 0$ are $y_*(x, t)$ and $y^*(x, t)$. If x, t is a point of continuity for $u(x, t)$ the triangle is a single characteristic x, t-segment. In virtue of theorem 5, two characteristic triangles either lie outside of each other or one encloses the other. Let x_1, t_1 denote an arbitrarily given point in $t > 0$. We assert that on every level $t = t_2 > t_1$ there exists one and only one point $x = x_2$ whose characteristic triangle contains the given point x_1, t_1; it then contains the whole triangle belonging to x_1, t_1. To see this, consider an arbitrary point x, t_2. We infer from lemma 1,c that there exist values of x for which the corresponding characteristic triangle lies on the right of the point x_1, t_1 and that for

$x \to -\infty$ the triangle ultimately always lies on the left of that point. The values x of the first kind, therefore, have a greatest lower bound x_2. It follows from lemma 1,b that the right side of the triangle belonging to x_2, t_2 does not lie on the left of x_1, t_1. Its left side cannot lie on the right of x_1, t_1 because the same lemma would then lead to a contradiction of the definition of x_2. Hence x_2, t_2 satisfies the requirement. The uniqueness of this point on $t = t_2$ easily follows from the non-intersection property of the triangles. In this way a unique curve $x = x(t)$, $x_1 = x(t_1)$, is defined for all $t \geqq t_1$. It is readily seen that the triangles belonging to the points $x(t)$, t form, in the sense of inclusion, an increasing family of triangles. The continuity of $x(t)$ for $t \geqq t_1$ is geometrically evident.

As t increases, $y_*(x(t), t)$ never increases and $y^*(x(t), t)$ never decreases. If the curve $x = x(t)$ contains a point x_2, t_2 of continuity for $u(x, t)$, it obviously is an x, t-characteristic in the interval $t_1 \leqq t \leqq t_2$. These facts prove the first assertion of

Theorem 7. *Each point* x_1, t_1 *in* $t > 0$ *uniquely determines a curve* $x = x(t)$, $x_1 = x(t_1)$, *for all* $t \geqq t_1$ *continuous and such that the characteristic triangles (triangular areas) belonging to the points of the curve form an increasing family of sets. At every* $t \geqq t_1$

$$(52) \qquad \lim_{\substack{t'=t+0 \\ t''=t+0}} \frac{x(t'') - x(t')}{t'' - t'} = \frac{1}{2}\,[u(x - 0, t) + u(x + 0, t)]; \qquad x = x(t).$$

For any $t > t_1$

$$(53) \qquad \lim_{\substack{t'=t-0 \\ t''=t-0}} \frac{x(t'') - x(t')}{t'' - t'}$$

exists, but it may have a different value.

It suffices to prove (52) at $t = t_1$. Let $x' = x(t')$, $x'' = x(t'')$, $t'' > t' > t_1$, and

$$y_{,} = y_*(x', t'), \quad y' = y^*(x', t'); \qquad y_{,,} = y_*(x'', t''), \quad y'' = y^*(x'', t'');$$

$$y_1 = y_*(x_1, t_1), \quad y^1 = y^*(x_1, t_1).$$

These numbers satisfy the inequalities

$$(54) \qquad y_{,,} \leqq y_{,} \leqq y_1 \leqq y^1 \leqq y' \leqq y''.$$

From lemma 3 and from what was said about the monotonicity of the two functions $y_*(x(t), t)$, $y^*(x(t), t)$ it follows that both functions of t are continuous on the right in t. Hence one infers that

$$(55) \qquad y_{,,} \to y_1, \; y'' \to y^1 \quad \text{as} \quad t'' \to t_1.$$

If the curve has an initial segment which is characteristic, the assertion is trivial. We therefore suppose that $y_* < y^*$ for $t > t_1$. This implies that

$$y_{,,} \leqq y_{,} < y' \leqq y''. \tag{56}$$

According to the original definition of y_* and y^* by means of the function (8), the following inequalities hold for all values of y

$$\int_{y_,}^{y} I'\, d\eta \geqq 0, \qquad \int_{y_{,,}}^{y} I''\, d\eta \geqq 0 \tag{57}$$

where

$$I' = u_0(\eta) + \frac{\eta - x'}{t'}, \qquad I'' = u_0(\eta) + \frac{\eta - x''}{t''}.$$

The inequalities (57) stay valid if the lower limits of integration are replaced by y' and y'', respectively. We also have

$$\int_{y_,}^{y'} I'\, d\eta = 0, \qquad \int_{y_{,,}}^{y''} I''\, d\eta = 0. \tag{58}$$

If the first of these equations is subtracted from the second, the following equation results:

$$\int_{y_{,,}}^{y'} I''\, d\eta + \int_{y_,}^{y''} I'\, d\eta + \int_{y_,}^{y''} (I'' - I')\, d\eta = 0.$$

From (57), where $y_,$ is to be replaced by y', it is inferred that the third term of the last relation is $\leqq 0$. If this term is evaluated and divided by $y'' - y_,$ which, in virtue of (56), is certainly not zero, the inequality

$$t'(x'' - x') \geqq \left(x' - \frac{y_, + y''}{2}\right)(t'' - t')$$

is finally obtained. Interchanging the two equations (58), one arrives at another inequality

$$t''(x' - x'') \geqq \left(x'' - \frac{y_{,,} + y'}{2}\right)(t' - t'').$$

From the last two inequalities we infer the desired result using (55) and the continuity of $x(t)$. The value of (52) follows from (49).[6] That (53) exists follows from the monotonicity of y_* and y^* along $x = x(t)$. On the left, however, these functions need not be continuous in t. Theorem 7 is hereby completely proved.

A curve $x = x(t)$ which, for any of its points, has the property stated in theorem 7 is called a *line of discontinuity* for $u(x, t)$ if $y_* < y^*$ holds along it. While, a discontinuity line is uniquely determined by any of its points in the direction of increasing t this need not be so in the opposite direction. Two lines of discontinuity can, as t increases, merge at some moment. A discontinuity line, $t > t_0$, always has a definite initial point. This is readily seen from the

[6]This value of the slope of the discontinuity line was derived by Burgers—see footnote 1(a)—on much more general grounds. The differentiability of $x(t)$, proven here, was taken for granted by Burgers.

property of the characteristic triangles if $t_0 > 0$. If $t_0 = 0$ the additional (and easily proved) fact must be used that

$$\lim_{t=0} y_*(x, t) = \lim_{t=0} y^*(x, t) = x$$

holds uniformly in every finite x-interval. A discontinuity line may be called complete if it is not part of another such line. Again by considering the triangles one can prove that a line of discontinuity can be completed in at least one way. If no differentiability conditions are imposed on the initial values, it may well happen that a discontinuity line possesses infinitely many completions and that, moreover, the discontinuity lines are everywhere dense. Conditions under which this does not occur are mentioned below. It is, however, generally true that the complete lines of discontinuity are only denumerable in number. This follows from the fact stated above that on $t =$ const. there are only denumerably many points of discontinuity and from the fact that two such lines must differ at some rational value of t if they do not coincide.

The differentiability assumptions mentioned about $u_0(y)$ are these: Suppose that the line of y can be subdivided into intervals which may accumulate only at ∞ such that $u_0'(y)$ is monotonic in the interior of each interval and bounded if the interval is finite. $u_0(y \pm 0)$ exists in all points of division. u_0' may have intervals of constancy which, however, may accumulate at ∞ only. Under these assumptions about the initial values the following statements are true (proofs are omitted): An arbitrary rectangle $a \leqq x \leqq b$, $0 \leqq t \leqq T$ can contain points of but a finite number of different lines of discontinuity. Only a finite number of starting points of discontinuity lines can lie in such a rectangle. On a finite segment $t < T$ of a discontinuity line there can be only finitely many points in which other discontinuity lines merge with it. Only in these points, by the way, can the two derivatives (52) and (53) have different values.

Lines of discontinuity "in general" do originate in $t > 0$ even if $u_0'(y)$ is continuous throughout. The only exceptional case is the one in which $u_0'(y) \geqq 0$ everywhere. These statements are proved in the next section. The starting point of a discontinuity line that originates in $t > 0$ is "in general" a point of continuity of $u(x, t)$. We do not go into the details back of this statement and we confine ourselves to proving a typical converse statement. *A point x_1, t_1 of continuity of $u(x, t)$ in which either the left or right derivative $\partial u/\partial x$ exists and equals $-\infty$ is the starting point of a discontinuity line.* This is readily inferred from the following fact: Let x_1, t_1 be a point on the characteristic x, t-segment belonging to a point x_2, t_2 in $t > 0$ which is a point of continuity for $u(x, t)$. The inequality

$$\frac{u(x', t_1) - u(x_1, t_1)}{x' - x_1} \geqq -\frac{1}{t_2 - t_1} \tag{59}$$

is then fulfilled by values x' arbitrarily near x_1 on the left as well as on the right. To prove this fact we observe that the characteristic x, t-segments belonging to the points x'', t_2 converge toward the segment belonging to x_2 , t_2 as $x'' \to x_2$. Denote by x', t_1 the point of intersection of such a segment with the line $t = t_1$. The value of $u(x, t)$ on such a segment equals its slope dx/dt which is equal to $(x'' - x')/(t_2 - t_1)$. The difference quotient in (59) therefore equals

$$\frac{(x'' - x') - (x_2 - x_1)}{(t_2 - t_1)(x' - x_1)} = \frac{1}{t_2 - t_1}\left[-1 + \frac{x'' - x_2}{x' - x_1}\right].$$

Since the x, t-segments cannot intersect, the second term in the square bracket is always positive. Hence (59) must hold as stipulated above.

6. *The function $u(x, t)$ and the initial values $u_0(y)$*

The connection between the characteristic segments of the preceding section which were defined by the surface $u = u(x, t)$, $t > 0$, and the characteristics that are directly defined by the initial curve is a matter that still needs some clarification. The explicit rule that connects $u(x, t)$ with the initial values is an accidental feature of Burgers' equation which has no analogue in other problems of similar nature. As in the preceding section we endeavor to give the results a form as independent as possible of that special feature.

We now suppose that the limits $u_0(y \pm 0)$ exist for every y, and we replace the condition at infinity (6) by the somewhat stronger one that

$$u_0(y) = 0(y), \qquad |y| \to \infty. \tag{60}$$

We assign a characteristic to each value of y for which $u_0(y - 0) \leqq u_0(y + 0)$ by the formula

$$x = y + vt, \qquad u = v$$

where v may have any value that satisfies

$$u_0(y - 0) \leqq v \leqq u_0(y + 0).$$

In particular, the single characteristic $x = y + u_0(y)t$, $u = u_0(y)$ is assigned to a value y of continuity for $u_0(y)$. No characteristic is assigned to a value for which $u_0(y - 0) > u_0(y + 0)$. These characteristics we call I-characteristics. The x, t, I-characteristics are their x, t-projections.

Theorem 4 now implies that every characteristic segment of the surface as defined by (50,) and (50′) is part of an I-characteristic. It remains to examine the converse question, if a given I-characteristic has an initial segment in common with the surface $u = u(x, t)$. This need not be true if the initial values satisfy no other local conditions besides the ones stated above. The form of such an additional differentiability condition that is most convenient for our purposes is a non-intersection condition or a field condition. *If an initial segment $0 < t < t_1$ of an x, t, I-characteristic is not met by any other x, t, I-characteristic then the surface contains the entire segment $0 < t < t_1$ of the I-characteristic.* This is easily seen on considering an arbitrary point x, t on that segment and a characteristic x, t-segment (in the sense of (50)) which belongs to this point. This latter segment is, as stated above, part of an x, t, I-characteristic. The hypothesis of the italicized statement clearly implies that this x, t, I-characteristic contains the original segment. Hence the latter must up to (x, t), and since this point was arbitrary on the segment, wholly lie on the surface.

The simplest and the only case in which no discontinuity appears is the one in which $u_0(y)$ is a nowhere decreasing function of y. In this case the x, t, I-characteristics never intersect and fill the entire semiplane $t > 0$. The surface $u(x, t)$ is formed by the I-characteristics.

In the following statement the hypothesis is a purely local one. *Suppose that the initial point $t = 0$ of an x, t, I-characteristic possesses a neighborhood, $t > 0$, in which it is never met by any other x, t, I-characteristic that starts in this neighborhood. Then the original x, t-characteristic has an initial segment in common with the surface.* To see this observe that the points of intersection with the x, t, I-characteristics that start outside of the neighborhood cannot come arbitrarily close to $t = 0$. This follows from the conditions of continuity for $u_0(y)$ and from the condition at infinity (60). The rest follows from the preceding italicized statement.

Suppose that, at $y = y_0$, $u_0(y)$ makes an upward jump. Consider an x, t, I-characteristic that belongs to this point and is not one of the two extreme x, t-characteristics of the bundle belonging to y_0. We assert that this x, t-characteristic has an initial segment in common with the surface. Indeed, it is not difficult to show that a neighborhood of $x = y_0$, $t = 0$ exists in which the preceding italicized statement becomes applicable.

An I-characteristic either has an initial segment in common with the surface or its initial point on $t = 0$ is the starting point of a discontinuity line. Suppose first that the initial point y_0 of that characteristic is a point of continuity for $u_0(y)$. Then no other x, t, I-characteristic starts at y_0. Hence it follows that $y_*(x, t)$ and $y^*(x, t)$ can never attain y_0 as a value because, otherwise, an x, t-characteristic (50) would begin at y_0 and because such a characteristic is always part of an I-characteristic. The curve $x(t)$ that starts at y_0 must, therefore, be a discontinuity line.[7] If u_0 is discontinuous at y_0 the argument is somewhat more involved; we omit the proof in this case.

We finally mention a criterion for the occurrence of discontinuities. *If two different x, t, I-characteristics intersect in $t > 0$, at least one of the two closed finite segments contains an interior point or a starting point of a line of discontinuity.* If no discontinuity line starts at the initial point of such a segment, we apply the preceding italicized statement. The segment therefore contains a subsegment which is a characteristic x, t-segment in the sense of the preceding section. If the longest subsegment of this kind is shorter than the original segment, then a discontinuity line starts at its end point. If this is the case for neither

[7]Our definition of the curve $x(t)$ through a given point x_0, t_0 defines it uniquely if $t_0 > 0$. This need not be the case if $t_0 = 0$ (consider, for instance, the case where $u_0(y)$ makes an upward jump at $y = x_0$). In the special case considered here we are, however, justified in speaking of "the" curve $x(t)$ issuing from $x = y_0$, $t = 0$.

original segment then both these segments must be x, t-projections of two surface-characteristics. Since their slopes are different the (constant) values of u on them differ. Hence u is discontinuous at the point of intersection. This completes the proof of the statement.

A point $y = y_0$ *on* $t = 0$ *where* $u_0(y - 0) > u_0(y + 0)$ *is the starting point of a discontinuity line for* $u(x, t)$. To see this we remember that no x, t, I-characteristic issues from such a point. As every characteristic x, t-segment is part of an x, t, I-characteristic neither $y_*(x, t)$ nor $y^*(x, t)$ can attain y_0 as a value. Hence a definite curve $x(t)$ starts from that point and this curve must be a discontinuity line.[7]

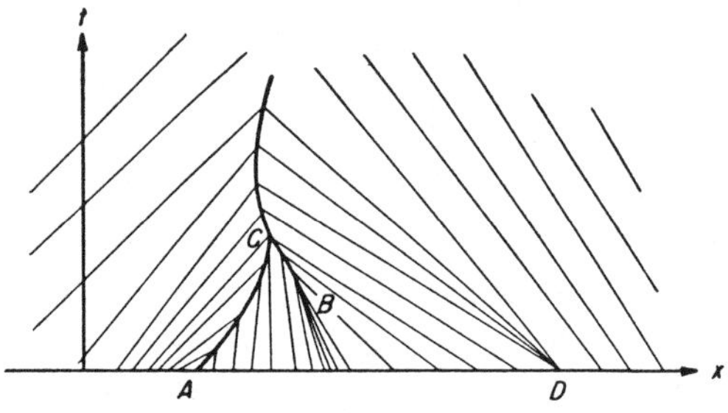

FIG. 1.

The figure shows an example where the initial values have a downward jump at A and an upward jump at D. A line of discontinuity starts at A. Another such line originates at B. The two lines merge at C.

7. *Behavior of* $u(x,t)$ *as* $t \to \infty$.

In section 3 we solved the problem how $u(x, t; \mu)$ behaves as $t \to \infty$ while $\mu > 0$ stays fixed. We refer to the formulation of the result given close to the end of that section. Right at the end it was stated how the result behaves if $\mu \to 0$. In hydrodynamics, the position of the problem in this order precisely corresponds to the problem of turbulent flow at high Reynolds numbers. This is the problem of the statistical long run behavior of a liquid of small viscosity. There is no reason to expect that the reversal of the order of the two limit processes leads to the same result. In the case of Burgers' equation one can easily show that the result is actually different.

We again assume the integral

$$M = \int_{-\infty}^{\infty} u_0(y)\, dy$$

to be finite. We show that the limit function

$$\lim_{t \to \infty} \sqrt{t}\, u(\bar{x}\sqrt{t}, t) \tag{61}$$

exists and that it may differ from the limit function mentioned at the end of section 3. Now, upon introducing new variables

$$x = \bar{x}\sqrt{t}, \qquad y = \bar{y}\sqrt{t}, \qquad z = \bar{z},$$

$$\sqrt{t}\, u(x, t) = \bar{u}(\bar{x}, t)$$

the geometrical rule that determines $u(x, t)$ from the curve $z = G(0, y, t)$ of (35) and (8) becomes this: If $\bar{y} = \bar{y}(\bar{x}, t)$ is the $\bar{y}$-coordinate of the point $\bar{y}, \bar{z}$ (we need only consider the cases with one such point) which the curve

$$\bar{z} = \int_{-\infty}^{\bar{y}\sqrt{t}} u_0(\eta)\, d\eta + \frac{\bar{y}^2}{2}$$

has in common with the lower support line of the slope $d\bar{z}/d\bar{y} = \bar{x}$ then $\bar{u} = \bar{u}(\bar{x}, t) = \bar{x} - \bar{y}$. The replacement of the lower limit 0 of integration by $-\infty$ is permitted because it only shifts the curve in the direction of $\bar{z}$. For large values of t the integral part differs little from 0 for $y < -\epsilon$ and from M for $y > \epsilon$. But the range of its values in the small interval $| y | < \epsilon$ is important for our purposes. Letting

$$N = \min_{\eta} \int_{-\infty}^{\eta} u_0(\eta)\, d\eta \tag{62}$$

we readily see that the limit curve which we have to consider is

$$\bar{z} = \begin{cases} \bar{y}^2/2 \quad , & \bar{y} < 0, \\ N \quad , & \bar{y} = 0, \\ M + \bar{y}^2/2, & \bar{y} > 0, \end{cases} \tag{63}$$

and that the limit (61) ($= \lim_{t=\infty} \bar{u}(\bar{x}, t)$) equals $\bar{x} - \bar{y}$ where $\bar{y}$ belongs to the point which the discontinuous curve (63) has in common with its lower support line of the slope $\bar{x}$. The curve is obtained by shifting the right half of the parabola $\bar{z} = \bar{y}^2/2$ by the amount M in the direction of $\bar{z}$ and by adding the isolated point $\bar{y} = 0, \bar{z} = N$. We have not added the analogous point corresponding to the maximum in (62). This point always lies in the convex hull of the curve and is therefore unimportant. This is not necessarily true for the other point, because

$$N \leqq \min (0, M). \tag{64}$$

A tangent to the continuous part of (63) has a slope equal to the $\bar{y}$-coordinate of the point of tangency. For two critical values

$$\bar{y}_1 = -\sqrt{-2N}, \qquad \bar{y}_2 = \sqrt{2(M - N)} \tag{65}$$

the tangent goes through the isolated point $\bar{y} = 0, \bar{z} = N$. Obviously, for any value of the slope $\bar{x}$ outside of the interval formed by (65), $\bar{u} = 0$. For any other value of $\bar{x}$ we have $\bar{u} = \bar{x}$ since the support line goes through that isolated point. This rule furnishes our limit function (61), $\bar{u} = \bar{u}(\bar{x})$,

$$\bar{u} = \begin{cases} 0, & \bar{x} < \bar{y}_1 , \\ \bar{x}, & \bar{y}_1 < \bar{x} < \bar{y}_2 , \\ 0, & \bar{x} > \bar{y}_2 . \end{cases}$$

Only in the exceptional case where the equality sign prevails in (64) does this function coincide with the function mentioned at the end of section 3.

Back of this discrepancy is the following fact (the proof of which is omitted here): We know that the moment

$$\int_{-\infty}^{\infty} u(x, t)\, dx$$

is an "integral" of Burgers' differential equation (1), $\mu > 0$. It is, essentially, the only such integral. After the passage to the limit $\mu \to 0$ it remains an integral. In this limit case, however, a new integral appears,

$$\min_x \int_{-\infty}^{x} u(x, t)\, dx.$$

8. *Integral form of the differential equations*

We first show that $u(x, t)$ satisfies the relation

$$\int_a^b u(\xi, t)\, d\xi = F(b, y_*(b, t), t) - F(a, y_*(a, t), t) \tag{66}$$

for all a and b. By virtue of lemma 2 this relation is valid if y_* is replaced by y^*. The relation holds if the initial values merely satisfy (6).

To prove it we use definition (8) of F and the definition of y^* and y_*. Letting

$$y_* = y_*(x, t), \qquad y_*' = y_*(x', t)$$

we have

$$\begin{aligned} F(x, y_*, t) - F(x', y_*', t) &= [F(x, y_*, t) - F(x, y_*', t)] \\ &\quad + [F(x, y_*', t) - F(x', y_*', t)] \\ &= [F(x, y_*, t) - F(x', y_*, t)] \\ &\quad + [F(x', y_*, t) - F(x', y_*', t)]. \end{aligned}$$

In the first relation the first term on the right is $\leqq 0$; in the second relation the second term is $\geqq 0$. By (8) this means that the quantity

$$F(x, y_*(x, t), t) - F(x', y_*(x', t), t)$$

lies between the limits

$$(x - x')\,\frac{x + x' - 2y_*(x', t)}{2t}, \qquad (x - x')\,\frac{x + x' - 2y_*(x, t)}{2t}.$$

In virtue of (49) these limits equal

$$\frac{(x - x')^2}{2t} + (x - x')u(x' - 0, t), \qquad -\frac{(x - x')^2}{2t} + (x - x')u(x - 0, t)$$

respectively. Identifying x, x' with successive points of a subdivision of the the interval (a, b) and adding the resulting inequalities, (66) becomes obvious.

We use (66) to prove that

$$\text{(67)} \qquad \lim_{t=0} \int_0^x u(\xi, t)\, d\xi = \int_0^x u_0(\eta)\, d\eta$$

holds uniformly in every finite x-interval, in other words, that the initial condition (9) is satisfied not only by the solution of (1) belonging to the initial values u_0 but also by the limit function obtained by letting $\mu \to 0$.

Obviously, $\min_y F \leqq F(x, x, t)$ and hence

$$\int_0^{y_*(x,t)} u_0(\eta)\, d\eta \leqq F(x, y_*(x, t), t) \leqq \int_0^x u_0(\eta)\, d\eta.$$

(67) follows from (66) and from the fact (stated in the second half of §5) that $y_*(x, t) \to x$ as $t \to 0$ holds uniformly in every x-interval.

We now turn to our main object and we assert: A solution u of (1), $\mu > 0$, which is regular in the semiplane $t > 0$ satisfies the relation (the double integrals are extended over the semiplane)

$$\text{(68)} \qquad \iint u f_t \, dx\, dt + \iint \frac{u^2}{2} f_x \, dx\, dt + \mu \iint u f_{xx} \, dx\, dt = 0$$

for each $f(x, t)$ such that f, f_x, f_t, f_{xx} are continuous in $t > 0$ and such that $f = 0$ holds outside some circle lying entirely within $t > 0$. This fact trivially results on multiplying (1) by f and on integrating over $t > 0$.

We are, however, more interested in the converse question. We consider (68), together with the requirement that it hold for arbitrary f as stipulated, as the primary form of the problem of defining u. Let us call a function u that solves (68) in this sense a generalized solution of (1) in $t > 0$. We assert: a generalized solution of (1) in $t > 0$ with the property that u, u_x, u_t, u_{xx} are continuous in $t > 0$ is a solution of (1) regular in $t > 0$. This fact also follows from (68) if we perform the integrations by parts backwards and observe that f is arbitrary within the specified function class.

The following theorem is deeper. *A generalized solution of* (1) *in* $t > 0$ *which is merely measurable and quadratically integrable in every closed rectangle in* $t > 0$ *coincides almost everywhere with a regular solution of* (1) *in* $t > 0$. This shows that nothing is gained by enlarging the function space to which u is confined. Later we will see that this is quite different in the limit case $\mu \to 0$. Only the idea of a proof of the theorem is briefly indicated here. It makes use of the additional supposition that for $|x|$ large, $u = e^{o(x^2)}$ holds uniformly in every t-interval. Appropriate approximations by functions f to the well known fundamental solution of $v_t + v_{xx} = 0$, the adjoint to the heat equation, lead to a quadratic integral equation for $u(x, t)$ which is satisfied almost everywhere. From this equation bounds can be derived for the difference quotients of the function $\bar{u}$ that solves the equation throughout and that coincides with u almost everywhere. In this way one can prove the differentiability properties required for u to render the transformation of (68) back to (1) trivial.

We now consider the limit case $\mu \to 0$. By a generalized solution u of (32), i.e. of (1), $\mu = 0$, in $t > 0$ we mean a function u that is measurable and quadratically integrable in every closed rectangle in the open semiplane $t > 0$ and that satisfies the relation

$$\iint \left[ug_t + \frac{u^2}{2} g_x \right] dx\, dt = 0, \tag{69}$$

where g is an arbitrary function of class C' in $t > 0$ that vanishes outside some circle lying entirely in $t > 0$. We assert: *Every limit function* u *obtained according to section* 4 *from the solution of* (1) *as* $\mu \to 0$ *is a generalized solution of* (1), $\mu = 0$. This can now be readily proved. From the results of section 4 it is inferred that the solutions of (1), for fixed initial values $u_0(x)$, converge almost everywhere in $t > 0$ as $\mu \to 0$. From the convergence theorem 3 it furthermore follows that every point x, t in $t > 0$ has a neighborhood in which the solutions of (1) stay uniformly bounded as $\mu \to 0$. In consequence of these facts the passage to the limit $\mu \to 0$ in (68) may be performed in the integrands (f being kept fixed). Thus it follows that (69) must hold for all functions f. That it must hold for all functions g, even if the class C' is replaced by the wider class of Lipschitz functions, follows from well known approximation theorems.

It is desirable to complete the picture and to show conversely that the limit function u constructed in section 4 is the only generalized solution of (1), $\mu = 0$, that satisfies the initial condition (67) uniformly in every finite x-interval. This uniqueness theorem is very likely to hold true, but we did not succeed in deriving the explicit formulae of section 4 directly from (69) and the initial condition.

The concept of the generalized solution of (1), $\mu = 0$, appears to furnish the simplest direct characterization of the limit case $\mu \to 0$ in the solutions of (1), whereas the case $\mu = 0$ of (1) in the literal sense cannot completely describe it.

The Partial Differential Equation $u_t + uu_x = \mu u_{xx}$, *Comm. on Pure and Appl. Math.* **3 (1950), 201-230.**

Commentary

Peter Lax

The equation*

$$u_t + uu_x = \mu u_{xx} \tag{1}$$

has been studied by J M Burgers (1948) as a scalar, one-dimensional model of the Navier-Stokes equation, although Bateman has already pointed out the analogy in 1915. In this paper Hopf determines all solutions whose initial values do not grow too fast as x goes to $\pm\infty$, say $u(x, 0)$ is integrable. Denote by $U(x, t)$ the integral of $u(x, t)$ from $-\infty$ to x; integrating (1) gives

$$U_t + \frac{1}{2}U_x^2 = \mu U_{xx} , \tag{2}$$

which can be linearized by introducing as new variable $\varphi = \exp(-U/2\mu)$:

$$\varphi_t = \mu\varphi_{xx} \tag{3}$$

In his 1951 paper Julian Cole uses the same transformation.Curiously, this linearization had already been observed by A.R. Forsyth in 1906, as an exercise in volume vi of his treatise on differential equations.

The solutions of (3) can be expressed in terms of the initial value of u and the heat conduction kernel; the corresponding solution of (1) can be expressed via the inverse of the transformation (2):

$$u = -2\mu\varphi_x/\varphi \tag{4}$$

Hopf bases his study on this explicit form for solutions of (1). He shows that as $t \to \infty$, $u(x, t)$ exhibits a rapid transition on the scale of $\sqrt{\mu}$.

More interesting is the limit $\mu \to 0$; Hopf shows that $u(x, t, \mu)$ tends to a limit $u(x, t)$, except along a set of curves -called shocks- across which u is discontinuous. At points of continuity

$$\lim_{\mu \to 0} u(x, t; \mu) = (x - y(x, t))/t , \tag{5}$$

where $y(x, t)$ is the unique point where

$$F(x, y, t) = U(x, 0) + (x - y)^2/2t \tag{6}$$

attains its minimum as function of y. The curves of discontinuity are the points x, t where the minimum is attained for two different values of y.

*Note that there is a misprint in the original title.

Hopf shows that $u(x,t,\mu)$ tends to the limit $u(x,t)$ boundedly and in the L^1 norm, and consequently the limit is a solution in the weak sense of the conservation law

$$u_t + (u^2/2)_x = 0 \; . \tag{7}$$

Today we would say that (7) is satisfied in the sense of distributions. Hopf derives the Rankine-Hugoniot jump condition across shocks:

$$s = \frac{u_\ell + u_r}{2} \; , \tag{8}$$

where s is the shock speed, u_ℓ and u_r are the values of u on the left and right side of the shock.

Hopf shows that the moment

$$\int_{\mathbb{R}} u(x,t)dx \tag{9}$$

is an invariant for all solutions of (1), and for all weak solutions of (7), that is, it is independent of t. He points out that the functional

$$\min_x \int_{-\infty}^{x} u(y,t)dy \tag{10}$$

is another invariant. It is not hard to show, that these two are the only invariant functionals that are continuous in the L^1 norm, see Lax (1969).

Hopf derives a one-sided Lipschitz condition for limits of solutions of (1). At the end of his paper Hopf surmises that all weak solutions of (7) are limits of solutions of (1). This is clearly false, for rarefaction shocks violate the one-sided Lipschitz condition. In a subsequent paper (1969), Hopf shows by an elegant argument that all limits of solutions of (1) satisfy Oleinik's entropy condition, which also excludes rarefaction shocks.

It is no exaggeration to call Hopf's paper epoch making; Math Reviews counts 539 references to it. I can only discuss a few of these.

Lax (1954) extends Hopf's analysis of the limit of solutions (1) to limits of solutions of the difference equation corresponding to

$$u_k^{n+1} = (u_{k-1}^n + u_{k+1}^n)/2 - [f(u_{k+1}^n) - f(u_{k-1}^n)]\frac{\Delta t}{2\Delta x} \tag{11}$$

as Δx and Δt tend to zero. For the case

$$f(u) = \log(a + b\exp u), \tag{12}$$

equation (11) can be linearized by the discrete analogue of the transformation (2), and solved explicitly. The limit satisfies in the weak sense the conservation law

$$u_t + f(u)_x = 0 \; , \tag{13}$$

and can be characterized by the explicit formula similar to (5) and (6):

$$u(x,t) = b\left(\frac{x - y(x,t)}{t}\right) \tag{14}$$

where $y(x,t)$ is the point where

$$F(x,y,t) = U(x,0) + tg\left(\frac{x-y}{t}\right) \tag{15}$$

assumes its minimum. Here $g(v)$ is the Legendre transform of $f(u)$, and $b(v) = dg/dv$. It is not hard to show, see Lax (1957),that (14) and (15) furnish a weak solution of (13) for any convex function $f(u)$.For concave $f(u)$ there is an analogous theory; for functions $f(u)$ which are neither convex nor concave, there is no analogue of the invariant functional (9).

The convergence of solutions of the parabolic equation

$$u_t + f(u)_x = \mu u_{xx} \tag{16}$$

to weak solutions of (13) was proved by Ladyzhenskaya (1956), and by Oleinik (1957).

Oleinik (1959) has given an entropy condition that singles out the physically relevant solutions of (13), valid even when $f(u)$ is neither concave nor convex. She has shown that such solutions are uniquely determined by their initial values.Another proof of uniqueness, based on L^1 contraction, has been given by Keyfitz-Quinn (1971).

It is observed in Lax (1954) that for nowhere linear functions f, the operator that relates values at time t to the initial values of solutions of the nonlinear equation (13) is compact. For f convex (or concave) this follows from the explicit formula (14),(15); for general nonlinear f it can be deduced from the convergence of solutions of the parabolic equation (16) to those of (13). Another proof has been furnished by Luc Tartar (1983). Ron DiPerna (1985) has extended this result to 2×2 systems that are genuinely nonlinear. An extension to $N \times N$ genuinely nonlinear systems would be extremely useful.

For a survey of modern developments in this area see the monographs of Smoller (1983), and of Dafermos (2000). Curiously, although motivated by the fluid dynamical analogy, Hopf had no interest in the subsequent extension of his results to hyperbolic systems of conservation laws.

References

BATEMAN, H., "Some recent researches on the motion of fluids," *Monthly Weather Review* **43** (1915), 163–170.

BURGERS, J. M., "A mathematical model illustrating the theory of turbulence," *Advances in Applied Mech.* **1** (1948), 171–1999.

COLE, J.D., "On a quasilinear parabolic equation occurring in areodynamics," *Quarterly of Applied Math.* **9** (1951),226–236.

DAFERMOS, C.M., *Hyperbolic Conservation laws in Continuum Physics*, vol. 325 Grundlehren, Springer Verlag, 2000.

DIPERNA, R. J., "Compensated compactness and general systems of conservation laws," *Trans. AMS* **292** (1985), 383–420.

FORSYTH, A. R., *Differential equations*, vol vi, Cambridge Univ. Press, 1906, Dover reprint, 1959.

HOPF, E., "On the right weak solution of the Cauchy problem for a quasilinear equation of first order," *J. of Math. and Mech.* **19** (1969), 483–487.

KEYFITZ-QUINN, B., "Solutions with shocks; an example of an L^1 contraction semigroup," *CPAM* **24** (1971), 125–132.

LADYZHENSKAYA, O.A., "On the construction of solutions of quasilinear hyperbolic equations as a limit of solutions of the corresponding parabolic equation when the viscosity coefficient tends to zero," *Dokl. Akad. Nauk. SSSR*, **111** (1956), 291–294.

LAX, P.D., "Weak solutions of nonlinear hyperbolic equations and their numerical construction," *CPAM* **7** (1954), 159–193.

LAX, P.D., "Hyperbolic systems of conservation laws," *CPAM* **10** (1957), 537–566.

LAX, P.D., "Invariant functionals of nonlinear equations of evolution," *Proc. of the International Conference on Functional Analysis and Related Topics*, Tokyo, (1969), 240-251.

OLEINIK, O.A., "Discontinuous solutions of nonlinear differential equations," *USP. MAT. NAuk.* (NS) **12** (1957), 3–73; AMS Trans. Ser. 2, 26, 95–172.

OLEINK, O.A., "Uniqueness and stability of the generalized solutions of the Cauchy problem for a quasilinear equation." *Usp. Mat. Nauk.*, 14 (1959), 165–170; AMS Transl., Ser. 2, **33** (1964), 285–290.

SMOLLER, J., *Shock waves and reaction-diffusion equation*, vol. 258 Grundlehren, Springer Verlag, 1983.

TARTAR, L., "The compensated compactness method applied to systems of conservation laws," *Systems of Nonlinear Partial Differential Systems*, 263–285, ed. J.M. Ball, D Reidel, 1983.

Sonderabdruck aus „Mathematische Nachrichten“
4. Band, Heft 1–6, Sept. 1950/Febr. 1951
Akademie-Verlag, Berlin

Über die Anfangswertaufgabe für die hydrodynamischen Grundgleichungen.

Erhard Schmidt zu seinem 75. Geburtstag gewidmet.

Von Eberhard Hopf in Bloomington, Ind. (USA.).

(Eingegangen am 10. 7. 1950.)

§ 1. Einführung.

Die Punkte des n-dimensionalen Raumes seien mit x bezeichnet; $x_1, x_2, \ldots, x_n$ seien die Koordinaten in einem festen cartesischen Koordinatensystem. $dx = dx_1 dx_2 \cdots dx_n$ bezeichne das Volumenelement im x-Raum. $u(x, t)$ sei ein zeitlich veränderliches, in einer offenen Menge $\hat{G}$ des x-t-Raumes erklärtes Vektorfeld mit den Komponenten u_i in jenem Koordinatensystem. Im folgenden wird nicht verlangt, daß $\hat{G}$ zusammenhängend ist, und wir werden nur der Kürze der Benennung halber von Gebieten sprechen. Gebiete im x-Raum werden mit G, solche im x-t-Raum mit $\hat{G}$ bezeichnet. Die Divergenzfreiheit eines in einem x-t-Gebiete $\hat{G}$ stetig x-differenzierbaren Vektorfeldes $u(x, t)$ ist durch das Erfülltsein der Differentialgleichung

$$\text{(1.1)} \qquad \operatorname{div} u = \frac{\partial u}{\partial x_\nu} = 0$$

gekennzeichnet. Wir benützen durchweg die übliche Summationsvorschrift ohne Gebrauch des Summenzeichens. Man kennt aber auch eine differentiallose Kennzeichnung der Divergenzfreiheit. Wir sagen, daß eine in einem x-t-Gebiete $\hat{G}$ definierte, skalare oder vektorwertige Funktion $v(x, t)$ in $\hat{G}$ zur Klasse N gehört, falls $v \equiv 0$ außerhalb einer passenden kompakten Teilmenge dieses Gebietes gilt. Die Funktionen dieser im folgenden oft gebrauchten Klasse verschwinden also in einem Randstreifen von $\hat{G}$. Jene Kennzeichnung ist dann: Für die Divergenzfreiheit in $\hat{G}$ eines in $\hat{G}$ stetig x-differenzierbaren Feldes $u(x, t)$ ist es notwendig und hinreichend, daß

$$\text{(1.2)} \qquad \iint\limits_{\hat{G}} u_i \frac{\partial h}{\partial x_i}\, dx\, dt = 0$$

für jede in $\hat{G}$ eindeutige und mitsamt ihren x-Ableitungen stetige Funktion $h(x, t)$ gilt, die in $\hat{G}$ zur Klasse N gehört. Diese Tatsache ist eine Folge des Gaußschen Integralsatzes, der wegen $h \in N$ in $\hat{G}$ anwendbar ist, und des Fundamental-lemmas der Variationsrechnung. Wenn man das skalare Produkt zweier Vektorfelder $v(x, t)$ und $w(x, t)$ in $\hat{G}$ einführt,

$$\iint_{\hat{G}} v_i w_i \, dx \, dt,$$

so kann man sagen: Divergenzfreiheit in $\hat{G}$ eines in $\hat{G}$ stetig x-differenzierbaren Feldes u bedeutet, daß u in $\hat{G}$ zum Gradientenfeld einer jeden in $\hat{G}$ eindeutigen und mitsamt ihren x-Ableitungen stetigen Funktion der Klasse N in $\hat{G}$ orthogonal ist[1]).

Folgendes Gegenstück zu dieser Tatsache ist hier von Interesse: Dafür, daß ein in $\hat{G}$ stetiges Feld $h'(x, t)$ mit den Komponenten h'_i das Gradientenfeld $h'_i = \frac{\partial h}{\partial x_i}$ einer in $\hat{G}$ eindeutigen und mitsamt ihren x-Ableitungen stetigen Funktion $h(x, t)$ ist, ist es notwendig und auch hinreichend, daß es in $\hat{G}$ zu jedem in $\hat{G}$ stetig x-differenzierbaren und divergenzfreien Felde der Klasse N orthogonal ist.

Die Notwendigkeit ist wieder eine Folge des Integralsatzes. Daß die Bedingung hinreicht, ergibt sich aus folgenden Überlegungen. Die Betrachtung von Feldern der Form $w(x, t) = \varphi(t) w(x)$ mit skalarem φ lehrt zunächst, daß man sich auf die entsprechende Behauptung für x-Gebiete G beschränken kann. Es gelte also

$$\int_G w_i h'_i \, dx = 0$$

für jedes glatte divergenzfreie Feld $w(x)$ der Klasse N in G. Die Behauptung folgt, wenn man zeigen kann, daß die Zirkulation des Feldes h'

$$\int_{\mathfrak{C}} h'_i \, dx_i = \int_{\mathfrak{C}} h'_s \, ds$$

längs eines beliebigen geschlossenen Weges $\mathfrak{C}$ in G verschwindet. Man sieht leicht, daß man das nur für stetig gekrümmte Wege ohne Selbstüberkreuzungen zu beweisen braucht. Man erschließt das Verschwinden durch geeignete Wahl von Feldern w. Zu beliebig klein vorgegebenem $\varepsilon > 0$ gibt es nun stets eine in G glatte und divergenzfreie Strömung $w(x)$ mit folgenden Eigenschaften. w ist nur in einer geschlossenen Röhre der Dicke $< \varepsilon$ um den Weg $\mathfrak{C}$ von Null verschieden. Auf jedem ebenen, $\mathfrak{C}$ senkrecht durchsetzenden Röhrenquerschnitt bildet der Vektor w mit der Normalenrichtung (Richtung von $\mathfrak{C}$ im Querschnitt) einen Winkel $< \varepsilon$. Der Querschnittsfluß von w, der wegen der Divergenzfreiheit vom speziellen Querschnitt unabhängig ist, ist gleich 1. Diese Tatsache reicht zum Beweise des Verschwindens der Zirkulation längs $\mathfrak{C}$ aus. Wir betrachten ein solches Feld $w(x)$, das zu einem vorgegebenen, aber genügend klein gewählten

---

[1]) Die Formulierung der Begriffe im x-t-Raum statt im x-Raum ist für unser Problem vorteilhafter.

Anwendungen der Theorie des Hilbertschen Raumes auf Probleme der Potentialtheorie und der mathematischen Hydrodynamik findet man in folgenden Arbeiten: O. Nikodym, Sur un théorème de M. S. Zaremba concernant les fonctions harmoniques. J. Math. pur. appl., Paris, Sér. IX **12** (1933), 95—109; J. Leray, Sur le mouvement d'un liquide visqueux emplissant l'espace. Acta math., Uppsala **63** (1934), 193—248; H. Weyl, The method of orthogonal projection in potential theory. Duke math. J. **7** (1940), 411—444.

$\varepsilon > 0$ gehört. Bezeichnet man mit dF das Hyperflächenelement auf jenen Röhrenquerschnitten und wählt man die Bogenlänge s längs $\mathfrak{C}$ als Parameter quer zu den Schnitten, so kann man in der Röhre das Volumenelement dx in der Form $\varrho(x)dF\,ds$ schreiben, wo ϱ in der Umgebung von $\mathfrak{C}$ stetig und auf $\mathfrak{C}$ gleich 1 ist. Es gilt dann

$$\int h'_i w_i dx = \int \left[\int h'_w |w| \varrho dF\right] ds.$$

Ersetzt man hier die Komponente h'_w durch die Komponente h'_s, genommen im Schnittpunkt von $\mathfrak{C}$ mit dem Querschnitt, ersetzt man ferner $|w(x)|$ durch die Komponente $w_s(x)$, gebildet in der Normalenrichtung von dF, und ersetzt man ϱ durch 1, so wird aus dem rechten Integral

$$\int h'_s \left[\int w_s\, dF\right] ds = \int h'_s ds,$$

d. h. die Zirkulation. Auf Grund der angegebenen Eigenschaften des Feldes w läßt sich indessen leicht beweisen, daß der durch diese Ersetzung begangene Fehler mit ε gegen Null strebt. Damit ist die Behauptung bewiesen.

Die Grundgleichungen von Navier-Stokes für die Bewegung einer homogenen inkompressiblen Flüssigkeit lauten

$$\frac{\partial u_i}{\partial t} + u_\alpha \frac{\partial u_i}{\partial x_\alpha} = -\frac{\partial p}{\partial x_i} + \mu \frac{\partial^2 u_i}{\partial x_\beta \partial x_\beta}, \tag{1.3}$$

worin μ eine positive Konstante, den kinematischen Viskositätskoeffizienten, bedeutet und

$$\operatorname{div} u = 0$$

ist. Es sei $u(x, t)$, $p(x, t)$ eine Lösung in einem x-t-Gebiete $\hat{G}$, die in $\hat{G}$ mitsamt den in den Gleichungen vorkommenden Derivierten u_t, u_x, u_{xx} stetig sein möge. Wir führen nun ein neues zeitabhängiges und in $\hat{G}$ divergenzfreies Vektorfeld $a = a(x, t)$ ein. Es soll in $\hat{G}$ von der Klasse $\overline{N}$ und genügend glatt sein: a und die Ableitungen a_t, a_x, a_{xx} seien in $\hat{G}$ stetig. Sonst werden dem Felde a keine Beschränkungen auferlegt. Wegen $a \in N$ in G und wegen $u_\alpha \frac{\partial u_i}{\partial x_\alpha} = \frac{\partial u_i u_\alpha}{\partial x_\alpha}$ ist

$$\iint_{\hat{G}} a_i \frac{\partial u_i}{\partial t}\, dx\, dt = -\iint_{\hat{G}} \frac{\partial a_i}{\partial t} u_i\, dx\, dt,$$

$$\iint_{\hat{G}} a_i u_\alpha \frac{\partial u_i}{\partial x_\alpha}\, dx\, dt = -\iint_{\hat{G}} \frac{\partial a_i}{\partial x_\alpha} u_\alpha u_i\, dx\, dt,$$

$$\iint_{\hat{G}} a_i \frac{\partial^2 u_i}{\partial x_\beta \partial x_\beta}\, dx\, dt = -\iint_{\hat{G}} \frac{\partial a_i}{\partial x_\beta} \frac{\partial u_i}{\partial x_\beta}\, dx\, dt = \iint_{\hat{G}} \frac{\partial^2 a_i}{\partial x_\beta \partial x_\beta}\, dx\, dt$$

und wegen $\operatorname{div} a = 0$ und $a \in N$

$$\iint_{\hat{G}} a_i \frac{\partial p}{\partial x_i}\, dx\, dt = 0.$$

Man findet somit, daß das Feld $u(x, t)$ die folgenden Bedingungen erfüllt:

$$\iint_{\hat{G}} \frac{\partial a_i}{\partial t} u_i\, dx\, dt + \iint_{\hat{G}} \frac{\partial a_i}{\partial x_\alpha} u_\alpha u_i\, dx\, dt + \mu \iint_{\hat{G}} \frac{\partial^2 a_i}{\partial x_\beta \partial x_\beta} u_i\, dx\, dt = 0 \tag{1.4}$$

gilt für jedes in $\hat{G}$ genügend glatte Feld $a(x, t)$ mit den Eigenschaften

$$\operatorname{div} a = 0 \text{ in } \hat{G}, \qquad a \in N \text{ in } \hat{G}. \tag{1.5}$$

Hierzu kommt noch die Divergenzfreiheit des Feldes u:

$$\iint\limits_{\hat{G}} \frac{\partial h}{\partial x_i} u_i \, dx \, dt = 0, \qquad h \in N \text{ in } \hat{G}, \tag{1.6}$$

gilt für jede in G genügend glatte Funktion der angegebenen Klasse. Die Grundgleichungen sind damit auf die Form von Gleichungen zwischen linearen Funktionaloperatoren willkürlicher Felder und Funktionen a, h gebracht. Das Wesentliche dabei ist, daß das unbekannte Feld $u(x, t)$, von welchem diese Operatoren noch abhängen, ohne Ableitungen in ihnen auftritt.

Wir müssen uns noch davon überzeugen, daß man von den im eben erklärten Sinne verstandenen Gleichungen (1.4) und (1.6) wieder auf die Differentialform der Gleichungen zurückkommt, wenn man sich auf genügend glatte Lösungsfelder u in $\hat{G}$ beschränkt. Wir wissen bereits, daß unter dieser Voraussetzung (1.6) wieder zu $\operatorname{div} u = 0$ in $\hat{G}$ zurückführt. Für genügend glattes u darf man alle benützten partiellen Integrationen rückgängig machen. Es folgt dann, daß

$$\iint\limits_{\hat{G}} a_i \left\{ \frac{\partial u_i}{\partial t} + u_\alpha \frac{\partial u_i}{\partial x_\alpha} - \mu \frac{\partial^2 u_i}{\partial x_\beta \partial x_\beta} \right\} dx \, dt$$

für jedes glatte Feld $a(x, t)$ der Form (1.5) gelten muß. Mit Hilfe des weiter oben bewiesenen Satzes folgt daraus, daß die geschweiften Klammern die partiellen Ableitungen $\frac{\partial p}{\partial x_i}$ einer in $\hat{G}$ eindeutigen Funktion $p(x, t)$ sein müssen, daß also die Differentialgleichungen der Bewegung in $\hat{G}$ gelten müssen. Man sieht, daß die obige Integralform der Gleichungen die physikalische Forderung der Eindeutigkeit des Druckes genau trifft.

Es liegt nahe, die allgemeine mathematische Theorie auf der Integralform der Gleichungen aufzubauen. Dann ist es aber zweckmäßig, sich von der künstlichen Beschränkung auf glatte Lösungsfelder u zu befreien. Das Auftreten der beiden quadratischen Formen

$$\int u_i u_i \, dx, \qquad \int \frac{\partial u_i}{\partial x_\beta} \frac{\partial u_i}{\partial x_\beta} dx$$

in der Energiegleichung führt ganz von selbst dazu, dem Problem einen Hilbertschen Raum der Vektorfelder zugrunde zu legen. Es bedeutet einen methodischen Vorteil, daß in diesem weiteren Rahmen die Differenzierbarkeitseigenschaften der Lösungen u Gegenstand einer Aufgabe werden, die von den Existenzproblemen gänzlich abgesondert studiert werden kann[2]).

Das übliche Anfangswertproblem der hydrodynamischen Grundgleichungen ist folgendes. Die Lösung $u(x, t)$ ist in einem vorgeschrieben bewegten Gebiete $G(t)$, $t \geqq 0$, des x-Raumes zu bestimmen, wenn $u(x, 0)$ in $G(0)$ vorgeschrieben ist (zusammen mit einer passend formulierten Bedingung des stetigen Anschlusses

---

[2]) Vgl. hierzu die Behandlung quadratischer Variations- und linearer Differentialprobleme mit den Methoden des Hilbertschen Raumes bei R. Courant und D. Hilbert, Methoden der mathematischen Physik, Bd. 2. Berlin 1937, Kap. VII.

für $t \to 0$) und wenn die Randwerte von u am Rande von $G(t)$, $t > 0$, gegeben sind (in passend formuliertem Sinne des Anschlusses). J. LERAY hat diesem Problem in den frühen dreißiger Jahren drei umfangreiche Arbeiten gewidmet[3]). Diese Untersuchungen haben Leray bereits zur Heranziehung der Methoden des Hilbertschen Raumes und zur integralen Auffassung der Gleichungen in drei Dimensionen gezwungen[3a]). Gelöst hat Leray die Existenzfrage für alle $t > 0$ in seinen drei Arbeiten in den folgenden Fällen: zwei Dimensionen, a) $G =$ ganze Ebene, Nebenbedingung endlicher kinetischer Energie, b) $G =$ festes Oval, Randwerte Null; c) G ist der ganze dreidimensionale Raum, Nebenbedingung endlicher kinetischer Energie. Die bemerkenswerten Untersuchungen, die Leray der Differenzierbarkeitsfrage widmet, weisen auf einen merkwürdigen Unterschied zwischen den Dimensionenzahlen $n = 2$ und $n > 2$ hin. Während im ersteren Falle, wenigstens wenn G die ganze Ebene ist, der Beweis der unbeschränkten stetigen Differenzierbarkeit der Lösung gelingt, versagen für $n \geqq 3$ die Beweismethoden, die man als die natürlichen ansehen sollte. Auch bei beliebiger Glattheit aller vorgeschriebenen Daten gelang es nicht, die Glattheit der Lösung für alle Zeiten zu beweisen. Merkwürdig ist auch das Versagen der Eindeutigkeitsbeweise in drei Dimensionen. Diese Fragen sind immer noch nicht befriedigend geklärt. Es ist schwer zu glauben, daß die Anfangswertaufgabe zäher Flüssigkeiten für $n = 3$ mehr als eine Lösung haben könnte, und der Erledigung der Eindeutigkeitsfrage sollte mehr Aufmerksamkeit geschenkt werden. Daß indessen bei nichtlinearen partiellen Differentialproblemen die Anzahl der unabhängigen Veränderlichen auf die lokalen Eigenschaften der Lösungen einen wesentlichen Einfluß hat, geht auch aus anderen neueren Untersuchungen hervor.

Wir lassen in der vorliegenden Arbeit, die ebenfalls dem Anfangswertproblem gewidmet ist und in der wir von der integralen Auffassung der Gleichungen als der primären Form derselben ausgehen, die Frage der Differenzierbarkeit und auch der Eindeutigkeit beiseite. Auf diese Dinge sowie auf den (in unserem Rahmen nicht leichten) Beweis der Energiegleichung hoffen wir in späteren Mitteilungen zurückzukommen. Der Hauptpunkt der Arbeit ist, daß die Konstruktion der angenäherten Lösungen, die in Lerays Arbeiten einen so breiten Raum einnimmt, hier durch einen einfachen Prozeß ersetzt wird, der sich auch auf viel weitere Klassen partieller Differentialprobleme anwenden läßt. Auch darauf hoffen wir später zurückzukommen. Die Methode ermöglicht die Lösung der Anfangswertaufgabe für alle $t > 0$ in beträchtlicher Allgemeinheit, doch kommt es uns in dieser ersten Mitteilung mehr auf die Hervorhebung der methodischen Grundgedanken als auf die Allgemeinheit der Resultate an. Wir beschränken uns hier auf den Fall, in dem das x-Gebiet G zeitlich fest, aber sonst völlig beliebig ist

---

[3]) J. LERAY, a) Étude de diverses équations intégrales non linéaires et de quelques problèmes que pose l'Hydrodynamique. J. Math. pur. appl., Paris, Sér. IX **12** (1933), 1—82; b) Essay sur les mouvements plans d'un liquide visqueux que limitent des parois. J. Math. pur. appl., Paris, Sér. IX **13** (1934), 331—418; c) l. c. Fußnote [1]) S. 214.

[3a]) Lange vorher hatte schon C. W. OSEEN seinen bekannten hydrodynamischen Untersuchungen eine Form der Grundgleichungen zugrunde gelegt, die von zweiten Ableitungen frei ist. Der Existenzbeweis gelang ihm jedoch nur für hinreichend kleine Zeiten. Vgl. sein Werk: Hydrodynamik (Leipzig 1927).

und wo u verschwindende Randwerte besitzt. Die Randbedingung wird mit Begriffen des Hilbertschen Raumes definiert, weit genug, um die Lösbarkeit, und eng genug, um die Eindeutigkeit der Lösung, jedenfalls in zwei Dimensionen, zu gewährleisten[4]). In der reinen Existenztheorie wird die Anzahl der Raumdimensionen keine Rolle spielen.

§ 2. Die Funktionenklasse H'. Lösungen der Klasse H'.

Unter der Klasse H bezüglich eines x-t-Gebietes $\hat{G}$ verstehen wir die Klasse aller reellen, in diesem Gebiete definierten und meßbaren Funktionen $f(x, t)$ mit endlicher Norm

$$\iint_{\hat{G}} f^2 dx\, dt.$$

H ist ein reeller Hilbertscher Raum. Begriffe wie schwache und starke Konvergenz in $\hat{G}$ werden im folgenden stets hinsichtlich dieser Norm verstanden. Wir erinnern daran, daß eine Folge von Funktionen f aus H in $\hat{G}$ gegen eine Funktion f^* aus H schwach konvergiert, wenn erstens die Normen aller f unter einer festen Schranke bleiben und wenn zweitens

$$\iint_{\hat{G}} fg\, dx\, dt \to \iint_{\hat{G}} f^* g\, dx\, dt$$

für jede feste Funktion g gilt, die in $\hat{G}$ zu H gehört. Unter Beibehaltung der ersten Bedingung darf die zweite Bedingung dahin abgeschwächt werden, daß die Zahlenfolge

$$\iint_{\hat{G}} fg\, dx\, dt$$

für jedes feste g einer in H stark dichten Menge konvergiert. Es existiert dann eine und im wesentlichen nur eine schwache Grenzfunktion f^* in $\hat{G}$. Neben diesen Begriffen, denen hier ein x-t-Gebiet zugrunde liegt, werden wir die gleichen Begriffe für ein rein räumliches x-Gebiet G benützen müssen. Es wird dann die Norm

$$\int_G f^2 dx$$

zugrunde gelegt. Wir erinnern an die schwache Kompaktheit einer Folge von Funktionen mit gleichmäßig beschränkten Normen (Satz von F. Riesz).

Das folgende, von Leray ausgiebig benützte Kriterium für starke Konvergenz wird auch hier benötigt. Für eine in $\hat{G}$ schwach konvergente Folge von Funktionen f mit der Grenzfunktion f^* gilt

$$\overline{\lim} \iint_{\hat{G}} f^2 dx\, dt \geqq \iint_{\hat{G}} f^{*2} dx\, dt,$$

---

[4]) Wenn G der ganze x-Raum ist, geht die so gefaßte Randbedingung in die Bedingung endlicher kinetischer Energie und endlichen Dissipationsintegrals über.

Die Fassung der Randbedingung ist durch die Ausführungen bei R. Courant und D. Hilbert, Methoden der mathematischen Physik, Bd. 2. Berlin 1937, Kap. VII, § 1, 3. Abschn., nahegelegt.

wobei die Gleichheit dann und nur dann zutrifft, wenn $f \to f^*$ in $\hat{G}$ im starken Sinne gilt. Alle diese Dinge übertragen sich auf Vektorfelder u, v in $\hat{G}$, wenn man das skalare Produkt

$$\iint\limits_{\hat{G}} u_i v_i \, dx \, dt$$

und die entsprechende Norm zugrunde legt.

Lemma 2.1. *Konvergieren die Vektorfelder $u(x, t)$ in $\hat{G}$ schwach gegen ein Limesfeld $u^*(x, t)$, so gilt*

$$\overline{\lim} \iint\limits_{\hat{G}} u_i u_i \, dx \, dt \geqq \iint\limits_{\hat{G}} u_i^* u_i^* \, dx \, dt.$$

Dabei trifft das Gleichheitszeichen dann und nur dann zu, wenn die Konvergenz in $\hat{G}$ stark ist.

Wir benötigen wie Leray den Begriff der verallgemeinerten (rein räumlichen) x-Derivierten von Funktionen $f(x, t)$ und Feldern $u(x, t)$.

Definition 2.1. *Ein in einem x-t-Gebiete $\hat{G}$ definiertes $f(x, t)$ soll dann und nur dann zur Klasse H' in diesem Gebiete gerechnet werden, wenn es folgende Eigenschaften hat. f gehört in $\hat{G}$ zu H. Es existieren n ebenfalls in $\hat{G}$ zu H gehörende Funktionen $f_{/i}$ derart, daß die Relationen*

$$\iint\limits_{\hat{G}} h f_{/i} \, dx \, dt = -\iint\limits_{\hat{G}} \frac{\partial h}{\partial x_i} f \, dx \, dt \qquad \left(h \in N \text{ in } \hat{G}\right) \tag{2.1}$$

für jede Funktion $h(x, t)$, die in $\hat{G}$ mitsamt ihren x-Ableitungen stetig ist und zur Klasse N gehört, und für jedes $i = 1, 2, \ldots, n$ erfüllt sind.

Die Klasse H' enthält offensichtlich jedes $f(x, t)$, welches in $\hat{G}$ stetig x-differenzierbar ist derart, daß f und alle $\frac{\partial f}{\partial x_i}$ in $\hat{G}$ zu H gehören. Für ein solches f ist $\frac{\partial f}{\partial x_i} = f_{/i}$. Dies folgt aus dem Integralsatze und aus der Forderung, daß h zu N gehört, d. h. daß h außerhalb einer gewissen kompakten Teilmenge von $\hat{G}$ verschwindet. Offenbar sind, falls $f \in H'$ in $\hat{G}$, die verallgemeinerten x-Derivierten $f_{/i}$ in G bis auf die Werte in einer x-t-Nullmenge eindeutig bestimmt.

Lemma 2.2. *Konvergiert eine Folge von Funktionen f der Klasse H' in $\hat{G}$ schwach gegen f^* und sind für alle f die Ausdrücke*

$$\iint\limits_{\hat{G}} f^2 \, dx \, dt + \iint\limits_{\hat{G}} f_{/i} f_{/i} \, dx \, dt$$

gleichmäßig beschränkt, so gehört auch f^* zu H' in $\hat{G}$, und jede der x-Derivierten $f_{/i}$ konvergiert in $\hat{G}$ schwach gegen die entsprechende x-Derivierte $f^*_{/i}$.

Beweis. Jedes f erfüllt (2.1), worin h eine beliebige der dort zugelassenen Funktionen ist. Die rechten Seiten konvergieren gegen

$$-\iint\limits_{\hat{G}} \frac{\partial h}{\partial x_i} f^* \, dx \, dt.$$

Für festes h und i konvergieren also entlang der Folge der f die linken Seiten. Die zugelassenen Funktionen h in $\hat{G}$ liegen im Hilbertschen Raum H stark dicht, und nach Voraussetzung bleiben die H-Normen der $f_{/i}$ unter einer festen Schranke. Also ist für jedes feste i die Folge der $f_{/i}$ in $\hat{G}$ schwach konvergent. Bezeichnet f_i^* die Grenzfunktion, so folgt aus (2.1), daß

$$\iint_{\hat{G}} h f_i^* \, dx\, dt = -\iint_{\hat{G}} \frac{\partial h}{\partial x_i} f^* \, dx\, dt$$

für jedes zugelassene h und jedes i gilt. Nach Definition 2.1 gehört also f^* zu H' in $\hat{G}$, und wegen der Eindeutigkeit der x-Derivierten ist $f_i^* = f_{/i}^*$.

Ein Feld heißt von der Klasse H' in $\hat{G}$, wenn das für alle Komponenten zutrifft.

In der obigen Integralform der hydrodynamischen Grundgleichungen kommen keine Ableitungen von u vor. Es ist indessen zweckmäßig, an die Lösungen u eine schwache Differenzierbarkeitsforderung wie Zugehörigkeit zur Klasse H' zu stellen. Man darf dann für das Reibungsglied in (1.4)

$$\mu \iint_{\hat{G}} \frac{\partial^2 a_i}{\partial x_\beta \partial x_\beta} u_i \, dx\, dt = -\mu \iint_{\hat{G}} \frac{\partial a_i}{\partial x_\beta} u_{i,\beta} \, dx\, dt \tag{2.2}$$

schreiben.

Definition 2.2. *Ein Feld $u(x, t)$ heißt Lösung der Klasse H' der hydrodynamischen Gleichungen im x-t-Gebiete $\hat{G}$, wenn es folgende Bedingungen erfüllt:* a) *$u \in H'$ in $\hat{G}$.* b) *Divergenzfreiheit; die Relation* (1.6) *wird durch jede Funktion h befriedigt, die in $\hat{G}$ von der Klasse N und mitsamt ihren x-Ableitungen stetig ist.* c) *Bewegungsgleichungen; die Relation* (1.4) *wird durch jedes Feld $a(x, t)$ befriedigt, das in $\hat{G}$ von der Klasse N, divergenzfrei und mitsamt den Ableitungen a_t, a_x, a_{xx} stetig ist.*

Man beachte, daß unter der Voraussetzung a) auch das in u nichtlineare Glied in den Grundgleichungen (1.4) bei beliebigem zugelassenem Felde a ein wohldefiniertes Lebesguesches Integral ist. Dies ist bereits dann der Fall, wenn $u \in H$ in $\hat{G}$ ist.

Wegen a) ist die Bedingung der Inkompressibilität b) damit gleichbedeutend, daß

$$\operatorname{div} u \equiv u_{i,i} = 0$$

für fast alle x, t in $\hat{G}$ gilt[5]).

Wir denken uns in der Grundgleichung (1.4) alle Integranden außerhalb $\hat{G}$ durch Null erklärt. Die Integrale können dann über den ganzen x-t-Raum erstreckt werden. Mit dieser Verabredung gilt der folgende Satz, den wir hier noch beweisen wollen, obwohl er in dieser Arbeit nicht gebraucht wird.

---

[5]) Wenn man, was wir indessen in dieser Arbeit nicht tun, das Problem des Grenzüberganges $\mu \to 0$ in den hydrodynamischen Strömungen studiert, so verliert der mit dem Fall $\mu > 0$ wesentlich verknüpfte Funktionenraum H' seine Bedeutung. Selbstverständlich muß man sich dann an die ableitungsfreie Definition der Divergenzfreiheit halten.

Theorem 2.1. *Eine Lösung der Klasse H' in G erfüllt für fast alle Werte von τ die Gleichung*

$$\int\limits_{t=\tau} a_i u_i \, dx = \iint\limits_{t<\tau} \frac{\partial a_i}{\partial t} u_i \, dx \, dt + \iint\limits_{t<\tau} \frac{\partial a_i}{\partial x_\alpha} u_\alpha u_i \, dx \, dt - \mu \iint\limits_{t<\tau} \frac{\partial a_i}{\partial x_\beta} u_{i,\beta} \, dx \, dt. \tag{2.3}$$

Darin ist $a(x, t)$ ein beliebiges der unter Definition 2.2, c) zugelassenen Felder.

Beweis. Man bedenke, daß mit $a(x, t)$ auch $h(t) a(x, t)$ ein zugelassenes Feld ist, wenn $h(t)$ eine ganz beliebige, für alle t stetig differenzierbare Funktion bedeutet. Ersetzt man in der in der abgekürzten Form

$$\iint K[a, u] \, dx \, dt = \int\limits_{-\infty}^{\infty} \left\{ \int\limits_{t=\tau} K[a, u] \, dx \right\} d\tau = 0$$

geschriebenen Gleichung (1.4) a durch ha, so folgt, daß auch die Gleichung

$$\int\limits_{-\infty}^{\infty} h(\tau) \left\{ \int\limits_{t=\tau} K \, dx \right\} d\tau + \int\limits_{-\infty}^{\infty} h'(\tau) \left\{ \int\limits_{t=\tau} a_i u_i \, dx \right\} d\tau = 0 \tag{2.4}$$

befriedigt ist. Die geschweiften Klammern sind in $-\infty < \tau < \infty$ Lebesgue-integrable Funktionen von τ, die für alle großen $|\tau|$ verschwinden. Die Gültigkeit von (2.4) für willkürliches $h(\tau)$ mit stetigem $h'(\tau)$ ist bekanntlich damit gleichbedeutend, daß

$$\int\limits_{t=\tau} a_i u_i \, dx = \int\limits_{-\infty}^{\tau} \left\{ \int\limits_{t \text{ fest}} K \, dx \right\} dt = \iint\limits_{t<\tau} K \, dx \, dt$$

für fast alle τ gilt. Damit ist die Behauptung bewiesen.

In (2.3) ist die linke Seite nur für fast alle Werte von τ definiert, während die rechte Seite eine totalstetige Funktion von τ ist. Man kann jedoch beweisen: Eine Lösung der Klasse H' in $\hat{G}$ läßt sich stets in einer x-t-Nullmenge so abändern, daß das neue u die Relation (2.3) ausnahmslos, d. h. für jedes zugelassene Feld a und für jedes τ erfüllt. Doch gehen wir hier nicht weiter darauf ein.

§ 3. Die Randbedingung des Verschwindens. Die Anfangswertaufgabe.

Die Querschnitte $t = \text{const}$ des x-t-Gebietes $\hat{G}$ sind x-Gebiete $G(t)$. Wir müssen mit Hilfe von Begriffen des Hilbertschen Raumes der Randbedingung des Verschwindens einer Funktion $g(x, t)$ und eines Feldes $u(x, t)$ für alle t am Rande von $G(t)$ möglichst nahe kommen. Das kann man dadurch erreichen, daß man die Funktionen g aus Funktionen, die in $\hat{G}$ zur Klasse N gehören, durch geeigneten Grenzübergang gewinnt. Dabei ist es nötig, genügend wirksame Schranken für die rein räumlichen x-Derivierten der approximierenden Funktionen (nicht aber die t-Derivierte) mitzubenützen, damit das „Verschwinden" an den Rändern der x-Gebiete $G(t)$ im wesentlichen erhalten bleibt. Wir drücken die Randbedingung durch die Zugehörigkeit zu der folgenden Funktionenklasse $H'(N)$ aus.

Definition 3.1. *Eine Funktion $g(x, t)$ heißt von der Klasse $H'(N)$ in $\hat{G}$, wenn sie in $\hat{G}$ schwache Grenzfunktion einer Folge von Funktionen $\gamma(x, t)$ ist, welche in $\hat{G}$ zu N gehören und mitsamt ihren x-Ableitungen stetig sind und für welche die Ausdrücke*

$$\iint\limits_{\hat{G}} \gamma^2 dx\, dt + \iint\limits_{\hat{G}} \gamma_{/i}\gamma_{/i} dx\, dt \tag{3.1}$$

gleichmäßig beschränkt sind[6]).

Aus Lemma 2.2 folgt: Für ein gegebenes x-t-Gebiet G ist die Klasse $H'(N)$ in der Klasse H' enthalten.

Lemma 3.1. *$\hat{G}$ sei eine Zylindermenge $x \subset G$, $0 < t < T$. $g(x, t)$ sei in $\hat{G}$ schwacher Limes einer Folge von in $\hat{G}$ stetig x-differenzierbaren Funktionen $\gamma(x, t)$ folgender Art: Zu jedem γ gibt es eine kompakte Teilmenge des x-Gebietes G derart, daß γ verschwindet, wenn x außerhalb dieser Menge liegt; die Integrale* (3.1) *seien gleichmäßig beschränkt. Dann gehört g in $\hat{G}$ zur Klasse $H'(N)$*[7]).

Beweis. Man beachte den Unterschied zwischen der Klasse der in diesem Lemma zugelassenen γ und der engeren Klasse der γ von Definition 3.1. Zugehörigkeit von γ zu N im x-t-Gebiet $\hat{G}$ erfordert im gegenwärtigen Falle, daß γ auch genügend nahe bei $t = 0$ und $t = T$ verschwindet. Da aber in (3.1) nur die x-Ableitungen auftreten, ist dieser Unterschied belanglos. Wenn man die hiesigen γ durch Funktionen $\varphi(t)\gamma(x, t)$ ersetzt, wobei φ in $\langle 0, T\rangle$ stetig und

$$\varphi = \begin{cases} 0 & \text{für} \quad 0 < t < \varepsilon, \quad T - \varepsilon < t < T, \\ 1 & \text{für} \quad 2\varepsilon < t < T - 2\varepsilon \end{cases}$$

und sonst $0 < \varphi < 1$ ist ($\varepsilon \to 0$), so trifft Definition 3.1 mit den neuen $\tilde{\gamma} = \varphi\gamma$ zu. g gehört also zu $H'(N)$.

Lemma 3.2. *Die Relationen*

$$\iint\limits_{\hat{G}} g_{/i} f\, dx\, dt = -\iint\limits_{\hat{G}} g f_{/i} dx\, dt \qquad (i = 1, 2, \ldots, n)$$

werden durch jedes f der Klasse H' in $\hat{G}$ und durch jedes g der Klasse $H'(N)$ in $\hat{G}$ befriedigt.

Beweis. Nach Definition 2.1 gelten die Relationen für jedes der angegebenen f und für jedes γ, das in $\hat{G}$ stetig x-differenzierbar und von der Klasse N ist. Nach

[6]) Vgl. hierzu Courant-Hilbert, l. c. Fußnote [4]) S. 218. Die dort gegebene Definition der Randbedingung des Verschwindens ist nur scheinbar stärker als unsere. Nach dem Satze von S. Saks enthält die Folge der arithmetischen Mittel einer schwach konvergenten Folge eine stark konvergente Teilfolge. Aus diesem Satze und aus Lemma 2.2 folgt, daß zu jedem g aus $H'(N)$ eine Folge von Funktionen γ der obigen Art existiert derart, daß

$$\gamma \to g, \quad \gamma_{/i} \to g_{/i}$$

im starken Sinne gilt.

[7]) Wenn G der ganze x-Raum ist, so fällt die Klasse $H'(N)$ mit der Klasse H' zusammen. In diesem Falle liegen die zugelassenen γ im Funktionenraume H' im Sinne der Norm (3.1) stark dicht.

Definition 3.1 ist g schwacher Limes einer Folge derartiger γ mit gleichmäßig beschränkten Integralen (3.1). Nach Lemma 2.2 gilt daher neben $\gamma \to g$ auch $\gamma_{/i} \to g_{/i}$ schwach in $\hat{G}$. Die für f, γ geltenden Relationen gelten somit auch für f, g.

Zur zweckmäßigen Formulierung der Anfangsbedingung führen wir noch die Klasse $H(N)$ ein. Wir beschränken uns dabei auf den x-Raum und auf Felder $u(x)$, die in einem x-Gebiete G erklärt sind. Wenn man nur Funktionen $f(x)$ betrachtet, die in G zu beiden Klassen H und N gehören, so ist es klar, daß die stark abgeschlossene Hülle dieser Funktionenmenge mit H identisch ist. Dasselbe gilt von Vektorfeldern in G. Anders ist es jedoch, wenn man sich auf die in G divergenzfreien Felder beschränkt.

Definition 3.2. *Ein in G divergenzfreies Feld der Klasse H heißt von der Klasse $H(N)$, wenn es schwaches Limesfeld von Feldern ist, welche in G zu N gehören, zweimal stetig differenzierbar und divergenzfrei sind*[8]).

Man beweist leicht: Ist das Feld $u(x)$ in G divergenzfrei und von der Klasse $H(N)$ und ist die Funktion $\varphi(x)$ in G von der Klasse H', so gilt

$$\int_G u_i \varphi_{/i} \, dx = 0.$$

Zugehörigkeit eines divergenzfreien Feldes zu $H(N)$ ersetzt offenbar die Randbedingung des Verschwindens der Normalkomponente.

Wir können nunmehr den Existenzsatz für das hydrodynamische Anfangswertproblem formulieren.

Existenzsatz. *G sei ein beliebiges Gebiet des x-Raumes. Das Feld $U(x)$ sei in G divergenzfrei und von der Klasse $H(N)$, sonst aber beliebig. Dann gibt es ein für alle $t > 0$ in G definiertes Feld $u(x, t)$ mit folgenden Eigenschaften:* A) *In jedem x-t-Zylindergebiet $x \subset G$, $0 < t < T$ ist u Lösung der Klasse H' der hydrodynamischen Grundgleichungen (Definition 2.2).* B) *„Verschwinden der Randwerte" für $t > 0$: in jedem jener Zylindergebiete gehört u zu $H'(N)$.* C) *Anfangsbedingung: für $t \to 0$ konvergiert $u(x, t) \to U(x)$ stark in G.*

§ 4. Vereinfachung der Aufgabe. Der Annäherungsprozeß.

Zur Konstruktion der Lösung u der Anfangswertaufgabe für ein zeitlich festes x-Gebiet G gehen wir von der Gleichung

$$\int_G a_i u_i \, dx \Big|_{t=\tau'} - \int_G a_i u_i \, dx \Big|_{t=\tau}$$

(4.1)
$$= \int_\tau^{\tau'}\!\!\int_G \frac{\partial a_i}{\partial t} u_i \, dx \, dt + \int_\tau^{\tau'}\!\!\int_G \frac{\partial a_i}{\partial x_\alpha} u_\alpha u_i \, dx \, dt + \mu \int_\tau^{\tau'}\!\!\int_G \frac{\partial^2 a_i}{\partial x_\beta \partial x_\beta} u \, dx \, dt$$

aus.

---

[8]) Nach dem Satze von Saks ist es dann auch starkes Limesfeld ebensolcher Felder.

Lemma 4.1. *Das Feld $u(x, t)$ sei in G für alle $t > 0$ erklärt und gehöre in jedem Zylinderabschnitt $x \subset G$, $0 < t < T$ des x-t-Raumes zur Klasse H. Es befriedige die Gleichung* (4.1) *für alle $\tau' > \tau > 0$ und für jedes der folgenden Felder a: $a = a(x)$ ist in G zweimal stetig differenzierbar und es sei*

$$a = a(x), \qquad \operatorname{div} a = 0 \text{ in } G, \qquad a \in N \text{ in } G, \tag{4.2}$$

d. h. $a(x)$ verschwinde außerhalb einer passenden kompakten Teilmenge von G. Dann befriedigt u die Grundgleichung (1.4) *für den Halbzylinder $\hat{G}$: $x \subset G$, $t > 0$ und für jedes der dort zugelassenen Felder $a(x, t)$ (siehe die Bedingung c) in der Lösungsdefinition* 2.2).

Beweis. Schreibt man (4.1) in der abgekürzten Form

$$f(\tau') - f(\tau) = \int_{\tau}^{\tau'} g(t)\,dt,$$

so sieht man, daß die Gleichung

$$\int_0^\infty \varphi'(t)\, f(t)\,dt + \int_0^\infty \varphi(t)\, g(t)\,dt = 0$$

für jedes in $(0, \infty)$ stetig differenzierbare $\varphi(t)$, das für alle hinreichend kleinen und großen t verschwindet, erfüllt sein muß. Schreibt man diese Gleichung wieder voll aus, so erkennt man, daß die Gleichung (1.4) im besagten Halbzylinder durch jedes Feld $a = \varphi(t)a(x)$ befriedigt wird, worin $a(x)$ ein beliebiges der oben zugelassenen Felder (4.2) und $\varphi(t)$ eine beliebige der oben zugelassenen Funktionen ist. Nun kann aber jedes der in der Bedingung c) der Lösungsdefinition 2.2 zugelassenen $a(x, t)$ im Halbzylinder $\hat{G}$ so durch Summen von Feldern der speziellen Art approximiert werden, daß man in der Grundgleichung (1.4) Integration und Grenzübergang vertauschen kann. Man kann es z. B. immer so einrichten, daß die Konvergenz der Felder und ihrer Ableitungen bis zu einer vorgeschriebenen Ordnung in $\hat{G}$ gleichmäßig ist und daß die approximierenden Felder alle außerhalb einer passenden festen und kompakten Teilmenge von $\hat{G}$ verschwinden.

Es ist nunmehr klar, daß ein Feld $u(x, t)$, welches die Gleichung (4.1) in dem im Lemma geforderten Umfang löst, welches ferner im Halbzylinder divergenzfrei ist und welches in jedem Zylinderabschnitt zur Klasse H' gehört, in jedem Zylinderabschnitt der vollen Lösungsdefinition 2.2 genügt.

Eine noch besser geeignete Form der Grundgleichung erhält man auf Grund der folgenden Tatsache.

Lemma 4.2. *Es existiert eine Folge von in G zweimal stetig differenzierbaren und in G linear unabhängigen Feldern im Feldraum* (4.2)

$$a = a^{\nu}(x), \qquad \operatorname{div} a^{\nu} = 0 \text{ in } G, \qquad a^{\nu} \in N \text{ in } G \tag{4.3}$$

mit folgender Eigenschaft: Ein beliebiges in G zweimal stetig differenzierbares Feld der Form (4.2) *ist in G gleichmäßiges Limesfeld einer Folge von endlichen Linearkombinationen der Felder $a^{\nu}(x)$, mit gleichmäßiger Konvergenz auch der Ableitungen bis zur zweiten Ordnung in G. Bei vorgegebenem $a(x)$ kommen in dieser Approxi-*

mation nur solche Linearkombinationen zur Verwendung, welche außerhalb einer gewissen nur von a abhängigen kompakten Teilmenge von G den Wert Null haben.

Auf Grund dieser Tatsache ist es klar, daß ein Feld $u(x, t)$, welches in jedem Zylinderabschnitt von der Klasse H ist und welches die Grundgleichung (4.1) für alle $\tau' > \tau > 0$ und für jedes Feld a der obigen Folge erfüllt, dies automatisch für alle oben zugelassenen Felder (4.2) tut. Zusammenfassend können wir sagen, daß die Grundgleichungen (1.4) für die gegenwärtige Aufgabe vollständig durch die Gleichungen (4.1) mit (4.3) ersetzt werden können

Im Funktionenraum der divergenzfreien Vektorfelder a ist (4.1), (4.3) eine affine Koordinatendarstellung der hydrodynamischen Grundgleichungen. Das affine System von Koordinatenvektoren (4.3) kann durch eine eineindeutige lineare Transformation einfacher Art in ein neues verwandelt werden, das im Sinne der Bilinearform

$$\int_G v_i w_i \, dx$$

orthonormal ist. Man darf zusätzlich voraussetzen, daß die Folge (4.3) diese Bedingung erfüllt:

$$\int_G a_i^\lambda a_i^\nu \, dx = \delta_{\lambda\nu} . \tag{4.4}$$

Lemma 4.3. *Das Orthonormalsystem der Felder $a^\nu(x)$ ist im Feldraum der in G divergenzfreien Felder $U(x)$ der Klasse $H(N)$ vollständig.*

Der Beweis ergibt sich aus Definition 3.2 und Lemma 4.2.

Das Annäherungsverfahren. Der k-te Annäherungsschritt besteht einfach darin, daß man von den unendlich vielen Grundgleichungen (4.1), (4.3) nur die ersten k betrachtet,

$$a = a^\nu(x) \qquad (\nu = 1, 2, \ldots, k), \tag{4.5}$$

und diese durch den Ansatz

$$u = u^k(x, t) = \sum_{\nu=1}^{k} \lambda_\nu(t) \, a^\nu(x) \tag{4.6}$$

mit noch zu bestimmenden skalaren Faktoren $\lambda_\nu = \lambda_\nu^k$ zu lösen sucht. Dieser Ansatz erfüllt wegen (4.3) von selbst die Bedingung der Divergenzfreiheit und die Randbedingung des Verschwindens:

$$\operatorname{div} u^k = 0 \text{ in } G, \qquad u^k \in N \text{ in } G. \tag{4.7}$$

Da nur differenzierbare $\lambda(t)$ in Frage kommen und da die zugelassenen Felder a nicht von t abhängen, dürfen die ersten k Gleichungen (4.1) in der Form

$$\int_G a_i \frac{\partial u_i}{\partial t} \, dx = \int_G \frac{\partial a_i}{\partial x_\alpha} u_\alpha u_i \, dx + \mu \int_G \frac{\partial^2 a_i}{\partial x_\beta \partial x_\beta} u_i \, dx \tag{4.8}$$

geschrieben werden. Wegen (4.4) stellen die k Gleichungen (4.8), (4.5) zusammen mit (4.6) ein gewöhnliches Differentialsystem

$$\frac{d\lambda_\nu}{dt} = F_\nu(\lambda_1, \ldots, \lambda_k) \qquad (\nu = 1, 2, \ldots, k) \tag{4.9}$$

für die λ dar, worin die rechten Seiten $F_\nu = F_\nu^k$ Polynome der λ mit konstanten Koeffizienten sind.

Die Gleichungen (4.8), (4.5), (4.6) oder die gleichbedeutenden Gleichungen (4.9) teilen nun mit den strengen hydrodynamischen Gleichungen die wichtige Eigenschaft, daß für die Lösungen die Energiegleichung

$$\frac{d}{dt}\frac{1}{2}\int_G u_i u_i\,dx = -\mu\int_G \frac{\partial u_i}{\partial x_\beta}\frac{\partial u_i}{\partial x_\beta}\,dx \tag{4.10}$$

gilt. Da nämlich die Gleichungen (4.8) für alle Felder (4.5) bestehen, so gelten sie auch für deren Linearkombination (4.6), $u = u^k$. Die Energiegleichung folgt in der üblichen Weise (und ohne Schwierigkeiten am Rande), weil wegen (4.7)

$$\int_G \frac{\partial u_i}{\partial x_\alpha} u_\alpha u_i\,dx = \int_G \frac{\partial K}{\partial x_\alpha} u_\alpha\,dx = 0 \qquad \left(K = \frac{1}{2}u_i u_i\right)$$

und

$$\int_G \frac{\partial^2 u_i}{\partial x_\beta \partial x_\beta} u_i\,dx = -\int_G \frac{\partial u_i}{\partial x_\beta}\frac{\partial u_i}{\partial x_\beta}\,dx \qquad (u = u^k)$$

ist. Aus (4.10) folgt, daß

$$\int_G u_i u_i\,dx = \lambda_1^2 + \cdots + \lambda_k^2; \qquad (u = u^k)$$

niemals zunimmt. Daraus schließen wir, daß jede bei $t = 0$ begonnene Lösung des Differentialsystems (4.9) für alle $t = 0$ existiert.

Das Annäherungsverfahren läßt sich formal sehr einfach folgendermaßen deuten. Man denke sich beide Seiten der Navier-Stokesschen Differentialgleichungen und die Lösung u formal nach dem Orthonormalsystem der Felder a^ν entwickelt: $u = \lambda_\nu a^\nu$. Man erhält dann rein formal ein System von unendlich vielen Differentialgleichungen erster Ordnung für die unendlich vielen skalaren Fourierkoeffizienten λ. Unser k-ter Schritt besteht dann einfach darin, daß man nur die ersten k dieser Gleichungen betrachtet und in ihnen alle Unbekannten mit den Indizes $\nu > k$ gleich Null setzt. Die Art und Weise, auf die wir im folgenden den Existenzansatz beweisen, liefert uns zugleich eine Aussage über die Konvergenzeigenschaften dieses einfachsten und nächstliegenden Annäherungsverfahrens.

Als Anfangswerte der $\lambda_\nu(t)$ bei $t = 0$ wähle man die Fourierkoeffizienten der Entwicklung des gegebenen Feldes $U(x)$ nach den a^ν. Während die Lösungen $\lambda(t)$ im k-ten Schritt im allgemeinen von k abhängen, sind diese Anfangswerte davon unabhängig. Nach der Voraussetzung $U \in H(N)$ in G und nach dem Vollständigkeitslemma 4.3 gilt

$$u_k(x, 0) \to U(x) \quad \text{stark in } G \quad (k \to \infty). \tag{4.11}$$

§ 5. Beweis des Existenzsatzes.

Wir fassen die im folgenden benötigten Eigenschaften der Felder der Folge $u^k(x, t)$ zusammen:

5a) Jedes $u^k(x, t)$ ist für $x \subset G$, $t \geqq 0$ zweimal stetig x-t-differenzierbar und divergenzfrei.

5b) $u^k(x, t)$ verschwindet, wenn x außerhalb einer nur von k abhängenden kompakten Teilmenge des x-Gebietes G liegt.

5c) $u^k(x, t)$ befriedigt die Gleichung (4.8) $(t \geqq 0)$ und die Gleichung (4.1) $(\tau' > \tau \geqq 0)$ in den k Fällen (4.3) $(\nu = 1, 2, \ldots, k)$.

5d) Die Integrale

$$\int_G u_i u_i \, dx, \qquad \int_0^T \int_G \frac{\partial u_i}{\partial x_\beta} \frac{\partial u_i}{\partial x_\beta} \, dx \, dt \qquad (u = u^k(x, t))$$

verbleiben unterhalb einer von k, t, T unabhängigen Schranke.

5e) Die Anfangswerte $u^k(x, 0)$ genügen der Limesrelation (4.11).

5d) folgt direkt aus der zeitlich integrierten Energiegleichung (4.10) in Verbindung mit (4.11).

Erster Schritt. Jedes Feld $a^\nu(x)$ ist in G stetig und nur in einer kompakten Teilmenge von G ungleich Null. Wendet man die erste Hälfte von 5d) auf die rechte Seite von (4.8) ($a = a^\nu$) an, indem man den in $u = u^k$ linearen Term mit Hilfe der Schwarzschen Ungleichung und den in u quadratischen Term mit Hilfe einer Absolutschranke für die Ableitungen von a abschätzt, so ergibt sich folgendes: Die rechte Seite von (4.8) ($a = a^\nu$, $u = u^k$, $k \geqq \nu$) ist bei festgehaltenem ν für alle k und t gleichmäßig beschränkt. Dasselbe gilt also von der linken Seite

$$\frac{d}{dt} \int_G a_i u_i \, dx.$$

Bei festgehaltenem ν genügen somit die Zeitfunktionen

$$\int_G a_i^\nu(x) u_i^k(x, t) \, dx$$

für alle $t \geqq 0$ einer von k unabhängigen Lipschitzbedingung. Sie bleiben außerdem für alle t und k gleichmäßig beschränkt. Nach einem bekannten Auswahlsatze gibt es also bei beliebigem, aber festgehaltenem ν eine Folge ganzer Zahlen k' derart, daß

$$\lim_{k' \to \infty} \int_G a_i^\nu(x) \, u_i^{k'}(x, t) \, dx \tag{5.1}$$

für jedes $t \geqq 0$ existiert, und zwar gleichmäßig in jedem endlichen t-Intervall. Die Folge der k' hängt vom Index ν ab, aber man kann die zum Index $\nu + 1$ gehörige Folge als Teilfolge der vorangehenden wählen. Mit Hilfe des Diagonalverfahrens läßt sich dann eine feste Folge ganzer Zahlen, die wir wieder mit k' bezeichnen, bilden derart, daß die eben gemachte Limesaussage für jedes feste $\nu = 1, 2, \ldots$ zu Recht besteht. Mit dieser Folge k' wird im folgenden operiert.

Zweiter Schritt. Wir beweisen jetzt, daß die Folge der Felder $u^{k'}(x, t)$ für jedes feste $t \geqq 0$ im x-Gebiete G schwach konvergiert. Zum Beweise fassen wir einen beliebig gewählten festen Wert t_0 von t ins Auge und beachten, daß nach 5d), erste Hälfte, die Folge dieser Felder ($t = t_0$) in G schwach kompakt ist. Die Behauptung wird bewiesen sein, wenn wir zeigen, daß jene Folge in G nur ein einziges schwaches Limesfeld besitzen kann. Sei $u^*(x, t_0)$ ein solches Limesfeld und sei k'' eine Teilfolge der k' (die Teilfolge wird von t_0 abhängen) derart, daß

$$\lim_{k'' \to \infty} \int_G w_i(x) \, u_i^{k''}(x, t_0) \, dx = \int_G w_i(x) \, u_i^*(x, t_0) \, dx$$

für jedes Feld $w(x)$ der Klasse H in G gilt. Im Falle $w = a^\nu$ ist aber der Wert der rechten Seite bereits durch den Limes (5.1) festgelegt. Sind u^* und u^{**} zwei

15*

schwache Limesfelder und ist v ihr Differenzfeld, so gilt daher

$$\int_G a_i^\nu v_i \, dx = 0$$

für jedes ν. Nach Definition 3.2 gehören die Felder u^*, u^{**}, also auch v zur Klasse $H(N)$ in G. Nach Lemma 4.3 spannen aber die Felder a^ν in G denselben Feldraum auf. Daraus ergibt sich $\int_G v_i v_i \, dx = 0$ und damit die Behauptung. Es gibt also ein in G für alle $t > 0$ wohlbestimmtes Feld $u^*(x, t)$ derart, daß

$$(5.2) \qquad \lim_{k' \to \infty} \int_G w_i(x) \, u_i^{k'}(x, t) \, dx = \int_G w_i(x) \, u_i^*(x, t) \, dx$$

für jedes Feld $w(x)$ ($w \in H$ in G) und für jedes $t > 0$ gilt. Das Feld u^* befriedigt die Bedingung B) des Existenzsatzes am Ende von § 3. Dies folgt aus 5b) und der zweiten Hälfte von 5d) durch Anwendung von Lemma 3.1. Man beweist leicht, daß $u^{k'} \to u^*$ auch in x und t ($0 < t < T$) schwach zutrifft.

Dritter Schritt. Beweis, daß das Feld $u^*(x, t)$ die Bedingung A) des Existenzsatzes erfüllt. u^* gehört in jedem Zylindergebiet $x \subset G$, $0 < t < T$ zur Klasse H', die ja Oberklasse von $H'(N)$ ist (wegen B) gehört es zu der letzteren Klasse). Nach den Ausführungen in der ersten Hälfte von § 4 brauchen wir nur zu zeigen, daß u^* die Gleichungen (4.1) für jedes $a = a^\nu$ und für alle $\tau' > \tau > 0$ befriedigt. Nach 5c) erfüllt $u = u^*$ diese Gleichungen für dieselben τ, τ' und für die ersten k' Felder a^ν. Wir halten jetzt τ, τ' und den Index ν fest und gehen zur Grenze $k' \to \infty$ über. Es ist klar, daß auf der linken Seite von (4.1) u durch u^* ersetzt werden darf. Dasselbe gilt vom dritten Integral auf der rechten Seite (das erste ist Null). Man bedenke, daß in

$$\int_\tau^{\tau'} \left[\int_G w_i(x) \, u_i^{k'}(x,t) \, dx \right] dt$$

das innere Integral wegen 5d), erste Hälfte, eine hinsichtlich k' gleichmäßig beschränkte Funktion von t ist und daß man auf das äußere t-Integral einen bekannten Lebesgueschen Konvergenzsatz anwenden kann. Daß man auch im zweiten Integral rechts in (4.1) Grenzübergang $k' \to \infty$ und Integration vertauschen darf, erfordert eine tiefere Überlegung, die von der zweiten Hälfte von 5d) Gebrauch macht. Wir benötigen dazu den folgenden Satz, den wir in § 6 nachträglich beweisen werden.

Lemma 5.1. *Die Folge der für $x \subset G$, $0 < t < T$ stetig x-differenzierbaren Funktionen $f^k(x, t)$ habe folgende Eigenschaften: Für jedes feste t gehört f^k im x-Gebiete G zur Klasse N. Für jedes feste t konvergieren die $f^k(x, t)$ in G schwach gegen eine Funktion $f^*(x, t)$. Die Integrale*

$$\int_G f^2(x, t) \, dx, \qquad \int_0^T \int_G f_{/i} f_{/i} \, dx \, dt \qquad (f = f^k)$$

bleiben hinsichtlich t und k gleichmäßig beschränkt. Dann konvergieren die f^k im x-t-Gebiete $x \subset QG$, $0 < t < T$ stark gegen f^*. Dabei ist Q ein beliebiger endlicher Quader des x-Raumes. Insbesondere gilt die Folgerung für G selbst, falls G beschränkt ist.

Wegen 5a), 5b), wegen des Resultates des zweiten Schrittes und wegen 5d) sind die Voraussetzungen des Lemmas bei den Komponenten der Feldfolge $u^{k'}(x, t)$ für beliebiges festes T erfüllt. Es folgt also, daß

$$\int_0^T \int_{QG} (u_i - u_i^*)(u_i - u_i^*)\, dx\, dt \qquad (u = u^{k'})$$

für $k' \to \infty$ nach Null strebt, wenn Q ein beliebiger endlicher Quader des x-Raumes ist. Damit kann man den Grenzübergang $k' \to \infty$ im zweiten Integral rechts in (4.1) ($a = a^\nu$, ν fest) rechtfertigen. Man bedenke, daß der a-Faktor des Integranden außerhalb einer festen kompakten Teilmenge C von G verschwindet. Wählt man $Q \supset C$ und $T > \tau'$, so liegt bei diesem Integral

$$\int_\tau^{\tau'} \int_{QG} (a_{i,\alpha} u_\alpha)(u_i)\, dx\, dt \qquad (a = a^\nu,\ u = u^{k'})$$

folgende Situation vor. Der erste Faktor des Integranden konvergiert im Integrationsgebiet schwach gegen $a_{i,\alpha} u_\alpha^*$, während der zweite Faktor stark gegen u_i^* konvergiert. Dies genügt bekanntlich zur Ausführung des Grenzüberganges $k' \to \infty$ unter dem Integralzeichen. Damit ist bewiesen, daß das Feld u^* die Gleichungen (4.1) für jedes Feld $a^\nu(x)$ und für alle positiven τ, τ' befriedigt. Die Bedingung A) des Existenzsatzes ist damit bis auf die Divergenzfreiheit verifiziert. Die letztere Eigenschaft ist indessen trivialerweise erfüllt, sogar für jedes feste $t > 0$.

Zur Vervollständigung des Beweises des Existenzsatzes braucht nur noch gezeigt zu werden, daß auch die Anfangsbedingung C) erfüllt ist. Aus der Energiegleichung (4.10) folgt

$$\frac{1}{2}\int_G u_i u_i\, dx \Big|_0 = \frac{1}{2}\int_G u_i u_i\, dx \Big|_T + \mu \int_0^T \int_G \frac{\partial u_i}{\partial x_\beta} \frac{\partial u_i}{\partial x_\beta}\, dx\, dt \tag{5.3}$$

für jedes Feld u unserer Folge. Die linke Seite strebt wegen (4.11) für $k' \to \infty$ nach

$$\frac{1}{2}\int_G U_i U_i\, dx.$$

Für $t = T$ konvergieren in G die Felder für $k' \to \infty$ schwach gegen u^*. Im x-t-Zylinderabschnitt gilt wegen Lemma 2.2 und 5d) schwach

$$u_{i,\beta}^{k'} \to u_{i,\beta}^*.$$

Durch Anwendung von Lemma 2.1 folgt daher aus (5.3) die Ungleichung

$$\frac{1}{2}\int_G U_i U_i\, dx \geqq \frac{1}{2}\int_G u_i^* u_i^*\, dx \Big|_T + \mu \int_0^T \int_G u_{i,\beta}^* u_{i,\beta}^*\, dx\, dt$$

für beliebiges $T > 0$. Insbesondere ist

$$\overline{\lim_{t \to 0}} \int_G u_i^* u_i^*\, dx \leqq \int_G U_i U_i\, dx.$$

Wenn man auf die letztere Ungleichung noch einmal Lemma 2.1 anwendet, erkennt man, daß die Anfangsbedingung C) erfüllt ist, was zu beweisen war.

Auf die Frage der starken Konvergenz bei festem t gehen wir hier nicht mehr ein.

§ 6. Beweis von Lemma 5.1.

Das Lemma ist mit dem Rellichschen Auswahlsatz eng verwandt und wird auch ähnlich bewiesen[9]).

Im voraus sei bemerkt, daß das Lemma wie auch der Satz von Rellich nicht für G selbst zu gelten braucht, wenn G unbeschränkt ist. Ein Gegenbeispiel ist der Fall, wo G der ganze x-Raum und

$$f^k(x, t) = f(x_1 + k, x_2, \ldots, x_n)$$

mit in G zu H' und N gehörigem f ist. In diesem Fall ist $f^* = 0$, aber es findet keine starke Konvergenz gegen Null statt[10]).

Der Beweis von Lemma 5.1 ergibt sich aus der Ungleichung von Friedrichs: Q sei ein endlicher Quader im x-Raum. Zu beliebig vorgegebenem $\varepsilon > 0$ existiert eine endliche Anzahl von in Q zu H gehörigen festen Funktionen $\omega_\nu(x)$ derart, daß die Ungleichung

$$\int\limits_Q f^2 dx \leqq \sum_\nu \Big[\int\limits_Q f\,\omega_\nu\,dx\Big]^2 + \varepsilon \int\limits_Q f_{/i} f_{/i}\,dx$$

durch jede in Q zu H' gehörende Funktion $f(x)$ befriedigt wird[11]).

Zum Beweise von Lemma 5.1 beachte man zunächst, daß bei festem t die Funktionen $f^k(x, t)$ des Lemmas in G stetig differenzierbar und von der Klasse N sind. Wenn man diese Funktionen außerhalb G durch Null erklärt, so bleibt diese Aussage gültig, wenn sie auf den ganzen x-Raum statt auf G bezogen wird. Bei festem t gehört insbesondere jede der Funktionen in jedem endlichen Quader Q des x-Raumes zur Klasse H'. Die Ergänzung der Funktionen und die letztere Aussage wurde durch jene Voraussetzung der Zugehörigkeit zu N ermöglicht. Diese Voraussetzung wird aber auch nur an dieser Stelle benützt. Wir geben nun einen Quader Q und eine Zahl $\varepsilon > 0$ beliebig vor und wählen die endlich vielen Hilfsfunktionen $\omega_\nu(x)$ so, daß in Q die Friedrichssche Ungleichung gilt. Wir wenden sie bei festem t auf die Funktionen

$$f(x, t) = f^k(x, t) - f^l(x, t) \tag{6.1}$$

---

[9]) Vgl. Courant-Hilbert, l. c. Fußnote [4]) S. 218. Im Satze von Rellich wird die Beschränktheit der x-Integrale der Quadrate der Derivierten vorausgesetzt. Unsere Beschränktheitsvoraussetzung betrifft lediglich das x-t-Integral und ist deshalb der Sachlage in unserem Problem besser angepaßt.

Leray beweist und benützt l. c. Fußnote [1]) S. 214, Lemma 2, ein dem Rellichschen Satze noch näher stehendes Lemma, das wie dieser Satz nur mit den x-Integralen arbeitet. Unser Konvergenzbeweis ist direkter.

[10]) Aus dem Lemma allein kann man daher die starke Konvergenz der angenäherten Felder $u(x, t)$ gegen $u^*(x, t)$ in den Zylinderabschnitten nur dann erschließen, wenn G beschränkt ist. Indessen trifft starke Konvergenz zweifellos bei beliebigem G zu. Leray hat sie bei seinen Approximationen im Falle, wo G der ganze x-Raum ist, mit Hilfe einer komplizierten Abschätzung der Energieverteilung über G deduziert. Wir hoffen, auf die stärkeren Konvergenzeigenschaften unserer Annäherungen später zurückzukommen.

[11]) Die ω_ν können als orthonormal in Q angenommen werden. Die Ungleichung stellt dann eine Abschätzung des Unterschiedes in der Besselschen Ungleichung dar. Den Beweis der Ungleichung findet man bei Courant-Hilbert, l. c. Fußnote [4]) S. 218, Kap. VII, § 3, Abschn. 1. Man überzeugt sich leicht, daß der Beweis, der dort in zwei Dimensionen geführt wird, auch in n Dimensionen funktioniert. Die Friedrichssche Ungleichung gilt nicht für beliebige beschränkte Gebiete.

an, die in Q gewiß zu H' gehören. Durch Integration nach t folgt, daß sämtliche Funktionen (6.1) die Ungleichung

$$\int_0^T\int_Q f^2\,dx\,dt \leqq \sum_\nu \int_0^T \left[\int_Q f\,\omega_\nu\,dx\right]^2 dt + \varepsilon \int_0^T\int_Q f_{/i} f_{/i}\,dx\,dt \tag{6.2}$$

befriedigen. Nach Voraussetzung (schwache Konvergenz bei festem t) ist

$$\lim_{\substack{k\to\infty\\ l\to\infty}} \int_Q f\,\omega_\nu\,dx = 0$$

für jedes feste t. Auf Grund der Beschränktheitsvoraussetzung (erste Hälfte) bleibt ferner die Funktion von t

$$\int_Q (f^k - f^l)\,\omega_\nu\,dx$$

bezüglich k, l gleichmäßig beschränkt. Daher strebt der erste Term rechts in (6.2) für $k \to \infty$, $l \to \infty$ nach Null. Nach Voraussetzung bleibt auch der Faktor von ε für die Funktionen (6.1) unterhalb einer festen Schranke. Aus

$$\overline{\lim_{\substack{k\to\infty\\ l\to\infty}}} \int_0^T\int_Q (f^k - f^l)^2\,dx\,dt \leqq c\,\varepsilon$$

folgt aber, da ε beliebig war, die starke Konvergenz unserer Folge im x-t-Gebiet $x \subset Q$, $0 < t < T$. Man überlegt sich leicht, daß die Grenzfunktion die im Wortlaut des Lemmas erwähnte Funktion $f^*(x, t)$ ist. Damit ist Lemma 5.1 bewiesen.

Über die Anfangswertaufgabe für die hydrodynamischen Grundgleichungen, ***Math. Nachr.*** **4 (1950), 213-231.**

Commentary

James B. Serrin

In 1951 Eberhard Hopf reopened the existence question for the initial-boundary value problem for incompressible Navier-Stokes equations, a problem left essentially untouched since the pioneering papers of Jean Leray in 1932-1934. Leray had established the existence and uniqueness of regular solutions *in space dimension 2 when the domain of the solution is the whole space*; and had also found regular solutions *in a finite time interval* $[0, T)$ for bounded domains in dimension $n = 3$, but there the problem had rested until Hopf's work.

Hopf's paper "Über die Anfangswertaufgabe für die hydrodynamischen Grundgleichungen" brought new life and cohesiveness to the problem, and has been a source of stimulation to authors ever since that time. In this commentary I wish to consider Hopf's work in some detail, especially since the paper itself, as a *mathematical contribution*, has been unduly neglected. The reasons for this lie, I believe, in the difficult German text, in the lengthy elaboration of differentiation theory in the first part of the paper (material now well known in the theory of partial differential equations), and finally in some troubles in reading the proof, which I shall return to later. At the same time, one cannot over-emphasize the far reaching nature of Hopf's fundamental result, that it applies in an arbitrary number of space dimensions, and the fact that the proof is carried out by simpler means than Leray's, simpler also than the later proofs which I know. In this respect, Hopf makes the following interesting comment (page 217): "Der Hauptpunkt der Arbeit ist, dass die Konstruktion der angenäherten Lösungen die in Lerays Arbeiten einen so breiten Raum einnimmt, hier durch einen einfachen Prozess ersetzt wird, des sich auch auf viel weitere Klassen partieller Differentialprobleme anwenden lässt."

The commentary will consist of three sections. First, the statement of Hopf's result and a short discussion of the proof, including an elegant but little known compactness lemma. I also emphasize several places which Hopf left without sufficient discussion. The second section concerns the exceptional case of two spatial dimenions, where a remarkable degree of completeness has subsequently been obtained through work of Prodi, Lions and Ladyzhenskaya.

The third section discusses various refinements of Hopf's work in the space dimensions $n = 3$ and $n = 4$ (and some additional considerations for $n \geq 5$), largely due to Ladyzhenskaya (1957).

There are any number of monographs describing the Navier-Stokes equations and the associated initial-boumdary value problem, excellent sources being Ladyzhenskaya's book (1967), Temam's treatise (2000) and the monograph of Sell and You (2002). Historically minded readers will also be rewarded by consulting Leray's original papers.

1. Hopf's fundamental result. Hopf's main existence theorem for the initial-boundary value problem is formulated for weak solutions

$$u = u(x,t) = (u_1(x,t), u_2(x,t), \cdots, u_n(x,t))$$

in a strict distributional sense. For the convenience of the reader we restate the result (Existenzsatz, page 223) in a more compact form. Here it is useful first to introduce the function space V:

Let G be an arbitrary domain of $\mathbf{R}^n$. By V we denote the closure of smooth divergence-free vector fields $\phi(x,t)$ in $\mathbf{R}^n$ with compact support in $\hat{G} = G \times (0,\infty)$, under the space-time norm

$$\int_0^\infty (|\phi|^2 + |D\phi|^2)\, dt,$$

where $|\phi|^2 = |\phi|^2(t) = \int_G \phi^2(x,t)\, dx$ and $D\phi$ denotes the gradient matrix of ϕ (notation obvious).

[The space V is essentially Hopf's space $H'(N)$. The definition of V encloses the notion of weakly divergence-free vector fields as well as the weak satisfaction of the imposed boundary condition $u = 0$. In his definitions Hopf uses weak rather than strong closures, although these are equivalent in the present case, as Hopf notices in the footnote on page 223.]

Main Existence Theorem (page 223). *Let G be an arbitrary domain in $\mathbf{R}^n$, and consider initial data $u_0(x)$ which is in the $L^2(G)$ closure of smooth divergence-free vector fields in $\mathbf{R}^n$, with compact support in G.*

Then there exists a velocity vector field $u = u(x,t)$ in V, such that the distribution identity (1.4) (*page* 215) *for the Navier-Stokes equations is satisfied. Moreover the initial data u_0 is attained in the sense that $|u(\cdot,t) - u_0(\cdot)| \to 0$ as $t \to 0$.*

The vector field u in the main theorem has two important properties not included in the statement, (*i*) the identity (1.4) can be replaced by (4.1), and (*ii*) the solution obeys an *energy inequality*

$$\frac{1}{2}|u|^2 + \nu \int_0^t |Du|^2 dt \le \frac{1}{2}|u_0|^2$$

at each time $t \ge 0$, where ν denotes the kinematic viscosity of the fluid in question. In fact, it is precisely (4.1) which Hopf actually proves in the paper, see the third part (Dritter Schritt) of the the demonstration, page 228. The energy inequality, finally, is included (in passing) at the last paragraph of the proof, page 229.

Turning to the main proof of the theorem, the application of the Galerkin method is particularly to be emphasized. Hopf in fact describes the technique as "dieses einfachsten und nächstliegenden Annäherungsverfahrens" (page 225). Hopf's preparation of the Galerkin method in Section 4 is exemplary: we would only add that the orthonormal construction preceding Lemma 4.3 on page 225 is in fact the well-known Gram-Schmidt procedure. [Two years before Hopf's work, Faedo had suggested application of the Galerkin method to the simpler case of linear evolutionary equations, but it was Hopf's use of the technique for *nonlinear* problems that cemented its importance in the context of partial differential equations.]

The proof itself (Section 5, pages 226–229) rests squarely on the basic (pre-energy) identity (4.10) and the subsequent relation

$$|u^{(k)}|^2 = \sum_{i=1}^{k} (\lambda_i^{(k)})^2$$

which appears a few lines later.

The whole treatment is masterful. It is worth noting that the ultimate step in the demonstration necessarily involves proving the (strong) compactness of the approximating family $u^{(k)}$; here Hopf proceeds in a beautiful way through Lemma 5.1 (page 228). [In fact, the present writer, in admiration of Hopf's lemma, gave a generalization of the result (1962), with perhaps a slightly more transparent proof.]

Altogether, Hopf's result and its proof are remarkable, though a small final point should be noted: the proof of the last part of the theorem, namely that the initial data u_0 is strongly attained in $L^2(G)$, requires for the validity of the argument (foot of page 229), the assertion of *weak continuity in time of the vector* $u(x,t)$. This however follows immediately from (4.1).

2. The special dimension $n = 2$. Prodi (1960) showed by not entirely obvious means that the energy inequality for $n = 2$ can be improved to an equality

$$|u|^2 = |u_0|^2 - \nu \int_0^t |Du|^2 dt.$$

Even more, from this it follows that $|u|^2$ is weakly differentiable (as a function of time) with L^1 derivative $-\nu|Du|^2$; hence, for *bounded* domains contained in some ball of radius R, one gets from Poincaré's inequality

$$\frac{d|u|^2}{dt} \le -\frac{2\nu}{R^2}|u|^2.$$

By integration we find that $|u|^2 e^{2\nu t/R^2}$ is a decreasing function, and in particular

$$|u|^2 \le |u_0|^2 e^{-2\nu t/R^2}.$$

Thus in turn $|u| \to 0$ as $t \to \infty$. This calculation of course applies in any case where the energy equality holds and the domain is bounded.

It is surprising that the exponential decay condition continues to be valid for arbitray dimensions even though the energy equality itself may no longer hold. In fact by using Hopf's pre-energy equality (4.10) together with the above calculation, one finds at once that the approximating energy $|u^k|$, see above, also exhibits the exponential decay found before. But then, since $|u|^2 \le |u^k|^2$ as in Hopf's proof, the desired conclusion follows at once.

It was also shown by Lions and Prodi (1959) that *Hopf's weak solution (for* $n = 2$*) is uniquely determined by its initial data* u_0. In their proof, the inequality

$$\int_0^t (u, v \cdot Dw)dt \le \text{Const.}\, |u|^{1/2}|v|^{1/2} \left(\int_0^t |Du|^2 \int_0^t |Dv|^2 \right)^{1/4} \left(\int_0^t |Dw|^2 \right)^{1/2} \quad (1)$$

plays a crucial role: (1) is in fact an easy consequence of Hölder's inequality together with the Sobolev interpolation inequality

$$|u|_4 \leq c|u|^{1/2}|Du|^{1/2}; \tag{2}$$

here $(\cdot,\cdot)$ denotes the L^2 inner product on G, and $|u|_s$ the standard $L^s(G)$ norm of u.

It seems highly unlikely that Hopf's weak solution should be a regular function of the spatial variable x; on the other hand, if the data u_0 is somewhat more regular, then this will in fact be the case (precisely, u will then be analytic in the spatial variable x at each time $t > 0$). This result is due to the combined work of Ladyzhenskaya (1957, 1959), Golovkin (1960), Ohyama (1960), Serrin (1962) and Kahane (1969). Ladyzhenskaya, first of all, showed that *if the initial data u_0 is in H^2, namely the closure under the norm*

$$|\varphi|^2 + |D\varphi|^2 + |D^2\varphi|^2 \tag{3}$$

of smooth divergence-free vector fields φ with compact support in G, then there are weak solutions $u(x,t)$ of the Navier-Stokes equatiions such that $|Du(\cdot,t)|$ and $|u_t(\cdot,t)|$ exist and are uniformly bounded, at least for initial time intervals $[0,\tau]$.

Here τ is any fixed time, $0 < \tau < T$, while $T < \infty$ depends only on the norm $\|u_0\|_H$. Serrin (1963) showed that one can in fact take $T = \infty$ here, and the demonstration is then completed through the work of the other authors cited. It is interesting that the inequality (1) for the triple product $(u, v\cdot Dw)$ is closely related to this proof.

The preceding results, that is, the energy equality and uniqueness and regularity, provide a satisfactory resolution of the initial-boundary value problem for $n = 2$ dimensions, in the classical sense of Hadamard. The situation when $n \geq 3$ is less complete, as we shall see, though important conclusions can still be drawn.

3. Dimensions $n \geq 3$. Here very little additional information can be deduced without first establishing further regularity of the solution vector $u(x,t)$. To unify the presentation, we begin by presenting a general result applying for all dimensions $n \geq 3$. In stating this result, we shall denote by $L^{s,s'}(\hat{G})$ $(\hat{G} = G \times (0,\tau))$ the function space defined by the condition

$$\int_0^\tau |u(\cdot,t)|_s^{s'}\,dt < \infty,$$

where $1 \leq s, s' < \infty$, with natural extensions to the values s, $s' = \infty$.

Theorem 1. *Let u be a weak solution of the Navier-Stokes equations in the sense of the Main Existence Theorem of Section 1. Suppose that $u \in L^{s,s'}(\hat{G})$ $(\hat{G} = G \times (0,\tau))$, where*

$$\frac{n}{s} + \frac{2}{s'} = 1. \tag{4}$$

Then the energy equality holds for all $t \leq \tau$. Moreover, if v is any solution satisfying the conditions of Hopf's theorem, and $u_0 = v_0$, then $u \equiv v$ for all $t \leq \tau$. (Serrin (1963, Theorems 5, 6).

The basic inequality

$$\int_0^\tau (u, v\cdot Dw)dt \leq \text{Const.}\, |u|_{s.s'}|Dw|_{2.2}\left(|v|_{2.\infty}^{2/s'}\ |Dv|_{2.2}^{n/s}\right)$$

plays a key role in the proofs when $n \geq 3$. In the same way as (1), this inequality arises from a straightforward application of the Hölder and Sobolev inequalities. (The space $L^{s,s'}$ was also used later by J.-L. Lions (1969).)

A direct connection with the two dimensional case is also worth noting: that is, for the Hopf solution in $n = 2$ one finds from (2) and the energy inequality that

$$|u|_{4.4} \leq \text{Const.}\, |u_0|^{1/2} \cdot |u_0|^{1/2}/\nu^{1/4} < \infty$$

on $G \times (0, \infty)$; the pair $(s, s') = (4, 4)$ moreover satisfies (4) when $n = 2$, thus providing another motivation for the uniqueness of weak solutions in this case.

The following general regularity result also holds.

Theorem 2. *Let the hypotheses of Theorem 1 be satisfied, except that (4) is replaced by*

$$\frac{n}{s} + \frac{2}{s'} < 1. \tag{5}$$

Then $u(x,t)$ is analytic in the space variable x for each t in $(0, \tau)$, and each spatial derivative is bounded in compact subregions of $\hat{G}$.

This result is due to Kahane (1969). The condition (4) precludes direct application of Theorem 2 to Hopf's weak solutions in any dimension $n \geq 3$, while the stronger condition (5) even disallows direct application when $n = 2$. It is consequently necessary once more to have stronger existence theorems.

Theorem 3. (Ladyzhenskaya (1957, 1959). *Let $n = 3$. Then for initial data $u_0 \in H^2$, there exists a time T, $0 < T < \infty$, and weak solutions $u(x,t)$ of the Navier-Stokes equations, such that $|Du(\cdot, t)|$ and $|u_t(\cdot, t)|$ are uniformly bounded for all times t in $[0, \tau]$ with $\tau < T$.*

This result is crucially different than the corresponding one in two dimensions, in that the solution may not satisfy the boundedness conclusions for all times $\tau < \infty$. A refinement of Ladyzhenskaya's result does show however that if $\|u_0\|_H$ is sufficiently small, then one can take $T = \infty$; see Serrin (1963, Theorem 2), and Fujita and Kato (1964).

With the help of Theorem 3 and Sobolev's inequality, one obtains for the solutions in question that, for $n = 3$,

$$|u|_6 \leq \text{Const.}\, |Du| \leq \text{Const.},$$

for $0 \leq t \leq \tau$, $\tau < T$. This in turn gives

$$|u|_{6.4} \leq \text{Const.}\, t^{1/4}, \quad 0 \leq t \leq \tau,$$

where the pair $(s, s') = (6, 4)$ now obeys (4). In consequence of Theorem 1, Ladyzhenskaya 's solution $u(x,t)$ then obeys the *energy equality* and is *unique* – for $0 \leq t < T$ – in the class of weak solutions obtained by Hopf; see Serrin (1963, Theorem 6). It should be noted as well that Ladyzhenskaya (1957) had obtained a preliminary version of the uniqueness of solutions, this being the foundation for the more general result stated here.

Also one has $|u|_{6.\infty} \leq \text{Const.}$, so Theorem 2 shows that Ladyzhenskaya's solution is analytic in the space variables at each fixed time $0 < t < T$.

Turning to higher dimensions, one can in fact extend Theorem 3 to the case $n = 4$, the conclusion being that *if the initial data u_0 is sufficiently small in the* *norm H^2, then $|Du(\cdot,t)|$ and $|u_t(0,t)|$ are uniformly bounded for all times $t \geq 0$.* (Serrin (1963); Theorem 3); see also Temam (2000, Chapter 3, Theorem 3.1) and J.-L. Lions (1959).)

Again with the help of the Sobolev inequality, one has for $n = 4$,

$$|u|_4 \leq \text{Const.}\,|Du| \leq \text{Const.}$$

for all $t \geq 0$, that is $|u|_{4,\infty} \leq$ Const. for $t \geq 0$. The pair $(s, s') = (4, \infty)$ satisfies (4) when $n = 4$. Thus by Theorem 1 the solution again obeys the *energy equality* and is *unique* in the class of weak solutions obtained by Hopf.

The pair $(4, \infty)$ on the other hand does *not* satisfy (5) when $n = 4$ (though it comes close). Consequently, when $n = 4$ one does not have a regularity theorem *for solutions which are known to exist.*

For $n \geq 5$ the result of Theorem 1 is of course still valid, but a corresponding existence theorem is unavailable; in other words, we know the existence but *not* the uniqueness and regularity of weak solutions, and we know uniqueness and regularity of suitably restricted solutions, but *not* the existence of such solutions.

A pivotal question of intense interest, finally, is the behavior of solutions for $n = 3$ when the initial data is not small enough to allow $T = \infty$; see Theorem 3 and the remark following it (explicit bounds are stated in Serrin (1963)). In such cases, even for smooth data, a regular solution may exist only for some sufficiently small time interval, after which it loses both its regularity and uniqueness. It would be fruitless to speculate here on whether such situations can actually occur, but the reader can be referred to Temam's monograph (2000), page 345, and to the recent review article of J.-Y. Chemin (2000). At ths same time, one may express the belief that irregularity and non-uniqueness do in fact occur for the three dimensional Navier-Stokes equations for appropriate initial data u_0. On the other hand, Hopf writes on page 217: "Es ist schwer zu glauben, dass die Anfangswertaufgabe zäher Flussigkeiten für $n = 3$ mehr als eine Lösung haben könnte, und der Erledigung der Eindeutigkeitsfrage sollte mehr Aufmerksamkeit geschenkt werden."

Of course, the validity of the Navier-Stokes equations themselves may be questioned in cases where velocity gradients become excessively large, a point raised by Temam in the comments noted above. Thus if one assumes that stresses grow superlinearly with respect to velocity gradients (not just linearly as for the Navier-Stokes equations), it becomes believable that uniqueness and regularity will persist for all time even in dimension 3, or for that matter, in arbitrary dimensions. A result to this effect has in fact been established in a little known but perceptive paper of Ladyzhenskaya!

Acknowledgement. The author wishes to thank Prof. Norman Meyers for a number of helpful conversations concerning this work.

References

CHEMIN, J.-Y., "Jean Leray and Navier-Stokes." *Gazette des Mathématiciens* **84** (Supplement) (2000), pp. 71-82.

FAEDO, S., "Un nuove metodo per l'analisi esistenziale e quantitativa dei problemi di propagazione." *Annali Scuola Norm. Sup. Pisa*, Ser 3 **1** (1949), pp. 1-41.

FOIAS, C., "Une remarque sur l'unicité des solution des équations de Navier-Stokes en dimension n." *Bull. Soc. Math. France* **89** (1961), pp. 1-8.

FUJITA, H., AND T. KATO, "On the Navier-Stokes initial value problem." *Archive Rational Mech. Analysis* **16** (1964), pp. 269-315.

GALDL, G. P., *An Introduction to the Mathematical Theory of the Navier-Stokes Equations.* New York: Springer, 1994.

ITO, S., "The existence and uniqueness of regular solution of non-stationary Navier-Stokes equation." *J. Faculty Science, Univ. Tokyo,* **9** (1961), pp. 103-140.

KAHANE, C., "On the spatial analyticity of solutions of the Navier-Stokes equations." *Archive Rational Mech. Analysis* **33** (1969), pp.386-405.

KISELEV, A.A., AND O. A. LADYZHENSKAYA, "On existence and uniqueness of the solution of the non-stationary problem for a viscous incompressible fluid." *Izvestiya Akad. Nauk SSSR* **21** (1957), pp. 655-680.

LADYZHENSKAYA, O.A., "Solution "in the large" of non-stationary boundary value problem for the Navier-Stokes system with two space variables." *Comm. Pure Applied Math.* **12** (1959), pp. 427-433.

LADYZHENSKAYA, O. A., *Mathematical problems in the dynamics of viscous incompressible fluids.* Moscow 1961.

LERAY, J., "Etude de diverses équations intégrales non-linéares et de quelques problèmes que pose l'hydrodynamique." *Journ. Math. Pures Appl.* **12** (1933), pp.1-82.

LERAY, J., "Essai sur les mouvements plans d'un liquide visqueux que limitent des parois." *J. Math. Pures Appl.* **13** (1934), pp. 331-418.

LERAY, J., "Sur le mouvement d'un liquide visqueux emplissant l'espace." Acta Math. **63** (1934), pp. 193-248.

LION, J.-L., "Quelques résultats d'existence dans les équations aux dérivées partielles nonlinéaires." *Bull. Soc. Math. France* **87** (1959), pp. 245-273.

LIONS, J.-L., *Quelques methodes de resolution des problèmes aux limites non linéaires.* Paris: Dunod 1969.

LIONS, J. L., AND G. PRODI, "Un theoreme d'existence et unicite dans les equations de Navier-Stokes en dimension 2." *C. R. Acad. Sci. Paris* **248** (1959), pp. 3519-3521.

PRODI, G., *Un teorema di unicitá per le equazioni di Navier-Stokes.*

PRODI, G., "Qualche risultato riguardo alle equazioni di Navier-Stokes nel caso bidimensionable." *Rend. Sem. Mat. Padova* **30** (1960), pp. 1-15.

SELL, G, AND Y. YOU, *Dynamics of Evolutionary Equations.* New York: Springer 2002. See especially Chapter 6.

SERRIN, J., "Strong convergence in a product space." *Proc. Amer. Math. Soc.* **13** (1962). pp. 651-655.

SERRIN, J., "On the interior regularity of weak solutions of the Navier-Stokes equations." *Arch. Rational Mech. Analysis.* **9** (1962), pp. 187-195.

SERRIN, J., "The initial value problem for the Navier-Stokes equations." In *Nonlinear Problems*, ed. R. E. Langer, pp. 69-98. Univ. Wisconsin Press, 1963.

TEMAM, R., *Navier-Stokes Equations.* Third edition, North-Holland, 1984; revised and with extensive additional comments, Providence: AMS Chelsea Publ., 2000.

Hamilton's Theory and Generalized Solutions of the Hamilton-Jacobi Equation

E. D. CONWAY AND E. HOPF*

1. Introduction. Let $f(t, x, v)$ be of class C'' and let

$$f_{vv}(t, x, v) > 0 \quad \text{and} \quad f_v(t, x, \pm\infty) = \pm\infty,$$

respectively, for all values of the real variables t, x, v. Let $g(t, x, z)$ be the conjugate function figuring in the Legendre transformation and its inverse

$$z = f_v(t, x, v), \qquad v = g_z(t, x, z),$$

$$f(t, x, v) + g(t, x, z) = vz,$$

(with t, x as parameters). $g(t, x, z)$ is then defined and also of class C'' for all values of t, x, z.

The classical theory of Hamilton (see the respective chapters in [1] and [2]) furnishes smooth solutions of the Hamilton-Jacobi equation

$$J_t + f(t, x, J_x) = 0, \tag{1.1}$$

by means of the associated variational problem

$$\delta \int_{t_0}^{t_1} g(\tau, x(\tau), \dot{x}(\tau))\, d\tau = 0,$$

in the following direct way. Consider the endpoint t_1, $x_1 = x(t_1)$ fixed and the other endpoint $(t_0, x(t_0))$ variable on a fixed curve γ (of class C'') in the plane of (t, x). The solutions $x(\tau)$ of this variational problem, for various fixed endpoints (t_1, x_1) are the extremals (extremal = stationary curve for fixed end points) that meet γ transversally. Suppose that these transversal extremals form a field in some onesided neighborhood N of γ. Then, for any point (t, x) inside of N, the expression

$$J(t, x) = \int_{t_0}^{t} g(\tau, x(\tau), \dot{x}(\tau))\, d\tau,$$

* The second author wishes to express his indebtedness to the Office of Naval Research for a research contract at Indiana University.

Journal of Mathematics and Mechanics, Vol. 13, No. 6 (1964).

evaluated along the unique field extremal $x(\tau)$ from $(t, x = x(t))$ down to its foot point (t_0, x_0) on γ represents a smooth solution in N of the Hamilton-Jacobi equation (1.1) with initial values zero on γ.

A simple generalization furnishes a smooth solution of the Hamilton-Jacobi equation with prescribed initial values (of class C'' along γ)

$$J = \varphi(t_0, x_0) \quad \text{on} \quad \gamma.$$

The correspondingly modified variational problem is

$$\delta\left[\int_{t_0}^{t} g(\tau, x(\tau), \dot{x}(\tau))\, d\tau + \varphi(t_0, x_0)\right] = 0, \tag{1.2}$$

again with $(t, x = x(t))$ fixed and with $(t_0, x_0 = x(t_0))$ variable on γ, and the transversality condition becomes a generalized one in which φ figures. Again, if and as far as the extremals starting on γ "transversally" form a field the expression

$$J(t, x) = \int_{t_0}^{t} g(\tau, x(\tau), \dot{x}(\tau))\, d\tau + \varphi(t_0, x_0)$$

evaluated as above represents the required solution.

The classical theory shows that it is the smoothness of the solution of the initial value problem for (1.1) that is closely tied up with the restrictive field condition. The aim of the present paper is a global extension of Hamilton's theory that furnishes a solution in a t, x-domain wider than such a field. Of course this will be a solution in a generalized sense. We impose a drastic restriction on f or, equivalently, on the associated variational problem:

Assumption. *Any two points (t_0, x_0) and (t_1, x_1), $t_0 < t_1$, can be joined by a unique extremal $x(\tau)$. Among arbitrary absolutely continuous curves $x(\tau)$ joining these the extremal furnishes the absolute minimum of $\int_{t_0}^{t_1} g\, d\tau$.*

(The second part could be replaced by the (already stated) condition that $f_{vv} > 0$ plus certain growth conditions on f for large $|v|$, uniformly with respect to t, x. The existence of the absolute minimum could then be inferred from known theorems concerning this minimum.)

It follows that for this unique extremal is the absolute minimum attained.

We first describe the general trend of this paper (in rather vague terms). Under the assumption made above it is reasonable to look for the minimum of the expression considered in (1.2) and thus we define a function

$$J(t, x) = \min\left[\int_{t_0}^{t} g(\tau, x(\tau), x(\tau))\, d\tau + \varphi(t_0, x_0)\right], \tag{1.3}$$

with respect to all piecewise smooth curves $x(\tau)$ joining $(t, x = x(t))$ to some point $(t_0, x_0 = x(\tau_0))$ variable along the initial curve γ. Certain precautions have to be taken, however.

If γ is the line $t_0 = 0$ or part of it the point (t, x) is to be restricted to the upper half plane $t > 0$. If t were < 0 (1.3) would in general be $-\infty$; consider, for instance, the case in which $f = v^2/2$ (and, hence $g = z^2/2$) and $\varphi = 0$. If, more generally, γ is of the form $t_0 = t_0(x_0)$ then, for similar reasons, we have to restrict the points (t_0, x_0) figuring in (1.3) to the part of γ on which $t_0 < t$ (this part may vary with t, x). If the minimum in (1.3) is understood in this way then $J(t, x)$ is well-defined in the half-plane $t > \inf t_0(x_0)$. Under certain additional conditions on f and on the initial data γ, φ it is possible to deduce the following properties of $J(t, x)$ in this half plane. J is finite, locally Lipschitzian and it satisfies almost everywhere the Hamilton-Jacobi-equation (1.1); furthermore, it has limit values φ on γ. (In unpublished notes written in 1959–1960 the second author had carried out part of this program.) Also, within a field of "transversal" extremals abutting on γ on the appropriate side, (1.3) coincides with the smooth solution mentioned above.

Of some interest is the close connection between the Hamilton-Jacobi-equation (1.1) and the quasilinear equation

$$v_t + [f(t, x, v)]_x = 0 \tag{1.4}$$

(conservation law). The transformation

$$v = J_x \tag{1.5}$$

carries C'' — solutions of (1.1) into C' — solutions of (1.4). Conversely, to every C' — solution v of (1.4) in a $t - x$-rectangle R, (1.5) determines a C'' — solution J of (1.1) in R. This connection is shown by us to remain valid between locally Lipschitzian solutions of (1.1) and locally bounded weak solutions of (1.4). Weak solutions of (1.4) have received some attention in the literature of the past 15 years (E. Hopf [5], case $f = v^2/2$, P. Lax [6], case $f = f(v)$, A. Douglis [3], same case, O. A. Oleinik [7], more general $f(t, x, v)$ and others). In her penetrating and comprehensive work Oleinik solves the weak initial value problem for (1.4) with γ an interval on $t_0 = 0$. Her assumptions on f go in the same general direction as ours. Her main result about this particular problem is its solution by an explicit formula. This formula is not easy to understand but it becomes clear when viewed from the variational background, through its connection with the Hamilton-Jacobi-equation (1.1) the initial values for J on $t = 0$ being

$$J(0, x_0) = \varphi(x_0) = \int v(0, x_0)\, dx_0 . \tag{1.6}$$

Oleinik's formula is nothing but (1.3), with $x(\tau)$ restricted to extremals (this makes no difference in (1.3)).

In order to avoid the many complicated details of the more general theory we have confined the discussion (by the first author) to the simple yet sufficiently typical case

$$f = f(v), \qquad f'' > 0, \qquad f'(\pm\infty) = \pm\infty, \qquad \text{respectively,}$$

and, accordingly, $g = g(z)$, $g'' > 0$. In this case the extremals are straight lines. Our assumption stated above is satisfied, and the part concerning the absolute minimum follows from Jensen's inequality for convex functions.

Suppose that γ is the entire line $t_0 = 0$ and that the initial values $\varphi(x_0)$ of J are, say, continuous and of at most linear growth for large $|x_0|$. Then the formula (1.3), with $x(\tau)$ extremals = straight line segments, becomes

$$J(t, x) = \min_{x_0} \left[tg\left(\frac{x - x_0}{\tau}\right) + \varphi(x_0) \right], \tag{1.7}$$

and it represents a Lipschitzian solution of

$$J_t + f(J_x) = 0$$

in all of $t > 0$ with limit values $\varphi(x)$ on $t = 0$. Therefore, for given bounded and measurable initial values $v(0, x_0)$ on $t_0 = 0$, (1.6), (1.7), (1.5) represents a weak solution $v(t, x)$ in $t > 0$ of

$$v_t + [f(v)]_x = 0,$$

with weak limit values $v(0, x)$ on $t = 0$. This explicit solution had been found before (E. Hopf [5], special case $f = v^2/2$, P. Lax [6], general f). Most of our results on the weak solutions are found already in the papers of the authors name above but it seemed worthwhile to derive them by methods of calculus of variations.

The present paper also deals with the case where γ is the broken line $t = 0$, $x \geqq 0$ plus $x = 0$, $t \geqq 0$ (called mixed initial and boundary value problem). The weak solution of this problem appears to be new.

There is the intricate question of uniqueness for the initial value problem, and we must mention what is known about it. If γ is a finite or semi-infinite interval on $t = 0$ the *weak* solution is never unique, not for any initial values of J or v, and not even in any $t - x$-domain abutting on γ. However, even if γ is all of $t = 0$ then, for certain initial values of v, the weak solution is not unique (P. Lax [6]). A striking example is obtained in the following way. We start from a general fact (proved in this paper): If J_1, J_2 are Lipschitzian solutions of the Hamilton-Jacobi-equation (1.1) so are

$$\min(J_1, J_2), \qquad \max(J_1, J_2).$$

The equation

$$J_t + \tfrac{1}{2}J_x^2 = 0$$

has the three solutions 0, $2(x - t)$, $2(-x - t)$. Hence, the maximum of the second and third, or $2(|x| - t)$ and

$$J = \min(2|x| - 2t, 0) = \begin{cases} 2|x| - 2t, & |x| < t \\ 0, & |x| \geqq t \end{cases}$$

are Lipschitzian solutions. J has the same limit values (zero) on $t = 0$ as the solution zero. Consequently,

$$u_t + uu_x = 0$$

has more than one weak solution in $t > 0$ with weak limit values zero on $t = 0$.

In the case where γ is the whole line $t = 0$. Oleinik, in her remarkable work [7], has obtained weak solutions with given initial values by different methods. Firstly, by a finite difference scheme (suggested by Lax) and secondly, by the limit passage $\epsilon \to 0$ in the solutions $v(t, x; \epsilon)$ of the parabolic equation ($\epsilon > 0$)

$$v_t + [f(t, x, v)]_x = \epsilon v_{xx} ,$$

with fixed initial values $v(0, x)$. Thirdly, by the already mentioned explicit formula. Oleinik has also, in the same memoir, established an additional condition on the solution which fixes the weak solution v of the initial-value problem (we repeat, for the case where γ is all of $t = 0$) uniquely. The solutions obtained by the first two approaches satisfy this condition and, consequently, they coincide. That the weak solution furnished by the variational approach, *i.e.* by the formula (1.5) with arbitrary curves $x(t)$, also satisfies Oleinik's condition is most easily proved by means of simple minimum arguments. We leave the verification to the reader.

In section 2 we set up the variational problem for the initial value problem in the case where γ is a finite part of the line $t = 0$ and we show that a minimum exists. In the next section we show that this minimum is a Lipschitz continuous solution of the Hamilton-Jacobi equation having prescribed initial values. In section 4 we show the connection between weak solutions of the quasilinear equation and Lipschitzian solutions of the Hamilton-Jacobi equation. Using this relationship we obtain weak solutions of the initial value problem from the solutions of the Hamilton-Jacobi equation obtained in sections 2 and 3.

In sections 5 through 9 we give analogous treatments of a boundary value problem and a mixed initial—and boundary-value problem.

In section 10 we outline the treatment of these problems for the case where γ consists of unbounded line segments.

I. The Initial Value Problem

2. Let D be the rectangle,

$$D = \{(t, x): \quad 0 \leqq t \leqq T \quad \text{and} \quad 0 \leqq x \leqq L\}. \tag{2.1}$$

Let $\varphi(x)$ be a Lipschitz continuous function on the closed interval $[0, L]$. We search for a function, $J(t, x)$, which is Lipschitz continuous in D, satisfies the Hamilton-Jacobi equation,

$$J_t + F(J_x) = 0, \tag{2.2}$$

almost everywhere in D and satisfies the initial condition,

$$\lim_{t \to 0} J(t, x) = \varphi(x). \tag{2.3}$$

Throughout this paper the function f is assumed to satisfy the following two conditions.

A. The function f is defined and twice continuously differentiable on the entire real line. Moreover, for all real v we have

$$f''(v) > 0. \tag{2.4}$$

B. The range of f' is the entire real line.

Because of these two conditions, f' maps $(-\infty, +\infty)$ in a one-to-one manner onto $(-\infty, +\infty)$ so that we may employ the Legendre transformation defined by

$$z = f'(v) \tag{2.5}$$

$$g(z) = zv - f(v). \tag{2.6}$$

The function g so defined is twice differentiable over the entire real line and

$$g''(z) > 0, \tag{2.7}$$

for all real z. Moreover,

$$v = g'(z) = g'(f'(v)). \tag{2.8}$$

i.e. g' and f' are inverse mappings.

Lemma 2.1

$$\lim_{\|z\| \to \infty} \frac{g(z)}{|z|} = +\infty. \tag{2.9}$$

Proof. This follows from the fact that g' is an increasing function on the entire real line and has $(-\infty, +\infty)$ as its range.

For $(t, x) \in D$, $t > 0$, let $A(t, x)$ be the set of all (single-valued) absolutely continuous functions, $w(\tau)$, defined on $0 \leqq \tau \leqq t$ such that $0 \leqq w(\tau) \leqq L$ and $w(t) = x$. Each such function represents a curve lying in D and joining (t, x) to some point of the line, $t = 0$.

Let $H[t, x]$ be the function defined on $A(t, x)$ whose value at $w \in A(t, x)$ is given by

$$\int_0^t g(\dot{w}(\tau))\, d\tau + \varphi(w(0)). \tag{2.10}$$

Lemma 2.2. *For each $(t, x) \in D$, $t > 0$, the functional $H[t, x]$ attains a minimum value in $A(t, x)$.*

Proof. Let $w(\tau)$ be any absolutely continuous function on $t_1 \leqq \tau \leqq t_2$, such that $w(t_1) = x_1$ and $w(t_2) = x_2$. Then from the integral form of Jensen's inequality ([4], theorem 204) it follows that

$$\frac{1}{t_2 - t_1}\int_{t_1}^{t_2} g(\dot{w}(\tau))\,d\tau \geqq g\left(\frac{1}{t_2 - t_1}\int_{t_1}^{t_2} \dot{w}(\tau)\,d\tau\right) = g\left(\frac{x_2 - x_1}{t_2 - t_1}\right).$$

Moreover, equality can occur only when

$$\dot{w}(z) \equiv \frac{x_2 - x_1}{t_2 - t_1}.$$

Hence,

$$\int_{t_1}^{t_2} g(\dot{w}(\tau))\,d\tau > (t_2 - t_1)g\left(\frac{x_2 - x_1}{t_2 - t_1}\right) = \int_{t_1}^{t_2} g(\dot{w}_0(\tau))\,d\tau, \tag{2.11}$$

for all $w \neq w_0$,

where

$$w_0(\tau) = x_1 + \left(\frac{x_2 - x_1}{t_2 - t_1}\right)(\tau - t_1).$$

Therefore, for any element $w \varepsilon A\ (t, x)$ such that $w(0) = y$, we have

$$\int_0^t g(\dot{w}(\tau))\,d\tau + \varphi(y) \geqq tg\left(\frac{x - y}{t}\right) + \varphi(y), \tag{2.12}$$

and equality can only occur when $\dot{w}(\tau) = (x - y)/t$. Now the right hand side of (2.12) is a continuous function of y for $0 \leqq y \leqq L$. We can therefore define

$$J(t, x) = \min_{0 \leqq y \leqq L}\left\{tg\left(\frac{x - y}{t}\right) + \varphi(y)\right\}. \tag{2.13}$$

(This function has been used before by P. Lax. *Cf.* [6].) From (2.12) it is clear that $J(t, x)$ is a lower bound for the functional $H[t, x]$ in $A(t, x)$. But it is also a minimum value since if y_0 is any value of y for which the minimum in (2.13) is attained, then

$$J(t, x) = tg\left(\frac{x - y_0}{t}\right) + \varphi(y_0).$$

But this is precisely the value of $H[t, x]$ for the function,

$$w(\tau) = y_0 + \left(\frac{x - y_0}{t}\right)\tau,$$

which is clearly in $A(t, x)$. *q.e.d.*

Let an extremal be any element of $A(t, x)$ for which $H[t, x]$ attains its minimum value. Then every extremal is a straight line segment joining (t, x) to $(0, y)$ where y is some value, $0 \leqq y \leqq L$. More specifically,

Corollary 2.2. If w is an extremal then

$$w(\tau) = y + \left(\frac{x - y}{t}\right)\tau,$$

for some y, $0 \leqq y \leqq L$.

This follows immediately from the exceptional case of Jensen's inequality.

We shall let $y_*(t, x)$ and $y^*(t, x)$ be the infimum and supremum respectively of those values of y for which the minimum in (2.13) is attained. Since g and φ are continuous, we have

$$J(t, x) = tg\left(\frac{x - y_*(t, x)}{t}\right) + \varphi(y_*(t, x)), \tag{2.14a}$$

and

$$J(t, x) = tg\left(\frac{x - y^*(t, x)}{t}\right) + \varphi(y^*(t, x)). \tag{2.14b}$$

Clearly,

$$0 \leqq y_*(t, x) \leqq y^*(t, x) \leqq L. \tag{2.15}$$

3. The main result of this section is embodied in Theorem 1.

Lemma 3.1. Let (t, x), $t > 0$, be in D and let y be the base point of any extremal from (t, x). i.e. the line joining (t, x) to $(0, y)$ is an extremal. Let (t_1, x_1), $t_1 > 0$, be any other point on this extremal. Then $y_*(t_1, x_1) = y = y^*(t_1, x_1)$ and

$$J(t, x) = (t - t_1)g\left(\frac{x - y}{t}\right) + J(t_1, x_1). \tag{3.1}$$

From this lemma we obtain directly the following corollary.

Corollary 3.1. Two extremals from distinct points of D cannot cross (i.e. intersect with different slopes) in the interior of D.

Proof of Lemma 3.1. Let y_1 be the base point for any extremal from (t_1, x_1). Let γ be the broken-line curve formed by the segments joining (t, x) to (t_1, x_1) and (t_1, x_1) to $(0, y_1)$. The functional (2.10) evaluated for this curve results in the value

$$(t - t_1)g\left(\frac{x - y}{t}\right) + t_1 g\left(\frac{x_1 - y_1}{t_1}\right) + \varphi(y_1) = (t - t_1)g\left(\frac{x - y}{t}\right) + J(t_1, x_1).$$

But from (2.13) it follows that

$$J(t_1, x_1) \leqq t_1 g\left(\frac{x_1 - y}{t_1}\right) + \varphi(y).$$

Therefore, since

$$\frac{x_1 - y}{t_1} = \frac{x - y}{t},$$

we see that the functional value at γ is not greater than

$$(t - t_1)g\left(\frac{x - y}{t}\right) + t_1 g\left(\frac{x - y}{t}\right) + \varphi(y) = J(t, x).$$

Therefore, γ is an extremal. But then γ must be a straight line segment so that $y = y_1$. Since y_1 was the base point of an arbitrary extremal from (t_1, x_1), this completes the proof of the lemma.

Lemma 3.2. *$J(t, x)$ is bounded from above and below in D.*

Proof. From (2.13) we obtain

$$J(t, x) \leqq tg(0) + \varphi(x) \leqq T\,|g(0)| + k, \tag{3.2}$$

where k is the supremum of $|\varphi(x)|$, $0 \leqq x \leqq L$. On the other hand, g has a minimum value, g_0, so that from (2.14) we obtain

$$J(t, x) \geqq tg_0 + \varphi(y_*(t, x)) \geqq -T\,|g_0| - k. \tag{3.3}$$

Lemma 3.3. *$J(t, x)$ is Lipschitz continuous in any closed set contained in the interior of D.*

Proof. Let Q be any closed rectangle contained in the interior of D. It is sufficient to prove that J is Lipschitz continuous in Q. To this end let

$$K = \sup_{(t,x)\in Q} \max \left\{ \left| \frac{x - y_*(t, x)}{t} \right|, \left| \frac{x - y^*(t, x)}{t} \right| \right\}.$$

K is finite since t is uniformly bounded away from zero providing (t, x) is in Q. Clearly, the slope of any extremal issuing from a point of Q has magnitude not greater than K. We now let the number A, $A > K$, be large enough so that if P_1 and P_2 are any two points of Q having the same t-co-ordinate, if L_K is a line drawn from P_2 in the direction of decreasing t with slope having magnitude K, and finally if L_A is the line whose slope has magnitude A and is drawn from P_1 so as to intersect the line segment L_K, then the point of intersection of L_A and L_K lies in the interior of D. Such an A exists because Q is a positive distance from the boundary of D.

Now let

$$M_1 = \sup_{|t| \leqq K} |g(t)|,$$

$$M_2 = \max\{|g(A)|, |g(-A)|\}.$$

We first show that $J(t, x)$ is Lipschitz continuous with respect to the variable x in Q. Let (t, x_1) and (t, x_2) be in Q. To be definite let x_1 be greater than x_2. Let E be any extremal from (t, x_2). Let P_K be the line segment of slope K joining (t, x_2) to the line $t = 0$. Let P_A be the line segment of slope A joining (t, x_1) with the line $t = 0$. Let (t_1, ζ_1) be the point of intersection of P_A with P_K. Let (t_2, ζ_2) be the point of intersection of P_A with E. Let y_2 be the base point of E.

Because of the definition of A and K we have

$$t > t_2 \geqq t_1 > 0.$$

Hence, if we let p be the slope of E, then, as a consequence of the definition of J, we have

$$\begin{aligned} J(t, x_1) &< (t - t_2)g(A) + t_2 g(p) + \varphi(y_2) \\ &= (t - t_2)g(A) + J(t, x_2) - (t - t_2)g(p) \\ &\leqq J(t, x_2) + (t - t_2)(M_1 + M_2) \\ &\leqq J(t, x_2) + (t - t_1)(M_1 + M_2). \end{aligned}$$

But from the similarity of triangles

$$(t - t_1) = B(x_1 - x_2),$$

where B depends only upon A and K. Therefore, if we let $B' = B(M_1 + M_2)$ we have

$$J(t, x_1) - J(t, x_2) \leqq B' \, |x_1 - x_2|.$$

Upon reversing our procedure by drawing the line segment P_{-K} (slope is $-K$) from (t, x_1) and the line P_A from (t, x_2) while considering an extremal from (t, x_1) we arrive in the same way at the inequality

$$J(t, x_2) - J(t, x_1) \leqq B' \, |x_1 - x_2|.$$

Therefore

$$|J(t, x_1) - J(t, x_2)| \leqq B' \, |x_1 - x_2|. \tag{3.4}$$

We now show that J is Lipschitz continuous with respect to the variable t in Q. Let (t_1, x) and (t_2, x) be in Q and let $t_1 \leqq t_2$. Let (t_1, x_1) be the point of intersection of the line $t = t_1$ with some arbitrary extremal from (t_2, x). Let this extremal have slope p. Then

$$J(t_1, x) - J(t_2, x) = J(t_1, x) - J(t_1, x_1) - (t_2 - t_1)g(p),$$

where we have used corollary 3.1. Therefore, using (3.4) we have

$$|J(t_1, x) - J(t_2, x)| \leqq B' \, |x - x_1| + M_1 \, |t_1 - t_2|.$$

However

$$\left| \frac{x - x_1}{t_2 - t_1} \right| = |p| \leqq K,$$

so that if $B'' = B'K + M_1$ we have

$$|J(t_1, x) - J(t_2, x)| \leqq B'' \, |t_1 - t_2|. \tag{3.5}$$

Then, of course, if $\bar{B}$ is the larger of B' and B'' we have from the triangle inequality

$$|J(t_1, x_1) - J(t_2, x_2)| \leqq \bar{B} \, (|t_1 - t_2| + |x_1 - x_2|), \tag{3.6}$$

for any points (t_1, x_1) and (t_2, x_2) in Q.

The following lemma is due to P. Lax [6] although his proof differs from ours.

Lemma 3.4. *For fixed t, $0 < t \leqq T$, the functions $y_*(t, \cdot)$ and $y^*(t, \cdot)$ are non-decreasing functions of x. $y_*(t, \cdot)$ is left continuous and $y^*(t, \cdot)$ is right continuous. Moreover, the two functions have the same points of discontinuity and except at these countably many points of discontinuity the two functions are equal.*

Proof. Let $0 \leqq x_1 < x_2 \leqq L$. If $y_*(t, x_2) < y^*(t, x_1)$ an extremal for (t, x_1) would intersect an extremal for (t, x_2) in the interior of D. Corollary 3.1 assures us that this cannot be. Therefore,

$$y_*(t, x_1) \leqq y^*(t, x_1) \leqq y_*(t, x_2) \leqq y^*(t, x_2). \tag{3.7}$$

This inequality demonstrates the non-decreasing nature of the two functions. But since g, φ and J are continuous, inequality (3.7) also implies:

$$\begin{aligned} y_*(t, x) &= y_*(t, x - 0) = y^*(t, x - 0), \\ y^*(t, x) &= y^*(t, x + 0) = y_*(t, x + 0), \end{aligned} \tag{3.8}$$

for all x, $0 \leqq x \leqq L$. From these relations it follows that the two functions have the same points of discontinuity, and except at these points, the two functions are equal.

Lemma 3.5. *If (t, x) is in the interior of D, then*

$$J_x^+(t, x) = \lim_{h \to 0+} \frac{J(t, x + h) - J(t, x)}{h} = g'\left(\frac{x - y^*(t, x)}{t}\right), \tag{3.9}$$

$$J_x^-(t, x) = \lim_{h \to 0-} \frac{J(t, x + h) - J(t, x)}{h} = g'\left(\frac{x - y_*(t, x)}{t}\right). \tag{3.10}$$

Proof. Let h be positive. From (2.13) we have

$$J(t, x + h) \leqq tg\left(\frac{x + h - y^*(t, x)}{t}\right) + \varphi(y^*(t, x)),$$

Therefore, using the mean-value theorem we have

$$\begin{aligned} \frac{1}{h} [J(t, x + h) - J(t, x)] &\leqq \frac{t}{h} \left[g\left(\frac{x + h - y^*(t, x)}{t}\right) \right. \\ &\quad \left. - g\left(\frac{x - y^*(t, x)}{t}\right) \right] = g'\left(\frac{x + \xi(h) - y^*(t, x)}{t}\right), \end{aligned} \tag{3.11a}$$

where $0 \leqq \xi(h) \leqq h$.

Similarly we have

$$J(t, x) \leqq tg\left(\frac{x - y^*(t, x+h)}{t}\right) + \varphi(y^*(t, x + h)),$$

so that

$$\frac{J(t, x + h) - J(t, x)}{h} \geqq g'\left(\frac{x + \zeta(h) - y^*(t, x + h)}{t}\right), \tag{3.11b}$$

where $0 \leqq \zeta(h) \leqq h$.

Since $y^*(t, \cdot)$ is right continuous, (3.11a) and (3.11b) together imply (3.9). Equation (3.10) is proved in the same way.

Lemma 3.6. *Let $(\bar{t}, \bar{x})$ be in the interior of D. If $y_*(\bar{t}, \cdot)$ is continuous at $x = \bar{x}$ then y_* is continuous as a function of (t, x) at $(\bar{t}, \bar{x})$.*

Proof. Let

$$m = \liminf_{k, h \to 0} y_*(t + k, x + h), \qquad M = \limsup_{k, h \to 0} y_*(t + k, x + h).$$

Choose any sequence (t_n, x_n) converging to $(\bar{t}, \bar{x})$ such that $y_*(t_n, x_n)$ converges to m. Clearly,

$$J(t_n, x_n) = t_n g\left(\frac{x_n - y_*(t_n, x_n)}{t_n}\right) + \varphi(y_*(t_n, x_n)).$$

But J, g and φ are continuous. Therefore,

$$J(\bar{t}, \bar{x}) = \bar{t} g\left(\frac{\bar{x} - m}{\bar{t}}\right) + \varphi(m).$$

We can show in the same way that

$$J(\bar{t}, \bar{x}) = \bar{t} g\left(\frac{\bar{x} - M}{\bar{t}}\right) + \varphi(M).$$

But then we must have

$$y_*(\bar{t}, \bar{x}) \leqq m \leqq M \leqq y^*(\bar{t}, \bar{x}).$$

But $\bar{x}$ is a point of continuity of $y^*(\bar{t}, \cdot)$ so that from lemma 3.4 we obtain

$$y_*(\bar{t}, \bar{x}) = m = M = y^*(\bar{t}, \bar{x}).$$

Lemma 3.7. *For any t, $0 < t \leqq T$, $J_t(t, x)$ exists and the equation*

$$J_t(t, x) = -f\left(g'\left(\frac{x - y_*(t, x)}{t}\right)\right) \tag{3.12}$$

is valid for all x except the countably many values x at which $y_*(t, \cdot)$ is discontinuous.

Proof. Let h be positive. From (2.13) it follows that

$$J(t + h, x) - J(t, x) \leqq (t + h) g\left(\frac{x - y_*(t, x)}{t + h}\right) - t g\left(\frac{x - y_*(t, x)}{t}\right)$$

$$\leqq h g\left(\frac{x - y_*(t, x)}{t}\right) - \left(\frac{x - y_*(t, x)}{t}\right) g'\left(\frac{x - y_*(t, x)}{t + \sigma(h)}\right) h,$$

where we have used the mean value theorem and $0 \leqq \sigma(h) \leqq h$. Similarly

$$J(t+h, x) - J(t, x) \geqq (t+h)g\left(\frac{x - y_*(t+h, x)}{t+h}\right) - tg\left(\frac{x - y_*(t+h, x)}{t}\right)$$

$$\geqq hg\left(\frac{x - y_*(t+h, x)}{t}\right) - h\left(\frac{x - y_*(t+h, x)}{t+h}\right)g'\left(\frac{x - y_*(t+h, x)}{t+\theta(h)}\right),$$

$$0 \leqq \theta(h) \leqq h.$$

Since $y_*(t, \cdot)$ is continuous at x, y_* is continuous at (t, x) (lemma 3.6). Therefore, from the above inequalities we obtain

$$\lim_{h \to 0+} \frac{J(t+h, x) - J(t, x)}{h} = g\left(\frac{x - y_*(t, x)}{t}\right) - \left(\frac{x - y_*(t, x)}{t}\right)g'\left(\frac{x - y_*(t, x)}{t}\right)$$

$$= -f\left(g'\left(\frac{x - y_*(t, x)}{t}\right)\right).$$

This argument works equally well when h is negative so that lemma 3.7 is seen to be true.

So far we have not made use of the fact that φ is Lipschitz continuous. Only the continuity of φ has been used, and indeed, continuity is all that is needed to prove that

$$\lim_{t \to 0} J(t, x) = \varphi(x). \tag{3.16}$$

(This can be proved using lemma 2.1). However, since $J(t, x)$ is Lipschitz continuous, the choice of Lipschitz continuous φ seems natural. Moreover, if φ is Lipschitz continuous, then the slopes of extremals are uniformly bounded in D as is stated in the next Lemma. (3.16) is an immediate consequence of this fact.

Lemma 3.8. *Let*

$$\rho_1 \leqq \varphi'(x) \leqq \rho_2 \tag{3.17}$$

for almost all x in $[0, L]$. *Then for (t, x) in D, $t > 0$, we have*

$$f'(\rho_1) \leqq \frac{x - y^*(t, x)}{t} \leqq \frac{x - y_*(t, x)}{t} \leqq f'(\rho_2), \tag{3.18}$$

providing

$$tf'(\rho_1) < x < L + tf'(\rho_2). \tag{3.19}$$

If $0 \leqq x \leqq tf'(\rho_1)$ then

$$0 \leqq \frac{x - y^*(t, x)}{t} = \frac{x - y_*(t, x)}{t} = \frac{x}{t} \leqq f'(\rho_1). \tag{3.20}$$

If $L \geqq x \geqq L + tf'(\rho_2)$ *then*

$$0 \geqq \frac{x - y^*(t, x)}{t} = \frac{x - y_*(t, x)}{t} = \frac{x - L}{t} \geqq f'(\rho_2). \tag{3.21}$$

Before proving this lemma we remark that such numbers, ρ_1 and ρ_2 exist inasmuch as φ is Lipschitzian. We also see that if $f'(\rho_1)$ is negative and $f'(\rho_2)$ is positive then (3.19) is valid for all (t, x) in D. From this and the fact that f' and g' are inverse functions we obtain the following corollary of the above lemma.

Corollary 3.8. *If*

$$|f'(\varphi'(x))| \leqq \omega, \tag{3.22}$$

for almost all x *in* $[0, L]$, *then*

$$\left|\frac{x - y_*(t, x)}{t}\right| \leqq \omega \quad \text{and} \quad \left|\frac{x - y^*(t, x)}{t}\right| \leqq \omega. \tag{3.23}$$

Proof of Lemma 3.8. Let $\omega_1 = f'(\rho_1)$ and $\omega_2 = f'(\rho_2)$. We first prove (3.18) and we do so indirectly. Assume that, contrary to (3.18), there is a point (t_1, x_1) in D, $t_1 > 0$, such that $x_1 - y^*(t_1, x_1) < t_1\omega_1$ even though $x_1 > \omega_1 t_1$. If we let $y_1 = y^*(t_1, x_1)$ and $y_2 = x_1 - \omega_1 t_1$ then this assumption is equivalent to

$$y_1 > y_2 > 0. \tag{3.24}$$

On the interval, $0 \leqq y \leqq L$, we define the function

$$S(y) = t_1 g\left(\frac{x_1 - y}{t_1}\right) + \rho_1(y - y_1) + \varphi(y_1).$$

Now

$$S'(y) = -g'\left(\frac{x_1 - y}{t_1}\right) + \rho_1,$$

and

$$S''(y) = \frac{1}{t_1} g''\left(\frac{x_1 - y}{t_1}\right) > 0.$$

Moreover,

$$S'(y_2) = -g'(\omega_1) + \rho_1 = -g'(f'(\rho_1)) + \rho_1 = 0.$$

Therefore, $S(y_2)$ is the minimum value for S in $[0, L]$ and

$$S(y_2) < S(y_1) = J(t_1, x_1). \tag{3.25}$$

But

$$J(t_1, x_1) \leqq t_1 g\left(\frac{x_1 - y_2}{t_1}\right) + \varphi(y_2),$$

and

$$\varphi(y_2) = \varphi(y_1) + \int_{y_1}^{y_2} \varphi'(x)\, dx \leqq \varphi(y_1) + \rho_1(y_2 - y_1),$$

as is seen from (3.17) and (3.24). Therefore,

$$J(t_1, x_1) \leqq t_1 g\left(\frac{x_1 - y_2}{t_1}\right) + \rho_1(y_2 - y_1) + \varphi(y_1) = S(y_2),$$

which contradicts (3.25) and shows that our assumption (3.24) cannot hold. This proves the first inequality in (3.18).

The second inequality in (3.18) is proved in the same way. Assume that for some (t_1, x_1) in D, $t_1 > 0$, we have

$$L > y_2 > y_1,$$

where $y_2 = x_1 - \omega_2 t_1$ and $y_1 = y_*(t_1, x_1)$. On the interval $[0, L]$ we define

$$S(y) = t_1 g\left(\frac{x_1 - y}{t_1}\right) + \rho_2(y - y_1) + \varphi(y_1).$$

In this case also, $S(y_2)$ is the minimum value for S in $[0, L]$ and $S(y_2) < S(y_1) = J(t_1, x_1)$. But again,

$$J(t_1, x_1) \leqq t_1 g\left(\frac{x_1 - y_2}{t_1}\right) + \varphi(y_2)$$

$$\leqq t_1 g\left(\frac{x_1 - y_2}{t_1}\right) + \rho_2(y_2 - y_1) + \varphi(y_1) = S(y_2),$$

which is the desired contradiction. This completes the proof of (3.18).

We also prove (3.20) indirectly by assuming that for some (t, x) in D, $y^*(t, x)$ is positive even though $0 \leqq x \leqq tf'(\rho_1)$. But since $y^*(t, x)$ is positive there must exist a point (t_0, x_0), $t_0 > 0$, on the line segment (extremal) joining (t, x) to $(0, y^*(t, x))$ such that $f'(\rho_1)t_0 < x_0 < L + t_0 f'(\rho_2)$. But then it follows from (3.18) that

$$f'(\rho_1) \leqq \frac{x_0 - y^*(t_0, x_0)}{t_0}.$$

But lemma 3.1 implies that $y^*(t_0, x_0) = y^*(t, x)$ so that

$$f'(\rho_1) \leqq \frac{x_0 - y^*(t_0, x_0)}{t_0} = \frac{x - y^*(t, x)}{t},$$

which contradicts our original assumption that $x \leqq tf'(\rho_1)$.

Relation (3.21) is proved in exactly the same way.

Lemma 3.9. *If* $0 \leqq x_0 \leqq L$ *then*

$$\lim_{(t,x)\to(0,x_0)} y_*(t, x) = x_0 = \lim_{(t,x)\to(0,x_0)} y^*(t, x), \tag{3.26}$$

and

$$\lim_{(t,x)\to(0,x_0)} J(t, x) = \varphi(x_0), \tag{3.27}$$

where the limits are taken over points (t, x) *in* D, $t > 0$.

Proof. Let ω be the supremum of $|f'(\varphi'(x))|$ for x in $[0, L]$. Then, from (3.23) it follows that

$$x - \omega t \leqq y_*(t, x) \leqq y^*(t, x) \leqq x + \omega t, \tag{3.28}$$

for all (t, x) in D, $t > 0$. Equation (3.26) follows immediately from (3.28) From (3.23) it also follows that

$$tg\left(\frac{x - y_*(t, x)}{t}\right)$$

converges to zero as (t, x) converges to $(0, x_0)$. Therefore, (3.27) follows from (3.26) and the continuity of φ and the lemma is proved.

We can now summarize the result of several of the preceding lemmas in the following theorem.

Theorem 1. *Let* φ *be a Lipschitz continuous function on* $[0, L]$. *Then the function* $J(t, x)$, *defined as the minimum of the functional described in* §2, *is Lipschitz continuous in* D. *For every* t, $0 < t \leqq T$, $J(t, x)$ *satisfies the Hamilton-Jacobi differential equation*

$$J_t(t, x) + f(J_x(t, x)) = 0, \tag{3.29}$$

for all x *save the (at most) countably many points where* $y_*(t, \cdot)$ *fails to be continuous.* $J(t, x)$ *has the initial values* $\varphi(x)$.

Except for the fact that J is Lipschitz continuous in D, this theorem follows directly from lemmas 3.3, 3.5, 3.7, and 3.9. Lemma 3.3 assures us only that J is locally Lipschitz continuous in the interior of D. The fact that J does satisfy a Lipschitz condition throughout D is due to the fact that the slopes of the extremals are bounded in D. Indeed,

$$J_x(t, x) = g'\left(\frac{x - y_*(t, x)}{t}\right),$$

and

$$J_t(t, x) = -f\left(g'\left(\frac{x - y_*(t, x)}{t}\right)\right),$$

at almost every point of D. Therefore (3.23) assures us that J_x and J_t are bounded in D. This of course implies that J is Lipschitz continuous in D.

4. In this section we consider the quasi-linear equation,

$$u_t + [f(u)]_x = u_t + f'(u)u_x = 0. \tag{4.1}$$

As is the case throughout this paper, the functions f and g and the rectangle D are as defined in section 2.

Definition. *A function $u(t, x)$ is termed a weak solution of* (4.1) *in D if it is measurable and bounded in every compact subset of the interior of D and if it satisfies*

$$\iint_D [u(t, x)w_t(t, x) + f(u(t, x))w_x(t, x)]\, dt\, dx = 0, \tag{4.2}$$

for every test function w. By test function we mean any function satisfying a Lipschitz condition in D and vanishing outside of some compact subset of the interior of D.

Although this is the most common definition of weak solutions others have been used including one based upon associated solutions of (1.1) (*cf.* [10]). However, these two definitions are equivalent as is stated in the following theorem.

Theorem 2. *A function $u(t, x)$ is a weak solution of* (4.1) *in D if, and only if, there exists a function $J(t, x)$ which is Lipschitz continuous in every compact subset of the interior of D and satisfies the following relations almost everywhere in D.*

$$J_t + f(J_x) = 0, \tag{4.3}$$

$$u(t, x) = J_x(t, x). \tag{4.4}$$

This theorem is really a corollary of the following basic theorem due to Haar and Schauder, [8].

Theorem. *Let G be a closed rectangle in the $t - x$ plane. Let $u(t, x)$ and $v(t, x)$ be defined in G. Then u and v will be measurable and bounded in G and will satisfy*

$$\iint_G [u(t, x)w_t(t, x) + v(t, x)w_x(t, x)]\, dt\, dx = 0, \tag{4.5}$$

for all functions $w(t, x)$ which are Lipschitz continuous in D and vanish on the boundary of G if, and only if there exists a function $J(t, x)$ which is Lipschitz continuous in G and such that

$$u(t, x) = J_x(t, x) \quad \text{and} \quad v(t, x) = J_t(t, x), \tag{4.6}$$

at almost every point of G.

Remark. The theorem is valid for more general sets G, but we need only assume G to be a rectangle. (For this case the proof of the theorem is simplified.)

Proof of Theorem 2. Assume first that there does exist a function $J(t, x)$ satisfying (4.3) and (4.4). Then $u(t, x)$ is certainly bounded and measurable in every compact subset of D^0, the interior of D. To see that (4.2) is satisfied let $w(t, x)$ be any Lipschitz continuous function whose support, S, is a compact subset of D^0. Choose G to be any closed rectangle such that

$$S \subset G \subset D^0.$$

Clearly w vanishes on the boundary of G. Hence, from the Haar-Schauder theorem where

$$v(t, x) = f(u(t, x)),$$

we obtain

$$\iint_G [w_t u + w_x f(u)] \, dt \, dx = 0,$$

from which follows (4.2) inasmuch as w vanishes outside of G.

Now assume that $u(t, x)$ is a weak solution of (4.1) in D. Let G_n be an ascending sequence of closed rectangles each of which is contained in D^0 and such that D^0 is the limit of this ascending sequence. Let $v(t, x) = f(u(t, x))$. Then u and v are measurable and bounded in each G_n. Moreover,

$$\iint_{G_n} [w_t u + w_x v] \, dt \, dx = 0,$$

for all Lipschitz continuous functions $w(t, x)$ which vanish on the boundary of G_n, providing we define $w(t, x)$ to be identically zero outside of G_n. This of course follows from (4.2). Therefore, according to the Haar-Schauder theorem, there exists a function $J_n(t, x)$ which is Lipschitz continuous in G_n and such that

$$u(t, x) = J_{nx}(t, x),$$

and

$$v(t, x) = -J_{nt}(t, x),$$

or

$$J_{nt}(t, x) + f(J_{nx}(t, x)) = 0,$$

almost everywhere in G_n. Now, by redefining each J_n on a set of measure zero in G_n, we can be sure that for each n,

$$J_{n+1}(t, x) = J_n(t, x), \qquad (t, x) \in G_n .$$

We now define $J(t, x)$ in D^0 as the limit of J_n; *i.e.*, if (t, x) is in D^0 then (t, x) is in G_n for some n. We then define $J(t, x)$ as $J_n(t, x)$. Clearly $J(t, x)$ is Lipschitz continuous in every compact subset of D^0 since every compact subset of D^0 is in G_n for some n. Also, J satisfies the Hamilton-Jacobi equation almost everywhere in each G_n. Therefore it satisfies it almost everywhere in D. Finally $u(t, x) = J_x(t, x)$ almost everywhere in D.

Corollary 2. *A necessary and sufficient condition that $u(t, x)$ be a bounded weak solution of (4.1) is that there exists a function $J(t, x)$ which is Lipschitz continuous in D and satisfies (4.3) and (4.4) almost everywhere in D.*

Let $u_0(x)$, $0 \leqq x \leqq L$, be bounded and measurable. Define the function φ by

$$\varphi(x) = \int_0^x u_0(y)\, dy, \qquad 0 \leqq x \leqq L. \tag{4.7}$$

clearly, φ is Lipschitz continuous on $[0, L]$ and therefore determines the functions J and y_* defined in section 2. We define the function $u(t, x)$ by

$$u(t, x) = g'\left(\frac{x - y_*(t, x)}{t}\right), \tag{4.8}$$

for (t, x) in D, $t > 0$.

From lemmas 3.4 and 3.5 it follows that

$$u(t, x) = J_x(t, x), \tag{4.9}$$

at almost every point of D. Moreover, as is stated in theorem 1, J is Lipschitz continuous in D and satisfies (4.3) almost everywhere in D. The first part of the following theorem now follows immediately from Corollary 2.

Theorem 3. *Let $u_0(x)$ be bounded and measurable on $[0, L]$. Then the function $u(t, x)$ defined by (4.8) is a bounded weak solution of (4.1) in D satisfying the initial condition*

$$\lim_{t \to 0} \int_0^x u(t, y)\, dy = \int_0^x u_0(y)\, dy, \tag{4.10}$$

for all x, $0 \leqq x \leqq L$.

Relation (4.10) follows from lemma 3.9 since for each fixed $t > 0$ the equation (4.9) is valid for almost all x, $0 \leqq x \leqq L$, as is clear from lemma 3.4 and 3.5.

Inasmuch as weak solutions of (4.1) are not uniquely determined by their initial values (*cf.* [7]) it is of interest to determine what properties are possessed by the weak solution $u(t, x)$.

Corollary 3a. *For each fixed t, $0 < t \leqq T$, $u(t, x)$ is continuous at all but a countable number of points (t, x) in D.*

Corollary 3b. *If (t, x) is in the interior of D, then*

$$u(t, x) = u(t, x - 0) \geqq u(t, x + 0), \tag{4.11}$$

Corollary 3c. *There exists a positive number A such that*

$$\frac{u(t, x_1) - u(t, x_2)}{x_1 - x_2} \leqq \frac{A}{t}, \tag{4.12}$$

for all points (t, x_1) and (t, x_2), $t > 0$, in D.

Corollary 3d. *Let ρ_1 and ρ_2 be the essential infimum and supremum respectively of $u_0(x)$ so that*

$$\rho_1 \leqq u_0(x) \leqq \rho_2 \,, \tag{4.13}$$

for almost every x, $0 \leqq x \leqq L$. Then for (t, x) in D, $t > 0$, we have:

i)
$$\rho_1 \leqq u(r, x) \leqq \rho_2$$
$$\text{providing} \quad tf'(\rho_1) \leqq x \leqq L + tf'(p_2),$$

ii)
$$0 \leqq x \leqq tf'(\rho_1) \Rightarrow u(t, x) = g'\left(\frac{x}{t}\right),$$

iii)
$$L \geqq x \geqq L + tf'(\rho_2) \Rightarrow u(t, x) = g'\left(\frac{x-L}{t}\right).$$

Proof. Corollary 3a follows from the corresponding properties of $y_*(t, x)$ as expressed in lemmas 3.4 and 3.6. From (3.8) we see that

$$u(t, x) = u(t, x - 0),$$

and

$$u(t, x + 0) = g'\left(\frac{x - y^*(t, x)}{t}\right).$$

Since $y_*(t, x) \leqq y^*(t, x)$, Corollary 3b follows from the fact that g' is an increasing function.

Let h be positive and let (t, x) and $(t, x + h)$ be in D. Applying the mean value theorem we obtain.

$$u(t, x + h) - u(t, x) = \frac{g''(\theta)}{t} [h + y_*(t, x) - y_*(t, x + h)],$$

where θ lies between

$$\frac{x - y_*(t, x)}{t} \quad \text{and} \quad \frac{x + h - y_*(t, x + h)}{t}.$$

But $y_*(t, x) - y_*(t, x + h)$ is non-positive so that

$$\frac{u(t, x + h) - u(t, x)}{h} \leqq \frac{g''(\theta)}{t}.$$

Corollary 3c follows immediately from this inequality upon defining

$$A = \sup_{|\theta| \leqq \omega} g''(\theta),$$

where ω is the supremum of

$$\left| \frac{x - y_*(t, x)}{t} \right|.$$

That this supremum is finite follows from Corollary 3.8.

Finally, Corollary 3d is nothing more than a paraphrase of lemma 3.8, where we have used the fact that g' and f' are inverse mappings.

The property expressed in Corollary 3c, the so called "strong entropy condition," is important, for it is for weak solutions satisfying this condition that uniqueness theorems have been proved by O. Oleinik [7]. She has also shown that such solutions can be obtained as limits of solutions of related second order parabolic equations. This is the so called "viscosity method."

It should be noted that the reliance upon weak convergence in the statement of the initial condition is due only to the extreme latitude that we have allowed in the choice of the initial data $u_0(x)$. In classical situations, there will be convergence in the conventional sense as is shown in the following corollary.

Corollary 3e. *If $u_0(x)$ is continuous at $x = \bar{x}$, $0 < \bar{x} < L$, then*

$$\lim_{t \to 0} u(t, \bar{x}) = u_0(\bar{x}). \tag{4.14}$$

Proof. Choose $a > 0$ small enough so that $\bar{x} - a$ and $\bar{x} + a$ are in $[0, L]$. From relation (3.18) of lemma 3.8 it follows that

$$\bar{x} - a < y_*(t, \bar{x}) < \bar{x} + a,$$

for all sufficiently small t. Therefore, for these values of t we have

$$J(t, \bar{x}) = \min_{\bar{x}-a \leqq y \leqq \bar{x}+a} \left\{ tg\left(\frac{\bar{x} - y}{t}\right) + \varphi(y) \right\}.$$

Let $r(a)$ and $R(a)$ be the essential infimum and supremum respectively of $u_0(y)$ for $\bar{x} - a \leqq y \leqq \bar{x} + a$. Then from Corollary 3d applied to the interval $[\bar{x} - a, \bar{x} + a]$ rather than to $[0, L]$ we see that

$$r(a) \leqq u(t, \bar{x}) \leqq R(a), \tag{4.15}$$

for all sufficiently small values of t, $t > 0$.

But u_0 is continuous at $\bar{x}$ so that

$$\lim_{a \to 0} r(a) = u_0(\bar{x}) = \lim_{a \to 0} R(a).$$

This relation, together with (4.15), implies (4.14) as was to be shown.

II. The Boundary Value Problem

5. Let the functions f and g and the rectangle D be as defined in section 2. Let $\psi(t)$, $0 \leqq t \leqq T$, be absolutely continuous. We search for a function $K(t, x)$ which is at least locally Lipschitz continuous in the interior of D, satisfies the Hamilton-Jacobi equation,

$$K_t + f(K_x) = 0), \tag{5.1}$$

almost everywhere in D and which satisfies the boundary condition,

$$\lim_{x \to 0} K(t, x) = \psi(t). \tag{5.2}$$

We again define such a function as the minimum value of the functional $H[t, x]$ which is in this case defined on curves joining (t, x) to the left lateral boundary of D. However, in contradistinction to the initial value problem restrictions that have nothing to do with smoothness must be imposed upon $\psi(t)$.

For (t, x) in the interior of D define $B(t, x)$ to be the set of all absolutely continuous functions $\omega(\tau)$, defined on $s \leqq \tau \leqq t$ for some s, $0 \leqq s < t$, satisfying

$$w(t) = x, \qquad w(s) = 0, \qquad 0 \leqq w(\tau) \leqq L,$$

for $s \leqq \tau \leqq t$. $H[t, x]$ is defined on $B(t, x)$ as the functional assigning the value

$$\int_s^t g(\dot{w}(\tau))\, d\tau + \psi(s)$$

to $w \in B(t, x)$ where w is defined on $[s, t]$.

Lemma 5.1. *For each (t, x) in the interior of D, the functional $H[t, x]$ assumes a minimum value in $B(t, x)$.*

Proof. Let s be any number such that $0 \leqq s < t$. Let w be any element of $B(t, x)$ defined on $[s, t]$. Then, just as in the proof of lemma 2.2, it follows from the integral form of Jensen's inequality that

$$(t - s)g\left(\frac{x}{t - s}\right) \leqq \int_s^t g(\dot{w}(\tau))\, d\tau. \tag{5.3}$$

Moreover, equality can occur only if

$$\dot{w}(\tau) \equiv \frac{x}{t - s}, \qquad s \leqq \tau \leqq t.$$

Now define the function $K(t, x)$ by

$$K(t, x) = \min_{0 \leqq s < t} \left\{(t - s)g\left(\frac{x}{t - s}\right) + \psi(s)\right\}. \tag{5.4}$$

That this minimum exists follows from the continuity of the functions and lemma 2.1. From (5.3) it follows that $K(t, x)$ is a lower bound for $H[t, x]$ over $B(t, x)$. But it is also a minimum since

$$(t - s)g\left(\frac{x}{t - s}\right) + \psi(s)$$

is the value of the functional for

$$w(\tau) = \left(\frac{x}{t - s}\right)(\tau - s),$$

which is certainly a member of $B(t, x)$.

Again let an extremal be an element of $B(t, x)$ for which the functional assumes its minimum value.

Corollary 5.1. *If w is an extremal then*

$$w(\tau) = \left(\frac{x}{t - s}\right)(\tau - s),$$

for some s, $0 \leqq s < t$. i.e. An extremal can only be a straight line joining (t, x) and a point $(s, 0)$, $s < t$, on the lateral boundary of D.

This of course follows immediately from the exceptional case of the Jensen inequality.

We now let $s^*(t, x)$ and $s_*(t, x)$ be the infimum and supremum respectively, of those values of s for which the minimum in (5.4) is attained. Clearly,

$$K(t, x) = (t - s^*(t, x))g\left(\frac{x}{t - s^*(t, x)}\right) + \psi(s^*(t, x)), \tag{5.5a}$$

and

$$K(t, x) = (t - s_*(t, x))g\left(\frac{x}{t - s_*(t, x)}\right) + \psi(s_*(t, x)). \tag{5.5b}$$

Also

$$0 \leqq s^*(t, x) \leqq s_*(t, x) < t \leqq T. \tag{5.6}$$

6. In this section we study the function, $K(t, x)$.

Lemma 6.1. a) *$K(t, x)$ is bounded from below in D. It is bounded from above in any subset of D in which t is uniformly bounded away from zero.*

b) *If x is positive, then*

$$\lim_{t \to 0} K(t, x) = +\infty. \tag{6.1}$$

Proof. Let g_0 be the minimum value of g. (Remember that $g'' > 0$ and g' assumes all real values.) From (5.4) it follows that, for all (t, x) in D,

$$K(t, x) \geqq -T\,|g_0| - e, \tag{6.2}$$

where

$$e = \max_{0 \leqq s \leqq T} |\psi(s)|. \tag{6.3}$$

Now

$$0 \leq \left(\frac{L - x}{L}\right)t < t,$$

for all positive $x \leq L$. Therefore, if x is positive, it follows from (5.4) that

$$K(t, x) \leqq \left[t - \left(\frac{L - x}{L}\right)t\right]g\left(\frac{L}{t}\right) + \psi\left(\frac{L - x}{L}\, t\right) \tag{6.4}$$
$$\leqq t\left|g\left(\frac{L}{t}\right)\right| + e.$$

But in the subset of points (t, x) in D for which $t \geqq \rho > 0$ the right hand side of (6.4) is bounded from above. We have therefore proven a). To prove b) we notice from (5.5) that for fixed positive x,

$$K(t, x) \geqq x\left(\frac{t - s_*(t, x)}{x}\right)g\left(\frac{x}{t - s_*(t, x)}\right) - e.$$

But as t tends to zero, $t - s_*(t, x)$ also tends to zero. Hence, by (2.9), the right hand side of the above inequality tends to plus infinity as t tends to zero.

Lemma 6.2. *Let the line joining (t, x) to $(s, 0)$ be an extremal for (t, x). Let (t_1, x_1) be any point on this extremal, $(t_1 > s)$. Then*

$$s_*(t_1, x_1) = s = s^*(t_1, x_1),$$

and

$$K(t, x) = (t - t_1)g\left(\frac{x}{t - s}\right) + K(t_1, x_1). \tag{6.5}$$

Proof. This lemma is proved in the same way as is lemma 3.1. Let $s_1 = s_*(t_1, x_1)$. Let γ be the path formed by the segments joining (t, x) to (t_1, x_1) and (t_1, x_1) to $(s_1, 0)$. The functional, $H(t, x)$, evaluated for γ is given by

$$(t - t_1)g\left(\frac{x}{t - s}\right) + (t_1 - s_1)g\left(\frac{x_1}{t_1 - s_1}\right) + \psi(s_1),$$

which is not greater than

$$(t - t_1)g\left(\frac{x}{t - s}\right) + (t_1 - s)g\left(\frac{x_1}{t_1 - s}\right) + \psi(s) = K(t, x).$$

Hence, γ is an extremal for the point (t, x). But all extremals are straight lines. Therefore,

$$s^*(t_1, x_1) = s.$$

Similarly,

$$s^*(t_1, x_1) = s.$$

Equation (6.5) is an immediate consequence of this.

From lemma 6.2 we obtain immediately,

Corollary 6.3. *Two extremals from distinct points in D cannot cross in the interior of D.*

Lemma 6.4. *The function $K(t, x)$ is Lipschitz continuous in any closed set contained in the interior of D.*

The proof of this lemma is exactly the same as that of lemma 3.3 so there is no need to repeat it. However, the proof does depend upon the fact that if R is a closed set contained in the interior of D then the absolute values of the slopes of extremals for points of R have a uniform bound. To see that this is true assume that it is false. Then there would be a sequence $\{t_n, x_n, s_n\}$ such that (t_n, x_n) is in R, the line joining (t_n, x_n) and $(s_n, 0)$ is an extremal, and yet

$$\frac{x_n}{t_n - s_n}$$

tends to infinity as n increases. But since R is closed and contained in the interior of D there is a positive number δ such that $x_n > \delta$ for all n. Therefore

$$K(t_n, x_n) \geqq \delta\left(\frac{t_n - s_n}{x_n}\right) g\left(\frac{x_n}{t_n - s_n}\right) - e,$$

for all sufficiently large n. But then from lemma 2.1 it follows that K is not bounded in R. This of course contradicts lemma 6.1. Hence, the slope of extremals are indeed bounded in R.

Lemma 6.5. *For fixed x, $x > 0$, the functions $s_*(\cdot, x)$ and $s^*(\cdot, x)$ are non-decreasing functions of t. They have the same countable set of points of discontinuity and, except at these points of discontinuity, the two functions are equal. Moreover, the relations*

$$\begin{cases} s_*(t, x) = s_*(t + 0, x) = s^*(t + 0, x) \\ s^*(t, x) = s^*(t - 0, x) = s_*(t - 0, x) \end{cases} \tag{6.6}$$

are valid for each t, $0 < t \leqq T$.

Lemma 6.6. *For fixed t, $0 < t \leqq T$, the functions $s_*(t, \cdot)$ and $s^*(t, \cdot)$ are non-increasing functions of x. They have the same points of discontinuity and, except at these countably many points, the two functions are equal. Moreover, for $0 < x < L$ the following relations are valid.*

$$\begin{cases} s_*(t, x) = s_*(t, x - 0) = s^*(t, x - 0) \\ s^*(t, x) = s^*(t, x + 0) = s_*(t, x + 0). \end{cases} \tag{6.7}$$

These two lemmas are proved in the same way as is their counterpart in section 3, lemma 3.4.

Lemma 6.7. *If (t, x) is in the interior of D then*

$$K_x^+(t, x) \equiv \lim_{h \to 0+} \frac{K(t, x + h) - K(t, x)}{h} = g'\left(\frac{x}{t - s^*(t, x)}\right), \tag{6.8}$$

$$K_x^-(t, x) \equiv \lim_{h \to 0-} \frac{K(t, x + h) - K(t, x)}{h} = g'\left(\frac{x}{t - s_*(t, x)}\right). \tag{6.9}$$

Proof. Let h be positive and assume $x + h < L$. Then because of (5.4) we have

$$K(t, x + h) \leqq (t - s^*(t, x))g\left(\frac{x + h}{t - s^*(t, x)}\right) + \psi(s^*(t, x)).$$

Hence,

$$\frac{K(t, x + h) - K(t, x)}{h}$$

$$\leqq \left(\frac{t - s^*(t, x)}{h}\right)\left[g\left(\frac{x + h}{t - s^*(t, x)}\right) - g\left(\frac{x}{t - s^*(t, x)}\right)\right] \tag{6.10}$$

$$= g'\left(\frac{x + \xi(h)}{t - s^*(t, x)}\right), \qquad 0 \leqq \xi(h) \leqq h.$$

Similarly,

$$K(t, x) \leqq (t - s^*(t, x + h))g\left(\frac{x}{t - s^*(t, x + h)}\right) + \psi(s^*(t, x + h),$$

so that

$$\frac{K(t, x + h) - K(t, x)}{h} \geqq g'\left(\frac{x + \zeta(h)}{t - s^*(t, x + h)}\right), \qquad 0 \leqq \zeta(h) \leqq h, \tag{6.11}$$

Equation (6.8) follows from (6.10), (6.11) and (6.7).

Equation (6.9) is verified in exactly the same manner.

Lemma 6.8. *If (t, x) is in the interior of D then*

$$K_t^+(t, x) \equiv \lim_{h \to 0+} \frac{K(t + h, x) - K(t, x)}{h} = -f\left(g'\left(\frac{x}{t - s_*(t, x)}\right)\right), \tag{6.12}$$

$$K_t^-(t, x) \equiv \lim_{h \to 0-} \frac{K(t + h, x) - K(t, x)}{h} = -f\left(g'\left(\frac{x}{t - s^*(t, x)}\right)\right). \tag{6.13}$$

Proof. Let h be positive and let $(t + h, x)$ be in D. (5.4) implies that

$$K(t + h, x) \leqq [t + h - s_*(t, x)]g\left(\frac{x}{t + h - s_*(t, x)}\right) + \psi(s_*(t, x)).$$

Hence, letting

$$\Delta(h) = \frac{K(t + h, x) - K(t, x)}{h},$$

we have

$$h\Delta(h) \leqq g\left(\frac{x}{t + h - s_*(t, x)}\right)h$$

$$+ [t - s_*(t, x)]\left\{g\left(\frac{x}{t + h - s_*(t, x)}\right) - g\left(\frac{x}{t - s_*(t, z)}\right)\right\},$$

or

$$(6.14)\quad \Delta(h) \leqq g\left(\frac{x}{t+h-s_*(t,x)}\right) - \left[\frac{x}{t+h-s_*(t,x)}\right] g'\left(\frac{x}{t+\xi(h)-s_*(t,x)}\right),$$

where

$$0 \leqq \xi(h) \leqq h.$$

Now since $s_*(t, x) < t$ and $s_*(t + 0, x) = s_*(t, x)$ it follows that for all h small enough $s_*(t + h, x) < t$. For these values of h we can again apply (5.4) to obtain

$$K(t, x) \leqq [t - s_*(t+h, x)] g\left(\frac{x}{t - s_*(t+h, x)}\right) + \psi(s_*(t+h, x)).$$

Therefore,

$$\Delta(h) \geqq g\left(\frac{x}{t+h-s_*(t+h,x)}\right) - \left(\frac{x}{t+h-s_*(t+h,x)}\right) g'\left(\frac{x}{t+\zeta(h)-s_*(t+h,x)}\right),$$

where $0 \leqq \zeta(h) \leqq h$. Now letting h approach zero in this inequality and in (6.14) we obtain (*cf.* (6.6)).

$$(6.15)\qquad K_t^+(t, x) = g\left(\frac{x}{t - s_*(t,x)}\right) - \left(\frac{x}{t-s_*(t,x)}\right) g'\left(\frac{x}{t-s_*(t,x)}\right).$$

If we now recall (2.6) and (2.8) we see that (6.12) is an immediate consequence of (6.15). Equation (6.13) is verified in exactly the same manner. *q.e.d.*

At this point we have shown that $K(t, x)$ is locally Lipschitz continuous in the interior of D and satisfies the Hamilton-Jacobi equation almost everywhere in D. Furthermore, the only property of the function ψ that was needed to show this was continuity. However, unless ψ satisfies more stringent requirements it is not generally true that

$$(6.16)\qquad \lim_{x \to 0} K(t, x) = \psi(t).$$

Moreover, the difficulty that arises has nothing to do with smoothness properties of ψ.

There are several ways to see this. For example, let $\psi(t)$ be continuously differentiable in $[0, T]$. $K(t, x)$ is the minimum, over the interval $0 \leqq s < t$, of the function

$$(6.17)\qquad (t - s) g\left(\frac{x}{t-s}\right) + \psi(s).$$

Now the derivative with respect to the variable s of this function is given by

$$(6.18)\qquad f\left(g'\left(\frac{x}{t-s}\right)\right) + \psi'(s),$$

where we have made use of (2.6). If this derivative does not vanish for some value of s in $(0, t)$ then the minimum value of (6.17) occurs for $s = 0$. In this case,

$$K(t, x) = tg\left(\frac{x}{t}\right) + \psi(0),$$

so that

$$\lim_{x \to 0} K(t, x) = K(t, 0) = tg(0) + \psi(0), \tag{6.19}$$

which will not in general be equal to $\psi(t)$. Therefore, in order to insure (6.16) being valid we must restrict ψ in such a way that (6.18) vanishes. This of course requires that $-\psi'(s)$ assume values in the range of f.

A slightly more refined argument shows that $-\psi'(s)$ must be in the range of f for all values of s, $0 < s \leqq T$. We are thus led (at least in the case where $\psi \in C^1$) to the necessary condition,

$$-\psi'(t) \geqq m,$$

$0 < t \leqq T$, where m is the minimum value of f. But this condition is also sufficient and in the case of ψ being absolutely continuous we have the following lemma.

Lemma 6.9. *If $\psi(t)$ is absolutely continuous on $[0, T]$ and if*

$$-\psi'(t) \geqq m, \qquad a.a. t \in [0, T], \tag{6.20}$$

then

$$\lim_{x \to 0} K(t, x) = \psi(t), \tag{6.21}$$

for all $t \in (0, T]$.

Proof. Since $s_*(t, x)$ is a decreasing function of x and since $s_*(t, x) < t$ we see that the limit

$$\lim_{x \to 0} s_*(t, x) = \bar{s}(t)$$

exists for all t and $\bar{s}(t) \leqq t$.

Case 1: $\bar{s}(t) = t$.

As in the proof of lemma 6.1 we see that

$$K(t, x) \leqq \left[t - \left(\frac{L - x}{L}\right)t\right]g\left(\frac{L}{t}\right) + \psi\left(\left(\frac{L - x}{L}\right)t\right).$$

But if g_0 is the minimum value of g then it follows from (5.5b) that

$$K(t, x) \geqq (t - s_*(t, x))g_0 + \psi(s_*(t, x)).$$

Since we are dealing with the case where $\bar{s}(t) = t$, equation (6.21) follows immediately from these two inequalities.

Case 2: $\bar{s}(t) < t$.

In this case,

$$\lim_{x \to 0} K(t, x) = (t - \bar{s})g(0) + \psi(\bar{s}) \tag{6.22}$$

$$= -m(t - \bar{s}) + \psi(\bar{s}), \tag{6.23}$$

where we have used the fact that $g(0) = -m$ and have written $\bar{s}$ for $\bar{s}(t)$. We shall prove that

$$\psi'(\tau) = -m, \qquad a.a. \tau \in (\bar{s}, t) \tag{6.24}$$

so that,

$$\psi(t) = -m(t - \bar{s}) + \psi(\bar{s}).$$

Then of course, (6.21) will follow from (6.23).

Define the set $E(\mu)$ by

$$E(\mu) = \{\tau : t > \tau > \bar{s} \text{ and } \psi'(\tau) < -\mu\}.$$

Because of (6.20) we can prove (6.24) by showing that for all $\mu > m$, the set $E(\mu)$ has measure zero. We show this indirectly. Assume that for some $\mu > m$ the set $E(\mu)$ has positive measure. Let

$$\text{measure } E(\mu) = \delta > 0.$$

From (6.22) we see that for any ϵ, $\epsilon > 0$, we can find $x_1 > 0$, such that

$$x_1 > x > 0 \Rightarrow |K(t, x) - (t - \bar{s})g(0) - \psi(\bar{s})| < \epsilon\delta. \tag{6.25}$$

But $x > 0 \Rightarrow s_*(t, x) \leqq \bar{s}$ so that we can choose x_1 so that

$$|g(\gamma_1) - g(0)| < \epsilon\delta, \qquad \gamma_1 = \frac{x_1}{t - s_*(t, x_1)}, \tag{6.26}$$

is satisfied as well as (6.25).

Now choose s_0, $t > s_0 > \bar{s}$, so that $t - s_0 < 1$ and

$$\text{measure } \{E(\mu) \cap [\bar{s}, s_0]\} \geqq \frac{\delta}{2}. \tag{6.27}$$

Then

$$\psi(s_0) = \psi(\bar{s}) + \int_{\bar{s}}^{s_0} \psi'(\tau)\, d\tau < \psi(\bar{s}) - \frac{\mu\delta}{2} - m\left(s_0 - \bar{s} - \frac{\delta}{2}\right). \tag{6.28}$$

Now let

$$x_0 = \gamma_1(t - s_0) \quad \text{so that} \quad \gamma_1 = \frac{x_0}{t - s_0}.$$

Then

$$K(t, x_0) \leqq (t - s_0)g(\gamma_1) + \psi(s_0).$$

Upon using (6.26) and (6.28) we obtain

$$K(t, x_0) < (t - s_0)g(0) - m(s_0 - \bar{s}) + \psi(\bar{s}) - \frac{\delta}{2}(\mu - m) + \epsilon\delta.$$

But $g(0) = -m$ so that

$$K(t, x_0) < (t - \bar{s})g(0) + \psi(\bar{s}) - \frac{\delta}{2}(\mu - m) + \epsilon\delta. \tag{6.29}$$

However, since $0 < x_0 < x_1$ it follows from (6.25) that

$$K(t, x_0) > (t - \bar{s})g(0) + \psi(\bar{s}) - \epsilon\delta.$$

Combining this with (6.29) we obtain

$$4\epsilon > \mu - m.$$

But this relation is false for ϵ chosen small enough. Hence, for all $\mu > m$ the set $E(\mu)$ has measure zero. The proof of (6.24) and the lemma is complete.

Corollary 6.9. *If $\psi(t)$ is absolutely continuous and*

$$-\psi'(t) > m \text{ for a.a. } t \in [0, T],$$

then

$$\lim_{x \to 0} s_*(t, x) = t, \qquad 0 < t \leqq T. \tag{6.30}$$

This of course is an immediate consequence of case 2 in the above proof.

The following theorem is the main result of this section.

Theorem 4. *Let $\psi(t)$ be absolutely continuous on $[0, T]$ and satisfy*

$$\psi'(t) \leqq -m, \qquad m = \text{minimum of } f,$$

for almost all points $t \in [0, T]$. Then the function $K(t, x)$ defined in section 5 is locally Lipschitz continuous in the interior of D and for each t, $T \geqq t > 0$, satisfies the Hamilton-Jacobi equation,

$$K_t(t, x) + f(K_x(t, x)) = 0,$$

at all but a countable set of values of x. Moreover,

$$\lim_{x \to 0} K(t, x) = \psi(t)$$

is satisfied for each t, $T \geqq t > 0$.

This theorem is a summary of lemmas 6.4, 6.6, 6.7, 6.8 and 6.9.

If in addition to the conditions specified in theorem 4, $\psi(t)$ is Lipschitz continuous in $[0, T]$ then we see that

$$-M \leqq \psi'(t) \leqq -m, \tag{6.31}$$

for some $M > m$. Because of the properties of f as delineated in section 2, the fact that M is greater than m implies that for some number ζ we have,

$$f(\zeta) = M \quad \text{and} \quad f'(\zeta) = \sigma > 0. \tag{6.32}$$

Moreover, ζ is uniquely determined by M.

Corollary 4. *If $\psi(t)$ is Lipschitz continuous and satisfies* (6.31) *at almost every point of $[0, T]$ then, in addition to possessing the properties expressed in theorem 4, the function $K(t, x)$ satisfies a uniform Lipschitz condition in the wedge-shaped subset of D defined by*

$$0 \leqq t \leqq T, \qquad 0 \leqq x \leqq \sigma t. \tag{6.33}$$

Moreover, if (t, x), $t > 0$, is in D and $x \geqq \sigma t$ then

$$K(t, x) = tg\left(\frac{x}{t}\right) + \psi(0), \tag{6.34}$$

i.e. $s_*(t, x) = 0$.

Proof. Because of lemma 6.2 this corollary is an immediate consequence of the fact that

$$0 \leqq x < \sigma t \Rightarrow \frac{x}{t - s_*(t, x)} \leqq \sigma, \tag{6.35}$$

which we shall now prove.

The proof is indirect and is a variation of the proof of lemma 3.8. If (6.35) were false then for some (t_1, x_1), $0 < x_1 < \sigma t_1$, we would have

$$\frac{x_1}{t_1 - s_*(t_1, x_1)} > \sigma.$$

If we let $s_1 = s_*(t_1, x_1)$ and $s_2 = t_1 - x_1/\sigma$ then this situation is expressed by,

$$s_1 > s_2 > 0. \tag{6.36}$$

Now define the function $y(s)$ by

$$y(s) = (t_1 - s)g\left(\frac{x_1}{t_1 - s}\right) - M(s - s_1) + \psi(s_1),$$

for $0 \leqq s \leqq s_1$. An elementary calculation results in

$$y'(s) = f\left(g'\left(\frac{x_1}{t_1 - s}\right)\right) - M,$$

and

$$y''(s) = \frac{x_1^2}{(t - s)^3}\, g''\left(\frac{x_1}{t_1 - s}\right) > 0.$$

Therefore, since $y'(s_2) = f(g'(\sigma)) - M = f(\zeta) - M = 0$ we see that $y(s_2)$ is the minimum value of y in $[0, s_1]$. Hence,

$$y(s_2) < y(s_1) = K(t_1, x_1). \tag{6.37}$$

But on the other hand it follows from (5.4) that

$$K(t_1, x_1) < (t_1 - s_2)g\left(\frac{x_1}{t_1 - s_2}\right) + \psi(s_2),$$

and from (6.31) we have

$$\psi(s_2) \leqq \psi(s_1) - M(s_2 - s_1).$$

Therefore,

$$K(t_1, x_1) \leqq (t - s_2)g\left(\frac{x_1}{t_1 - s_2}\right) - M(s_2 - s_1) + \psi(s_1) = y(s_2),$$

which contradicts (6.37). This shows that (6.35) is correct.

7. In this section we show how the ideas developed in the previous two sections can be applied to the problem of obtaining weak solutions of the quasilinear equation,

$$v_t + \frac{\partial}{\partial x}[f(v)] = 0, \tag{7.1}$$

having prescribed boundary values.

Let $\bar{v}(t)$, $0 \leqq t \leqq T$, be bounded and measurable and define the function $\psi(t)$, $0 \leqq t \leqq T$, by

$$\psi(t) = \psi_0 - \int_0^t f(\bar{v}(\tau))\, d\tau, \tag{7.2}$$

where ψ_0 is any constant. It is clear that ψ is a Lipschitz continuous function and that

$$\psi'(t) \leqq -m \quad \text{for} \quad a.a.t \in [0, T], \tag{7.3}$$

where m, as before, is the minimum value of f. If we use ψ to define, as in section 5, the function $K(t, x)$ then it follows from corollary 4 that K is Lipschitz continuous in the set,

$$\{(t, x): 0 \leqq t \leqq T \quad \text{and} \quad 0 \leqq x \leqq \omega t\}, \tag{7.4}$$

for any $\omega > 0$. Moreover, K satisfies the Hamilton-Jacobi equation almost everywhere in D and if we define the function v by

$$v(t, x) = g'\left(\frac{x}{t - s_*(t, x)}\right), \tag{7.5}$$

then v is equal to K_x almost everywhere in D. It therefore follows from theorem 2 that $v(t, x)$ is a weak solution of (7.1) in D and that v is bounded in the set (7.4) for any $\omega > 0$.

Since $K(t, x) \to \psi(t)$ as $x \to 0$ and from lemmas 6.5 and 6.8 it follows that

$$\lim_{x\to 0} \int_{t_1}^{t_2} f(v(t, x))\, dt = \int_{t_1}^{t_2} f(\bar{v}(t))\, dt, \tag{7.6}$$

for all t_1, t_2 such that $0 < t_1 < t_2 < T$.

On the basis of (7.6) we cannot say very much regarding the relationship of $\bar{v}(t)$ to the limit values (in some sense of convergence) of $v(t, x)$. This is partly due to the fact that f is not one-to-one on the real line. However, f is one-to-one on the interval $\{z : z \geqq g'(0)\}$ since $f'(g'(0)) = 0$ and f' is strictly increasing. Moreover,

$$v(t, x) > g'(0), \tag{7.7}$$

because $s_*(t, x) < t$ and g' is increasing. Therefore, the values of $\bar{v}$ and v appearing in (7.6) will be in an interval on which f is one-to-one if we impose the following restriction upon $\bar{v}$.

$$\bar{v}(t) \geqq g'(0) \quad \text{for} \quad t \in [0, T]. \tag{7.8}$$

But even with the restriction (7.8) we cannot conclude from (7.6) that

$$\int_{t_1}^{t_2} v(t, x)\, dt \to \int_{t_1}^{t_2} \bar{v}(t)\, dt. \tag{7.9}$$

To see this notice that $v(t, x)$ and $\bar{v}(t)$ are bounded providing $t \geqq \rho > 0$ for some ρ. Therefore, (7.6) and (7.9) imply weak convergence in $L_2[\rho, T]$. (Since all functions concerned are bounded (uniformly) the particular choice of L_p is not significant.) But it follows from a theorem due to P. Lax (section 5 of reference [9]) that weak convergence of both $v(t, x)$ to $\bar{v}(t)$ and $f(v(t, x))$ to $f(\bar{v}(t))$ can occur only when $v(t, x)$ in fact converges strongly to $\bar{v}(t)$. Because of this we will be content, for the time being, with the result embodied in (7.6).

Theorem 5. *Let $\bar{v}(t)$, $0 \leqq t \leqq T$, be measurable and bounded, and satisfy*

$$\bar{v}(t) \geqq g'(0) \tag{7.8}$$

for almost all t, $0 \leqq t \leqq T$.

Then the function $v(t, x)$, defined by (7.5), is a weak solution of (7.1) in D which satisfies the following conditions.

(A) (7.7) $\qquad (t, x) \in D$ *and* $x > 0 \Rightarrow v(t, x) > g'(0)$,

(B) (7.6) $\qquad t_2 > t_1 > 0 \Rightarrow \lim_{x\to 0} \int_{t_1}^{t_2} f(v(t, x))\, dt = \int_{t_1}^{t_2} f(\bar{v}(t))\, dt.$

(C) *If ω is any positive number then $v(t, x)$ is bounded in the set,*

$$\{(t, x): 0 \leqq t \leqq T \quad \text{and} \quad 0 \leqq x \leqq \omega t\}. \tag{7.10}$$

(D) *For each fixed* t, $0 < t < T$, $v(t, x)$ *is continuous for all but a countable set of points* (t, x) *in* D. *For each fixed* x, $0 < x \leqq L$, $v(t, x)$ *is continuous for all but a countable set of points* (t, x) *in* D,

(E) (7.11) $$v(t, x) = v(t, x - 0) \geqq v(t, x + 0).$$

Proof. We have seen that the function $v(t, x)$ defined by (7.5) is a weak solution of (7.1) in D. Also, v satisfies conditions (A) and (B). From corollary 4 (*cf.* (6.35)) it follows that since $\bar{v}$ (and therefore ψ') is bounded, the quantity

$$\frac{x}{t - s_*(t, x)}$$

is bounded in any set defined as in (7.10). (C) is an immediate consequence of this fact. Condition (D) follows from lemmas 6.5 and 6.6. Lemma 6.6, together with the fact that g' is increasing, imply (E).

By imposing further restrictions upon $\bar{v}$ we can replace (B) by a more convenient type of convergence. For example, if $\bar{v}$ is Riemann integrable then $v(\cdot, x)$ will converge almost everywhere to $\bar{v}(\cdot)$. This follows from the following corollary.

Corollary 5. *If* $\bar{v}$ *is continuous at* $\bar{t}$, $0 < \bar{t} \leqq T$, *then*

(7.12) $$\lim_{x \to 0} v(\bar{t}, x) = \bar{v}(\bar{t}).$$

Proof. We treat two cases separately.

Case 1.

$$\lim_{x \to 0} s_*(\bar{t}, x) = \bar{s} < \bar{t}.$$

From equation (6.24) it follows that in this case

$$f(\bar{v}(t)) = m,$$

for almost all t, $\bar{s} \leqq t < \bar{t}$. From this and the continuity of $\bar{v}$ at E it follows that

$$f(\bar{v}(\bar{t})) = m,$$

or equivalently,

$$\bar{v}(\bar{t}) = g'(0).$$

But since $\bar{s} < t$ it follows that

$$\lim_{x \to 0} v(t, x) = \lim_{x \to 0} g'\left(\frac{x}{t - s_*(\bar{t}, x)}\right) = g'(0),$$

so that (7.12) is verified.

Case 2.

(7.13) $$\lim_{x \to 0} s_*(\bar{t}, x) = \bar{t}.$$

From (7.13) it follows that for any a, $t > a > 0$, we have

$$\bar{t} > s_*(\bar{t}, x) > \bar{t} - a,$$

for all sufficiently small x. Since extremals cannot cross, we have, for these values of x, the relation

$$K(\bar{t}, x) = \min_{\bar{t}-a \leqq s < \bar{t}} \left\{(\bar{t} - s)g\left(\frac{x}{\bar{t} - s}\right) + \psi(s)\right\}, \tag{7.14}$$

where ψ is given by (7.2). Let $k(a)$ and $K(a)$ be the essential infimum and essential supremum respectively of $\bar{v}(t)$ for $\bar{t} - a \leqq t < \bar{t}$. From (7.8) we obtain

$$f(k(a)) \leqq f(\bar{v}(t)) \leqq f(K(a)), \tag{7.15}$$

for almost all t, $\bar{t} - a \leqq t < \bar{t}$. As is shown below, these inequalities imply that

$$k(a) \leqq v(\bar{t}, x) \leqq K(a), \tag{7.16}$$

for all sufficiently small x.

But $\bar{v}$ is continuous at $\bar{t}$ so that,

$$\lim_{a \to 0} k(a) = \bar{v}(\bar{t}) = \lim_{a \to 0} K(a).$$

From these relations and (7.16) it follows that

$$\lim_{x \to 0} v(\bar{t}, x) = \bar{v}(\bar{t}),$$

as was to be shown.

We now need only show that (7.15) implies (7.16). We give a detailed proof of only the left side of (7.16) since the proof of the right side is quite similar. The proof of the right side is in fact identical to the proof of (6.35).

Because of the definition of v and the fact that f' and g' are inverse functions, the left side of (7.16) follows from

$$\gamma = f'(k(a)) \leqq \frac{x}{\bar{t} - s_*(\bar{t}, x)}. \tag{7.17}$$

To see that (7.17) is satisfied for all sufficiently small x assume the contrary; *i.e.* assume that for some $\bar{x}$,

$$\gamma > \frac{\bar{x}}{\bar{t} - s_*(\bar{t}, \bar{x})},$$

even though $s_*(\bar{t}, \bar{x}) > \bar{t} - a$. Letting $s_1 = s_*(\bar{t}, \bar{x})$ and $s_2 = \bar{t} - \bar{x}/\gamma$, these two conditions may be written as

$$t - a < s_1 < s_2 < \bar{t}.$$

Now define

$$y(s) = (\bar{t} - s)g\left(\frac{\bar{x}}{\bar{t} - s}\right) + \psi(s_1) - f(k)(s - s_1),$$

for $\bar{t} - a < s < \bar{t}$. Elementary calculation gives

$$y'(s) = f\left(g'\left(\frac{\bar{x}}{\bar{t} - s}\right)\right) - f(k),$$

and

$$y''(s) = \frac{\bar{x}^2}{(\bar{t} - s)^3} g''\left(\frac{\bar{x}}{\bar{t} - s}\right) > 0.$$

Therefore, since $y'(s_2) = f(g'(f'(k))) - f(k) = 0$ we see that

$$y(s_2) < y(s_1) = K(\bar{t}, \bar{x}).$$

On the other hand, it follows from (7.14) that

$$K(\bar{t}, \bar{x}) \leqq (\bar{t} - s_2)g\left(\frac{\bar{x}}{\bar{t} - s_2}\right) + \psi(s_2). \tag{7.18}$$

But from (7.15) and (7.2) it follows that

$$\psi(s_2) = \psi(s_1) - \int_{s_1}^{s_2} f(\bar{v}(t))\, dt \leqq \psi(s_1) - f(k)(s_2 - s_1).$$

Therefore,

$$K(\bar{t}, \bar{x}) \leqq (\bar{t} - s_2)g\left(\frac{\bar{x}}{\bar{t} - s_2}\right) + \psi(s_1) - f(k)(s_2 - s_1) = y(s_2),$$

which contradicts (7.18). This proves that (7.17) is valid for all sufficiently small x.

III. The Mixed Problem

8. Again the functions f and g and the rectangle D are as in section 2. Let $\varphi(x)$, $0 \leqq x \leqq L$, and $\psi(t)$, $0 \leqq t \leqq T$, be Lipschitz continuous. We search for a function $M(t, x)$ which is Lipschitz continuous in D, satisfies the Hamilton-Jacobi equation

$$M_t + f(M_x) = 0 \tag{8.1}$$

almost everywhere in D and satisfies the following initial and boundary conditions.

$$\lim_{t \to 0} M(t, x) = \varphi(x), \tag{8.2}$$

$$\lim_{x \to 0} M(t, x) = \psi(t). \tag{8.3}$$

Because of the considerations in section 6 we also assume that

$$-\psi'(t) \geqq m, \tag{8.4}$$

for almost all t, $0 \leqq t \leqq T$. Also, since M is to be continuous, it is natural to assume that

$$\psi(0) = \varphi(0). \tag{8.5}$$

This assumption is largely one of convenience however, and most of what we say shall be valid in situations in which it is not satisified.

The desired solution is again obtained from Hamilton's variational procedure.

Consider the curve S consisting of the two segments forming the base and left lateral boundary of D. We employ the following parameterization of S: the set of points $(t(z), x(z))$ such that

$$-T \leqq z \leqq L, \tag{8.6}$$

$$\begin{aligned} x(z) &= 0 \quad \text{if} \quad -T \leqq z \leqq 0, \\ x(z) &= z \quad \text{if} \quad 0 \leqq z \leqq L, \end{aligned} \tag{8.7}$$

and

$$\begin{aligned} t(z) &= -z \quad \text{if} \quad -T \leqq z \leqq 0, \\ t(z) &= 0 \quad \text{if} \quad 0 \leqq z \leqq L. \end{aligned} \tag{8.8}$$

On the interval $-T \leqq z \leqq L$ we define the function $\theta(z)$ as follows.

$$\begin{aligned} \theta(z) &= \psi(t(z)) = \psi(-z) \quad \text{for} \quad -T \leqq z \leqq 0, \\ \theta(z) &= \varphi(x(z)) = \varphi(z) \quad \text{for} \quad 0 \leqq z \leqq L. \end{aligned} \tag{8.9}$$

Clearly, θ is Lipschitz continuous on $[-T, L]$.

Let (t, x) where $t > 0$, $x > 0$ be in D. For any z_0, $-t < z_0 \leqq L$, let $C(t, x; z_0)$ be the set of all absolutely continuous functions $w(\tau)$ defined for

$$t(z_0) \leqq \tau \leqq t,$$

and subject to the following conditions.

$$0 \leqq w(\tau) \leqq L, \tag{8.10}$$

$$w(t) = x, \qquad w(t(z_0)) = x(z_0).$$

Finally, let $C(t, x)$ be the union of all sets $C(t, x; z_0)$ for $-t < z_0 \leqq L$. We see that each member of $C(t, x)$ is a curve joining (t, x) to the curve S.

$H[t, x]$ is defined on $C(t, x)$ as the functional assigning the value

$$\int_{t(z_0)}^{t} g(\dot{w}(\tau))\, d\tau + \theta(z_0) \tag{8.11}$$

to the element $w \in C(t, x; z_0) \subset C(t, x)$.

Lemma 8.1.

a) *For each (t, x) in the interior of D, the functional $H[t, x]$ attains a minimum value in $C(t, x)$.*

b) *Every extremal is a straight line joining* (t, x) *to* S.

c) *Extremals cannot cross in the interior of* D.

d) *If* $M(t, x)$ *is the minimum value of* $H[t, x]$ *then*

$$M(t, x) = \min_{-t<z\leq L} \left\{(t - t(z))g\left(\frac{x - x(z)}{t - t(z)}\right) + \theta(z)\right\}. \tag{8.12}$$

Since this lemma is proved in exactly the same way as are its counterparts in Chapters I and II we shall not bother to give the details.

If we let $z_*(t, x)$ and $z^*(t, x)$ be the infimum and supremum respectively of those values of z for which the minimum in (8.12) is attained then we have

$$-t < z_*(t, x) \leqq z^*(t, x) \leqq L, \tag{8.13}$$

and

$$\begin{aligned} M(t, x) &= (t - t(z_*))g\left(\frac{x - x(z_*)}{t - t(z_*)}\right) + \theta(z_*(t, x)), \\ M(t, x) &= (t - t(z^*))g\left(\frac{x - x(z^*)}{t - t(z^*)}\right) + \theta(z^*(t, x)). \end{aligned} \tag{8.14}$$

Now to the function φ we can associate, as in Chapter I, the function $J(t, x)$ and to ψ we can associate, as in Chapter II, the function $K(t, x)$. It is clear that at all points (t, x) of D, $t > 0$ and $x > 0$, we have the following useful relation:

$$M(t, x) = \min \{J(t, x), K(t, x)\}. \tag{8.15}$$

Since both J and K are Lipschitz continuous in any compact set contained in the interior of D, the following lemma follows immediately from (8.15).

Lemma 8.2. *M is Lipschitz continuous in any compact subset of the interior of D.*

Since both J and K are bounded from below in D and since J is also bounded from above it follows from (8.15) that,

Lemma 8.3. *M is bounded from above and below in D.*

Lemma 8.4. *The function M satisfies the Hamilton-Jacobi equation* (8.1) *almost everywhere in D.*

This is an immediate consequence of the following more general lemma.

Lemma 8.5. *Let* $F(t, x, u, p, q)$ *be continuous and let* $u(t, x)$ *and* $v(t, x)$ *be Lipschitzian functions satisfying*

$$F(t, x, u, u_t, u_x) = 0 \quad \text{and} \quad F(t, x, v, v_t, v_x) = 0,$$

almost everywhere in some open set D. Then the function

$$w(t, x) = \min \{u(t, x), v(t, x)\}$$

is a Lipschitzian function satisfying

$$F(t, x, w, w_t, w_x) = 0, \tag{8.16}$$

almost everywhere in D.

Proof. It is easy to see that w is Lipschitzian. Therefore, if D_1 is the set of those points of D at which the partial derivatives of all three functions u, v and w exist and at which u and v satisfy the differential equation (8.16) then the set of points of D which are not in D_1 has measure zero. We prove this lemma by showing that w satisfies (8.16) at every point of D_1.

If $(\bar{t}, \bar{x})$ is a point of D_1 such that $u(\bar{t}, \bar{x}) > v(\bar{t}, \bar{x})$ then $w = v$ in some neighborhood of $(\bar{t}, \bar{x})$. But then $w_x(\bar{t}, \bar{x}) = v_x(\bar{t}, \bar{x})$ and $w_t(\bar{t}, \bar{x}) = v_t(\bar{t}, \bar{x})$ so that w satisfies (8.16) at $(\bar{t}, \bar{x})$. The same argument works when $v(\bar{t}, \bar{x}) > u(\bar{t}, \bar{x})$.

Now consider a point $(\bar{t}, \bar{x})$ of D_1 at which u and v are equal. We then have

$$u(\bar{t}, \bar{x}) = w(\bar{t}, \bar{x}) = v(\bar{t}, \bar{x}). \tag{8.17}$$

Let $\{x_n\}$ be any sequence of points converging to $\bar{x}$. We can certainly find a subsequence $\{y_n\}$ such that either $w(\bar{t}, y_n) = u(\bar{t}, y_n)$ for all n or $w(\bar{t}, y_n) = v(\bar{t}, y_n)$ for all n. Without loss of generality we assume the former. Then because of (8.17) we have

$$\frac{w(\bar{t}, y_n) - w(\bar{t}, \bar{x})}{y_n - \bar{x}} = \frac{u(\bar{t}, y_n) - u(\bar{t}, \bar{x})}{y_n - \bar{x}},$$

so that $w_x(\bar{t}, \bar{x}) = u_x(\bar{t}, \bar{x})$. Now if for some sequence $\{z_n\}$ converging to $\bar{x}$ we have $w(\bar{t}, z_n) = v(\bar{t}, z_n)$ then by forming differential quotients with this sequence we obtain $w_x(\bar{t}, \bar{x}) = v_x(\bar{t}, \bar{x})$. If there is no such sequence then $v(\bar{t}, x) > u(\bar{t}, x)$ for all $x \neq \bar{x}$ in some interval about $\bar{x}$. But then, because of (8.17), we must have

$$v_x^+(\bar{t}, \bar{x}) \geqq u_x^+(\bar{t}, \bar{x}) = u_x^-(\bar{t}, \bar{x}) \geqq v_x^-(\bar{t}, \bar{x}).$$

But $v_x(\bar{t}, \bar{x})$ exists so that $v_x(\bar{t}, \bar{x}) = u_x(\bar{t}, \bar{x})$. We see then that if (8.17) is satisfied,

$$u_x(\bar{t}, \bar{x}) = w_x(\bar{t}, \bar{x}) = v_x(\bar{t}, \bar{x}).$$

In the same way we show that

$$u_t(\bar{t}, \bar{x}) = w_t(\bar{t}, \bar{x}) = v_t(\bar{t}, \bar{x}).$$

From these relations we see that w satisfies (8.16) at $(\bar{t}, \bar{x})$ and the lemma is proved.

From the above proof and our knowledge of J and K we see immediately that

$$M_x(t, x) = g'\left(\frac{x - x(z_*(t, x))}{t - t(z_*(y, x))}\right), \tag{8.18}$$

almost everywhere in D.

Lemma 8.6. *The function M is Lipschitz continuous in D.*

Proof. Since both ψ and φ satisfy Lipschitz conditions it follows that J is Lipschitz continuous in D (Theorem 1) and that K is Lipschitz continuous in the set

$$\{t, x: \ 0 \leqq t \leqq T \ \text{ and } \ 0 \leqq x \leqq \omega t\},$$

for any positive ω. (Corollary 4).

Therefore, the lemma will be proved if there exists a positive number ω such that

$$0 < t \leqq T \ \text{ and } \ x > \omega t \Rightarrow M(t, x) = J(t, x) < K(t, x).$$

But this is so because if not, then there would be a sequence (t_n, x_n) such that

$$\frac{x_n}{t_n} > n, \qquad n = 1, 2, \cdots,$$

even though $K(t_n, x_n) \leqq J(t_n, x_n)$. Now from corollary 4 it follows that for all sufficiently large n, $s_*(t_n, x_n) = 0$ so that

$$K(t_n, x_n) = t_n g\left(\frac{x_n}{t_n}\right) + \psi(0).$$

But $\psi(0) = \varphi(0)$ so that

$$t_n g\left(\frac{x_n}{t_n}\right) + \varphi(0) \leqq J(t_n, x_n),$$

for all sufficiently large n. Hence, it follows from (2.13) that $y_*(t_n, x_n) = 0$ so that

$$\frac{x_n - y_*(t_n, x_n)}{t_n} = \frac{x_n}{t_n} > n,$$

which contradicts corollary 3.8. Therefore, for some ω, $\omega > 0$, we see that $x > \omega t \Rightarrow J(t, x) < K(t, x)$ and the lemma is proved.

Now from lemma 6.1 and from (3.27) it follows that

$$\lim_{t \to 0} M(t, x) = \lim_{t \to 0} J(t, x) = \varphi(x), \tag{8.18}$$

for $0 \leqq x \leqq L$. It is not necessarily true however, that

$$\lim_{x \to 0} M(t, x) = \psi(t). \tag{8.19}$$

We do know that

$$\lim_{x \to 0} K(t, x) = \psi(t),$$

and that

$$\lim_{x \to 0} J(t, x) = tg\left(-\frac{y_*(t, 0)}{t}\right) + \varphi(y_*(t, 0)).$$

Therefore, for (8.19) to be valid it is both necessary and sufficient that φ and ψ satisfy the following compatibility condition.

$$\psi(t) \leqq tg\left(-\frac{y_*(t, 0)}{t}\right) + \varphi(y_*(t, 0)). \tag{8.20}$$

The main results of this section may be summarized as follows.

Theorem 6. *Let $\varphi(x)$, $0 \leqq x \leqq L$, and $\psi(t)$, $0 \leqq t \leqq T$, be Lipschitz continuous functions such that:*

(i) $\psi'(t) \geqq -m$ *for almost all* t, $\quad 0 \leqq t \leqq T$.

(ii) $\psi(0) = \varphi(0)$.

(iii) $\psi(t) \leqq tg\left(-\frac{y_*(t, 0)}{t}\right) + \varphi(y_*(t, 0))$, $\quad 0 \leqq t \leqq T$,

where y_* is determined from φ as in section 2. Then the function $M(t, x)$, defined as in lemma 8.1, is Lipschitz continuous in D and satisfies the Hamilton-Jacobi equation

$$M_t + f(M_x) = 0,$$

almost everywhere in D. Moreover, M has initial value φ and boundary value ψ in the sense of (8.2) and (8.3).

Remarks. In the light of the proof of lemma 8.5 and by examining the behavior of the extremals when $M = J = K$ it is easy to see that for each t, $0 < t \leqq T$, the partial derivatives of M fail to exist at only countably many values of x.

If condition (8.5) is not satisfied then we define $\theta(0)$ as the smaller of $\varphi(0)$ and $\psi(0)$. Even though θ is discontinuous the minimum of the functional (8.11) will still exist. Moreover, the function M will be as in theorem 6 except that it will be only locally Lipschitzian since (8.5) was essential to the proof of lemma 8.6.

9. The main result of this section is embodied in the following theorem.

Theorem 7. *Let $u_0(x)$, $0 \leqq x \leqq L$, and $\bar{u}(t)$, $0 \leqq t \leqq T$, be bounded measurable functions satisfying the following two conditions:*

$$\bar{u}(t) \geqq g'(0), \qquad 0 < t \leqq T. \tag{9.1}$$

$$-\int_0^t f(\bar{u}(\tau))\, d\tau \leqq tg\left(-\frac{y_*(t, 0)}{t}\right) + \int_0^{y_*(t,0)} u_0(\zeta)\, d\zeta, \tag{9.2}$$

where the function y_* is determined from $\varphi(x) = \int_0^x u_0(\zeta)\, d\zeta$ as in section 2.

Then there exists a weak solution u of

$$u_t + (f(u))_x = 0 \tag{9.3}$$

in D. This solution is bounded in D and satisfies the following relations:

$$\lim_{t\to 0}\int_0^x u(t,\zeta)\,d\zeta = \int_0^x u_0(\zeta)\,d\zeta, \qquad 0 < x < L, \tag{9.4}$$

$$\lim_{x\to 0}\int_0^t f(u(\tau,x))\,d\tau = \int_0^t f(\bar u(\tau))\,d\tau, \qquad 0 < t \leqq T; \tag{9.5}$$

$$u(t,x) = u(t,x-0) \geqq u(t,x+0); \tag{9.6}$$

$$u_0 \ \textit{continuous at}\ \ x_0 \Rightarrow \lim_{t\to 0} u(t,x_0) = u_0(x_0); \tag{9.7}$$

$$\bar u \ \textit{continuous at}\ \ t_0 \Rightarrow \lim_{x\to 0} u(t_0\,,x) = \bar u(t_0). \tag{9.8}$$

Proof. The functions

$$\varphi(x) = \int_0^x u_0(\zeta)\,d\zeta, \qquad 0 \leqq x \leqq L,$$

and

$$\psi(t) = -\int_0^t f(\bar u(\tau))\,d\tau, \qquad 0 \leqq t \leqq T,$$

are Lipschitzian over their respective domains. Moreover, from the definition of these functions and from (9.2) it follows that hypotheses (i), (ii) and (iii) of theorem 6 are satisfied. Therefore, there is associated with these functions, as in section 8, a function $M(t, x)$ which is Lipschitz continuous in D and satisfies the Hamilton-Jacobi Equation almost everywhere there. If we define $u(t, x)$ by

$$u(t,x) = g'\left(\frac{x - x(z_*(t,x))}{t - t(z_*(t,x))}\right), \tag{9.9}$$

then u is equal to M_x almost everywhere and therefore is a bounded weak solution of (9.3) in D. (Corollary 2).

From the remarks made immediately following theorem 6 it follows that on each line, t = constant, u is equal to M_x at all but countably many points. Therefore, (9.4) follows immediately from (8.18).

Since u is bounded in D it is clear that the limit in (9.5) exists for all t, $0 < t \leqq T$. That the limit is as given in (9.5) follows from the fact that $M(t, x) \to \psi(t)$ and that $f(u(t, x))$ is equal to $-M_t(t, x)$ at almost every point of D.

It follows from the non-crossing of extremals in exactly the same way as in lemmas 3.4 and 6.6 that $z_*(t, x)$ is a non decreasing function of x and that $z_*(t, x) = z_*(t, x - 0) \leqq z_*(t, x + 0)$. Since g' is monotonically increasing, relation (9.6) follows immediately from this and the definitions of $x(z)$ and $t(z)$.

Relations (9.7) and (9.8) follow from Corollaries 3 and 5.

Remark. If we give up the requirement that the weak solution be bounded in D then we can dispense with condition (9.2). We merely outline the procedure to be followed.

Define $\varphi(x)$ as before but define ψ by

$$\psi(t) = \psi_0 - \int_0^t f(\bar{u}(\tau))\, d\tau,$$

where ψ_0 is a constant chosen so that condition (8.20) is satisfied. This can always be done by taking ψ_0 small enough. *E.g.* if μ_0 is the minimum value of $J(t, 0)$, $0 \leqq t \leqq T$, then take

$$\psi_0 = \min_{0 \leqq t \leqq T} \{\mu_0 + mt\}.$$

Since $f(\bar{u}) \geqq m$, this choice of ψ_0 clearly insures that (8.22) is satisfied.

Now of course, $\psi(0) \neq \varphi(0)$. Hence, from the remarks made immediately following theorem 6, the function M is not Lipschitz continuous throughout all of D. Therefore, M_x and therefore u will not be bounded in D. Notice that u will be bounded in any set obtained from D by deleting a neighborhood of the origin.

In this case (9.4) and (9.5) must be replaced by

$$\lim_{t \to 0} \int_{x_1}^{x_2} u(t, \zeta)\, d\zeta = \int_{x_1}^{x_2} u_0(\zeta)\, d\zeta, \qquad 0 < x_1, \quad x_2 \leqq L,$$

and

$$\lim_{x \to 0} \int_{t_1}^{t_2} f(u(\tau, x)\, d\tau = \int_{t_1}^{t_2} f(\bar{u}(\tau))\, d\tau, \qquad 0 < t_1 < t_2 \leqq T.$$

IV. Unbounded Domains

10. Let $\varphi(x)$, $-\infty < x < +\infty$, be Lipschitzian over the real line so that

$$k_1 < \varphi'(x) < k_2 \tag{10.1}$$

for all real x. We search for a solution of the Hamilton-Jacobi Equation in the infinite strip

$$\{(t, x): 0 \leqq t \leqq T, -\infty < x < +\infty\}, \tag{10.2}$$

having φ as its values on the line $t = 0$.

We define the functional (2.10) on the set of all absolutely continuous functions on $[0, t]$ having the value x at t. Then, because of Jensen's inequality, the minimum

$$J(t, x) = \min_{-\infty < y < +\infty} \left\{ tg\left(\frac{x - y}{t}\right) + \varphi(y) \right\}, \tag{10.3}$$

the existence of which follows from (2.9) and (10.1), is a minimum value of the functional. If we again define $y_*(t, x)$ and $y^*(t, x)$ as the infimum and supremum of the values of y for which the minimum in (10.3) is obtained then the representation (2.14) is valid. We also see, just as in section 3, that extremals cannot cross in the interior of the strip (10.2).

If we now restrict (t, x) to any bounded subset of the strip then $y_*(t, x)$ and $y^*(t, x)$ will lie within a bounded interval of the line $t = 0$. Since the arguments of Chapter I apply equally well to any bounded interval we see that $J(t, x)$ is a locally Lipschitzian solution of the Hamilton-Jacobi Equation,

$$J_t + f(J_x) = 0. \tag{10.4}$$

Moreover,

$$J_x(t, x) = g'\left(\frac{x - y_*(t, x)}{t}\right), \tag{10.5}$$

almost everywhere in the strip, and

$$\lim_{t \to 0} J(t, x) = \varphi(x). \tag{10.6}$$

It also follows from lemma 3.8 and (10.1) that

$$f'(k_1) \leqq \frac{x - y^*(t, x)}{t} \leqq \frac{x - y_*(t, x)}{t} \leqq f'(k_2), \tag{10.7}$$

for all points (t, x) of the strip (10.1). This is easily seen by applying lemma 3.8 to an interval $[a, b]$ chosen large enough so that x, $y_*(t, x)$ and $y^*(t, x)$ are contained in the interval and such that

$$a + tf'(k_1) < x < b + tf'(k_2).$$

Finally it follows from (10.7), (10.5) and (10.4) that the partial derivatives of J are bounded in the strip so that J is actually Lipschitz on the strip (10.1).

We summarize these results in the following theorem.

Theorem 8. *Let $\varphi(x)$ be Lipschitzian on the real line. Then, for fixed (t, x), $t > 0$, Hamilton's functional assumes a minimum value in the set of all absolutely continuous curves joining (t, x) to the line $t = 0$. This minimum value, $J(t, x)$, is Lipschitzian in the strip* (10.1) *and satisfies the Hamilton-Jacobi Equation almost everywhere in this strip. $J(t, x)$ has φ as its initial values in the sense of* (10.6).

Now let $u_0(x)$, $-\infty < x < +\infty$, be bounded and measurable. Then the function

$$\varphi(x) = \int_0^x u_0(\zeta)\, d\zeta$$

is Lipschitzian and therefore determines the functions J and y_* as above. We define the function $u(t, x)$ by

$$u(t, x) = g'\left(\frac{x - y_*(t, x)}{t}\right), \tag{10.8}$$

for (t, x), $t > 0$. From (10.7) it follows that $u(t, x)$ is bounded in the strip (10.1) and has the same upper and lower bounds as u_0. From (10.5) and theorem 2 it follows that

$$(10.9) \qquad \iint [w_t u + w_x f(u)]\, dt\, dx = 0,$$

for all Lipschitzian functions w having compact support in the strip (10.1). The integral in (10.9) is taken over this strip. These results are summarized in the first part of the following theorem.

Theorem 9. *Let $u_0(x)$ be bounded and measurable on the real line. Then the function $u(t, x)$ defined by (10.8) is a bounded weak solution of (4.1) in the strip (10.1). The function $u(t, x)$ has the same upper and lower bounds as does $u_0(x)$ and satisfies the initial condition*

$$(10.10) \qquad \lim_{t \to 0} \int_0^x u(t, \zeta)\, d\zeta = \int_0^x u_0(\zeta)\, d\zeta,$$

for all x, $-\infty < x < +\infty$.

Relation (10.10) follows from (10.6) and the fact that for each fixed t, $u(t, x)$ is equal to $J_x(t, x)$ for almost all x.

Finally, we can prove just as in section 4 that for each fixed t, $u(t, x)$ is continuous for all save a countable set of values of x and that

$$(10.11) \qquad u(t, x) = u(t, x - 0) \geqq u(t, x + 0).$$

Moreover, if u_0 is continuous at $\bar{x}$ then

$$(10.12) \qquad \lim_{t \to 0} u(t, \bar{x}) = u_0(\bar{x}).$$

We now consider the following mixed problem: Given two functions $\psi(t)$, $0 \leqq t$, and $\varphi(x)$, $0 \leqq x$, which are Lipschitz continuous on their respective half-lines of definition, we search for a function $M(t, x)$ which is Lipschitz continuous in the region

$$(10.13) \qquad \{(t, x) : 0 < t < \infty \quad \text{and} \quad 0 < x < \infty\},$$

satisfies the Hamilton-Jacobi Equation almost everywhere in this region and satisfies the following initial and boundary conditions.

$$(10.14) \qquad \lim_{t \to 0} M(t, x) = \varphi(x), \qquad x \geqq 0;$$

$$(10.15) \qquad \lim_{x \to 0} M(t, x) = \psi(t), \qquad t \geqq 0.$$

In light of section 8 we impose the following conditions on the function ψ:

$$(10.16) \qquad -\psi'(t) \geqq m, \qquad t \geqq 0,$$

$$(10.17) \qquad \psi(0) = \varphi(0),$$

where m is, as before, the minimum value of f. We again define M as the minimum value of Hamilton's functional which we define as follows. Let S be the curve consisting of the two half-lines bounding the region (10.13). Thus S is the set of all points $(t(z), x(z))$, $-\infty < z < +\infty$, where

$$x(z) = \begin{cases} 0 & \text{if} \quad z \leqq 0 \\ z & \text{if} \quad z \geqq 0 \end{cases},$$

$$t(z) = \begin{cases} -z & \text{if} \quad z \leqq 0 \\ 0 & \text{if} \quad z \geqq 0 \end{cases}.$$

We define the function $\theta(z)$, $-\infty < z < +\infty$, as follows:

$$\theta(z) = \begin{cases} \psi(t(z)) = \psi(-z) & \text{if} \quad z \leqq 0 \\ \varphi(x(z)) = \varphi(z) & \text{if} \quad z \geqq 0 \end{cases}.$$

For (t, x) in the region (10.13) we define $H[t, x]$ on the set of all absolutely continuous functions $w(\tau)$, $t(z) \leqq \tau \leqq t$, such that $w(t) = x$ and $w(t(z)) = x(z)$ where z is any number such that $z > -t$. To such a function $H[t, x]$ assigns the value

$$\int_{t(z)}^{t} g(\dot{w}(\tau))\, d\tau + \theta(z).$$

Just as in the earlier chapters we see that this functional attains a minimum value for each (t, x) and that this minimum is attained only for linear functions w. Moreover, this minimum value is precisely

$$M(t, x) = \min_{-t<z<+\infty} \left\{ (t - t(z))g\left(\frac{x - x(z)}{t - t(z)}\right) + \theta(z) \right\}. \tag{10.18}$$

That this latter minimum exists follows from (2.9) and the boundedness of the derivative of θ over the real line.

Theorem 10. *Let $\varphi(x)$, $0 \leqq x < \infty$, and $\psi(t)$, $0 \leqq t < \infty$, be Lipschitz continuous and satisfy* (10.16) *and* (10.17). *Then for each point (t, x) in region* (10.13) *Hamilton's functional assumes a minimum value $M(t, x)$. $M(t, x)$ is Lipschitz continuous throughout region* (10.13) *and satisfies the Hamilton-Jacobi Equation almost everywhere there. M satisfies* (10.14) *for all positive x and will satisfy* (10.15) *if and only if*

$$\psi(t) \leqq tg\left(-\frac{y_*(t, 0)}{t}\right) + \varphi(y_*(t, 0)), \tag{10.19}$$

for all positive t. The function y_* is obtained by solving the initial value problem for the Hamilton-Jacobi Equation with φ as initial data.

The proof of this theorem is quite similar to that of prior theorems and need not be given in any detail. It depends upon the solution, in the positive quadrant, of the initial value problem where the initial data φ is given on the half-line, $x \geqq 0$. This is accomplished in much the same way as is done in the proof of theorem 8.

Now let $u_0(x)$, $x \geqq 0$, and $\bar{u}(t)$, $t \geqq 0$, be bounded and measurable. Then the functions

$$\varphi(x) = \int_0^x u_0(\zeta)\, d\zeta, \tag{10.20}$$

and

$$\psi(t) = -\int_0^t f(\bar{u}(\tau))\, d\tau \tag{10.21}$$

are Lipschitz continuous on the intervals $x \geqq 0$ and $t \geqq 0$ respectively. Moreover, φ and ψ satisfy relations (10.16) and (10.17). We can therefore apply the previous theorem to obtain the function M which is a Lipschitz continuous solution in (10.13) of the Hamilton-Jacobi Equation. If we now define the function $u(t, x)$ by

$$u(t, x) = g'\left(\frac{x - x(z_*(t, x))}{t - t(z_*(t, x))}\right), \tag{10.22}$$

then u is bounded and measurable in (10.13) and satisfies

$$u(t, x) = M_x(t, x),$$

almost everywhere there. The following theorem is now proved in much the same was as is theorem 7 or it can be obtained as a consequence of that theorem.

Theorem 11. *Let $u_0(x)$, $x \geqq 0$, and $\bar{u}(t)$, $t \geqq 0$, be bounded and measurable functions and let $\bar{u}(t) \geqq g'(0)$. Let the functions φ and ψ, defined as in* (10.20) *and* (10.21), *satisfy relation* (10.19).

Then the function $u(t, x)$ defined as in (10.22) *is a weak solution in the positive quadrant of the quasi-linear equation,*

$$u_t + (f(u))_x = 0.$$

Moreover, u is bounded and satisfies the following conditions:

$$\lim_{t \to 0} \int_0^x u(t, \zeta)\, d\zeta = \int_0^x u_0(\zeta)\, d\zeta, \qquad x > 0;$$

$$\lim_{x \to 0} \int_0^t f(u(\tau, x))\, d\tau = \int_0^t f(\bar{u}(\tau))\, d\tau \qquad t > 0;$$

$$u(t, x) = u(t, x - 0) \geq u(t, x + 0);$$

$$u_0 \ \textit{continuous at} \ x_0 \Rightarrow \lim_{t \to 0} u(t, x_0) = u_0(x_0);$$

$$\bar{u} \ \textit{continuous at} \ t_0 \Rightarrow \lim_{x \to 0} u(t_0, x) = \bar{u}(t_0).$$

References

[1] Akhiezer, N., *The Calculus of Variations*, Blaisdell Pub., New York, 1962.

[2] Courant, R. & Hilbert, D., *Methods of Mathematical Physics*, Vol. II, Interscience, New York, 1962.

[3] Douglis, A., An ordering principle and generalized solution of certain quasi-linear partial differential equations, *Comm. Pure Appl. Math.*, **XII** 87–112.

[4] Hardy, G., Littlewood, J. & Polya, G., *Inequalities*, Cambridge Univ. Press, 1934.

[5] Hopf, E., The partial differential equation $u_t + uu_x = \mu u_{xx}$, *Comm. Pure Appl. Math.*, III (1950) 201-230.

[6] Lax, P. D., Hyperbolic systems of conservation laws, *Comm. Pure Appl. Math.*, **X** (1957) 537–566.

[7] Oleinik, O. A., On discontinuous solutions of nonlinear differential equations, *Uspehi Matem. Nauk.*, **12** (1957) 3–73. English translation in *American Mathematical Society Translations* ser. 2, **26** 95–172.

[8] Schauder, J., Über die Umkehrung eines Satzes aus der Variationsrechnung, *Acta Szeged*, **4** (1928) 38–50.

[9] Lax, P. D., Weak solutions of non-linear hyperbolic equations and their numerical computation, *Comm. Pure Appl. Math.*, **7** (1954) 159–193.

[10] Douglis, A., The continuous dependence of generalized solutions of non-linear partial differential equations upon initial data, *Comm. Pure Appl. Math.*, **XIV** (1961) 267–284.

Courant Institute of Mathematics
Indiana University

Hamilton's theory and generalized solutions of the Hamilton-Jacobi equation, *J. Math. & Mech.* **13 (1964), 939-986.**

Commentary

Cathleen S. Morawetz

The last work in this part and Hopf's last lengthy work is a joint work with his student, E. Conway. In the fifties, when Eberhard Hopf wrote his impressive work on the Burger's equation (see Selecta p. 159), he had privately extended some of it. This paper pursued some of the basic ideas in a different way and included Hopf's earlier extensions.

In the non-linear Hamilton-Jacobi equation, there is no viscosity and no entropy as meant in current terminology. The equation, for example, $u_t + uu_x = 0$ is regarded as the variational equation for a functional and among its many solutions satisfying the given data, one picks out the one that minimizes the functional. Several interesting examples are shown for nonuniqueness without some additional condition. These are cited by P. L. Lions on p. 48 of his book *Generalized solutions of Hamilton-Jacobi equations* (Pitman, 1982). Lions and others approach the problem by adding artificial viscosity and going to the limit of zero viscosity to obtain unique solutions.

The approach of Conway and Hopf has not been pursued in the later literature, but one can reasonably ask whether for some nonlinear systems, where there is no natural notion of entropy, would this method be viable. One thinks here of possible financial problems or equations in physics arising from some loose averaging procedure.

Part II

Papers on Ergodic Theory

Edited by
Yakov G. Sinai

ABDRUCK
AUS DEN BERICHTEN DER MATHEMATISCH-PHYSISCHEN KLASSE DER SÄCHSISCHEN AKADEMIE DER WISSENSCHAFTEN ZU LEIPZIG.
XCI. BAND.

SITZUNG VOM 11. NOVEMBER 1939.

Statistik der geodätischen Linien in Mannigfaltigkeiten negativer Krümmung.

Von

Eberhard Hopf.

Einleitung.

Statistische Untersuchungen über den Gesamtverlauf der geodätischen Linien auf Flächen $\mathfrak{F}$ negativer Krümmung sind erst neueren Datums. Sie waren bei der raschen, im letzten Jahrzehnt erfolgten Entwicklung der Ergodentheorie, der Lehre vom statistischen Verlauf der Lösungen gewöhnlicher Differentialsysteme[1]), unausbleiblich. Um die Begriffsbildungen und Sätze dieser Theorie anzuwenden, betrachtet man die geodätische Strömung, d. h. die stationäre Strömung im Raum Ω der gerichteten Linienelemente der Fläche, die durch die Bewegung längs den geodätischen Linien mit der Geschwindigkeit Eins entsteht. Die Strömung besitzt ein invariantes Volummaß m in Ω. Sein Element ist durch $dm = do d\vartheta$ definiert, wo do das Oberflächen- und $d\vartheta$ das Winkeldifferential auf $\mathfrak{F}$ bedeuten.

Die Fläche sei vollständig (Unendlichkeitspostulat von Koebe), d. h. die Geodätischen seien auf $\mathfrak{F}$ unbegrenzt fortsetzbar. Damit ist auch die Strömung für alle Zeiten in Ω definiert.

Was nun Flächen negativer Krümmung K vor anderen Flächen auszeichnet, ist vor allem die starke Unstabilität ihrer Geodätischen. Der Normalabstand n einer Geodätischen von einer infinitesimal benachbarten Geodätischen genügt längs derselben der Variationsgleichung

$$\frac{d^2 n}{d s^2} + K n = 0 .$$

Verläuft K auf $\mathfrak{F}$ zwischen festen negativen Grenzen, so wächst $n(s)$ in mindestens einer Richtung wie eine Exponentialfunktion. Dieses Auseinanderlaufen der Geodätischen führt zu der Vermutung, daß die Geodätischen, wenn es irgendwie begünstigt wird (z. B. auf Flächen mit hyper-

1) Vgl. meine Schrift „Ergodentheorie", Ergebnisse der Mathematik und ihrer Grenzgebiete V 2. Berlin, Springer, 1937. Im folgenden wird sie kurz ERG genannt.

bolischen Öffnungen), im allgemeinen auf $\mathfrak{F}$ ins Unendliche laufen, während sie, wenn dies nicht begünstigt wird (z. B. auf geschlossenen Flächen), im allgemeinen jeden Teil des Linienelementraumes Ω mit volumproportionaler Verweilzeit durchlaufen, d. h. in Ω gleichverteilt sind.

Das Hauptergebnis dieser Arbeit besteht in der präzisen Fassung und im Beweis dieser Vermutung. Unter der Voraussetzung, daß K zwischen festen negativen Grenzen verläuft und eine beschränkte Richtungsableitung besitzt, werden die Flächen in zwei Klassen verteilt. Für die erste Klasse, welche u. a. alle geschlossenen $\mathfrak{F}$ mit negativem K umfaßt, wird die Ergodizität der geodätischen Strömung bewiesen. Ergodizität bedeutet, wenn $\mathfrak{F}$ endlichen Flächeninhalt hat: Das Zeitmittel einer Funktion in Ω längs einer Stromlinie ist im allgemeinen gleich dem Raummittel. Die Gleichverteiltheit der allgemeinen Stromlinie in Ω ist damit gleichbedeutend. Die Ergodizität einer geodätischen Strömung kann auch in einer Form ausgesprochen werden, die von Nullmengen keinen Gebrauch macht. Eine derartige integrale Formulierung des Satzes, die sich durch größere Anschaulichkeit auszeichnet, ist folgende[1]):

Man denke sich auf $\mathfrak{F}$ zur Zeit $t = 0$ einen Haufen von Punkten verschiedenster Anfangsgeschwindigkeiten. Die Verteilung der Anfangsphasen, d. h. der Anfangslagen und -geschwindigkeiten (nicht nur der Anfangsrichtungen) sei kontinuierlich, sonst aber willkürlich. Denkt man sich die Punkte unter Beibehaltung ihrer Geschwindigkeiten längs den Geodätischen bewegt, so verteilen sich im Lauf der Zeit die Punkte sowohl nach Lage als nach Richtung gleichförmig in $\mathfrak{F}$.

Demgegenüber ist die geodätische Strömung auf allen Flächen zweiter Klasse vom dissipativen Typus. Die Stromlinien verlaufen, abgesehen von einer m-Nullmenge von solchen, in Ω ins Unendliche.

Diese Sätze werden im zweiten Kapitel bewiesen. Das erste Kapitel ist ausschließlich den Flächen und Räumen konstanter negativer Krümmung gewidmet. Wenn ihnen in dieser Arbeit ein so breiter Raum eingeräumt wird, so hat das zwei Gründe. Erstens, daß der wesentliche Punkt der Beweismethode in seiner ganzen Einfachheit klar hervortritt, während bei veränderlicher Krümmung mühsame differentialgeometrische Rechnungen vorausgeschickt werden mußten, bis er in Erscheinung tritt. Ich hoffe, die Lektüre dem Leser nicht zu erschweren, wenn ich von vornherein in n Dimensionen beginne. Der zweite Grund ist, daß für konstante Krümmung bei einer

1) Vgl. ERG, § 10, wo in gleicher Weise ein räumliches euklidisches Billardproblem betrachtet wird.

Unterklasse der ersten Klasse, bei den Räumen endlichen Gesamtvolumens, mehr als Ergodizität, nämlich der Mischungscharakter der geodätischen Strömung bewiesen wird:

Gleichförmige Verteilung des Punkthaufens nach Lage und Richtung kommt bereits dann zustande, wenn die Punkte alle dieselbe Geschwindigkeit Eins besitzen und anfänglich nach Lage und Richtung kontinuierlich verteilt waren.

Die bisher erzielten Resultate sind kürzlich von G. A. Hedlund zusammenfassend dargestellt worden[1]). Dort findet man Hinweise auf die Arbeiten von Artin, Birkhoff, Hadamard, Hedlund, E. Hopf, Koebe, Löbell, Morse, Myrberg, J. Nielsen, Tuller und anderer Forscher. Wir begnügen uns hier mit folgenden Angaben über das bisher Bewiesene im Vergleich zu den Methoden und Resultaten der vorliegenden Arbeit. Man hatte schon vor längerer Zeit bemerkt, daß sich auf Flächen negativer Krümmung die Geodätischen durch unendliche Symbolfolgen nach Art der Dezimalbrüche charakterisieren lassen. Mit Hilfe dieser eleganten Darstellung konnte man leicht das Vorkommen spezieller Typen, z. B. geschlossener Geodätischer von vorgegebenem Typ und quasiergodischer, d. h. in Ω überall dichter, Geodätischer, nachweisen. Hingegen bietet die Verknüpfung der Symbole mit dem Volummaß m, deren Herstellung für die Lösung der statistischen Probleme erforderlich wäre, neue Schwierigkeiten. Hedlund konnte den Ergodizitätsbeweis auf diesem Wege für spezielle Flächen mit konstantem K erbringen. Für alle Flächen mit endlicher Oberfläche und $K = \text{const}$ konnte ich den Beweis mit potentialtheoretischen Hilfsmitteln führen. Erst die hier entwickelte geometrische Methode der asymptotischen Geodätischen führt wesentlich weiter. Bei ihrer Anwendung auf Flächen mit variablem $K < 0$ sieht man, daß sie an der wesentlichen Stelle, nämlich der erwähnten Unstabilität der Geodätischen, einsetzt. Diese Unstabilität tritt nun nicht nur bei Flächen mit durchweg negativem K auf. Ich hoffe, in einer späteren Mitteilung, diesen Punkt schärfer als hier herauszuheben und in dieser Richtung einen Schritt weiterzugehen. Das schwierige Problem der Geodätischen auf Flächen überall positiver Krümmung kann, wie zu erwarten, mit diesen Methoden nicht behandelt werden. Für Flächen mit veränderlichem $K < 0$ hatte Hedlund vor kurzem bewiesen: Auf Flächen mit hyperbolischen Öffnungen ist die geodätische Strömung dissipativ, auf Flächen mit endlicher Oberfläche ist

1) G. A. Hedlund, The dynamics of geodesic flows, Bull. Am. Math. Soc. **45** (1939), S. 241. Da in diesem Bericht die Originalarbeiten vollständig zitiert sind, habe ich mich im folgenden vielfach auf bloße Namensnennung beschränken können.

fast jede Geodätische quasiergodisch. Die Voraussetzungen über K sind die gleichen wie hier. Seine Sätze sind also in unseren enthalten.

Der Mischungssatz ist im Falle $K = \text{const}$ und in zwei Dimensionen jüngst von Hedlund bewiesen worden. Unabhängig davon erschien etwas später mein potentialtheoretischer Beweis des Mischungscharakters im weiteren Sinne[1]). Die Analyse des Hedlundschen Beweises ergab indessen die größere Tragweite und Verallgemeinerungsfähigkeit seiner geometrischen Methode der Horozykel. Sie wird im folgenden vereinfacht dargestellt und auf n Dimensionen ausgedehnt. Der Ergodizitätssatz wird hier ebenso wie bei Hedlund benutzt. Es zeigt sich, daß als wesentliches Beweishilfsmittel neben dem Linienelementraum auch der Raum der orthogonalen Zweibeine herangezogen werden muß.

Inhaltsübersicht.

I. Kapitel. Mannigfaltigkeiten konstanter negativer Krümmung 265

§ 1. Die Grundmannigfaltigkeit und ihr Linienelementraum 265

§ 2. Die geodätische Strömung. Asymptotische Geodätische 267

§ 3. Die Mannigfaltigkeiten $\mathfrak{M}$ und ihre Linienelementräume 269

§ 4. Die fundamentale Einteilung der Mannigfaltigkeiten in zwei Klassen . . 270

§ 5. Dissipativität der geodätischen Strömung bei der zweiten Klasse . . . 272

§ 6. Beispiel einer $\mathfrak{M}_2$ erster Art und zweiter Klasse 275

§ 7. Ergodizität der geodätischen Strömung bei der ersten Klasse 277

§ 8. Anwendung auf automorphe Potentiale 281

§ 9. Hedlunds Horozykelströmung in zwei Dimensionen 282

§ 10. Mischungscharakter der geodätischen Strömung bei endlichem $V(\mathfrak{M}_2)$. 285

§ 11. Ausdehnung auf $n > 2$ Dimensionen 287

II. Kapitel. Zweidimensionale Mannigfaltigkeiten veränderlicher negativer Krümmung 292

§ 12. Abbildung auf die universelle Überlagerungsfläche 292

§ 13. Asymptotische Geodätische 294

§ 14. Fortsetzung. Die Koordinatentransformation 299

§ 15. Beweis des Dissipativitäts- und Ergodizitätssatzes 303

---

1) E. Hopf, Beweis des Mischungscharakters ..., Sitzungsber. d. Preuß. Akad. d. Wiss., Phys.-math. Klasse 1938 XXX.

I. Kapitel. Mannigfaltigkeiten konstanter negativer Krümmung.

§ 1. Die Grundmannigfaltigkeit und ihr Linienelementraum.

Wir gehen von der klassischen Konstruktion der in Frage stehenden Mannigfaltigkeiten durch einen nichteuklidisch metrisierten Überlagerungsraum und eine Gruppe von Deckbewegungen derselben aus. Das Innere $x_i x_i < 1$ der Einheitshyperkugel des $(x_1, x_2, \ldots, x_n)$-Raumes, $n \geqq 2$, wird mit der Bogenmetrik

$$d s^2 = \frac{4 d x_i d x_i}{(1 - x_i x_i)^2} \tag{1. 1}$$

versehen. Wegen der Proportionalität zur Euklidischen Bogenmetrik ist die induzierte Winkelmetrik die Euklidische. Das induzierte Volumdifferential ist

$$d V = \frac{2^n d x_1 \ldots d x_n}{(1 - x_i x_i)^n} \tag{1. 2}$$

In dieser hyperbolischen Geometrie sind die Bewegungen, d. h. die starren, ds ungeändert lassenden Transformationen bekanntlich identisch mit denjenigen Möbiusschen Kugelverwandtschaften des Euklidischen $(x_1, x_2, \ldots, x_n)$-Raumes, bei welchen $x_i x_i \leqq 1$ in sich übergeht. Sie lassen alle Größen und Beziehungen invariant, die aus ds^2 allein ableitbar sind und von der Wahl lokaler Koordinaten nicht abhängen, nämlich Winkel, Volumina, Krümmungstensor, infinitesimale Parallelverschiebung und Eigenschaft einer Linie, geodätisch zu sein, sowie alle diese Dinge in Teilräumen.

Versteht man unter einem orthogonalen n-Bein einen mit n numerierten, aufeinander senkrechten Richtungen versehenen Punkt (den Trägerpunkt des n-Beins), so gilt: Zu irgend zwei orthogonalen n-Beinen in $x_i x_i < 1$ gibt es eine und nur eine Bewegung, welche das erste in das zweite überführt. Bei gleichem Drehsinn der n-Beine ist die Bewegung eine eigentliche, bei entgegengesetztem eine uneigentliche, d. h. mit einer Spiegelung verbundene.

Die geodätischen Linien sind im $(x_1, x_2, \ldots, x_n)$-Raum die Innenbögen der auf $x_i x_i = 1$ orthogonalen Kreise. Denn: Kugeln beliebiger Dimensionszahl $< n$ gehen durch eine Kugelverwandtschaft wieder in solche Kugeln über. Wegen der Winkeltreue gehen also Orthogonalkreise in ebensolche über. Eine Bewegung, welche zwei Punkte in $x_i x_i < 1$ festläßt, hat sämtliche Punkte des durch sie gehenden Orthogonalkreises zu Fixpunkten. Diese sind auch die einzigen Punkte, die bei allen jenen Bewegungen fest bleiben. Andererseits ist ein nicht zu langer Teilbogen einer geodätischen Linie die eindeutig bestimmte kürzeste Verbindung seiner Endpunkte. Eine beliebige Bewegung, welche diese beiden Punkte festläßt, muß daher den Teilbogen Punkt für Punkt mit sich zur Deckung bringen. Daraus folgt die Behauptung. Eine gerichtete Geodätische ist durch ihre beiden unendlich fernen Punkte auf $x_i x_i = 1$ bestimmt.

Die zu (1. 1) gehörige Krümmung ist konstant und gleich minus Eins. Denn: Die Partialkrümmungen in einem festen Punkte sind als Flächenkrümmungen der geodätischen Flächen durch diesen Punkt definiert. Dabei ist eine geodätische Fläche durch die einparametrige Schar derjenigen von diesem Punkte ausgehenden Geodätischen definiert, welche in diesem Punkte ein zweidimensionales Flächenelement berühren. Bringt man durch eine Bewegung dieses Element in den Nullpunkt $x_1 = x_2 = \ldots = x_n = 0$ und in die Ebene $x_3 = \ldots = x_n = 0$, so fällt die geodätische Fläche ganz in diese Ebene hinein. Deren Krümmung berechnet sich aus (1. 1), $n = 2$, zu minus Eins. Aus dem eben ausgeführten folgt auch, daß die geodätischen Flächen die zweidimensionalen Orthogonalkugeln sind.

Im Riemannschen Raume läßt sich die infinitesimale Parallelverschiebung einer Richtung im Punkte x bei infinitesimaler Verschiebung dx dieses Punktes folgendermaßen definieren. Man verbinde x und $x + dx$ durch eine Geodätische. Gegebene Richtung in x und Anfangsrichtung der Geodätischen bestimmen in x ein zweidimensionales Flächenelement und damit eine geodätische Fläche durch x. Sie enthält die Geodätische x, $x + dx$. Die Richtung in $x + dx$, welche in der geodätischen Fläche liegt und mit jener Geodätischen denselben Winkel einschließt wie die gegebene Richtung, ist dann die parallel verschobene Richtung. Bekanntlich ist die Parallelverschiebung eine involutorische und winkeltreue Operation[1]).

Wir gehen nunmehr zum $(2n - 1)$-dimensionalen Raum der orientierten Linienelemente e mit Trägerpunkten in $x_i x_i < 1$ über. Man betrachte zwei unendlich benachbarte Linienelemente und bezeichne mit ds die Distanz ihrer Trägerpunkte und mit $d\chi$ den Winkel, welchen die in den zweiten Trägerpunkt parallel verschobene erste Richtung mit der zweiten einschließt. Dann definiert wegen der Bewegungsinvarianz der Parallelverschiebung

$$d\sigma^2 = ds^2 + d\chi^2 \tag{1. 3}$$

ein bewegungsinvariantes Bogendifferential im Linienelementraum. Nach obigem hängt $d\sigma$ von der Reihenfolge, in der man beide Elemente nimmt, nicht ab. Damit ist auch ein regulärer, bewegungsinvarianter Entfernungsbegriff $\sigma(e_1, e_2)$ im Linienelementraum definiert. $\sigma(e_1, e_2)$ ist die untere Grenze

1) Die geometrische Charakterisierung folgt direkt aus der analytischen, wenn man örtlich geodätische Koordinaten einführt. Zur Parallelverschiebung vgl. etwa H. Weyl, Raum, Zeit, Materie, § 11. Im übrigen sind die angeführten Haupteigenschaften der Parallelverschiebung, so wie sie oben definiert wird, in dem hier betrachteten Fall leicht direkt zu erkennen.

der $\int d\sigma$ für alle e_1 mit e_2 verbindenden Wege. Das bewegungsinvariante Volumdifferential im Linienelementraum wird durch

$$dm = dV d\omega \tag{1. 4}$$

gegeben, wo dV das Volumdifferential (1. 2) in $x_i x_i < 1$ und $d\omega$ das $(n-1)$-dimensionale Oberflächendifferential auf der Richtungskugel der Richtungen von einem Punkte x bedeutet. In $n = 2$ Dimensionen ist $d\omega$ das Winkeldifferential. $d\omega$ ist wegen der Winkeltreue bewegungsinvariant. Im übrigen ist (1. 4) das durch (1. 3) induzierte Volumdifferential. Es ändert sich nur um einen Zahlfaktor, wenn man in (1. 3) $d\chi^2$ mit einem positiven Zahlfaktor versieht.

§ 2. Die geodätische Strömung. Asymptotische Geodätische.

Betrachtet man die Bewegung längs den geodätischen Linien mit der Geschwindigkeit $\frac{ds}{dt} = 1$ im Raum der orientierten Linienelemente e, so erhält man eine stetige stationäre Strömung in diesem Raum, d. h. eine einparametrige Gruppe $\mathfrak{T}_t e$, $\mathfrak{T}_t \mathfrak{T}_s = \mathfrak{T}_{t+s}$ von stetigen eineindeutigen Abbildungen dieses Raumes auf sich. Sie heißt geodätische Strömung. Bei festem t ordnet $\mathfrak{T}_t$ jedem Linienelement dasjenige Linienelement zu, das man nach der Zeit t durch die geodätische Bewegung des ersten Elementes erhält. Bei festem Linienelement e beschreibt $\mathfrak{T}_t e$ die durch dasselbe gehende Stromlinie, in anderen Worten die auf der durch e bestimmten Geodätischen liegenden Linienelemente. Die geodätische Strömung läßt bekanntlich das Volumdifferential $dm = dV d\omega$ im Linienelementraum invariant. dm erzeugt im Linienelementraum ein sowohl strömungs- als auch bewegungsinvariantes Volummaß m im Lebesgueschen Sinne.

Für das Folgende ist es von Vorteil, neue Koordinaten für die orientierten Linienelemente einzuführen. Ein solches Linienelement bestimmt eine und nur eine gerichtete geodätische Linie. Ihre unendlichfernen Punkte $-\infty$, $+\infty$ seien auf $x_i x_i = 1$ die Punkte π_1, π_2. Die Entfernung $\int ds$ des Trägerpunktes vom Euklidischen Halbierungspunkt $x^0(\pi_1, \pi_2)$ des Orthogonalbogens $\overrightarrow{\pi_1 \pi_2}$ sei $s (s \gtreqless 0)$. Durch Angabe von π_1, π_2, s ist ein Linienelement völlig charakterisiert. π_1, π_2 laufen unabhängig auf $x_i x_i = 1$ und s nimmt unabhängig davon beliebige reelle Werte an. Der Übergang von den Euklidischen Koordinaten des Linienelementes zu den neuen Koordinaten wird ebenso wie der umgekehrte Übergang durch analytische Funktionen dargestellt. Mehr brauchen wir im Folgenden nicht zu wissen. Die geodätische Strömung wird durch die einfache Formel

$$\mathfrak{T}_t(\pi_1, \pi_2, s) = (\pi_1, \pi_2, s + t) \tag{2. 1}$$

dargestellt. Das Volumdifferential dm erhält in den neuen Koordinaten die Form

$$dm = \varrho(\pi_1, \pi_2)\, d\omega_1\, d\omega_2\, ds, \tag{2.2}$$

wo die $d\omega_i$ Volumdifferentiale auf der Hyperkugel $x_i x_i = 1$ sind. Daß $\varrho = \varrho(\pi_1, \pi_2, s)$ von s unabhängig ist, folgt aus der Strömungsinvarianz von dm und ds gegenüber (2. 1). Die genaue Form von $\varrho(\pi_1, \pi_2)$ ist für uns unwichtig. Wichtig ist nur, daß ϱ eine positive und stetige Funktion ist.

Haben zwei Geodätische (π_1, π_2) und (π_1', π_2) den gleichen positiv-unendlichen Punkt, so nennen wir sie positiv asymptotisch zueinander. Entsprechend werden zwei Geodätische als negativ asymptotisch definiert. Im folgenden Lemma, das die Grundlage unserer Ausführungen bildet, ist mit $\sigma(e_1, e_2)$ die bewegungsinvariante Entfernung zweier Linienelemente e_1, e_2 im Sinne von (1. 3) bezeichnet. Ferner ist $\mathfrak{T}_t e = e_t$ gesetzt.

Hauptlemma 2. 1. Sind die durch zwei Linienelemente e, e' bestimmten Geodätischen zueinander positiv asymptotisch, so gibt es eine Zahl a derart, daß

$$\sigma(e_{t+a}, e_t') \to 0, \qquad t \to +\infty$$

gilt. Gleiches gilt für $t \to -\infty$, wenn die Geodätischen negativ asymptotisch zueinander sind.

Beweis. Man bilde $x_i x_i < 1$ durch eine Kugelverwandtschaft auf den Halbraum $y_1 > 0$ des $(y_1, y_2, \ldots, y_n)$-Raumes so ab, daß der Punkt π_2 in den unendlichfernen Punkt des neuen Raumes fällt. Dann gehen die beiden Geodätischen in zwei parallele, auf $y_1 = 0$ senkrechte Halbgerade G, G' über. Die invariante Längenmetrik wird bekanntlich

$$ds^2 = \frac{dy_i\, dy_i}{y_1^2}. \tag{2.3}$$

Die transformierten Linienelemente e auf G, e' auf G' sind nun durch ihre y_1-Koordinaten y_1, y_1' charakterisiert. Setzt man

$$a = \int_{y_1'}^{y_1} \frac{du}{u} = \ln y_1 - \ln y_1',$$

so haben die Trägerpunkte der Elemente e_{t+a} und e_t' die gleiche y_1-Koordinate η, wobei

$$\int_{y_1'}^{\eta} \frac{du}{u} = t, \qquad \eta = y_1' e^t$$

ist. Ihre nichteuklidische Entfernung strebt für $t \to +\infty$ nach Null. Gleichzeitig wird der sie verbindende Orthogonalkreisbogen euklidisch immer gerader. Die Entfernung $\sigma(e_{t+a}, e_t')$ ist nun nicht größer als die Länge $\int d\sigma$ irgendeines Verbindungsweges die beiden Linienelemente. Wählt man als Weg erstens den Orthogonalbogen für den Trägerpunkt und nichteuklidische Parallelverschiebung der Richtung längs dem Bogen ($d\sigma = ds$) — die Richtung bleibt dabei in der G und G' enthaltenden Ebene (geodätische Fläche) —, und zweitens die noch erforderliche Drehung der Richtung im Trägerpunkt des zweiten Elementes ($d\sigma = |d\chi|$), so ergibt sich die zu beweisende Behauptung.

§ 3. Die Mannigfaltigkeiten $\mathfrak{M}$ und ihre Linienelementräume Ω.

$\mathfrak{G}$ sei eine eigentlich diskontinuierliche, d. h. von infinitesimalen Transformationen freie Gruppe hyperbolischer Bewegungen. Sie enthält entweder endlich oder abzählbar unendlich viele Bewegungen, wie aus der eigentlichen Diskontinuität folgt. Ferner folgt, daß die voneinander verschiedenen, einem Punkte x in $x_i x_i < 1$ kongruenten Punkte sich in $x_i x_i < 1$ nirgends häufen können. Die Fixpunkte der von der Identität verschiedenen Bewegungen liegen auf abzählbar vielen, höchstens $(n - 1)$-dimensionalen Teilräumen und können daher $x_i x_i < 1$ nicht ausfüllen. $\mathfrak{G}$ besitzt immer einen Fundamentalbereich. Einen besonders einfachen Fundamentalbereich $B = B(x^0)$ erhält man bekanntlich folgendermaßen. Sein Inneres ist die Menge aller Punkte x in $x_i x_i < 1$, die den Ungleichungen

$$s(x, x^0) < s(x, Sx^0)$$

genügen, wobei x^0 ein fester Punkt ist, der kein Fixpunkt sein darf, und wo S alle von der Identität verschiedenen Substitutionen von $\mathfrak{G}$ durchläuft. Da durch $s(x, x^0) = s(x, x^1)$ eine $(n - 1)$-dimensionale Orthogonalkugel definiert wird, besteht der Rand von B aus endlich oder abzählbar unendlich vielen solcher Kugelstücke, eventuell auch aus Punkten auf der unendlichfernen Kugel $x_i x_i < 1$. B ist geodätisch konvex. Fixpunkte können nur auf dem Rande eines Fundamentalbereichs auftreten.

Eine Mannigfaltigkeit der in Rede stehenden Art wird dadurch gebildet, daß man alle einem Punkte x in $x_i x_i < 1$ kongruenten Punkte identifiziert und als einen Punkt p auffaßt. Umgekehrt läßt sich unter weiten Voraussetzungen eine vollständige Riemannsche Mannigfaltigkeit konstanter negativer Krümmung stets auf diese Weise erzeugen. Durch die Identifikationsvorschrift ist auch eine Zusammenfassung kongruenter, gerichteter Linienelemente e zu einem Linienelement P auf $\mathfrak{M}$ gegeben. Der Raum aller Linienelemente P wird mit Ω bezeichnet. Aus der geodätischen Strömung im e-Raum wird nun die geodätische Strömung

$$\mathfrak{T}_t P = P_t, \quad \mathfrak{T}_s \mathfrak{T}_t = \mathfrak{T}_{s+t}$$

in Ω.

Die infinitesimalen bewegungsinvarianten Maßverhältnisse in der Grundmannigfaltigkeit $x_i x_i < 1$ und in ihrem e-Raum übertragen sich direkt auf $\mathfrak{M}$ und Ω. Unter der Entfernung $s(p, p')$ zweier Punkte auf $\mathfrak{M}$ versteht man die Minimaldistanz irgendeines der kongruenten Repräsentantenpunkte x von p von irgendeinem der Repräsentanten x' von p'. Analog wird die Entfernung $\sigma(P, P')$ in Ω definiert. Das Volummaß V auf $\mathfrak{M}$ wird so definiert. Eine Menge M auf $\mathfrak{M}$ heißt V-meßbar, wenn die Menge aller, irgendeinem p von $\mathfrak{M}$ zugeordneten Punkte x in $x_i x_i < 1$ V-meßbar ist. Die letztere Menge besitzt stets eine meßbare Repräsentantenmenge (z. B. ist der Durchschnitt mit einem Fundamentalbereich eine solche Menge). Ihr V-Maß heißt V-Maß von M. Es ist, wie man auf Grund der Bewegungsinvarianz von V beweist, unabhängig von der Wahl der Repräsentantenmenge. Analog werden m-Meßbarkeit und m-Maß in Ω definiert. m ist natürlich strömungsinvariant.

Zwei gerichtete Geodätische in $\mathfrak{M}$ heißen zueinander positiv-asymptotisch, wenn sie im Überlagerungsraum $x_i x_i < 1$ je ein Urbild besitzen, auf welche die Definition im früheren Sinne zutrifft. Man betrachte zwei Punkte p, p' in $\mathfrak{M}$. x, x' seien zwei bestimmte Repräsentanten in $x_i x_i < 1$. Zu jeder gerichteten Geodätischen durch x gibt es genau eine zu ihr positiv-asymptotische gerichtete Geodätische durch x' und umgekehrt. Die dadurch definierte eineindeutige Zuordnung der Richtungen in x zu denen in x' wird durch analytische Funktionen vermittelt. Die Abbildung der beiden Richtungskugeln in x und x' aufeinander ist daher meßbar im Sinne des Maßes $\int d\omega$, insbesondere gehen Nullmengen in Nullmengen über. Man hat damit eine meßbare Beziehung zwischen den Richtungen in p und p' auf $\mathfrak{M}$. Sie hängt durchaus von der Wahl der Repräsentanten dieser Punkte ab.

Wegen $\sigma(P, P') \leqq \sigma(e, e')$, wo e bzw. e' irgendwelche Repräsentantenelemente von P bzw. P' sind, gilt das Hauptlemma erst recht auf der Mannigfaltigkeit $\mathfrak{M}$:

Hauptlemma 3. 1. Sind die durch zwei Linienelemente P, P' bestimmten gerichteten Geodätischen in $\mathfrak{M}$ zueinander positiv-asymptotisch, so gibt es eine Zahl a derart, daß

$$\sigma(P_{t+a}, P'_t) \to 0, \qquad t \to +\infty$$

gilt. Gleiches gilt für $t \to -\infty$, wenn sie negativ-asymptotisch sind.

§ 4. Die fundamentale Einteilung der Mannigfaltigkeiten in zwei Klassen.

Wir sagen, ein geodätischer Halbstrahl in $\mathfrak{M}$ laufe ins Unendliche in $\mathfrak{M}$, wenn — mit vom Anfangspunkt gezählter Bogenlänge λ — $s(p_\lambda, p_0) \to \infty$ für $\lambda \to \infty$ gilt. Man beachte, daß allgemein $s(p_\lambda, p_\mu) \leqq |\lambda - \mu|$ ist. Hat ein Halbstrahl die Eigenschaft, daß für $\lambda > \lambda_0$, $\mu > \lambda_0$ hier stets das Gleichheitszeichen gilt, so läuft er im Endverlauf „auf kürzestem Wege" ins Unendliche. Es ist jedoch zu bemerken, daß ein ins Unendliche laufender Halbstrahl dies nicht im Endverlauf auf kürzestem Wege zu tun braucht.

Läuft von zwei zueinander asymptotischen Halbstrahlen auf $\mathfrak{M}$ einer ins Unendliche, so tut es auch der andere. Dies folgt aus dem Hauptlemma und der Ungleichung $s(p, p') \leqq \sigma(P, P')$, wo p, p' die Trägerpunkte von P, P' sind. Man sieht nun leicht, daß die Menge derjenigen Richtungen in einem Punkte p_0, für welche die Halbstrahlen in $\mathfrak{M}$ ins Unendliche laufen, auf der Richtungskugel im Sinne des Maßes $\int d\omega$ meßbar ist. Hieraus und aus der Meßbarkeit der obigen Richtungsbeziehung in verschiedenen Punkten von $\mathfrak{M}$ folgt, daß jene Menge von Richtungen in p_0 entweder für jeden Punkt p_0 von $\mathfrak{M}$ das Maß Null oder für jedes p_0 positives Maß besitzt.

Definition. $\mathfrak{M}$ heißt von der ersten Klasse, wenn die von einem Punkte von $\mathfrak{M}$ ausgehenden, auf $\mathfrak{M}$ ins Unendliche laufenden geodätischen Halbstrahlen einer Nullmenge von Anfangsrichtungen entsprechen. Jeder Punkt von $\mathfrak{M}$ hat dann diese Eigenschaft. $\mathfrak{M}$ heißt von der zweiten Klasse, wenn $\mathfrak{M}$ nicht zur ersten Klasse gehört.

Über die Klassenzugehörigkeit kann zunächst folgendes gesagt werden: Eine geschlossene, orientierbare oder nicht orientierbare Mannigfaltigkeit $\mathfrak{M}$ ist selbstverständlich von der ersten Klasse. Dies ist überhaupt der Fall wenn $\mathfrak{G}$ einen mitsamt Rand ganz in $x_i x_i < 1$ gelegenen Fundamentalbereich besitzt. Dabei treten auch zahlreiche berandete $\mathfrak{M}$ auf. In diesem Falle ist das geodätische Problem ein nichteuklidisches Billardproblem. Später wird allgemeiner bewiesen, daß $\mathfrak{M}$ zur ersten Klasse gehört, wenn das nichteuklidische Totalvolumen $V(\mathfrak{M})$ endlich ist.

Es ist notwendig, hervorzuheben, wie sich die Klasseneinteilung zur klassischen Einteilung der Gruppen $\mathfrak{G}$ in die erste und zweite Art verhält. Man betrachte die Grenzpunktmenge von $\mathfrak{G}$ auf der unendlich fernen Kugel $x_i x_i = 1$. Ein Grenzpunkt von $\mathfrak{G}$ ist ein Häufungspunkt der verschiedenen, zu einem Punkt x^0 in $x_i x_i < 1$ kongruenten Punkte. Die Grenzpunktmenge ist bekanntlich $\mathfrak{G}$-invariant und vom Punkte x^0 unabhängig. Letzteres folgt aus der Bewegungsinvarianz von $s(x, x')$ und der Kleinheit der Euklidischen Distanz, wenn x oder x' nahe bei $x_i x_i = 1$ liegt. $\mathfrak{G}$ heißt von erster Art, wenn die Grenzpunktmenge $x_i x_i = 1$ ganz erfüllt, sonst von zweiter Art. Ist nun $\mathfrak{G}$ von der zweiten Art, so gehört $\mathfrak{M}$ zur zweiten Klasse. Denn die Komplementärmenge der Grenzpunktmenge auf $x_i x_i = 1$ ist dann offen, und daher entfernen sich die Punkte $x(s)$ eines Halbstrahls, der für $s \to \infty$ auf der Komplementärmenge endet, nichteuklidisch immer mehr von der Gesamtheit der Punkte, welche einen festen Punkt x^0 in $x_i x_i < 1$ kongruent sind. Im allgemeinen decken sich jedoch die Klassen mit den Arten nicht.

In dem wichtigen Falle, wo $\mathfrak{G}$ eine endliche Basis besitzt, mit anderen Worten wo $\mathfrak{M}$ von endlicher Zusammenhangsordnung ist, deckt sich hingegen die erste Klasse genau mit der ersten Art. Zur Illustration betrachten wir die zweidimensionalen $\mathfrak{M}_2$. B sei ein Fundamentalbereich von $\mathfrak{G}$, der von Orthogonalkreisbögen und evtl. Teilbögen von $x_1^2 + x_2^2 = 1$ in endlicher Gesamtzahl begrenzt ist. Je nach dem Fehlen oder Vorhandensein der letzteren Teilbögen ist $\mathfrak{G}$ von erster oder zweiter Art. Es braucht nur gezeigt zu werden, daß $\mathfrak{M}_2$ von erster Klasse ist, wenn $\mathfrak{G}$ zur ersten Art gehört. Ist letzteres der Fall, so hat B höchstens Spitzen in endlicher Anzahl auf $x_1^2 + x_2^2 = 1$. B hat dann einen endlichen nichteuklidischen Flächeninhalt, wie sich direkt mit Hilfe der Integralformel von Gauß-Bonnet ergibt. Man kann es auch so beweisen (dies gilt für $n \geqq 2$), daß man statt $x_i x_i < 1$ den Halbraum $y_1 > 0$ mit der Metrik $y^{-2} dy_i\, dy_i$ betrachtet und die Spitze ins Unendliche verlegt. Dann erhält man einen auf $y_1 = 0$ senkrechten Zylinderraum, der offenbar nur einen endlichen Beitrag zu $V(\mathfrak{M})$ liefert. Die Zugehörigkeit zur ersten Klasse bei endlichem $V(\mathfrak{M})$ wurde bereits erwähnt.

Anders ist es, wenn $\mathfrak{G}$ keine endliche Basis besitzt. Mit Hilfe der Gauß-Bonnetschen Formel beweist man, daß $\mathfrak{M}_2$ dann stets unendlichen Flächeninhalt besitzt. Es gibt Flächen erster Art, aber von der zweiten Klasse. Ein Beispiel wird in § 6 besprochen[1]).

§ 5. Dissipativität der geodätischen Strömung bei der zweiten Klasse.

Eine Menge vollständiger Geodätischer auf $\mathfrak{M}$ ist einmal durch eine strömungsinvariante Punktmenge im Linienelementraum Ω von $\mathfrak{M}$, andererseits auch durch eine $\mathfrak{G}$-invariante Menge von Punktepaaren π_1, π_2 auf $x_i x_i = 1$ darstellbar. Ich erinnere daran, daß m das strömungsinvariante Maß in Ω, und daß $\iint d\omega_1\, d\omega_2$ ein Volummaß im Raum jener Punktepaare ist. Sowohl hier als in § 6 benötigen wir das

---

1) Es wäre von Interesse, festzustellen, ob folgendes oder ein ähnliches Kriterium der Klassenzugehörigkeit allgemein zutrifft. V_a sei das Volumen des Kugelinneren $s(x, x^0) < a$. $V_a(\mathfrak{M}; p^0)$ sei hingegen das Volumen der Menge aller p auf $\mathfrak{M}$ mit $s(p, p^0) < a$. Dann ist $\mathfrak{M}$ von der ersten bzw. zweiten Klasse, wenn (p^0 fest)

$$\limsup_{a \to \infty} \frac{V_a(\mathfrak{M}; p^0)}{V_a}$$

Null bzw. positiv ist. Die Lage von p^0 ist in dieser Aussage belanglos.

Lemma 5. 1. Eine strömungsinvariante Punktmenge in Ω hat dann und nur dann das Maß $m = 0$, wenn die entsprechende Menge von Punktepaaren π_1, π_2 das Maß $\iint d\omega_1 \, d\omega_2 = 0$ besitzt.

Beweis. Die $\mathfrak{T}_t$-invariante Menge in Ω sei mit A bezeichnet. A' sei die Menge aller derjenigen Linienelemente mit Trägerpunkten in $x_i x_i < 1$, welche bei der Zusammenfassung kongruenter Elemente die Punkte von A liefern. Dann ist $m(A) = 0$ völlig gleichbedeutend mit $m(A') = 0$. A' ist gegenüber der geodätischen Strömung im Linienelementraum der Grundmannigfaltigkeit invariant. Charakterisiert man die Punkte von A' durch die Koordinaten π_1, π_2, s, so gehört wegen (2. 1) mit (π_1, π_2, s) auch jeder Punkt (π_1, π_2, s') zu A'. Aus (2. 2) folgt nunmehr unmittelbar, daß die entsprechende Menge der (π_1, π_2) das Maß $\iint \varrho \, d\omega_1 \, d\omega_2 = 0$ hat. Wegen $\varrho > 0$ folgt daraus die Behauptung.

Sind P, P' zwei Linienelemente mit den Trägerpunkten p, p' in $\mathfrak{M}$, so gilt

$$s(p, p') \leqq \sigma(P, P') \leqq s(p, p') + \pi.$$

Die rechte Ungleichung folgt, indem man die Richtung von P längs der kürzesten Verbindung von p mit p' parallel verschiebt und in p' die restliche Drehung in die Richtung von P' ausführt. Somit läuft ein geodätischer Halbstrahl dann und nur dann auf $\mathfrak{M}$ ins Unendliche, wenn die entsprechende Stromlinienhälfte $\mathfrak{T}_t P = P_t$ für $t \to +\infty$ in Ω keinen Häufungspunkt besitzt. Wir wollen in diesem Falle von einer für $t \to +\infty$ fliehenden Stromlinie sprechen. Von diesen Stromlinien gilt der folgende allgemeine Satz[1]):

$\mathfrak{T}_t P$ sei eine bezüglich (t, P) stetige stationäre Strömung in einem vollständigen Raume Ω. Sie lasse das Lebesguesche Maß m invariant. Ω sei Vereinigungsmenge abzählbar vieler kompakter Mengen endlichen m-Maßes. Dann sind die für $t \to +\infty$ fliehenden Stromlinien im allgemeinen, d. h. bis auf eine m-Nullmenge von Stromlinien, auch in Richtung $t \to -\infty$ fliehend, und umgekehrt.

Hauptsatz 5. 2. Ist $\mathfrak{M}$ von der zweiten Klasse, so laufen die von einem beliebigen festen Punkte p_0 von $\mathfrak{M}$ ausgehenden geodätischen Halbstrahlen im allgemeinen, d. h. abgesehen von einer Menge von Richtungen vom Maße $\int d\omega = 0$ auf der Richtungskugel in p_0, ins Unendliche.

Beweis. Etwas weniger besagt folgende Formulierung. Die geodätische Strömung ist vom dissipativen Typus, d. h. abgesehen von einer m-Nullmenge von Stromlinien sind alle Stromlinien für $t \to +\infty$ fliehend. Es genügt aber, das zu beweisen, denn wegen $m = \iint d\omega dV = \int [\int d\omega] \, dV$ würde daraus der

1) Vgl. E. Hopf, Zwei Sätze über den wahrscheinlichen Verlauf der Bewegungen dynamischer Systeme. Math. Annalen **103** (1930), S. 710. Oder: ERG § 13.

Satz für fast alle Punkte p von $\mathfrak{M}$, d. h. bis auf eine V-Nullmenge in $\mathfrak{M}$ folgen. Gilt er für einen Punkt, so gilt er indessen wegen der Richtungszuordnung vermittels asymptotischer Geodätischer in sämtlichen Punkten.

Was zu beweisen bleibt, kann nun wegen Lemma 5. 1 folgendermaßen ausgedrückt werden. Die Paare π_1, π_2 von Punkten auf $x_i x_i = 1$, welche nicht fliehende Geodätische ($t \to +\infty$) darstellen, bilden eine Menge vom Maß $\iint d\omega_1\, d\omega_2 = 0$. Der Beweis ist folgender: Aus der Zugehörigkeit von $\mathfrak{M}$ zur zweiten Klasse folgt zunächst, daß die Punktepaare, welche fliehenden Geodätischen ($t \to +\infty$) entsprechen, eine Menge von positivem $\iint d\omega_1\, d\omega_2$ bilden.

Ist nun die Geodätische π_1, π_2 für $t \to +\infty$ fliehend, so ist es auch jede zu ihr positiv asymptotische Geodätische π_1', π_2. Analoges gilt für $t \to -\infty$ bei gleichbleibendem ersten Punkt π_1. Bezeichnet man mit A_2 die Menge der π_2 für die $+\infty$-fliehenden Geodätischen, mit $\bar{A}_2$ ihr Komplement auf $x_i x_i = 1$, und analog mit A_1 die Menge der π_1 für die $-\infty$-fliehenden Geodätischen, sowie mit $\bar{A}_1$ ihr Komplement, so ist zu zeigen, daß die Punktepaarmengen $\bar{A}_1 \times \bar{A}_2$, $A_1 \times \bar{A}_2$ und $\bar{A}_1 \times A_2$ das Maß $\iint d\omega_1\, d\omega_2 = 0$ besitzen. Wie oben festgestellt wurde, ist das Maß $\iint d\omega_1\, d\omega_2$ der Menge $(A_1 + \bar{A}_1) \times A_2$ positiv. Also ist $\int d\omega_2$ von A_2 positiv. Nach dem oben zitierten Satze über fliehende Stromlinien ist nun $\bar{A}_1 \times A_2$ vom Maße $\iint d\omega_1\, d\omega_2 = 0$. Also hat $\bar{A}_1$ das Maß $\int d\omega = 0$. Da demnach A_1 positives Maß hat, folgt durch Wiederholung der Schlußweise nach Vertauschung der Indizes, daß $\bar{A}_2$ das Maß Null hat. Daraus ergibt sich die Richtigkeit der Behauptung und damit des Satzes.

Spezialfall der Gruppe zweiter Art mit endlicher Basis. In diesem Falle kann dem Hauptsatz eine schärfere Fassung gegeben werden. Wir beschränken uns, lediglich um die Betrachtung durchsichtiger werden zu lassen, auf $n = 2$ Dimensionen. B sei ein von endlich vielen Seiten begrenzter polygonaler Fundamentalbereich von $\mathfrak{G}$. Da $\mathfrak{G}$ von zweiter Art ist, hat der Rand von B Bögen mit dem Einheitskreis $x_1^2 + x_2^2 = 1$ gemein. Dann gilt:

Die auf $x_i x_i = 1$ gelegenen Randbögen von B und ihre $\mathfrak{G}$-Bilder füllen bis auf eine Menge vom Winkelmaß Null den Einheitskreis aus.

Daß in dem hier betrachteten Spezialfall aus dieser Behauptung der Hauptsatz folgt, ist trivial. Man kann aber auch umgekehrt die Behauptung aus dem Hauptsatz folgern. Die auf $x_i x_i = 1$ gelegenen Randbögen seien mit α_1, $\alpha_2, \ldots, \alpha_k$ bezeichnet. Wir zeigen dann, daß die von einem festen Punkte, etwa x^0, ausgehenden geodätischen Halbstrahlen, welche auf $\mathfrak{M}$ ins Unendliche laufen, notwendig in einem Punkte der α_i oder ihrer $\mathfrak{G}$-Bilder, oder andererseits in einer Spitze von B auf $x_i x_i = 1$ oder einem $\mathfrak{G}$-Bild einer solchen enden müssen. Zum Beweise dieser Tatsache nehmen wir zunächst an, daß keine Spitzen vorhanden sind. Ein auf $\mathfrak{M}$ ins Unendliche laufender Halbstrahl wird, wenn in B verfolgt, d. h. wenn beim Austreten aus B durch eine $\mathfrak{G}$-Substitution nach B zurückversetzt, zu einer Folge von Orthogonalkreisbögen in B mit Konvergenz des laufenden Punktes gegen den Rand auf $x_i x_i = 1$. Es gibt nun unter diesen Bögen einen letzten, mit dem Endpunkt auf einem der α_i. Denn die Bögen, welche diese Eigenschaft nicht besitzen, müssen gegen $\Sigma \alpha_i$ konvergieren, und dies ist nur dadurch möglich, daß die Endpunkte der sie enthaltenden vollständigen Orthogonalkreisbögen immer mehr gegeneinander rücken. Also muß einmal ein solcher Endpunkt in ein α_i hineinfallen, was zu beweisen war. Im allgemeinen Fall, wenn B auch Spitzen auf $x_i x_i = 1$ besitzt, kommt noch folgende, hier nur kurz anzudeutende Überlegung hinzu. Man kann annehmen, daß die den Halbstrahl repräsentierende Bogenfolge in B von einer Stelle ab ganz in den Spitzen verbleibt. Diese Spitzen bilden einen parabolischen Zykel. Die ihm entsprechende parabolische Öffnung von $\mathfrak{M}$ läßt sich stets auf die Pseudosphäre abwickeln, etwa indem man einen in einer Spitze von B mündenden Halbstrahl einer Meridiankurve der letzteren isometrisch zuordnet und die Zuordnung längs der orthogonalen Trajektorien der asymptotischen Halbstrahlen fortsetzt. Ein dabei im Sinne des Zykels durchlaufener Weg schließt sich auf $\mathfrak{M}$ genau. Die Abwicklung führt entweder zu einer vollen Umgebung des unendlich fernen Punktes der Pseudosphäre oder zu einer von zwei diametralen Meridiankurven begrenzten Hälfte derselben. In letzterem Falle geschieht die Fortsetzung der Geodätischen durch Spiegelung an den Randlinien, ein Fall, der aus dem ersten durch Identifizieren diametraler Punkte hervorgeht. Die Rechnung lehrt, daß die Meridiankurven die einzigen, auf der Pseudosphäre ins Unendliche laufenden Geodätischen sind. Jede andere auf den unendlich fernen Punkt zulaufende Geodätische windet sich um die Fläche unter abnehmendem Schnittwinkel mit den Breitenkreisen, kehrt einmal um und läuft wieder zurück. Damit ist auch der Spitzenfall erledigt und der Satz bewiesen.

§ 6. Beispiel einer $\mathfrak{M}_2$ erster Art und zweiter Klasse.

Wie kompliziert der Verlauf der geodätischen Linien auf einer Mannigfaltigkeit zweiter Klasse sein kann, lehrt folgende Betrachtung, in welcher wir uns auf $n = 2$ Dimensionen beschränken. Für eine Mannigfaltigkeit $\mathfrak{M}_2$ der ersten Art ist nach einem Satze von Hedlund die dreidimensionale geodätische Strömung permanent transitiv, d. h. irgendein offener, von der Strömung mitgeführter Teil von Ω verteilt sich allmählich überall dicht in Ω. Trotzdem kann dabei die Strömung dissipativ sein, d. h. die allgemeine Stromlinie eine fliehende sein. Dieser merkwürdige Fall präsentiert sich bei jeder $\mathfrak{M}_2$ erster Art und zweiter Klasse. Folgendes Beispiel liefert eine derartige $\mathfrak{M}_2$.

μ sei eine perfekte, nirgends dichte Menge auf $x_1^2 + x_2^2 = 1$ mit positivem Winkelmaß. Ihre Komplementärmenge $\bar{\mu}$ besteht dann aus unendlich vielen offenen Bögen von $x_1^2 + x_2^2 = 1$ ohne gemeinsame Endpunkte. Jede Umgebung eines beliebigen Punktes von μ enthält vollständige Bögen von $\bar{\mu}$. Die beiden Endpunkte eines Bogens von $\bar{\mu}$ bestimmen einen Orthogonalkreis. B' sei der von allen diesen Orthogonalkreisen und von μ berandete Bereich in $x_1^2 + x_2^2 < 1$. Man richte es so ein, daß B' bezüglich eines festen Durchmessers von $x_i x_i = 1$ symmetrisch ist und daß die Endpunkte desselben zu μ gehören. Dann paare man jeden Orthogonalkreisbogen mit dem symmetrisch gelegenen so, daß symmetrische Punkte einander entsprechen. Diese Paarung bestimmt eine hyperbolische Substitution, und diese erzeugen wiederum eine Gruppe. Statt dessen könnte man auch jeden Punkt mit sich selbst paaren. Die erzeugenden Substitutionen wären dann Spiegelungen. In einem Falle entsteht eine unberandete, im anderen eine berandete Fläche (Billardproblem). Zunächst ist jedenfalls zu zeigen, daß B' ein echter Fundamentalbereich der erzeugten Gruppe ist. Klar ist, daß zwei zu verschiedenen Substitutionen gehörige Gruppenbilder von B' keinen inneren Punkt gemein haben. Also ist die Gruppe eigentlich diskontinuierlich. Man konstruiere nun nach der Vorschrift am Anfang von § 3 den Fundamentalbereich $B(x^0)$, $x^0 = 0$, von $\mathfrak{G}$ (statt $x^0 = 0$ könnte man irgendeinen Punkt des Symmetriedurchmessers nehmen). Dann muß sich das Innere von $B(0)$ mit dem Inneren von B' decken. Denn jeder innere Punkt x von $B(0)$ liegt nach Definition von $B(0)$ dem Nullpunkt näher als alle zu x kongruenten Punkte $Sx \neq x$, eine Eigenschaft, die den Randpunkten von B' offenbar nicht zukommt. Da $B(0)$ zusammenhängend ist, muß demnach sein Inneres zu B' gehören, also, da $B(0)$ Fundamentalbereich ist, mit dem Inneren von B' identisch sein. Da der Rand von B' keine Bögen von $x_i x_i = 1$ enthält, ist $\mathfrak{G}$ von der ersten Art. $\mathfrak{M}$ ist offenbar von der zweiten Klasse. Denn die vom Nullpunkt ausgehenden Halbstrahlen mit Endpunkten auf μ verlaufen ganz in $B' = B(0)$. Diese Halbstrahlen laufen, da jeder ihrer Punkte zum Nullpunkt näher liegt als zu den $\mathfrak{G}$-Bildern desselben, auf $\mathfrak{M}$ ins Unendliche[1]).

1) Zur Konstruktion von Gruppen erster Art und zweiter Klasse sei schließlich noch folgendes bemerkt. Für die Zugehörigkeit zur zweiten Klasse ist es vermutlich nicht notwendig, daß die auf $x_i x_i = 1$ liegenden Randpunkte des Fundamentalbereichs B' positives Winkelmaß besitzen. Selbst wenn diese Randpunkte nur aus abzählbar vielen Spitzen mit einem einzigen Häufungspunkt bestehen, genügt es wahrscheinlich, daß die Berandung von B' zwischen dem Einheitskreis und einer im Häufungspunkt diesen Kreis innen mit genügend hoher Ordnung berührender Kurve verläuft. Dagegen wird man Zugehörigkeit zur ersten Klasse erwarten, wenn etwa, von $x_1 = x_2 = 0$ gesehen, die Winkeldistanz zwischen zwei benachbarten Spitzen auf einer Seite des Grenzpunktes, etwa der n-ten und der $(n+1)$-ten wie a^n, $a < 1$, nach Null geht.

§ 7. Ergodizität der geodätischen Strömung bei der ersten Klasse.

Ist das Totalvolumen $V(\mathfrak{M})$ von $\mathfrak{M}$ endlich, so gehört $\mathfrak{M}$ zur ersten Klasse. Dann ist nämlich das strömungsinvariante Volumen $m(\Omega)$ des Linienelementraumes ($V(\mathfrak{M})$ mal dem Oberflächeninhalt der $(n-1)$-dimensionalen Einheitskugel) endlich, und daher gilt von der geodätischen Strömung der Poincaré-sche Wiederkehrsatz. Die fliehenden Stromlinien bilden also in Ω eine m-Nullmenge, was die Zugehörigkeit zur ersten Klasse zur Folge hat.

Wir beschränken uns zunächst auf diesen wichtigsten Fall $V(\mathfrak{M}) < \infty$. Von der geodätischen Strömung gilt dann der Birkhoffsche Ergodensatz[1]):

Ist die Funktion $f(P)$ in Ω m-summierbar, so existieren auf allen Stromlinien, bis auf solche, welche in Ω eine Punktmenge vom m-Maß Null bilden, die auf der Stromlinie konstanten Zeitmittel

$$f_+^*(P) = \lim_{T=+\infty} \frac{1}{T}\int_0^T f(P_t)\,dt, \qquad f_-^*(P) = \lim_{T=+\infty} \frac{1}{T}\int_{-T}^0 f(P_t)\,dt. \tag{7.1}$$

Jedes der beiden $f^*(P)$ ist wieder summierbar und genügt für jede beschränkte, m-meßbare und $\mathfrak{T}_t$-invariante Funktion $\varphi(P)$ der Gleichung

$$\int_\Omega f^*\varphi\,dm = \int_\Omega f\varphi\,dm.$$

Durch sie sind die Zeitmittel im wesentlichen bestimmt, insbesondere gilt fast überall in Ω

$$f_+^*(P) = f_-^*(P).$$

Die Strömung heißt ergodisch, wenn $f^*(P)$ fast überall konstant ist, und zwar bei beliebigem $f(P)$. Für die Ergodizität reicht es aus, die Konstanz des Zeitmittels für solche Funktionen $f(P)$ zu beweisen, welche im Sinne des Distanzbegriffs

$$\int_\Omega |f-g|\,dm \tag{7.2}$$

im Raume aller m-summierbaren $f(P)$ überall dicht liegen. Im ergodischen Falle ist fast überall

$$f^*(P) = \frac{1}{m(\Omega)}\int_\Omega f\,dm.$$

Im Gegensatz zum Hauptsatz über die Mannigfaltigkeiten zweiter Klasse gilt nun

Hauptsatz 7. 1. Hat $\mathfrak{M}$ endliches Totalvolumen, so ist die geodätische Strömung in Ω ergodisch. Genauer gilt: Ist $f(P)$ in Ω gleichmäßig stetig

1) Vgl. ERG § 14.

(im Sinne der Distanz σ) und summierbar, und ist p ein willkürlicher Punkt in $\mathfrak{M}$, so gilt

$$\lim_{T=\infty} \frac{1}{T}\int_0^T f(P_t)\,dt = \frac{1}{m(\Omega)}\int_\Omega f\,dm$$

für alle Linienelemente P in p, abgesehen von einer Nullmenge von Richtungen in p.

Beweis. Ist $\mathfrak{M}$ geschlossen, so genügt es, von $f(P)$ Stetigkeit zu verlangen. Der Beweis des Hauptsatzes 7. 1 ist nun völlig analog zum Beweise des Hauptsatzes 5. 2. Der einzige Unterschied besteht in der Heranziehung des Satzes von der Gleichheit der Zeitmittel in Vergangenheit und Zukunft anstatt des Satzes von der Gleichzeitigkeit des Fliehcharakters in Vergangenheit und Zukunft. Der Tatsache, daß zwei zueinander im positiven Sinne asymptotische Geodätische für $t \to +\infty$ beide fliehend sind, wenn es eine ist, entspricht hier zunächst die Tatsache, daß für zwei Elemente P und P', wo P auf der ersten, P' auf der zweiten dieser zwei Geodätischen liegt,

$$\lim_{T=+\infty} \frac{1}{T}\int_0^T f(P'_t)\,dt = \lim_{T=+\infty} \frac{1}{T}\int_0^T f(P_t)\,dt$$

ist, sobald einer der Limites existiert. Denn aus dem Hauptlemma und aus der gleichmäßigen Stetigkeit von $f(P)$ folgt, wenn a eine passende Zahl ist,

$$\lim_{t=+\infty} \left\{ f(P'_t) - f(P_{t+a}) \right\} = 0\,.$$

Gleiches gilt, wenn P, P' negativ asymptotische Geodätische bestimmen, von den rückwärtigen Zeitmitteln. Um die Analogie vollständig zu machen, betrachten wir diejenigen Geodätischen, für welche

$$f_+^*(P) \geqq \alpha \tag{7. 3}$$

bei festem, aber beliebig wählbarem α besteht. α sei so gewählt, daß sie in Ω eine Menge positiven m-Maßes bilden. Die Menge der Geodätischen mit der Eigenschaft

$$f_-^*(P) \geqq \alpha \tag{7. 4}$$

kann sich von ihr nur um eine m-Nullmenge unterscheiden und umgekehrt. Charakterisiert man die Geodätischen durch die Punktepaare π_1, π_2 auf $x_i x_i = 1$, so kann man wegen Lemma 5. 1 in allen Aussagen das m-Maß durch das Maß $\iint d\omega_1\, d\omega_2$ ersetzen. Die Situation ist nunmehr dieselbe wie im Beweis von Satz 5. 2. Nennt man A_2 die Menge der Punkte π_2, die durch (7. 3) charakterisiert sind, und A_1 die Menge der π_1, welche (7. 4) erfüllen, so folgt genau wie dort, daß abgesehen von einer Nullmenge sämtliche Geodätische die Ungleichung (7. 3) befriedigen müssen, sobald es eine Menge positiven m-Maßes tut. Da dies für beliebiges α gelten muß, so ergibt sich schließlich, daß $f_+^*(P)$ fast überall in Ω konstant ist, wenn $f(P)$ den angegebenen Voraussetzungen genügt. Der Rest des Beweises ist derselbe wie früher.

Im Beweise spielt die gleichmäßige Stetigkeit von $f(P)$ eine wesentliche Rolle. Man kann aber dem Satz nachträglich andere Formen geben, wenn man statt solcher $f(P)$ eine andere Funktionenklasse betrachtet. Zunächst kann man aus den gleichmäßig stetigen $f(P)$ eine Folge solcher f herausgreifen, welche im Raume dieser $f(P)$ im Sinne gleichmäßiger Konvergenz dicht liegen. Für diese Folge läßt sich die Nullmenge des Satzes für alle angehörenden f fest wählen. Daraus folgt leicht, daß man mit dieser festen Nullmenge bei sämtlichen gleichmäßig stetigen Funktionen auskommt.

Versteht man nun unter einem Normalbereich in Ω einen Teilbereich von Ω, dessen Rand eine m-Nullmenge ist, und nennt man eine Stromlinie gleichverteilt in Ω, wenn sie jeden Normalbereich in Ω mit volumproportionaler mittlerer Verweilzeit durchläuft, so kann man dem Hauptsatz folgende Form geben:

Ist $V(\mathfrak{M}) < \infty$, so sind die von einem beliebigen festen Punkte von $\mathfrak{M}$ ausgehenden Geodätischen, wenn man von einer Nullmenge von Anfangsrichtungen absieht, in Ω gleichverteilt.

Es genügt zu zeigen, daß im Sinne des Maßes m fast alle Stromlinien in Ω gleichverteilt sind. A sei ein Normalbereich in Ω, $h(P)$ seine charakteristische Funktion. Dann kann man zwei gleichmäßig stetige und summierbare Funktionen f_1 und f_2 derart finden, daß überall $f_1 \leqq h \leqq f_2$ und

$$\frac{1}{M(\Omega)} \int_{\Omega} (f_2 - f_1) dm < \varepsilon$$

gilt. Gehört P nun nicht der obigen universellen m-Nullmenge an, so streben in

$$\frac{1}{T} \int_0^T f_1(P_t) dt \leqq \frac{1}{T} \int_0^T h(P_t) dt \leqq \frac{1}{T} \int_0^T f_2(P_t) dt$$

(gilt für alle T) das linke und rechte Glied für $T \to \infty$ gegen Grenzwerte, die um weniger als ε differieren. Daraus folgt, daß die mittlere Verweilzeit von $P_t = \mathfrak{T}_t(P)$ in jedem beliebigen Normalbereich A gleich $m(A)/m(\Omega)$ ist, was zu beweisen war.

Ist $\mathfrak{M}$ von der ersten Klasse, aber nicht notwendig von endlichem Volumen, so bilden in Ω die fliehenden Stromlinien eine m-Nullmenge. Dann gilt des Verfassers Verallgemeinerung des Ergodensatzes[1]):

1) ERG § 14.

Sind $f(P)$ und $g(P) > 0$ in Ω m-summierbar, so existiert und gilt fast überall in Ω

$$\lim_{T=+\infty} \frac{\int_0^T f(P_t)\,dt}{\int_0^T g(P_t)\,dt} = \lim_{T=+\infty} \frac{\int_{-T}^0 f(P_t)\,dt}{\int_{-T}^0 g(P_t)\,dt} = f^*(P).$$

f^* genügt für jede beschränkte, meßbare und $\mathfrak{T}_t$-invariante Funktion $\varphi(P)$ der Beziehung

$$\int_\Omega g f^* \varphi\, dm = \int_\Omega f \varphi\, dm.$$

$f^* g$ ist ebenfalls summierbar.

Die Strömung heißt wieder ergodisch, wenn f^* für jedes f fast überall konstant ist. Dann ist fast überall

$$f^*(P) = \frac{\int_\Omega f\,dm}{\int_\Omega g\,dm}.$$

Mit derselben Schlußweise wie oben läßt sich dann, was nicht näher ausgeführt werden soll, folgende Erweiterung des Hauptsatzes beweisen:

Hauptsatz 7. 2. Ist $\mathfrak{M}$ von der ersten Klasse, so ist die geodätische Strömung ergodisch. Genauer: $f(P)$ und $g(P) > 0$ sind in Ω summierbar und mögen der folgenden Stetigkeitsvoraussetzung genügen.

$$\lim_{P' \to P} \frac{f(P') - f(P)}{g(P)} = \lim_{P' \to P} \frac{g(P') - g(P)}{g(P)} = 0$$

gelte in Ω gleichmäßig ($P' \to P$ ist im Sinne von $\sigma(P, P') \to 0$ verstanden). Ist p ein willkürlicher Punkt in $\mathfrak{M}$, so gilt, abgesehen von einer Nullmenge von Anfangsrichtungen, für alle Linienelemente P in p

$$\lim_{T=+\infty} \frac{\int_0^T f(P_t)\,dt}{\int_0^T g(P_t)\,dt} = \frac{\int_\Omega f\,dm}{\int_\Omega g\,dm}.$$

Für festes $g(P)$ bilden die zugelassenen $f(P)$ eine im Sinne von (7. 2) im Raume aller summierbaren f dichte Funktionenmenge.

Bei ergodischer Strömung ist die allgemeine Stromlinie relativ gleichverteilt, also gewiß überall dicht (die letztere Eigenschaft nennt man bekanntlich Quasiergodizität). Die Existenz einer quasiergodischen Stromlinie bedingt jedoch keineswegs Ergodizität der Strömung. Die geodätische Strömung auf einer $\mathfrak{M}_2$ erster Art und zweiter Klasse (§ 6) beweist dies in eklatanter Weise. Auf einer $\mathfrak{M}_2$ erster Art gibt es stets quasiergodische Geodätische, was allgemein zuerst von Koebe bewiesen wurde. Diese Tatsache ist auch eine Folge des in § 6 erwähnten Hedlundschen Satzes von der permanenten Transitivität der Strömung.

§ 8. Anwendung auf automorphe Potentiale.

Eine Strömung mit invariantem Maß m ist dann und nur dann ergodisch, wenn die Konstanten die einzigen meßbaren und strömungsinvarianten Funktionen (Integrale der Strömungsgleichungen) sind. Nach Lemma 5. 1 kann man also dem Hauptsatz 7. 2 folgende Fassung erteilen.

Satz 8. 1. Definiert die Gruppe $\mathfrak{G}$ von Kugelverwandtschaften eine Mannigfaltigkeit $\mathfrak{M}$ der ersten Klasse, so ist eine meßbare und automorphe, d. h. für alle S von $\mathfrak{G}$

$$f(S\pi_1, S\pi_2) \equiv f(\pi_1, \pi_2)$$

erfüllende Funktion der Punktepaare auf $x_i x_i = 1$ notwendig konstant (bis auf eine Nullmenge).

Natürlich gilt das erst recht von automorphen Funktionen von $\pi_1 = \pi$ allein. Ob der Satz auch von den automorphen Funktionen von mehr als zwei Punkten π_i gilt, ist mir nicht bekannt. In $n = 2$ und in $n = 3$ Dimensionen gilt er jedenfalls nicht von den automorphen $f(\pi_1, \pi_2, \pi_3)$. Dies liegt daran, daß in beiden Fällen eine $x_i x_i \leqq 1$ in sich überführende Kugelverwandtschaft durch die Angabe von drei verschiedenen Originalpunkten und drei verschiedenen Bildpunkten auf $x_i x_i = 1$ immer und eindeutig — für $n = 2$ wirklich eindeutig, für $n = 3$ bis auf eine ganz bestimmte Spiegelung eindeutig — bestimmt ist. Daraus folgt aber, daß die $\mathfrak{G}$-Bilder $f\pi_1, f\pi_2, f\pi_3$ eines festen Tripels π_1, π_2, π_3 nur solche Häufungstripel π_1', π_2', π_3' besitzen können, wo mindestens zwei π_i' zusammenfallen. Anderenfalls ergäbt sich ein Widerspruch gegen die eigentliche Diskontinuität von $\mathfrak{G}$. Es gibt also sicher nichtkonstante automorphe $f(\pi_1, \pi_2, \pi_3)$. Die Fragestellung soll indessen, da für die hier verfolgten Ziele ohne Belang, nicht weiter verfolgt werden.

Erwähnen möchte ich jedoch eine potentialtheoretische Anwendung des zweidimensionalen Falles, deren Resultat mir als Ausgangspunkt meines früheren Beweises für den Hauptsatz 7. 1 ($n = 2$) gedient hatte. Dieses Resultat kann hier unter gegen früher erweiterten Voraussetzungen ausgesprochen werden. Dabei bedienen wir uns der komplexen Schreibweise $z = x_1 + i x_2$.

Satz 8. 2. Definiert eine Grenzkreisgruppe in $|z| < 1$ eine Fläche $\mathfrak{M}_2$ erster Klasse, so ist eine in $|z| < 1$, $|z'| < 1$ definierte Funktion $u(z, z')$, welche erstens sowohl in z als in z' harmonisch, zweitens automorph —

$$u(Sz, Sz') \equiv u(z, z')$$

für alle S von $\mathfrak{G}$ —, drittens beschränkt ist, notwendig eine Konstante.

Beweis. Man braucht nur zu zeigen, daß ein beschränktes Potential $u(z, z')$ Grenzwerte $u(\zeta, \zeta')$ auf dem Torus $|\zeta| = |\zeta'| = 1$ in einem gleich anzugebenden Sinne besitzt, daß es durch $u(\zeta, \zeta')$ eindeutig bestimmt ist, und daß sich die Automorphie auf die Grenzwerte $u(\zeta, \zeta')$ überträgt. Der Fatousche Satz über die Existenz der Grenzwerte bei radialer Annäherung läßt sich auf harmonische Funktionen zweier Punkte übertragen. Für fast alle (ζ, ζ') auf dem Torus existiert

$$\lim_{r = r' = 1} u(r\zeta, r'\zeta') = u(\zeta, \zeta').$$

$u(z, z')$ ist durch die Toruswerte eindeutig bestimmt und durch das Poissonsche Doppelintegral

$$u(z, z') = \frac{1}{4\pi^2} \int\int_{|\zeta| = |\zeta'| = 1} u(\zeta, \zeta') \frac{1 - |z|^2}{|\zeta - z|^2} \frac{1 - |z'|^2}{|\zeta' - z'|^2} |d\zeta| \, |d\zeta'|$$

dargestellt. Annäherung längs zweier Orthogonalkreise anstatt zweier Radien ändert nichts an den Grenzwerten, da ein beschränktes Potential im nichteuklidischen Sinne gleichmäßig stetig ist, sogar einen beschränkten Gradienten besitzt. Also gilt

$$\lim_{r = r' = 1} u(Sr\zeta, Sr'\zeta') = u(S\zeta, S\zeta'),$$

woraus alles folgt.

Auch umgekehrt läßt sich aus Satz 8. 2 der Satz 8. 1, $n = 2$, erschließen. Satz 8. 2 ist also mit der Behauptung, daß die geodätische Strömung auf den Flächen $\mathfrak{M}_2$ der ersten Klasse ergodisch ist, völlig äquivalent.

§ 9. Hedlunds Horozykelströmung in zwei Dimensionen.

Wir beschränken uns in diesem Abschnitt auf $n = 2$ Dimensionen. Ein Horozykel ist definiert als orthogonale Trajektorie eines Feldes von zueinander asymptotischen Geodätischen. Er ist ein euklidischer Kreis in $x_i x_i < 1$, welcher $x_i x_i = 1$ in dem gemeinsamen ∞-Punkt derselben berührt. Ein gerichtetes Linienelement e in $x_i x_i < 1$ bestimmt genau einen Horozykel, welcher e senkrecht durchsetzt, und dessen ∞-Punkt mit dem Punkt $+\infty$ der durch e bestimmten Geodätischen zusammenfällt. Die Horozykelströmung

$$\mathfrak{H}_t e, \quad \mathfrak{H}_t \mathfrak{H}_s = \mathfrak{H}_{t+s}$$

wird dann dadurch definiert, daß e längs dem Horozykel bei stets senkrecht gehaltener Richtung um die Bogenlänge s verschoben wird. Um die noch verbleibende Unbestimmtheit in der Verschiebungsrichtung zu beseitigen, wird folgende Verabredung getroffen. e läßt sich auf zweierlei Weisen zu einem (orthogonalen) Zweibein ergänzen. Ist die Wahl getroffen, so soll die Verschiebungsrichtung mit der zweiten Richtung des Zweibeins identisch sein. Dementsprechend fassen wir von nun ab e als Symbol für ein Zweibein auf, auch wenn im folgenden vielfach der Ausdruck Linienelement beibehalten wird[1]). Die Horozykelströmung ist nunmehr vollständig definiert. Ihr enger Zusammenhang mit der geodätischen Strömung $\mathfrak{T}_t$ ist zuerst von Hedlund erkannt und zur Untersuchung der letzteren auf den Flächen $\mathfrak{M}_2$ herangezogen worden. Man verbinde den Trägerpunkt von e mit dem von $\mathfrak{H}_t e$ durch einen gerichteten geodätischen Bogen. Wird sein Anfangselement mit e' bezeichnet, so ist sein Endelement von der Form $\mathfrak{T}_s e'$, wo $s = s(t)$ von t allein abhängt und

$$s(t) \to \infty, \quad t \to \infty \tag{9, 1}$$

gilt. Versteht man unter der Operation $\mathfrak{D}_\alpha$ die positive Drehung eines Zweibeins um seinen Trägerpunkt mit dem Winkel α, so ist geometrisch klar (Schnitt zweier Kreise), daß man

$$e' = \mathfrak{D}_\alpha e, \qquad \mathfrak{H}_t e = \mathfrak{D}_\pi \mathfrak{D}_\alpha \mathfrak{T}_s e',$$

also

$$\mathfrak{H}_t e = \mathfrak{D}_\pi \mathfrak{D}_\alpha \mathfrak{T}_s \mathfrak{D}_\alpha e \tag{9. 2}$$

schreiben kann. Dabei ist $\alpha = \alpha(t)$ nur von t abhängig und

$$\alpha(t) \to 0, \qquad t \to \infty. \tag{9. 3}$$

Denn für $t \to \infty$ nähert sich der verbindende Orthogonalkreisbogen demjenigen, welcher im unendlich fernen Punkt des Horozykels endet ($\alpha = 0$). Daß s und α nur von t abhängen, sieht man, wenn man e durch eine Bewegung in eine feste Normallage bringt. Dabei bleiben t, s, α ungeändert.

Die Beziehungen, welche die Horozykelströmung definieren, sind bewegungsinvariant. Daher kann man, wenn die Fläche $\mathfrak{M} = \mathfrak{M}_2$ durch eine diskrete Bewegungsgruppe definiert ist, von der Horozykelströmung $P \to \mathfrak{H}_t(P)$ im Linienelementraum Ω von $\mathfrak{M}$ sprechen. Dabei gilt wieder (9. 2), und die darin vorkommenden Einzeltransformationen von Ω in sich sind sämtlich m-treu. Daraus folgt die m-Treue von $\mathfrak{H}_t$.

Wir beweisen nun das Hedlundsche

Lemma 9. 1. Ist $\mathfrak{M}_2$ von endlichem Flächeninhalt, so ist die Horozykelströmung in Ω ergodisch.

---

1) Es ist unzweckmäßig, sich auf eine bestimmte zweite Richtung festzulegen, da diese Festsetzung gegenüber uneigentlichen Bewegungen nicht invariant wäre, und da dann nichtorientierbare $\mathfrak{M}_2$ nicht erfaßt würden. Während sich aus diesem Grunde hier die Einführung der Zweibeine empfiehlt, wird sie sich für $n > 2$ als unumgänglich erweisen (§ 11).

Beweis. Nach dem Hauptsatz 7.1 ist die geodätische Strömung ergodisch, d. h. jede m-meßbare, $\mathfrak{T}_t$-invariante Funktion von P ist notwendig in Ω fast überall konstant. Dasselbe muß nun von den $\mathfrak{H}_t$-invarianten Funktionen $h(P)$,

$$h(\mathfrak{H}_t P) \equiv h(P)$$

bewiesen werden. Es kann $h(P)$ als beschränkt angenommen werden. Nach 9.2 ist überall in Ω

$$h(P) \equiv h(\mathfrak{D}_\pi \mathfrak{D}_\alpha \mathfrak{T}_s \mathfrak{D}_\alpha P),$$

wofür man auch

$$h(\mathfrak{D}_{-\alpha} \mathfrak{T}_{-s} P) \equiv h(\mathfrak{D}_\alpha \mathfrak{D}_\pi P)$$

schreiben kann. Also ist für beliebiges beschränktes $g(P)$

$$\int_\Omega h(\mathfrak{D}_{-\alpha} \mathfrak{T}_{-s} P) g(P) dm = \int_\Omega h(\mathfrak{D}_\alpha \mathfrak{D}_\pi P) g(P) dm,$$

woraus nach Ausführung m-treuer Substitutionen in den Integralen

$$\int_\Omega h(\mathfrak{D}_{-\alpha} P) g(\mathfrak{T}_s P) dm = \int_\Omega h(\mathfrak{D}_\alpha \mathfrak{D}_\pi P) g(P) dm \tag{9. 4}$$

folgt. Im Sinne mittlerer Konvergenz in Ω, d. h. im Sinne des Distanzbegriffs

$$\sqrt{\int_\Omega (f-g)^2 dm}$$

gilt nun die allgemeine Grenzbeziehung

$$\lim_{\alpha=0} h(\mathfrak{D}_\alpha P) = h(P).$$

Wegen der Schwarzschen Ungleichung folgt hieraus und aus 9.4 — man berücksichtige (9.1) und (9.3) —

$$\lim_{s=\infty} \int_\Omega h(P) g(\mathfrak{T}_s P) dm = \int_\Omega h(\mathfrak{D}_\pi P) g(P) dm. \tag{9. 5}$$

Da die $\mathfrak{T}_t$-Strömung ergodisch ist, gilt

$$\lim_{S=\infty} \frac{1}{S} \int_0^S g(\mathfrak{T}_s P) ds = \frac{1}{m(\Omega)} \int_\Omega g dm \tag{9. 6}$$

im Sinne mittlerer Konvergenz. Aus (9.5) und (9.6) folgt daher für jedes $g(P)$

$$\int_\Omega h dm \int_\Omega g dm = \int_\Omega dm \int_\Omega h(P) g(\mathfrak{D}_\pi P) dm.$$

$\mathfrak{D}_\pi$ kann man auch weglassen, denn man kann $g'(P) = g(\mathfrak{D}_\pi P)$ statt $g(P)$ betrachten,

$$\int_\Omega h dm \int_\Omega g dm = \int_\Omega dm \int_\Omega h g dm. \tag{9. 7}$$

Dies kann nur dann für willkürliches $g(P)$ gelten, wenn $h(P)$ fast überall konstant ist, was zu beweisen war. Umgekehrt folgt in gleicher Weise die Ergodizität der geodätischen Strömung aus der der Horozykelströmung.

§ 10. Mischungscharakter der geodätischen Strömung bei endlichem $V(\mathfrak{M}_2)$.

Eine stationäre, m-treue Strömung in Ω, $m(\Omega) < \infty$, heißt vom Mischungstypus, wenn sich ein beliebiger Teil von Ω allmählich im Sinne des m-Maßes in Ω gleichförmig verteilt, oder, was dasselbe ist, wenn für irgend zwei quadratisch summierbare Funktionen $f(P)$, $g(P)$

$$\lim_{t=\infty} \int_\Omega f(P_t)g(P)dm = \frac{1}{m(\Omega)} \int_\Omega f dm \int_\Omega g dm \tag{10. 1}$$

gilt. Gilt dies für $t \to +\infty$, so gilt es auch für $t \to -\infty$. Eine Strömung vom Mischungstyp ist notwendig ergodisch, aber nicht umgekehrt. Wir beweisen nun den von Hedlund herrührenden

Hauptsatz 10. 1. Hat $\mathfrak{M}_2$ endlichen Flächeninhalt, so ist die geodätische Strömung vom Mischungstyp.

Beweis. Ein Horozykel wird von allen Geodätischen, die mit ihm den unendlich fernen Punkt auf $x_1^2 + x_2^2 = 1$ gemein haben, senkrecht geschnitten. Zwei Horozykel mit gleichem unendlich fernen Punkt (parallele Horozykel) schneiden also auf jenen geodätischen Strecken gleicher Länge t aus. Man rechnet aus, daß zwei Bögen, welche auf den beiden Horozykeln durch ein Paar jener Geodätischer ausgeschnitten werden, das Längenverhältnis e^t besitzen. In Formeln ausgedrückt (s, t sind unabhängig):

$$\mathfrak{T}_{-t}\mathfrak{H}_s P \equiv \mathfrak{H}_{se^t}\mathfrak{T}_{-t}P. \tag{10. 2}$$

Genau so, wie wir früher für die Linienelemente e die Koordinaten (π_1, π_2, s) eingeführt hatten, in welchen die Geodätischen die Parallelen zur s-Achse werden, führen wir diesmal Koordinaten (η, l, s) ein, bei welchen die Horozykel eine analoge Rolle spielen. Wir bestimmen wie oben den Horozykel, der das Linienelement e senkrecht durchsetzt. z sei der Trägerpunkt von e, und z_0 der dem Nullpunkt $x_1 = x_2 = 0$ am nächsten gelegene Punkt auf dem Horozykel. η ist der unendlich ferne Punkt des letzteren auf $|\eta| = 1$, l die (mit Vorzeichen versehene) Entfernung zwischen z_0 und $x_1 = x_2 = 0$, und s die (mit Vorzeichen je nach Umlaufssinn versehene) Entfernung zwischen z und z_0 entlang dem Horozykel. Die Horozykelströmung wird einfach durch

$$\mathfrak{H}_t(\eta, l, s) = (\eta, l, s + t)$$

dargestellt. Das Maßelement dm im e-Raum hat die Form

$$dm = \gamma(\eta, l) \mid d\eta \mid dl\, ds, \tag{10. 3}$$

wo γ wegen der Strömungsinvarianz von dm und ds nicht von s abhängen kann. Aus der Bewegungsinvarianz von dm folgt, wenn man Drehungen um $x_1 = x_2 = 0$ und hyperbolische Verschiebungen mit $(-\eta, \eta)$ als Achse betrachtet,

$$\gamma = \text{const.} \tag{10. 4}$$

Zum Beweise des Hedlundschen Satzes genügt es, (10. 1) unter der Voraussetzung

$$\int_\Omega f dm = 0 \tag{10. 5}$$

zu beweisen, d. h.

$$\lim_{t=+\infty} \int_\Omega f(\mathfrak{T}_{-t}P)g(P)dm = 0 \tag{10. 6}$$

zu beweisen. Es genügt ferner, dies für beschränktes f und beschränktes, auf Ω gleichmäßig stetiges g zu zeigen. Man denke sich im Integral (10. 6) den Raum Ω etwa durch einen Fundamentalbereich der Bewegungsgruppe, in welchem jeder Punkt mit allen möglichen Richtungen gepaart ist, dargestellt. Der Kern des Beweises beruht auf folgender Vorstellung. Man denke sich im Integral (10. 6) gemäß (10. 3) und (10. 4) zuerst die Integration nach s ausgeführt. P wandert dabei senkrecht gehalten entlang einem Horozykel und $\mathfrak{T}_{-t}P$ läuft entlang dem parallelen Horozykel im Abstand t, und zwar sehr rasch durch einen langen Bogen, wenn t groß ist. Daher kann man in (10. 6) f durch seinen Mittelwert längs dem großen Horozykel ersetzen. Wegen der Ergodizität der Horozykelströmung und wegen (10. 5) ist aber dieser Mittelwert konstant gleich Null. Nun zur strengen Ausführung! Dabei machen wir von der Ungleichung Gebrauch, die besagt, daß

$$\int_a^b f(s)g(s)ds$$

sich von

$$\int_a^b \left[\frac{1}{\delta}\int_s^{s+\delta} f(\tau)d\tau\right] g(s)ds$$

um weniger als

$$(b-a)\cdot \|f\| \cdot \omega_\delta(g) + 2\delta \cdot \|f\| \cdot \|g\|$$

unterscheidet. $\|f\|$ ist eine obere Schranke für $|f|$ und $\omega_\delta(g)$ ist der Stetigkeitsmodul von $g(s)$, d. h. die obere Grenze von $|g(s') - g(s)|$ für $|s' - s| < \delta$.

Die Ungleichung ergibt sich mit Hilfe partieller Integration. Ersetzt man entsprechend das Integral in (10. 6) durch

$$\int_{\Omega}\left[\frac{1}{\delta}\int_0^{\delta} f(\mathfrak{T}_{-t}\mathfrak{H}_s P)ds\right]g(P)dm\,, \tag{10. 7}$$

so ist der begangene Fehler kleiner als $\|f\|\cdot\omega_\delta(g)$ mal

$$\iint (b-a)\,|\,d\eta\,|\,dl = m(\Omega)$$

plus $2\,\delta\cdot\|f\|\cdot\|g\|$ mal

$$\iint |\,d\eta\,|\,dl.$$

Dieses Integral ist endlich, wenn der Fundamentalbereich von $\mathfrak{G}$ mitsamt Rand in $x_i x_i < 1$ liegt. Ist dies nicht der Fall, so schneide man die auf $x_i x_i = 1$ liegenden Spitzen ab und lege statt Ω von vornherein das entsprechend verkleinerte Ω' zugrunde. Man kann erreichen, daß der begangene Fehler gleichmäßig unter eine beliebige positive Schranke fällt. Nachher kann man den obigen Fehler durch Wahl eines passenden δ gleichmäßig klein machen.

Es bleibt zu zeigen, daß (10. 7) bei festem $\delta > 0$ für $t \to +\infty$ nach Null strebt. Wegen Lemma 9. 1 und wegen (10. 5) ist nach v. Neumann

$$\lim_{S=\infty}\int_{\Omega}\left\{\frac{1}{S}\int_0^{S} f(\mathfrak{H}_s Q)ds\right\}^2 dm_Q = 0 \tag{10. 8}$$

oder

$$\lim_{t=+\infty}\int_{\Omega}\left\{\frac{1}{\delta}\int_0^{\delta} f(\mathfrak{H}_{se^t} Q)ds\right\}^2 dm_Q = 0\,.$$

Durch Ausführung der m-treuen Substitution $Q = \mathfrak{T}_{-t}P$ im Integral, $dm_P = dm$, wird hieraus

$$\lim_{t=+\infty}\int_{\Omega}\left\{\frac{1}{\delta}\int_0^{\delta} f(\mathfrak{H}_{se^t}\mathfrak{T}_{-t}P)ds\right\}^2 dm = 0\,.$$

Berücksichtigt man hier die fundamentale Beziehung (10. 2), so folgt die Richtigkeit der Behauptung und damit des Haupsatzes.

§ 11. Ausdehnung auf $n > 2$ Dimensionen.

Zweibeine seien mit b bezeichnet. b besteht aus einem Trägerpunkt x in $x_i x_i < 1$ und zwei orthogonalen Richtungen r_1 und r_2. Wir betrachten im folgenden den Fall $n = 3$. Die Verallgemeinerung auf $n > 3$ Dimensionen führt auf keine neuen Schwierigkeiten.

Die geodätische Strömung $\mathfrak{T}_t e$ im Linienelementraum erweitern wir zur geodätischen Strömung $\mathfrak{T}_t b$ im Zweibeinraum. Sie ist dadurch definiert, daß b längs der in Richtung r_1 startenden Geodätischen um die Länge t parallel verschoben wird. Nach § 1 läßt sich das auch so beschreiben. Man konstruiere die das Zweibein b berührende (zweidimensionale) Orthogonalkugel. Sie enthält die erwähnte Geodätische. Dann verschiebe man b längs derselben so, daß r_1 tangential bleibt und r_2 senkrecht dazu immer die Orthogonalkugel berührt.

Das bewegungsinvariante Volumelement im b-Raum ist

$$d\mu = dV\, d\Sigma, \tag{11. 1}$$

wo dV wieder das bewegungsinvariante Volumelement in $x_i x_i < 1$ und $d\Sigma$ das drehungsinvariante Volumelement im Richtungsraum der b mit festem Trägerpunkt bedeutet. Man beweist dann, daß $d\mu$ auch $\mathfrak{T}_t$-invariant ist. Jedem Zweibein $b = (x, r_1, r_2)$ denke man sich das Linienelement $e = (x, r_1)$ zugeordnet. Dann läßt sich $d\mu$ auf die Form

$$d\mu = dm\, d\varphi_2 \tag{11. 2}$$

bringen, wo dm die alte Bedeutung hat und $d\varphi_2$ die infinitesimale Winkelvariation der Richtung r_2 bei festem $e = (x, r_1)$ bedeutet. Führt man die Bezeichnung

$$\overline{f(b)} = \frac{1}{2\pi} \int_0^{2\pi} f(b)\, d\varphi_2 \tag{11. 3}$$

bei festem e ein, so gilt die Relation

$$\int f(b)\, d\mu = 2\pi \int \overline{f(b)}\, dm, \tag{11. 4}$$

wenn der Integrationsbereich links eine Zylindermenge im b-Raum ist, d. h. eine Menge, in welcher irgendein auftretendes e mit jeder Richtung r_2 gepaart vorkommt. Rechts wird dann über die Projektion in den e-Raum integriert. In $n \geqq 3$ Dimensionen ist $d\varphi_2$ durch das $(n-2)$-dimensionale Volumelement auf der $(n-2)$-dimensionalen r_2-Kugel zu ersetzen.

In der n-dimensionalen hyperbolischen Geometrie versteht man unter einem Horozykel eine orthogonale Trajektorie einer einparametrigen Schar von Orthogonalkreisen, welche auf einer und derselben zweidimensionalen Orthogonalkugel verlaufen und zueinander asymptotisch sind. Die Horozykel sind in $x_i x_i < 1$ die euklidischen Kreise, welche $x_i x_i = 1$ von innen berühren.

Ein Zweibein bestimmt eindeutig einen Horozykel in folgender Weise. Die Geodätische durch x in Richtung r_1 hat einen $+\infty$-Punkt auf $x_i x_i = 1$. Unter den Horozykeln, welche denselben unendlich fernen Punkt besitzen und durch x gehen, gibt es genau einen, der x in Richtung r_2 durchläuft. Dieser wird dem Zweibein $b = (x, r_1, r_2)$ zugeordnet. Es ist gut, neben dem zugeordneten Horozykel auch die Orthogonalkugel, welche ihn enthält, im Auge zu behalten. Es ist dies die Orthogonalkugel, welche das Zweibein berührt. Wir definieren nun die Horozykelströmung $\mathfrak{H}_t b$ dadurch, daß wir b in Richtung r_2 längs dem Horozykel um die Länge t verschieben, wobei r_2 immer tangential zum Horozykel bleiben und r_1 immer senkrecht dazu die Orthogonalkugel berühren soll (r_1 weist dann immer geodätisch zum unendlich fernen Punkt des Horozykels). Die $\mathfrak{H}_t$, $\mathfrak{T}_t$ definierenden Beziehungen sind bewegungsinvariant. Beide Strömungen sind durch die gleichen Beziehungen verknüpft wie in zwei Dimensionen. Man sieht das, indem man das Zweibein b durch eine Bewegung in den Nullpunkt bringt und die Richtungen r_i mit den Richtungen der x_i-Achsen zusammenfallen läßt, $i = 1, 2$. Die b berührende Orthogonalkugel wird dann die Ebene $x_3 = 0$ (bzw. $x_3 = \ldots = x_n = 0$), wie überhaupt alles hier zu Beschreibende sich in dieser Ebene abspielt. Man hat nur dafür zu sorgen, daß alles bewegungsinvariant beschrieben wird. Die gewünschten Relationen, nämlich (9. 2) mit (9. 1) und (9. 3), und (10. 2), können nunmehr direkt dem Falle $n = 2$ entnommen werden. Dabei sind e bzw. P durch b zu ersetzen. $\mathfrak{D}_\alpha b'$ bedeutet nunmehr eine Drehung des Zweibeins $b' = (x', r_1', r_2')$ in seiner eigenen Ebene um seinen Trägerpunkt. Der Drehsinn wird wie üblich in der (r_1', r_2')-Ebene aufgefaßt. In $n \geqq 3$ Dimensionen sollen dabei alle auf b senkrechten Richtungen einzeln festbleiben. Aus der μ-Invarianz der Einzeltransformationen folgt wieder:

Nicht nur die geodätische, sondern auch die Horozykelströmung im b-Raum läßt das Volummaß μ invariant.

Alle obigen Begriffe und Zusammenhänge denke man sich nun auf eine Mannigfaltigkeit $\mathfrak{M}_3$, bzw. $\mathfrak{M}_n$, übertragen. Die Zweibeine seien auch weiterhin mit b bezeichnet. Ωb sei der Raum dieser b. Die für das Folgende benötigte Verallgemeinerung von Lemma 9. 1 lautet

Lemma 11. 1. $V(\mathfrak{M})$ sei endlich. Dann hat die Horozykelströmung in Ωb dieselben invarianten (und μ-meßbaren) Funktionen wie die geodätische Strömung in Ωb. Insbesondere gilt: Ist $h(b)$ meßbar, beschränkt und $\mathfrak{H}_t$-invariant, so ist die Funktion $\overline{h(b)}$ von P in Ω bis auf eine m-Nullmenge konstant.

Beweis. $h(b)$ genüge den Voraussetzungen, insbesondere sei

$$h(\mathfrak{H}_t b) \equiv h(b)$$

in Ωb. Der Beweis ist bis zur Formel (9. 5) derselbe wie der von Lemma 9. 1, wenn dort P durch b, Ω durch Ωb, dm durch $d\mu$ ersetzt wird. Es gilt also

$$\lim_{s=\infty} \int_{\Omega b} g(\mathfrak{T}_s b) h(b) d\mu = \int_{\Omega b} h(b) g(\mathfrak{D}_\pi b) d\mu$$

für beliebiges beschränktes $g(b)$. Mittelt man bezüglich s und setzt man $F^*(b)$ für das $\mathfrak{T}_s$-Mittel einer Funktion $F(b)$, so folgt (die Operation $\mathfrak{D}_\pi$ ist μ-invariant)

$$\int_{\Omega b} g^*(b) h(b) d\mu = \int_{\Omega b} h(\mathfrak{D}_\pi b) g(b) d\mu . \tag{11. 5}$$

Wegen der für jedes $\mathfrak{T}_s$-invariante $\varphi(b)$ geltenden Beziehung

$$\int_{\Omega b} F^* \varphi d\mu = \int_{\Omega b} F \varphi d\mu \tag{11. 6}$$

und wegen der $\mathfrak{T}_s$-Invarianz von $g^*(b)$ ist die linke Seite in (11. 5) gleich

$$\int_{\Omega b} h^*(b) g(b) d\mu .$$

Da $g(b)$ beliebig ist, folgt

$$h(\mathfrak{D}_\pi b) = h^*(b), \qquad h(b) = h^*(\mathfrak{D}_\pi b)$$

fast überall in Ωb. Wegen $\mathfrak{D}_\pi \mathfrak{T}_t \equiv \mathfrak{T}_{-t} \mathfrak{D}_\pi$ und wegen der $\mathfrak{T}_t$-Invarianz von $h^*(b)$ folgt

$$h(\mathfrak{T}_t b) = h^*(\mathfrak{T}_{-t} \mathfrak{D}_\pi b) = h^*(\mathfrak{D}_\pi b) = h(b),$$

d. h. $h(b)$ ist im wesentlichen $\mathfrak{T}_t$-invariant. Ganz analog beweist man umgekehrt, daß aus der $\mathfrak{T}_t$-Invarianz die $\mathfrak{H}_t$-Invarianz folgt.

Die zweite Behauptung folgt aus der $\mathfrak{T}_t$-Invarianz von $h(b)$; denn die Funktion $\overline{h(b)}$ von P ist in Ω $\mathfrak{T}_t$-invariant, und in Ω ist die $\mathfrak{T}_t$-Strömung ergodisch. Damit ist der Beweis beendet.

Die für den Mischungsbeweis wichtige Konsequenz des Lemmas ist, daß für ein beliebiges m-meßbares und beschränktes $f(b) = f(P)$ die Formel

$$\lim_{S=\infty} \int_{\Omega} \left\{ \frac{1}{S} \int_0^S \overline{f(\mathfrak{H}_s b)}\, ds - \frac{1}{m(\Omega)} \int_{\Omega} f dm \right\}^2 dm = 0 . \tag{11. 7}$$

Die Mittelung $\overline{f}$ ist bezüglich der zweiten Richtung r_2 von b verstanden (nicht etwa von $\mathfrak{H}_s b$!). Zum Beweise bedenke man zunächst, daß man den Mittelungsstrich über das ganze Innere der geschweiften Klammer — es sei im Moment mit $\psi(b)$ bezeichnet — setzen kann. Wegen der Ungleichung $\overline{\psi}^2 \leqq \overline{\psi^2}$ würde (11. 7) aus dem v. Neumannschen Ergodensatz

$$\lim_{S=\infty} \int_{\Omega b} \left\{ \frac{1}{S} \int_0^S f(\mathfrak{H}_s b) ds - \hat{f}(b) \right\}^2 d\mu = 0$$

und aus (11. 4) folgen, wenn von dem $\mathfrak{H}_s$-Mittel $\hat{f}$ von f

$$\overline{\hat{f}(b)} = \text{const} = \frac{\int_\Omega f dm}{m(\Omega)}$$

gezeigt werden kann. Die Konstanz folgt aber aus Lemma 11. 1, und der Wert ergibt sich aus

$$\text{const} = \frac{1}{m(\Omega)} \int_\Omega \bar{\hat{f}} dm = \frac{1}{\mu(\Omega b)} \int_{\Omega b} \hat{f} d\mu .$$

Hier kann wegen (11. 6), mit $\hat{F}$ statt F^* und $\varphi = 1$, $\hat{f}$ durch f ersetzt werden. Man beachte dann $\bar{f} = f(P)$.

Hauptsatz 11. 2. Hat $\mathfrak{M}$ endliches Volumen $V(\mathfrak{M})$, so ist die geodätische Strömung in Ω vom Mischungstypus (Verallgemeinerung von Satz 10. 1).

Beweis. Es ist wieder (10. 6) unter gleichen Voraussetzungen wie beim Beweis von Hauptsatz 10. 1 zu beweisen, insbesondere unter der Voraussetzung (10. 5). Um wieder von der $\mathfrak{H}_s$-Strömung Gebrauch machen zu können, fassen wir f und g als Funktionen von b auf. In (10. 6) darf dann P durch b, Ω durch Ωb, dm durch $d\mu$ ersetzt werden. Man zeigt dann analog wie dort, daß das Integral in (10. 6) praktisch durch (10. 7) ersetzt werden darf, wenn man dort in derselben Weise jene Buchstaben ersetzt. Wegen $g(b) = g(P)$ kann dieses Integral

$$\int_\Omega \left[\frac{1}{\delta} \int_0^\delta \overline{f(\mathfrak{T}_{-t} \mathfrak{H}_s b)} ds \right] g(P) dm \tag{11. 8}$$

geschrieben werden. Über den Mittelungsstrich ist gleiches zu sagen wie bei der Formel (11. 7). Zur Rechtfertigung des Überganges zum neuen Integral (11. 8), und auch nur zu diesem Zweck, kann man wieder wie damals geeignete Horozykelkoordinaten der Zweibeine heranziehen. Die Ausführung der Einzelheiten soll hier unterbleiben[1]). Der Beweis, daß für $t \to \infty$ (11. 8) bei

1) Wesentlich ist nur, daß s als neue Koordinate auftritt. Drückt man das Volumelement $d\mu$ durch die neuen Koordinatendifferentiale aus, so folgt wieder, daß die Gewichtsfunktion von s unabhängig ist.

festem $\delta > 0$ nach Null strebt, verläuft ebenso wie im Falle (10. 7). Wesentlich ist die fundamentale Formel (10. 2). Nur muß statt (10. 8) die Formel (11. 7), wo (10. 5) zu berücksichtigen ist, benützt werden.

Ich möchte hier dem Leser die Bemerkung nicht vorenthalten, daß gerade die Verallgemeinerung des Integrals (10. 7), in welchem zum erstenmal im Mischungsbeweis die Horozykelströmung auftritt, auf $n \geqq 3$ Dimensionen die Einführung der Zweibeine notwendig macht. Ein Horozykel ist durch ein Zweibein, aber nicht durch ein Linienelement bestimmt.

II. Kapitel. Zweidimensionale Mannigfaltigkeiten veränderlicher negativer Krümmung.

§ 12. Abbildung auf die universelle Überlagerungsfläche.

Wir gehen von einer zweidimensionalen Riemannschen Mannigfaltigkeit aus, kurz Fläche $\mathfrak{F}$ genannt. Ihr ist eine positiv definite quadratische Differentialform

$$ds^2 = g_{ik}(x_1, x_2)\, dx_i\, dx_k, \qquad i = 1, 2;\ k = 1, 2 \tag{12. 1}$$

in eindeutiger, d. h. von der Wahl lokaler Koordinaten x_1, x_2 unabhängiger Weise eingeprägt. Um die folgenden Ausführungen nicht mit störenden Details zu beschweren, setzen wir die g_{ik} als unendlich oft stetig differenzierbar voraus. $\mathfrak{F}$ sei vollständig. Die Punkte auf $\mathfrak{F}$ werden mit p, die gerichteten Linienelemente mit P und der Raum der P mit Ω bezeichnet. Wie früher wird mit Hilfe infinitesimaler Parallelverschiebung durch

$$d\sigma^2 = ds^2 + d\chi^2 \tag{12. 2}$$

ein von der lokalen Koordinatenwahl unabhängiges Längendifferential in Ω definiert; $d\chi$ ist der Winkel, welchen die längs der geodätischen Verbindung $\overrightarrow{ds}$ bei festbleibendem Winkel mit $\overrightarrow{ds}$ verschobene erste Richtung im zweiten Punkt mit der zweiten Richtung bildet. Durch die untere Grenze der Längen aller Verbindungswege wird in $\mathfrak{F}$ eine Entfernung $s(p, p')$ und in Ω eine Distanz $\sigma(P, P')$ erklärt. Sie genügen den wohlbekannten Axionen. (12. 1) definiert in $\mathfrak{F}$ ein Oberflächendifferential do. Analog definiert (12. 2) in Ω ein Volumdifferential

$$dm = do\, d\vartheta, \tag{12. 3}$$

$d\vartheta$ = Winkeldifferential in einem Punkte von $\mathfrak{F}$. dm ist gegenüber der geodätischen Strömung $\mathfrak{T}_t(P)$ in Ω invariant.

Die Flächenkrümmung K sei überall negativ. In diesem Fall weiß man, daß es in der Gesamtheit derjenigen Wege von einem Punkt p zu einem

Punkte p' auf $\mathfrak{F}$, welche bei festgehaltenen p, p' stetig in einen bestimmten Weg deformierbar sind, eine und nur eine geodätische Verbindung gibt (Hadamard)[1]). Ihre Länge ist gleich $s(p, p')$.

Die im folgenden durchzuführenden Überlegungen werden übersichtlicher, wenn man statt in $\mathfrak{F}$ in der universellen Überlagerungsfläche $\tilde{\mathfrak{F}}$ operiert. Eine ein-vieldeutige und stetige Beziehung zwischen $\mathfrak{F}$ und $\tilde{\mathfrak{F}}$ kann bekanntlich folgendermaßen hergestellt werden. Man betrachte die von einem festen Punkte p_0 von $\mathfrak{F}$ ausgehenden geodätischen Halbstrahlen und bezeichne mit λ die längs denselben von p_0 gerechnete Bogenlänge, mit ϑ den Winkel des Strahls mit einer festen Richtung in p_0. Faßt man etwa $r = \frac{\lambda}{\lambda + 1}$, ϑ als Polarkoordinaten eines Punktes der Einheitskreisfläche $r < 1$ auf, so entspricht jedem Punkt von $r < 1$ genau ein Bildpunkt $p(\lambda, \vartheta)$ von $\mathfrak{F}$ und $\mathfrak{F}$ wird dabei erschöpft[2]). Dieses Entsprechen ist stetig. Jedem Punkt p von $\mathfrak{F}$ entspricht umgekehrt eine Menge von Urbildpunkten in $r < 1$. Diese Menge hat folgende Eigenschaft:

Zu jedem ϱ, $0 < \varrho < 1$, gibt es ein $\delta > 0$ derart, daß die in $r \leqq \varrho$ gelegenen Urbilder irgendeines Punktes von $\mathfrak{F}$ euklidisch um mehr als δ voneinander entfernt sind.

Wäre dies falsch, so gäbe es Punkte p in $s(p_0, p) \leqq \frac{\varrho}{1-\varrho}$ auf $\mathfrak{F}$, die sich in zwei verschiedenen Weisen geodätisch mit p_0 verbinden ließen, wobei die Anfangsrichtungen in p_0 beliebig wenig voneinander abweichen würden[3]). Das ist nicht möglich, da sich sonst die beiden Geodätischen unter Festhaltung von p und p_0 stetig ineinander deformieren lassen würden.

Hiernach können sich die Urbilder in $r < 1$ nur auf $r = 1$ häufen. Ist $\mathfrak{C}$ eine stetige Kurve in $\mathfrak{F}$, p' ein Punkt auf ihr, so gibt es nach der obigen Bemerkung durch jeden Urbildpunkt in $r < 1$ eine eindeutig bestimmte stetige Urbildkurve $\tilde{\mathfrak{C}}$. Aus demselben Grunde müssen zwei Wege von p nach p'

---

1) Die hier zugelassenen Flächen sind frei von Singularitäten, also auch konischen Punkten, wie sie in Kap. I vorkommen durften. Kommen sie vor, so kann es mehrere geodätische Verbindungen geben.

2) Hat $\mathfrak{F}$ die konstante Krümmung minus Eins, und setzt man $r = \mathfrak{Tg}\,\frac{\lambda}{2}$, dann wird die Abbildung automatisch zu einer Abwickelung auf $x_i x_i < 1$. ds^2 geht dabei in (1.1) über. Wir betonen ausdrücklich, daß der analytische Charakter der Abbildung im folgenden unwichtig ist. Nur ihre Stetigkeit wird zur übersichtlicheren Darstellung der Lagebeziehungen benötigt.

3) Oder wo die geodätischen Strecken $p_0 p$ ganz in beliebiger Nachbarschaft von p_0 verlaufen würden, was offenbar nur möglich ist, wenn die Strecken zusammenfallen.

auf $\mathfrak{F}$, die unter Festhaltung von p, p' ineinander deformierbar sind, in $r < 1$ bei gleichem Anfangspunkt Kurven mit gleichem Endpunkt entsprechen. Die Umkehrung hiervon ist trivialerweise richtig.

Durch die stetige Abbildung von $\mathfrak{F}$ auf $r < 1$ wird auch die Metrik (12. 1) in $r < 1$ stetig hineinverpflanzt. Nur in ihrem Sinne sprechen wir von Längen, Winkeln, Flächeninhalt und Geodätischen in $r < 1$. Die letzteren sind die Urbilder der Geodätischen von $\mathfrak{F}$. Aus dem Obigen folgt ferner:

Satz 12. 1. Zwei Geodätische können sich in $r < 1$ höchstens einmal schneiden. Zu irgend zwei Punkten von $r < 1$ gibt es eine und nur eine geodätische Verbindung.

Definiert man die Distanz zweier Punkte in $r < 1$ durch die untere Grenze der Längen aller Verbindungswege in $r < 1$, so ist die Länge der geodätischen Verbindung gleich der kürzesten Verbindung beider Punkte. Die Entfernung $s(p, p')$ zweier Punkte auf $\mathfrak{F}$ ist offenbar die kürzeste der Distanzen eines Urbildes von p von einem Urbild von p' in $r < 1$.

§ 13. Asymptotische Geodätische.

Führt man um einen willkürlichen Punkt in $r < 1$ geodätische Polarkoordinaten ϱ, ϑ ein, so gilt bekanntlich

$$ds^2 = d\varrho^2 + Y^2(\varrho, \vartheta)\, d\vartheta^2, \tag{13. 1}$$

und Y ist bei festem ϑ diejenige Lösung der Variationsgleichung längs eines Strahles $\vartheta = \text{const.}$

$$Y'' + KY = 0, \qquad Y' = \frac{\partial Y}{\partial \varrho}, \tag{13. 2}$$

welche die Anfangsbedingungen

$$Y(0, \vartheta) = 0, \qquad Y'_\varrho(0, \vartheta) = 1 \tag{13. 3}$$

erfüllt. Die Darstellung (13. 1) gilt überall in $r < 1$, da die geodätischen Radien nach Satz 12. 1 bis auf $\varrho = 0$ in $r < 1$ ein Feld bilden.

Wegen $K < 0$ ist stets $Y'' > 0$ und daher $Y' > 1$. Führt man die Voraussetzung ein

Voraussetzung A. Die Krümmung K verläuft zwischen festen negativen Grenzen

$$a^2 < -K < b^2,$$

so folgt aus (13. 2) und (13. 3)

$$Y'^2 - 1 = -\int_0^\varrho K d(Y^2) > a^2 Y^2,$$

also wegen $Y > \varrho$

$$a^2 < \frac{Y'}{Y} < \sqrt{b^2 + \varrho_0^{-2}}, \qquad \varrho \geqq \varrho_0 > 0. \tag{13. 4}$$

Ist nun G eine gerichtete Geodätische in $r < 1$ mit der Bogenlänge s als Parameter, und ist α der Winkel zwischen der Geodätischen von einem Punkt von G zum Pol $\varrho = 0$ und der Richtung $s \to +\infty$ auf G, $0 < \alpha < \pi$, so gilt die Gleichung

$$\frac{d\alpha}{ds} = \frac{Y'}{Y} \sin \alpha \text{ [1]} \tag{13. 5}$$

Aus ihr und aus (13. 4) (linke Ungleichung) folgt, daß α monoton von 0 nach π läuft, wenn G von $s = -\infty$ bis $s = +\infty$ durchlaufen wird.

Aus den Sturmschen Vergleichssätzen und aus Voraussetzung A folgt

$$\frac{1}{a} \operatorname{Sin} a\varrho < Y(\varrho) < \frac{1}{b} \operatorname{Sin} b\varrho. \tag{13. 6}$$

Zwei gerichtete Geodätische in $r < 1$ heißen positiv asymptotisch zueinander, wenn sich die Bogenlängenparameter, t auf der einen und t' auf der anderen, einander stetig und monoton wachsend so ordnen lassen, daß die Distanz des Punktes $p(t)$ auf der einen vom Punkte $p'(t')$ der anderen für $t \to +\infty$, oder, was dasselbe bedeutet, für $t' \to +\infty$ nach Null strebt. Zwei Geodätische auf $\mathfrak{F}$ heißen positiv asymptotisch, wenn sie je ein Urbild in $r < 1$ besitzen, auf welche die Definition zutrifft.

Satz 13. 1. Durch einen beliebigen Punkt von $r < 1$ gibt es eine und nur eine gerichtete Geodätische, welche zu einer gegebenen gerichteten Geodätischen in $r < 1$ positiv asymptotisch ist.

Beweis. p^*, G sei der gegebene Punkt bzw. die gegebene Geodätische in $r < 1$, $p(t)$ der laufende Punkt auf G(t = Bogenlänge auf G). Ist $\bar{p} = p(\bar{t})$ ein fester Punkt auf G, so ist wegen $s(\bar{p}, p(t)) = |\bar{t} - t|$

$$|\bar{t} - t| - s(p^*, \bar{p}) \leqq s(p^*, p(t)) \leqq |\bar{t} - t| + s(p^*, \bar{p}).$$

Also gilt $s(p^*, p(t)) \to \infty$ für $|t| \to \infty$, und daher gibt es ein geodätisches Lot von p^* auf G. $\bar{p}$ sei nunmehr der Treffpunkt des Lotes auf G. Man kann annehmen, daß p^* nicht auf G liegt. Nun betrachte man das geodätische

1) Die Gaußsche Gleichung der Geodätischen für das Längenelement (13.1) lautet

$$\frac{d\alpha}{d\vartheta} = Y'_\varrho = Y'.$$

Hier kann $Y\,d\vartheta = ds \sin \alpha$ gesetzt werden mit ds = Bogenelement auf G.

Dreieck D_t mit den Ecken p^*, $\bar{p}$, $p(t)$. Nach Satz 12. 1 treffen sich die geodätischen Seiten nur in den Eckpunkten. Der Winkel bei $\bar{p}$ ist ein rechter. $\alpha^*(t)$ sei der Innenwinkel in p^*, $\alpha(t)$ der in $p(t)$. Mit wachsendem $t > \bar{t}$ nimmt $\alpha^*(t)$ offenbar monoton zu. Da wegen $K < 0$ die Winkelsumme kleiner als π ist, folgt die Existenz von

$$\lim_{t=+\infty} \alpha^*(t) = \beta.$$

Der diesem Winkel β entsprechende geodätische Halbstrahl G^* durch p^* liefert die gesuchte Geodätische. Aus der an die Gaußsche Formel (13. 5) angeknüpften Bemerkung folgt $\alpha(t) \to 0$, $t \to +\infty$. Wendet man auf das Dreieck D_t die Gauß-Bonnetsche Integralformel an, so folgt durch Grenzübergang

$$\frac{\pi}{2} - \beta = \int K do, \tag{13. 7}$$

wo das Integral über D_∞ erstreckt ist. Man fälle nun von irgendeinem Punkte $p'(t')$ von G^* das Lot auf G. Es verläuft für hinreichend großes t' innerhalb D_∞. Ist $p(t)$ der Treffpunkt des Lotes auf G, so wächst t monoton und stetig mit t'; denn für verschiedene t' können sich die Lote nicht schneiden. Auf die vom Lot, von G und G^* begrenzte Fläche kann ebensogut wie auf D_∞ die Formel (13. 6) angewandt werden. Da für $t' \to +\infty$ das Integral nach Null strebt, konvergiert der Winkel bei $p'(t')$ gegen $\frac{\pi}{2}$. Es ist nun zu zeigen, daß für $t' \to +\infty$ die Lotlänge nach Null strebt. Man ziehe hierzu von allen Punkten eines Lotes die Asymptoten zu G, und denke sie sich in der obigen Weise approximiert durch die Halbstrahlen von einem entfernten Punkte von G. Bezeichnet man mit α den Winkel zwischen Asymptote und Lot, so folgt aus (13. 4), (13, 5) längs dem Lot

$$ds < \frac{d\alpha}{a^2 \sin\alpha}.$$

Hieraus folgt, da α für große t' nahe bei $\frac{\pi}{2}$ bleibt, die Behauptung. Daß im Sinne der ursprünglichen Definition die Asymptote durch p^* eindeutig bestimmt ist, folgt leicht mit Hilfe von (13. 1) mit p^* als Pol und (13. 6).

Versteht man nun unter $\alpha = \alpha(s)$ den Schnittwinkel einer gerichteten Geodätischen mit der positiven Richtung in einem Felde gerichteter, einander positiv asymptotischer Geodätischer, $0 \leqq \alpha \leqq \pi$, so gilt statt (13. 5) längs jener Geodätischen

$$\frac{d\alpha}{ds} = -z_0' \sin\alpha, \tag{13. 8}$$

wo $z'_0 = z'(0)$ ist und $z(t)$ die den Randbedingungen

$$z(0) = 1, \qquad z(+\infty) = 0 \tag{13. 9}$$

genügende Lösung der Variationsgleichung

$$z'' + Kz = 0 \tag{13. 10}$$

bedeutet. (13. 10) ist dabei entlang einer Feldgeodätischen gebildet, und die Bogenlänge t ist vom Schnittpunkt mit der in (13. 8) betrachteten Geodätischen an gerechnet.

Zum Beweise von (13. 8) muß in (13. 5) der Grenzübergang, nämlich die Verlegung des Poles $\varrho = 0$ ins Unendliche längs einer festen Geodätischen, vollzogen werden. Zunächst wird längs den geodätischen Radien $\vartheta = \text{const}$ die Bogenlänge t (statt ϱ) vom Schnittpunkt mit der in (13. 8) betrachteten Geodätischen an zum Pol $\varrho = 0$ hin gerechnet. Bezeichnet man nun mit $z = z(t; A)$ die Lösung von (13. 10), bezogen auf eine beliebige geodätische Strecke der Länge A, mit den Randbedingungen

$$z(0; A) = 1, \qquad z(A; A) = 0, \tag{13. 11}$$

so ist, wenn man speziell die Strecken auf jenen Halbstrahlen bis zu $\varrho = 0$ betrachtet, das Y'/Y in (13. 5) gleich $-z'_t(0; A)$. Zur Rechtfertigung des besagten Grenzüberganges genügt es nun, allgemein zu zeigen, daß bei festem Anfangselement der Strecke

$$\lim_{A=\infty} z'(0; A) = z'(0)$$

gleichmäßig bezüglich der Lage des Anfangselementes in $r < 1$ gilt, wenn $z(t)$ die Lösung mit den Randbedingungen (13. 9) bedeutet. Wir beweisen dies ausführlich, da von den dabei gemachten allgemeinen Feststellungen später wieder Gebrauch gemacht wird.

Ist $W(t)$ die Lösung von (13. 10) mit den Anfangsbedingungen

$$W(0) = 0, \qquad W'(0) = 1, \tag{13. 12}$$

so gilt wegen (13. 11)

$$zW' - Wz' = \text{const} = 1. \tag{13. 13}$$

Daraus folgt

$$z(t; A) = W(t) \int_t^A \frac{d\tau}{W^2(\tau)}. \tag{13. 14}$$

Da $W(\varrho)$ der Ungleichung (13. 6) genügt, gilt

$$\lim_{A=\infty} z(t;A) = W(t)\int_t^\infty \frac{d\tau}{W^2(\tau)} = z(t)$$

gleichmäßig in jedem Intervall $0:0 \leqq t \leqq \omega$ und gleichmäßig bezüglich des Anfangselementes der Geodätischen. Da $W(\tau)$ wächst, ist

$$z(t,A) < \int_t^A \frac{d\tau}{W(\tau)} < \int_t^\infty \frac{a\,d\tau}{\mathfrak{Sin}\,a\tau} \sim 2e^{-at}. \tag{13. 15}$$

$z(t;A)$ verläuft also unterhalb einer universellen integrablen Funktion. Wegen (13. 10) ist

$$z'(0;A) = z'(A;A) + \int_0^A Kz\,dt.$$

Aus den obigen Bemerkungen und aus der Tatsache, daß

$$z'(A,A) = -\frac{1}{W(A)}$$

für $A \to \infty$ gleichmäßig nach Null strebt, folgt dann die zu beweisende Behauptung.

Integriert man (13. 10) nach Multiplikation mit z', so folgt wegen $z' < 0$ und nach Voraussetzung A

$$a < -z_0' < b. \tag{13. 16}$$

z_0' läßt sich als Funktion des Linienelementes in $r < 1$ auffassen. Man betrachte dasselbe als Anfangselement eines geodätischen Halbstrahls. z_0' ist eine nicht nur beschränkte, sondern auch stetige Funktion des Elementes. Für jedes feste A hängt nämlich $z'(0;A)$ stetig vom Anfangselement ab.

Satz 13. 2. Zu zwei gerichteten Geodätischen in $r < 1$ gibt es genau eine gerichtete Geodätische in $r < 1$, welche zur ersten positiv, zur zweiten negativ asymptotisch ist.

Beweis. G_1, G_2 seien die gegebenen Geodätischen. Der Punkt p^* liege auf keiner von ihnen. Durch ihn gibt es einen zu G_1 positiv und einen zu G_2 negativ asymptotischen Halbstrahl. Man wähle eine Geodätische G durch p^* so, daß die Halbstrahlen in p^* auf verschiedenen Seiten von G münden. Vom laufenden Punkt $p(s)$ auf G geht ebenfalls eine positive Asymptote zu G_1 und eine negative Asymptote zu G_2 aus. $\alpha_1(s)$ bzw. $\alpha_2(s)$, $0 \leqq \alpha_i \leqq \pi$, seien die von ihnen mit der positiven Richtung auf G in $p(s)$ gebildeten Winkel. In p^* ist $0 < \alpha_i < \pi$. $\alpha_i(s)$ erfüllen je eine Gleichung (13. 8) mit (13. 16). Aus ihr folgt, daß $\alpha_i(s)$ monoton von $\alpha_i(-\infty) = 0$ bis $\alpha_i(+\infty) = \pi$ wächst. Daher muß einmal, und nur einmal, $\alpha_1(s) + \alpha_2(s) = \pi$ werden. Dann bilden aber die beiden Halbstrahlen zusammen eine Geodätische der verlangten Art. Gäbe es nun zwei verschiedene derartige Geodätische, so verbinde man einen Punkt der einen mit einem nicht auf ihr gelegenen Punkt der anderen. Ist G die verbindende Geodätische, so ergibt sich nach dem eben Gesagten ein Widerspruch. Damit ist der Satz bewiesen.

Die Bedeutung der beiden Sätze für unsere Aufgabe beruht darin, daß die gerichteten Geodätischen in $r < 1$ sich nunmehr durch zwei Winkelkoordinaten φ, ψ charakterisieren lassen. Man wähle einen festen Punkt p_0 in $r < 1$, etwa $r = 0$, und eine feste Richtung in ihm, etwa $\varphi = 0$. Dann ist φ der Anfangswinkel des zur Geodätischen negativ asymptotischen Halbstrahles durch p_0 und ψ der Anfangswinkel des positiv asymptotischen Strahls durch p_0. Damit haben wir erst erreicht, was bei den Flächen konstanter Krümmung sofort in die Augen sprang. Die Geodätischen in $r < 1$, welche eine feste Geodätische G_0 schneiden, sind auch durch den Bogenlängenparameter s des Schnittpunktes auf G_0 und durch den Schnittwinkel α charakterisiert. Für unsere Aufgabe ist es notwendig, die Meßbarkeit der Koordinatentransformation

$$(s, \alpha) \longleftrightarrow (\varphi, \psi)$$

zu beweisen. Das war im Fall konstanter Krümmung trivial. Aber im allgemeinen Fall ist es nicht trivial, da die Transformation auf einem Durchgang durch das Unendliche ($r = 1$) beruht, welcher das Studium ihres analytischen Charakters erschwert.

§ 14. Fortsetzung. Die Koordinatentransformation.

$P_t = \mathfrak{T}_t P$ ist dasjenige Linienelement auf der durch das Element P bestimmten Geodätischen, welches durch Verschieben von P längs derselben in der Richtung von P um die Bogenlänge t entsteht.

Hauptlemma 14. 1. Sind die durch zwei Linienelemente P, P' bestimmten Geodätischen auf $\mathfrak{F}$ positiv asymptotisch zueinander, so gibt es eine Zahl a derart, daß

$$\sigma(P_{t+a}, P'_t) \to 0, \qquad t \to +\infty$$

gilt. Sind sie negativ asymptotisch, so gilt dies für $t \to -\infty$.

Beweis. G, G' seien die beiden Geodätischen. $p(t)$ bzw. $p'(t')$ seien die Trägerpunkte von P_t bzw. $P'_{t'}$. Man fälle von p' das Lot auf G. Der Treffpunkt sei p. Zwischen t und t' in $p = p(t)$, $p' = p'(t')$ besteht dann eine stetige, monotone Beziehung. Beim Beweis von Satz 13. 1 ergab sich $l(t) \to 0$ für die

Länge des Lotes und $\beta \to \frac{\pi}{2}$ für den Winkel bei p', wenn $t \to +\infty$ strebt. Nun existiert

$$\lim_{t=+\infty} (t - t') = a .$$

In dem von G, G' und zwei Loten, $t = t_1$ und $t = t_2$, begrenzten Viereck ist nämlich $(t_1 < t_2)$

$$t_2' - t_1' - l(t_1) - l(t_2) < t_2 - t_1 < t_2' - t_1' + l(t_1) + l(t_2),$$

also

$$|(t_2 - t_2') - (t_1 - t_1')| < l(t_1) + l(t_2).$$

Daraus folgt die Behauptung. Nun ist

$$\sigma(P_t, P'_{t-a}) < \sigma(P_t, P'_{t'}) + \sigma(P'_{t'}, P'_{t-a}).$$

Obere Schranken der rechten Distanzen erhält man durch Parallelverschiebungen und eine Drehung, und durch Ausrechnen von $\int d\sigma$ längs dieser Wege im Linienelementraum. Wegen (12. 2) ist daher die zweite Distanz gleich

$$|t' - (t - a)|$$

und die erste kleiner als

$$l(t) + \left|\beta - \frac{\pi}{2}\right|,$$

woraus das Lemma folgt.

Wir wenden unsere Aufmerksamkeit nun der am Ende von § 13 erwähnten Koordinatentransformation zu. Zu diesem Zweck untersuchen wir zunächst die vom Linienelement abhängige Größe z_0' als Funktion des Richtungswinkels α bei festem Trägerpunkt. Es ist $z = z(t) = z(t, \alpha)$ die den Bedingungen (13. 9) genügende Lösung von (13. 10), $K = K(t, \alpha)$. Mit K hängt auch z stetig von t, α ab. Führt man um den festen Trägerpunkt Polarkoordinaten ein, so tritt an die Stelle von $Y(\varrho, \vartheta)$ in (13. 1) die Lösung $W(t, \alpha)$ von (13. 10), welche den Anfangsbedingungen (13. 12) genügt. Daher ist

$$\frac{\partial K}{\partial \alpha} = \frac{dK}{ds} W;$$

der erste Faktor ist die Richtungsableitung von K normal zum Strahl. Wir machen von nun ab die

Voraussetzung B. Die Richtungsableitung der Krümmung ist auf $\mathfrak{F}$ beschränkt,

$$\left|\frac{dK}{ds}\right| < C.$$

Statt $z(t, \alpha)$ sei zunächst die Lösung $z(t, \alpha; A)$ mit den Randbedingungen (13. 11) betrachtet. Aus (13. 10) folgt durch Differenzieren nach α die Differentialgleichung

$$u'' + Ku = -K'_\alpha z = -\frac{dK}{ds} zW, \tag{14. 1}$$

wo die Funktion

$$u = u(t) = z'_\alpha(t, \alpha; A)$$

den Randbedingungen $u(0) = u(A) = 0$ genügt. Dabei ist

$$u'_0 = u'(0) = \frac{\partial}{\partial \alpha} z'_t(0, \alpha; A). \tag{14. 2}$$

Multiplikation von (14. 1) mit z, von (13. 10) mit u und Subtraktion liefert

$$(u'z - z'u)' = -\frac{dK}{ds} z^2 W,$$

woraus wegen der Randbedingungen für u und z

$$u'_0 = \int_0^A \frac{dK}{ds} z^2 W dt \tag{14. 3}$$

folgt. Im Integranden hängt nur z von A ab. z genügt für alle α, A der Ungleichung (13. 15) und $W(\varrho)$ erfüllt (13. 6),

$$W(t) < \frac{1}{b} \operatorname{Sin} bt.$$

Da ferner in $z(t, \alpha; A) \to z(t, \alpha)$, $A \to \infty$, die Konvergenz bezüglich t, α gleichmäßig ist, folgt mit Rücksicht auf Voraussetzung B, daß $\lim_{A=\infty} u_0'$ gleichmäßig in α existiert. Wegen (14. 2) folgt also aus (14. 3) die Formel

$$\frac{\partial z'_0}{\partial \alpha} = \frac{\partial}{\partial \alpha} z'_t(0, \alpha) = \int_0^\infty \frac{dK}{ds} z^2 W dt,$$

wo nun $z = z(t, \alpha)$ zu setzen ist. Da der Integrand unter einer festen integrablen Funktion liegt und stetig von t und dem Anfangselement abhängt, gilt das

Lemma 14. 2. Die Winkelableitung von z'_0 bei festem Trägerpunkt existiert, ist beschränkt auf $\mathfrak{F}$ und hängt stetig vom Linienelement ab.

Man betrachte nun zwei Punkte in $r < 1$. Winkel beliebiger Richtungen mit einer festen Richtung seien im ersten Punkt mit β, im zweiten mit γ bezeichnet. Zieht man vom ersten Punkt aus einen Strahl, vom zweiten den zu ihm asymptotischen Strahl (Richtungszuordnung vermittels asymptotischer Geodätischer), so wird $\beta = \beta(\gamma)$, $\gamma = \gamma(\beta)$. Dann gilt das

Lemma 14. 3. Bei der Richtungszuordnung vermittels asymptotischer Geodätischer ist $\frac{d\gamma}{d\beta}$ vorhanden, stetig in β und von Null verschieden.

Beweis. Man verbinde die beiden festen Punkte durch eine Geodätische G, $0 \leqq s \leqq a$. Die Winkel β, γ seien von der Richtung von G ab gerechnet. Die beiden zueinander asymptotischen Geodätischen bestimmen ein ganzes Feld von zueinander asymptotischen Geodätischen. Ihr Schnittwinkel $\alpha(s)$ mit G erfüllt längs G die Gleichung (13. 8) mit $z_0' = z_0'(s, \alpha)$. Dabei ist

$$\alpha(0) = \beta, \qquad \alpha(a) = \gamma.$$

Nach Lemma 14. 2 hängen z_0' und seine partielle Ableitung nach α stetig von s und α ab. Damit ist die Voraussetzung für die Differenzierbarkeit der Lösung $\alpha = \alpha(s; \beta)$ nach dem Anfangswert β gegeben. Die Ableitung $v = \frac{\partial \alpha}{\partial \beta}$ genügt der linearen Differentialgleichung

$$\frac{dv}{ds} = -\frac{\partial}{\partial \alpha}(z_0' \sin \alpha)\, v, \qquad v(0) = 1.$$

Daraus folgt wegen $v(a) = \frac{d\gamma}{d\beta}$ die Behauptung. Ferner erkennt man: Verschiebt man den ersten Punkt längs G, so hängt $\frac{d\gamma}{d\beta}$ stetig von β und der Verschiebung ab.

Jetzt haben wir alles, was zur Untersuchung der am Schluß von § 13 definierten Koordinatentransformation

$$(\varphi, \psi) \longleftrightarrow (s, \alpha)$$

erforderlich ist, beisammen.

Lemma 14. 4. Die Funktionaldeterminante

$$\frac{\partial(\varphi, \psi)}{\partial(s, \alpha)}$$

ist für $\alpha \neq 0$ von Null verschieden. Die vier partiellen Ableitungen in ihr hängen stetig von s und α ab.

Beweis. α ist der Winkel der positiven Richtung $s \to +\infty$ auf der festen Geodätischen G_0 mit der positiven Richtung der willkürlichen Geodätischen im Schnittpunkt, $0 \leqq \alpha \leqq \pi$. Auf $\alpha(s, \varphi) = \text{const}$ kann der Satz über die Ableitung einer impliziten Funktion angewandt werden,

$$\frac{\partial \varphi}{\partial s}\Big/_{\alpha} = -\frac{\dfrac{\partial \alpha}{\partial s}\Big/_{\varphi}}{\dfrac{\partial \alpha}{\partial \varphi}\Big/_{s}} = -\frac{\partial \alpha}{\partial s}\Big/_{\varphi} \cdot \frac{\partial \varphi}{\partial \alpha}\Big/_{s},$$

wo die Indizes die festzuhaltenden Größen angeben. Der erste Faktor — die Änderung von α längs G_0 im Asymptotenfelde $\varphi = \text{const}$ — ist gleich

$$-z_0' \sin \alpha ,$$

was stetig von s und α abhängt, und wo $z_0' \neq 0$ ist. Der zweite Faktor ist hingegen die Winkelverzerrung bei der Richtungszuordnung vermittels asymptotischer Geodätischer. Von ihm gilt Lemma 14. 3 und die Schlußbemerkung im Beweise desselben. Also sind die vier besagten Ableitungen stetig in s und α. Die Funktionaldeterminante hat den Wert

$$\left.\frac{\partial \varphi}{\partial \alpha}\right/_s \cdot \left.\frac{\partial \psi}{\partial \alpha}\right/_s \cdot \left(\left.\frac{\partial \alpha}{\partial s}\right/_\psi - \left.\frac{\partial \alpha}{\partial s}\right/_\varphi\right).$$

Hier ist nun zu beachten, daß die beiden Ableitungen in der Klammer stets entgegengesetztes Vorzeichen besitzen. Ist, was hier angenommen war, $\varphi = \text{const}$ die Schar der zur willkürlichen Geodätischen positiv asymptotischen Geodätischen, so ist es die Schar $\psi = \text{const}$ im negativen Sinne. Daher ist im letzteren Falle nicht α, sondern $\pi - \alpha$ der Schnittwinkel, von welchem die Gleichung (13. 8) gilt. Daraus folgt das Lemma.

§ 15. Beweis des Dissipativitäts- und Ergodizitätssatzes.

Die geodätische Strömung $\mathfrak{T}_t P$ in Ω läßt das Volumelement $dm = do\, d\vartheta$ in Ω invariant. Die Abbildung von $\mathfrak{F}$ auf die Überlagerungsfläche $r < 1$ induziert eine Abbildung von Ω auf den Raum der Linienelemente in $r < 1$. Dabei wird das Lebesguesche Maß m in diesen Raum übertragen. Es wird dann invariant gegenüber den Decktransformationen, welche auf Grund von § 12 eindeutig definiert sind. Hat eine Menge von Punkten P in Ω das m-Maß Null, so ist die ganze Menge von Urbildelementen im Linienelementraum von $r < 1$ eine m-Nullmenge. Wir sind jetzt erst in der Lage, von der Charakterisierung der Geodätischen durch Winkelpaare (φ, ψ) das Analogon zum Lemma 5. 1 zu beweisen.

Lemma 15. 1. Eine strömungsinvariante Menge (Menge von Geodätischen) in Ω hat dann und nur dann das Maß $m = 0$, wenn die entsprechende Menge von Winkelpaaren (φ, ψ) das Maß $\iint d\varphi\, d\psi = 0$ besitzt.

Beweis. Die Menge von Geodätischen in Ω hat dann und nur dann das Maß $m = 0$, wenn es bei der Menge aller entsprechenden Geodätischen in $r < 1$ zutrifft. Oder: Wenn es bei der Teilmenge derjenigen dieser Geodätischen zutrifft, welche eine feste Geodätische G_0 in $r < 1$ schneiden. Letzteres folgt daraus, daß man abzählbar viele G_0 so angehen kann, daß eine beliebige Geodätische in $r < 1$ mindestens eine dieser G_0 schneidet. Man kann nun ein

Linienelement P in $r < 1$ durch die Koordinaten τ, s, α kennzeichnen. τ ist die Entfernung von P vom Schnittpunkt der durch P bestimmten Geodätischen mit der festen G_0, s ist die Bogenlänge längs G_0 und α der Schnittwinkel. Dann ist bekanntlich

$$dm = \sin\alpha \, d\tau \, ds \, d\alpha .$$

Eine $\mathfrak{T}_t$-invariante Menge von Linienelementen in $r < 1$ (Menge von Geodätischen) ist also dann und nur dann vom m-Maße Null, wenn die Fußelemente (s, α) eine Menge vom Maß $\iint ds \, d\alpha = 0$ bilden. Hieraus und aus Lemma 14. 4 folgt die zu beweisende Behauptung.

Die in § 4 definierte Klasseneinteilung der Mannigfaltigkeiten ist auf Grund von Lemma 14. 3 in genau derselben Weise auf die Flächen $\mathfrak{F}$ anwendbar.

Hauptsatz 15. 2. Ist die Fläche $\mathfrak{F}$ von der ersten Klasse, so ist die geodätische Strömung ergodisch. Ist $\mathfrak{F}$ von der zweiten Klasse, so ist die Strömung dissipativ. Die Hauptsätze 5. 2, 7. 1 und 7. 2 gelten wörtlich bei den Flächen $\mathfrak{F}$.

Beweis. Die Beweise der genannten Hauptsätze können wörtlich übertragen werden, wenn man $dV = do$, $d\omega = d\vartheta$ schreibt und an Stelle der unendlich fernen Punkte π_1, π_2 der Geodätischen die Winkel φ, ψ setzt. Statt $d\omega_1$, $d\omega_2$ schreibe man $d\varphi$, $d\psi$. Zwei Geodätische (φ, ψ) sind positiv (bzw. negativ) asymptotisch zueinander, wenn die ψ-Winkel (bzw. φ-Winkel) übereinstimmen.

Statistik der Lösungen geodätischer Probleme vom unstabilen Typus. II.

Von

Eberhard Hopf in Leipzig.

Einleitung.

$\mathfrak{F}$ sei eine zweidimensionale Riemannsche, d. h. mit positiv definitem Bogenelement ds,

$$ds^2 = g_{ik}(u_1; u_2)\, du_i\, du_k,$$

versehene Mannigfaltigkeit, kurz Fläche genannt. Sie wird dreimal stetig differenzierbar vorausgesetzt, d. h. die g_{ik} und die Parametertransformationen sollen dieser Bedingung genügen. $\mathfrak{F}$ sei *vollständig*. Ω sei der Raum der gerichteten Linienelemente auf $\mathfrak{F}$. Unter der geodätischen Strömung in Ω versteht man die in den Phasenraum Ω verlegte Bewegung längs den Geodätischen von $\mathfrak{F}$ mit der Geschwindigkeit $ds/dt = 1$. Das strömungsinvariante Volumelement in Ω ist

$$dm = do\, d\vartheta,$$

wo $d\vartheta$ das Winkel- und do das Flächenelement auf $\mathfrak{F}$ bedeuten. Damit ist auch ein strömungsinvariantes Lebesguesches Maß m in Ω definiert.

In seiner Abhandlung „Statistik der geodätischen Linien in Mannigfaltigkeiten negativer Krümmung“ [1]) hat der Verfasser den statistischen Gesamtverlauf der Geodätischen auf Flächen $\mathfrak{F}$ mit folgender Eigenschaft untersucht: *A*). *Die Krümmung K verläuft zwischen festen negativen Grenzen.* *B*). *Die Richtungsableitung dK/ds ist beschränkt.*

Die Gesamtheit dieser Flächen $\mathfrak{F}$ wurde in zwei fundamentale Klassen geteilt. $\mathfrak{F}$ gehört zur *ersten Klasse*, wenn die Anfangsrichtungen derjenigen von einem festen Punkte von $\mathfrak{F}$ ausgehenden geodätischen Halbstrahlen, welche auf $\mathfrak{F}$ in unendlicher Entfernung enden, eine Menge vom Winkelmaß Null bilden. Die *zweite Klasse* ist die zur ersten komplementäre Klasse. Zur ersten Klasse gehören offenbar alle geschlossenen Flächen negativer Krümmung. Durch Anwendung des Poincaréschen Wiederkehrsatzes auf die geodätische Strömung wurde geschlossen, daß ihr allgemeiner alle $\mathfrak{F}$ mit den Eigenschaften A und B und mit endlichem Flächeninhalte angehören.

---

[1]) Leipziger Berichte **91** (1939), S. 261—304. Im folgenden als Hopf I zitiert. In der Einleitung findet man auch eine von Nullmengen freie Formulierung von Satz 1.

Auf Grund eines neuen Gedankens, der Methode der asymptotischen Geodätischen, gelang u. a. der Beweis der beiden folgenden Hauptsätze.

Satz I. *Ist $\mathfrak{F}$ von der ersten Klasse, so ist die geodätische Strömung in Ω ergodisch. Hat $\mathfrak{F}$ endlichen Flächeninhalt, so bedeutet dies: Das Zeitmittel irgendeiner in Ω m-summierbaren Funktion ist längs jeder Stromlinie, abgesehen von gewissen Stromlinien, die in Ω eine m-Nullmenge bilden, gleich dem Raummittel über Ω. Oder: Die allgemeine Geodätische ist auf $\mathfrak{F}$ nach Fläche und Richtung gleichverteilt. Im Falle unendlichen Flächeninhalts ist diese Aussage in naheliegender Weise zu modifizieren.*

Satz II. *Ist $\mathfrak{F}$ von der zweiten Klasse, so ist die geodätische Strömung in Ω dissipativ: Die Stromlinien enden im allgemeinen im Unendlichen von Ω. Oder: Die allgemeine Geodätische endet auf $\mathfrak{F}$ in unendlicher Entfernung.*

Die Tragweite der Methode der asymptotischen Geodätischen reicht indessen über diese Anwendung auf Flächen negativer Krümmung weit hinaus. In Hopf I wurden bereits n-dimensionale Mannigfaltigkeiten konstanter negativer Krümmung untersucht. Aber auch auf viel allgemeinere Variationsprobleme mit einer unabhängigen Veränderlichen läßt sich die Methode anwenden. Dabei ist als Hilfsmittel die Finslersche Geometrie des Problems zu benutzen. Damit ist ein weites Feld von Differentialgleichungs-Problemen angedeutet, bei welchen es nunmehr möglich sein wird, den Gesamtverlauf der Lösungen im Sinne des ungenau messenden makroskopischen Beobachters zu bestimmen.

Die Probleme, auf welche die Methode anwendbar ist, haben eine gewisse Unstabilität der Lösungen miteinander gemein. Welche Rolle diese Unstabilität dabei spielt, wird in der vorliegenden Arbeit am Beispiel des geodätischen Problems für die folgenden Flächen $\mathfrak{F}$ gezeigt. Man betrachte diejenige Lösung $y(s)$ der längs irgendeines geodätischen Halbstrahls, $s \geqq 0$, gebildeten Variationsgleichung

$$\frac{d^2 y}{d s^2} + K y = 0,$$

welche den Anfangsbedingungen $y_0 = 0$, $y'_0 = 1$ genügt. Wir verlangen dann von der Fläche: A'). *K ist beschränkt. Die Geodätischen genügen gleichmäßig der Unstabilitätsbedingung $y'/y \geqq c > 0$* [2]). $B') = B$). *dK/ds ist beschränkt.* Alle Flächen mit den Eigenschaften A und B erfüllen diese Bedingungen. Darüber hinaus (Flächen mit $K \gtrless 0$) kann natürlich eine Bedingung, welche etwas über die ganze unendliche Geodätische voraussetzt,

[2]) Sie ist wesentlich enger als die von Morse und später von Hedlund zur Untersuchung des topologischen Verlaufs der Geodätischen eingeführten Unstabilitätsbedingungen. Vgl. M. Morse, Instability and transitivity. Journal de Mathém. **14** (1935), S. 49—71; G. A. Hedlund, Two-dimensional manifolds and transitivity. Annals

nur eine provisorische Bedeutung haben. Es ist wünschenswert und wohl auch möglich, sie durch Voraussetzungen finiter Natur zu ersetzen oder anzunähern. Über einen primitiven Schritt in dieser Richtung (§ 1) soll hier nicht hinausgegangen werden.

Im folgenden wird bewiesen: *Die Gesamtheit der obigen Flächen zerfällt in die beiden erwähnten Klassen. Die beiden Hauptsätze bleiben wörtlich bestehen.*

Einige Änderungen der Beweisdetails waren gegenüber Hopf I erforderlich. Aus diesem Grunde, und um anknüpfende Untersuchungen in den angedeuteten Richtungen zu erleichtern, ist der Beweisgang unter schärferer Hervorhebung der Hauptpunkte noch einmal vollständig dargestellt worden.

§ 1.

Eine direkte Bedingung dafür, daß $\mathfrak{F}$ der Unstabilitätsbedingung genügt.

Auf $\mathfrak{F}$ sollen endlich viele getrennte Bereiche B angebbar sein, welche alle Punkte mit $K \geqq 0$ enthalten. Es existiere eine endliche obere Schranke l für die Bogenlängen der in irgendeinem B enthaltenen geodätischen Segmente. L sei eine untere Schranke für die Längen der geodätischen Segmente außerhalb $\sum B$ und mit beiden Endpunkten auf dem Rande von $\sum B$. Ferner sei $K \leqq m^2$ überhaupt und $K \leqq -\mu^2$ in $\mathfrak{F} - \sum B$, $\mu \neq 0$. Bestehen dann die Ungleichungen

$$(1.1) \qquad m\,l < \frac{\pi}{2}, \quad m \operatorname{tg} m\,l < \mu\, \mathfrak{Tg}\, \mu\, L,$$

so genügt $\mathfrak{F}$ der Unstabilitätsbedingung A').

Der Beweis beruht auf den Sturmschen Vergleichssätzen. Die logarithmische Ableitung $u = y'/y$ ist eine Lösung der Riccatischen Gleichung

$$u' = -K - u^2,$$

$u(+0) = +\infty$.

---

of Math. **37** (1936), S. 534—542. Unsere Bedingung kann zweifellos gemildert werden. Z. B. ist die in ihr enthaltene Forderung $y' > 0$ unwesentlich. Man kommt sicher auch mit folgender Bedingung aus. Es gibt zwei positive Konstante C und c derart, daß

$$\frac{y(s')}{y(s)} \geqq C e^{c(s'-s)}, \qquad s' > s > 0$$

gilt. Genauere Angaben darüber findet der Leser im Text. Im Zusatz bei der Korrektur wird der Beweis unter dieser Bedingung nachgeliefert. — Durch unsere Bedingung wird das Vorkommen von Geodätischen ausgeschlossen, längs welchen $K \geqq 0$ gilt. Liegen auf einer geschlossenen $\mathfrak{F}$ mit $K \leqq 0$ die Punkte mit $K = 0$ auf endlich vielen glatten Kurven, unter denen ganze Geodätische auftreten können, so gilt vermutlich noch Satz I. Der Beweis hat eine zusätzliche Schwierigkeit zu überwinden, indem die Hilfssätze 3. 3 und 3. 4 von § 3 etwas modifiziert werden müssen. Der Haupthilfssatz 3. 5 bleibt vermutlich richtig.

$s = s_0$ sei eine Stelle, wo die Geodätische ΣB verläßt, $s = s_1$ die nächste Stelle, wo sie wieder in ΣB eintritt, und schließlich $s = s_2$ die darauf folgende Stelle des Wiederverlassens. Dann gelten die Ungleichungen

(1. 2) $\quad -K \geqq \mu^2, \quad s_0 \leqq s \leqq s_1; \quad -K \geqq -m^2; \quad s_1 - s_0 \geqq L, \quad s_2 - s_1 \leqq l.$

Es sei nun $v(s)$ die bei $s = s_1$ stetige Lösung der Hilfsgleichung

(1. 3) $\quad v' = \mu^2 - v^2, \quad s_0 \leqq s \leqq s_1; \quad v' = -m^2 - v^2, \quad s_1 \leqq s \leqq s_2,$

mit der Anfangsbedingung $v(s_0) = 0$. Dann gilt wegen (1, 2) und nach Sturm im ganzen Intervall $\langle s, s_2\rangle$ die Ungleichung $u \geqq v$, sobald sie im Anfangspunkt desselben erfüllt ist. Rechnet man v aus, so ergibt sich folgendes. Infolge der ersten der Ungleichungen (1. 1) bleibt v im Intervall $\langle s_1, s_2\rangle$ stetig. Man findet

$$v(s_2) \geqq \frac{\mu\, \mathfrak{Tg}\, \mu L - m \operatorname{tg} m l}{1 + \frac{\mu}{m}\, \mathfrak{Tg}\, \mu L \cdot \operatorname{tg} m l}.$$

Infolge der zweiten Ungleichung (1, 1) hat also $v(s_2)$ auf der ganzen Fläche eine positive untere Schranke. Bildet man schließlich diejenige Lösung $w(s)$ von (1, 3), für welche $w(s_0)$ gleich dieser Schranke a ist, so gilt, wie leicht zu sehen, $w(s) \geqq a$. Durch Betrachtung der aufeinanderfolgenden Austrittsstellen aus ΣB,

$$s_0 \leqq 0 < s_2 < s_4 < s_6 < \dots$$

ergibt sich dann wegen $u(0) = +\infty > w(0)$ sukzessiv $u(s) > a$ für alle $s \geqq 0$. Damit ist die Behauptung bewiesen.

Wir erwähnen noch eine finite Bedingung, die auch hinreicht, aber schwächer als die obige ist. Bereiche B und Zahlen L und l seien wie oben zugrunde gelegt. Hat jede geodätische Strecke der Länge $L + l$, deren erstes Stück der Länge L in $\mathfrak{F} - \Sigma B$ liegt, die Eigenschaft, daß die im Anfangspunkt verschwindende Lösung von $u' = -K - u^2$ auf der ganzen Strecke positiv ist, so ist die Unstabilitätsbedingung A') erfüllt. Hierbei ist $\mathfrak{F}$ geschlossen vorausgesetzt. Allgemein ist die Behauptung richtig, wenn etwas mehr als bloße Positivität verlangt wird.

Damit die Bedingung $L > 0$ überhaupt erfüllbar ist, muß jeder Bereich extremalkonvex sein. Dies ist der Fall, wenn B einfach zusammenhängend und von einer Linie positiver geodätischer Krümmung (bei Durchlaufung in einem bestimmten, in B festgesetzten Drehsinn) begrenzt ist. Haben alle B diese Eigenschaft, so ist die untere Grenze L die kleinste der folgenden Zahlen. Die erste ist die Kleinstentfernung verschiedener B voneinander. Jede der anderen ist die Minimallänge derjenigen Linien außerhalb ΣB und mit Endpunkten auf einem bestimmten B, welche bei festgehaltenen Endpunkten nicht auf $\mathfrak{F} - \Sigma B$ einem Randbogen dieses B homotop sind. Dies ergibt sich aus folgender Tatsache:

B sei ein von einer geschlossenen Kurve positiver geodätischer Krümmung (im obigen Sinne) begrenzter Bereich auf einer Fläche. Außerhalb *B* sei die Flächenkrümmung negativ. Dann kann ein geodätischer Bogen, welcher außerhalb *B* zwei Randpunkte von *B* verbindet, unter Festhaltung derselben nicht in einen Randbogen von *B* stetig deformiert werden.

Wäre dies falsch, so würde der Bogen zusammen mit jedem der beiden Randbögen von *B* auf der Fläche ein Gebiet beranden. Doppelpunkte auf dem ersten Bogen sind ausgeschlossen, sonst ergäbe die Anwendung der Formel von Gauß-Bonnet auf eine Schleife einen Widerspruch. Eines jener Gebiete liegt nun außerhalb *B*. Anwendung derselben Formel auf dieses Gebiet führt dann ebenfalls zum Widerspruch.

Der Sinn jeder Unstabilitätsbedingung der eingeführten Art ist, daß dem Vorkommen von $K > 0$ auf $\mathfrak{F}$ Beschränkungen auferlegt werden. Solche Beschränkungen sind aber notwendig, wenn die Hauptsätze auf $\mathfrak{F}$ und auf allen durch hinreichend kleine und gleichmäßig glatte Deformation von $\mathfrak{F}$ erhaltenen Flächen gelten sollen.

§ 2.

Verlauf der Geodätischen auf der universellen Überlagerungsfläche.

$\mathfrak{F}$ ist vollständig und zweimal stetig differenzierbar. Daher gibt es zwischen irgend zwei Punkten unter allen denjenigen Verbindungswegen, welche einer vorgegebenen Verbindung homotop sind, stets einen kürzesten. Er ist geodätisch[3]).

Wir bezeichnen mit ϑ den Anfangswinkel der von einem festen Punkte p_0 von $\mathfrak{F}$ ausgehenden geodätischen Halbstrahlen, $s \geqq 0$, und fassen $r = s/(s+1)$ und ϑ als Polarkoordinaten in der Kreisscheibe $r < 1$ auf. Damit ergibt sich eine Ein-viele-Korrespondenz zwischen den (nach obigem allen) Punkten p von $\mathfrak{F}$ und den Punkten von $r < 1$. Dabei variiert $p(r, \vartheta)$ stetig. Infolge der Unstabilitätsbedingung A') ist stets $y > 0$, $s > 0$, auf einem ganz beliebigen Halbstrahl, d. h. es gibt keine konjugierten Punkte auf $\mathfrak{F}$. Aus der letzteren Tatsache allein folgt nun die Unverzweigtheit jener Korrespondenz: Zu einem beliebigen Punkte in $r < 1$ gibt es eine Umgebung, welche nicht zwei verschiedene Punkte mit demselben Spurpunkt auf $\mathfrak{F}$ enthält (kongruente Punkte). Im entgegengesetzten Falle wäre nämlich, wie leicht zu sehen, der Grenzpunkt dieser Spurpunkte zu p_0 konjugiert. Aus der Unverzweigtheit ergibt sich in geläufiger Weise folgendes. Einer stetigen Linie $p(\lambda)$ in $\mathfrak{F}$ entspricht, wenn man einen bestimmten Bildpunkt von $p(\lambda_0)$

[3]) H. Hopf und W. Rinow, Über den Begriff der vollständigen differentialgeometrischen Fläche. Comm. Math. Helvet. **3** (1931), S. 209. Statt der dort vorausgesetzten Analytizität genügt die obige Voraussetzung.

in $r < 1$ vorgibt, genau eine stetige Linie durch denselben in $r < 1$. Zwei verschiedenen Wegen zwischen zwei Punkten auf $\mathfrak{F}$ mögen in $r < 1$ zwei Kurven mit gleichem Anfangspunkt entsprechen. Dann fallen ihre Endpunkte dann und nur dann zusammen, wenn die beiden Wege auf $\mathfrak{F}$ einander homotop sind.

Die Kreisscheibe stellt hiernach eine Verwirklichung der universellen Überlagerungsfläche $\tilde{\mathfrak{F}}$ von $\mathfrak{F}$ dar. Wir denken uns durch die Korrespondenz die Metrik ds^2 auf sie übertragen. In diesem Sinne sprechen wir von Längen, Winkeln, geodätischen Linien, infinitesimaler Parallelverschiebung usw. auf $\tilde{\mathfrak{F}}$. Da konjugierte Punkte nicht existieren, und da $\tilde{\mathfrak{F}}$ einfach zusammenhängend ist, bilden die von einem festen Punkte von $\tilde{\mathfrak{F}}$ ausgehenden Strahlen auf $\tilde{\mathfrak{F}}$ außerhalb dieses Punktes ein Feld von Geodätischen. Zwei verschiedene Geodätische können sich auf $\tilde{\mathfrak{F}}$ nur einmal schneiden. Zusammenfassend: *Zu irgend zwei Punkten von $\tilde{\mathfrak{F}}$ gibt es auf $\tilde{\mathfrak{F}}$ eine und nur eine kürzeste Verbindung; sie ist geodätisch. Von zwei Punkten von $\mathfrak{F}$ gilt dasselbe, wenn man sich auf Wege einer Homotopieklasse beschränkt.* Da auf $\tilde{\mathfrak{F}}$ ein geodätischer Bogen die kürzeste Verbindung seiner Endpunkte ist, folgt: *Auf $\tilde{\mathfrak{F}}$ konvergiert jeder geodätische Strahl gegen die Grenze von $\tilde{\mathfrak{F}}$.*

Die Entfernung zweier Punkte auf $\mathfrak{F}$ oder auf $\tilde{\mathfrak{F}}$ wird im folgenden mit $s(p_1, p_2)$ bezeichnet. Dabei wird in Zweifelsfällen stets angegeben, ob $\mathfrak{F}$ oder $\tilde{\mathfrak{F}}$ zugrunde gelegt ist. Offenbar ist $s(p_1, p_2)$ auf $\mathfrak{F}$ der Kleinstwert aller $s(\tilde{p}_1, \tilde{p}_2)$ für irgend zwei Bildpunkte $\tilde{p}_1, \tilde{p}_2$ auf $\tilde{\mathfrak{F}}$.

Wir bezeichnen mit Ω $(\tilde{\Omega})$ den Raum der gerichteten Linienelemente P auf $\mathfrak{F}$ $(\tilde{\mathfrak{F}})$. Die Korrespondenz zwischen $\mathfrak{F}$ und $\tilde{\mathfrak{F}}$ definiert eine analoge Korrespondenz zwischen Ω und $\tilde{\Omega}$. Eine Metrik wird in Ω und $\tilde{\Omega}$ durch

$$d\sigma^2 = ds^2 + d\chi^2 \tag{2. 1}$$

eingeführt. Dabei bedeutet ds die Entfernung der Trägerpunkte und $d\chi$ den Winkel, welchen die längs der geodätischen Verbindung ds unter festgehaltenem Winkel mit ihr verschobene erste Richtung im zweiten Punkte mit der zweiten Richtung bildet (Parallelverschiebung längs ds). Damit ist ein regulärer Entfernungsbegriff $\sigma(P_1, P_2)$ in Ω und $\tilde{\Omega}$ definiert. Auch von ihm gilt das über s Gesagte. Von irgend zwei Elementen P_1, P_2 und ihren Trägerpunkten p_1, p_2 gilt in $\tilde{\mathfrak{F}}$, $\tilde{\Omega}$ und auch in $\mathfrak{F}$, Ω

$$s(p_1, p_2) \leqq \sigma(P_1, P_2) \leqq s(p_1, p_2) + \pi. \tag{2. 2}$$

Die zweite Ungleichung ergibt sich, indem man das erste Element längs der geodätischen Verbindung von p_1 mit p_2 parallel verschiebt und dann durch eine Drehung mit dem zweiten zur Deckung bringt. Dieser Weg von P_1 nach P_2 besteht aus zwei Stücken; auf dem ersten ist $d\chi = 0$, auf dem zweiten $ds = 0$. Schließlich sei noch folgendes über das Lebesguesche Volummaß m, das durch $dm = do\,d\vartheta$ in Ω und $\tilde{\Omega}$ definiert ist, bemerkt. Carathéodorysche m-Meßbarkeit einer Menge M in Ω ist mit der m-Meßbarkeit der Menge $\tilde{M}$ aller Bildpunkte aller Punkte von M in $\tilde{\Omega}$ gleichbedeutend. Insbesondere können M und $\tilde{M}$ immer nur gleichzeitig m-Nullmengen sein. Zum Beweis beachte man, daß die Gruppe der Decktransformationen von $\tilde{\Omega}$ in sich, durch welche aus $\tilde{\Omega}$ wieder Ω entsteht, eigentlich diskontinuierlich ist. In $\tilde{\Omega}$ läßt sich ein m-meßbarer Fundamentalbereich B angeben, der mit seinen abzählbar vielen Kopien $\tilde{\Omega}$ einfach überdeckt. Es gilt allgemein $m(M) = m(\widetilde{BM'})$. Die Behauptung folgt, wenn man die Invarianz von m gegenüber der Gruppe beachtet.

§ 3.

Asymptotische Geodätische. Formulierung der Hilfssätze. Klasseneinteilung.

Zwei geodätische Halbstrahlen $\tilde{\mathfrak{F}}$ heißen zueinander *asymptotisch*, wenn mit einer passenden Zahl a auf $\tilde{\mathfrak{F}}$

$$(3.1) \qquad s\,(p_{t+a}, p'_t) \to 0, \quad t \to \infty$$

gilt; dabei bedeuten p_t bzw. p'_t den laufenden Punkt auf dem einen bzw. anderen Halbstrahl und t die Bogenlänge. Analoges definieren wir auf $\mathfrak{F}$ (mit dem Distanzbegriff auf $\mathfrak{F}$). Zwei Halbstrahlen in $\mathfrak{F}$ sind sicher in $\mathfrak{F}$ asymptotisch, wenn zwei geeignete Bildstrahlen in $\tilde{\mathfrak{F}}$ asymptotisch sind. Die Umkehrung hiervon wird im folgenden nicht gebraucht. Sind auf $\tilde{\mathfrak{F}}$ zwei Halbstrahlen zu einem dritten asymptotisch, so sind sie es offenbar zueinander.

Wir führen nun eine Reihe von Sätzen, die für den Beweis der Hauptsätze von großer Wichtigkeit sind, ohne Beweis an. Ihre Beweise werden der Übersichtlichkeit halber im letzten Teil der Arbeit dargestellt werden.

Zunächst: Die in der Strahlrichtung orientierten Linienelemente auf zwei asymptotischen Halbstrahlen haben die Eigenschaft

$$(3.2) \qquad \sigma\,(P_{t+a}, P'_t) \to 0, \quad t \to \infty.$$

Dies gilt in $\tilde{\Omega}$ und erst recht in Ω bei entsprechend aufgefaßtem σ.

Auf $\tilde{\mathfrak{F}}$ gelten zwei fundamentale Sätze.

Satz 3.1. *Auf $\tilde{\mathfrak{F}}$ geht von einem beliebig gegebenen Punkt ein und nur ein Halbstrahl aus, der zu einem beliebig gegebenen Halbstrahl asymptotisch ist.*

Ein Halbstrahl auf $\tilde{\mathfrak{F}}$ heißt positiv asymptotisch zu einer gerichteten Geodätischen auf $\tilde{\mathfrak{F}}$, wenn er es zu einem positiven Teilstrahl derselben ist. Zwei Strahlen (gerichtete Geodätische) auf $\tilde{\mathfrak{F}}$ sind zueinander positiv (negativ) asymptotisch, wenn positive (negative) Hälften derselben es sind.

Satz 3. 2. *Zu zwei Halbstrahlen auf $\tilde{\mathfrak{F}}$, die nicht zueinander asymptotisch sind, gibt es einen und nur einen Strahl auf $\tilde{\mathfrak{F}}$, zu welchem der erste positiv und der zweite negativ asymptotisch ist.*

Zufolge der Sätze 3. 1 und 3. 2 lassen sich auf $\tilde{\mathfrak{F}}$ die gerichteten Geodätischen = Strahlen durch zwei Winkelkoordinaten φ_1, φ_2 charakterisieren. Man lege auf $\tilde{\mathfrak{F}}$ einen festen Punkt p_0 und eine feste Richtung in demselben zugrunde. H_0 sei der von p_0 aus in dieser Richtung laufende Halbstrahl. Zu einem beliebigen Strahl S gehören dann zwei Winkel φ_1, φ_2; φ_1 (φ_2) ist der in p_0 gebildete Winkel von H_0 bis zu demjenigen, von p_0 ausgehenden Halbstrahl, welcher zu S negativ (positiv) asymptotisch ist. Die Winkel charakterisieren die „unendlich fernen Punkte" von S. Es ist stets $\varphi_1 \neq \varphi_2$. Sonst wären zwei Hälften von S zueinander asymptotisch, was nach Satz 3. 1 (Eindeutigkeit) unmöglich ist. Umgekehrt bestimmen nach Satz 3. 2 zwei voneinander verschiedene Winkel φ_1, φ_2 genau einen Strahl auf $\tilde{\mathfrak{F}}$. Aus dem Vorangehenden folgt: Zwei Strahlen in $\tilde{\mathfrak{F}}$ sind dann und nur dann zueinander positiv asymptotisch, wenn sie dieselbe φ_2-Koordinate, negativ asymptotisch, wenn sie dieselbe φ_1-Koordinate besitzen.

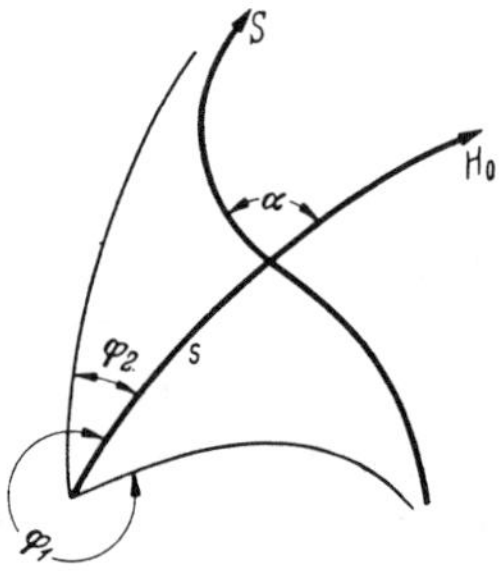

Fig. 1.

Diejenigen Strahlen in $\tilde{\mathfrak{F}}$, welche den Halbstrahl H_0 schneiden (sie schneiden ihn dann nur einmal), sind auch durch die Entfernung s des Schnittpunktes von p_0 und den Schnittwinkel α charakterisiert (Fig. 1). Die Parametertransformation

$$(\varphi_1, \varphi_2) \leftrightarrow (s, \alpha)$$

ist eineindeutig und hat die folgende Eigenschaft.

Hilfssatz 3. 3. *Die ersten partiellen Ableitungen von φ_1, φ_2 nach s, α sind stetig in s, α. Die Funktionaldeterminante der Transformation ist für $\alpha \neq 0, \pi$ von Null verschieden.*

Der Beweis dieses Satzes gründet sich auf den

Hilfssatz 3. 4. *Ordnet man vermittels asymptotischer Halbstrahlen die Richtungen in einem Punkte von $\tilde{\mathfrak{F}}$ den Richtungen in einem anderen Punkte von $\tilde{\mathfrak{F}}$ zu und sind β, γ ihre Winkel mit festen Richtungen in diesen Punkten, so ist $d\gamma/d\beta$ vorhanden und stetig. Zusatz: Läßt man den zweiten Punkt auf einem festen Strahl durch den ersten variieren, so hängt die Ableitung auch stetig von β und der Bogenlänge s auf dem Strahl ab. γ wird dabei vom Strahl an gerechnet.*

Man betrachte nun eine gegenüber der geodätischen Strömung in Ω invariante Punktmenge in Ω, m. a. W. eine Menge von Stromlinien in Ω. Jeder dieser Stromlinien entsprechen viele Stromlinien in $\tilde{\Omega}$. Jede der letzteren ist wiederum durch ein Winkelpaar φ_1, φ_2 charakterisiert.

Haupthilfssatz 3. 5. *Eine strömungsinvariante Punktmenge in Ω hat dann und nur dann das m-Maß Null, wenn die gesamte entsprechende Punktmenge auf dem (φ_1, φ_2)-Torus das Maß $\iint d\varphi_1 d\varphi_2$ Null besitzt.*

Die zwei Klassen von Flächen $\mathfrak{F}$. Wir sagen von einem Halbstrahl auf $\mathfrak{F}$, er ende auf $\mathfrak{F}$ im Unendlichen oder in unendlicher Entfernung, wenn seine Punkte p der Grenzbeziehung $s(p_0, p_t) \to \infty$, $t \to \infty$, genügen. Hat von zwei auf $\mathfrak{F}$ zueinander asymptotischen Halbstrahlen einer diese Eigenschaft, so hat sie auch der andere, wie man sofort einsieht. Die Anfangsrichtungen derjenigen, von einem festen Punkte p_0 von $\mathfrak{F}$ ausgehenden Halbstrahlen, welche auf $\mathfrak{F}$ im Unendlichen enden, bilden nun, wie leicht zu zeigen, eine im Sinne des gewöhnlichen Winkelmaßes meßbare Winkelmenge. Ist sie eine Nullmenge, so ist die in einem anderen Punkte p_1 von $\mathfrak{F}$ analog definierte Richtungsmenge ebenfalls eine Nullmenge. Dies folgt durch Richtungszuordnung vermittels asymptotischer Halbstrahlen aus dem Hilfssatz 3. 4; da die Ableitung von Null verschieden sein muß, gehen Nullmengen in Nullmengen über. Durch diese Tatsache ist die in der Einleitung beschriebene Klasseneinteilung gerechtfertigt.

Hat $\mathfrak{F}$ endlichen Flächeninhalt, so gehört $\mathfrak{F}$ zur ersten Klasse. Wegen der in Ω geltenden Ungleichung (2. 2) endet nämlich ein Halbstrahl in $\mathfrak{F}$ dann und nur dann im Unendlichen, wenn dasselbe bei der entsprechenden Stromlinienhälfte in Ω der Fall ist. Nach dem Poincaréschen Wiederkehrsatz bilden aber die Stromlinien, welche diese Eigenschaft haben, eine m-Nullmenge M von Punkten P in Ω. Bezeichnet man mit $\mu(p)$ das Winkelmaß der vom Punkte p von $\mathfrak{F}$ ausgehenden und im Unendlichen endenden Halbstrahlen, so gilt nach Fubini

$$m(M) = \int_{\mathfrak{F}} \mu(p)\, do.$$

Daraus folgt die Behauptung, da entweder stets $\mu = 0$ oder stets $\mu > 0$ ist.

§ 4.

Beweis der Hauptsätze.

Beweis von Satz I. Wir beschränken uns auf Flächen endlicher Oberfläche $O(\mathfrak{F})$. Dann ist auch $m(\Omega) = 2\pi O(\mathfrak{F})$ endlich. Von der geodätischen Strömung $\mathfrak{T}_t(P) = P_t$ gilt daher der Birkhoffsche Ergodensatz:

Ist die Funktion $f(P)$ in Ω m-summierbar, so existieren, wenn man von einer m-Nullmenge in Ω absieht, in jedem P die längs der Stromlinie konstanten Limites

$$f_2^*(P) = \lim_{T=\infty} \frac{1}{T} \int_0^T f(P_t)\, dt, \quad f_1^*(P) = \lim_{T=\infty} \frac{1}{T} \int_{-T}^0 f(P_t)\, dt.$$

Jedes der beiden Zeitmittel $f^*(P)$ ist summierbar und genügt für jede beschränkte, m-meßbare und strömungsinvariante Funktion $\varphi(P)$ der Gleichung

$$\int_\Omega f^* \varphi \, dm = \int_\Omega f \varphi \, dm.$$

Fast überall in Ω ist

$$f_1^* = f_2^*.$$

Die Strömung heißt ergodisch, wenn $f^*(P)$ für beliebiges $f(P)$ fast überall konstant ist. f^* ist dann gleich dem m-Mittel von f. Für die Ergodizität ist die Konstanz von f^* für solche f hinreichend, welche im Sinne der Distanz

$$\int_\Omega |f - g| \, dm$$

im Raume aller obigen f dicht liegen.

Wir können voraussetzen, daß $f(P)$ im Sinne der Distanz σ in Ω gleichmäßig stetig ist. Liegen nun P und P' auf zwei zueinander positiv asymptotischen Stromlinien, so gilt bei passendem a

$$\lim_{t=\infty} [f(P_{t+a}) - f(P'_t)] = 0.$$

Also müssen die Zukunftsmittel $f_2^*(P)$ und $f_2^*(P')$ einander gleich sein, wenn eins von ihnen existiert. Analoges gilt von den Vergangenheitsmitteln, wenn die Linien negativ asymptotisch sind. Wegen $f_1^* = f_2^*$ können sich die beiden Mengen

(4. 1) $$f_1^*(P) \geqq z$$

und

(4. 2) $$f_2^*(P) \geqq z$$

nur um m-Nullmengen unterscheiden. Die behauptete Konstanz folgt nun, wenn bei beliebigem z gezeigt wird, daß entweder die Menge (4. 2) oder ihre Komplementärmenge eine m-Nullmenge sein muß. Dieser Nachweis gelingt aufs einfachste, wenn man die den Stromlinienmengen (4. 1) und (4. 2) entsprechenden Punktmengen auf dem Torus $\Phi_1 \times \Phi_2$ betrachtet; mit Φ_i sind die beiden φ_i-Kreislinien bezeichnet. Nach Obigem hat die Menge (4. 1) die Eigenschaft, mit einem Punkte (φ_1, φ_2) auch alle Punkte (φ_1, φ_2') zu enthalten, also eine Produktmenge $A_1 \times \Phi_2$ zu sein. Ebenso ist (4. 2) von der Form $\Phi_1 \times A_2$. Nach dem Hauptnilfssatz 3. 5 können sich diese beiden Mengen von ihrem Durchschnitt $A_1 \times A_2$ nur um Mengen vom Maß $\iint d\varphi_1 \, d\varphi_2$ Null unterscheiden. In diesem Sinne sind also $\bar{A}_1 \times A_2$ und $A_1 \times \bar{A}_2$ Nullmengen, $\bar{A}_i = \Phi_i - A_i$. Ist nun (4. 2) keine m-Nullmenge, so ist A_2 keine Nullmenge auf Φ_2. Daraus folgt, daß $\bar{A}_1$ auf Φ_1 eine Nullmenge ist, und schließlich, daß $\bar{A}_2$ auf Φ_2 das Maß Null hat. Nach dem Haupthilfssatz ist also die Komplementärmenge von (4. 2) eine m-Nullmenge. Damit ist Satz I bewiesen.

Für gleichmäßig stetiges $f(P)$ ergibt sich aus obigem eine etwas schärfere Fassung des Satzes. Ist p ein beliebiger Punkt auf der Fläche, so ist im Sinne des Winkelmaßes für fast alle P in p das Zeitmittel f^* von f gleich dem m-Mittel von f.

Die Formulierung des Satzes im Falle, daß $\mathfrak{F}$ zur ersten Klasse gehört, aber unendliches $O(\mathfrak{F})$ hat, findet der Leser in Hopf I.

Beweis von Satz II. Aus der Zugehörigkeit von $\mathfrak{F}$ zur zweiten Klasse folgt, daß die für $t \to +\infty$ im Unendlichen von Ω endenden Stromlinien in Ω eine Menge positiven m-Maßes bilden. Nach einem allgemeinen Satze über solche Stromlinien kann sich diese Menge von der Menge derjenigen Stromlinien, welche für $t \to -\infty$ im Unendlichen enden, nur um m-Nullmengen unterscheiden. Läßt man nun die beiden Mengen an die Stelle der Mengen (4. 1) und (4. 2) im vorangehenden Beweis treten, so ist der Beweis wörtlich derselbe wie oben.

§ 5.

Beweis von Satz 3. 1.

Statt A' wird die schwächere Voraussetzung zugrunde gelegt, daß K beschränkt ist und daß die Lösung $y(s)$, $s \geqq 0$, von

$$\frac{d^2 y}{d s^2} + K y = 0, \tag{5.1}$$

welche $y(0) = 0$, $y'(0) = 1$ erfüllt, für beliebige Halbstrahlen den Ungleichungen

$$\frac{y(s_2)}{y(s_1)} > C e^{c(s_2 - s_1)}, \qquad s_2 > s_1 > 0, \tag{5.2}$$

mit festen $C > 0$, $c > 0$ genügt. Sie reicht für die Gültigkeit aller Sätze von § 3 aus; nur dem Beweis von Satz 3. 2 wurde A' zugrunde gelegt. Wahrscheinlich ist aber auch er unter der schwächeren Bedingung richtig.

Wir ziehen zunächst einige einfache Folgerungen. Durch Integration von (5. 1) über $(0, s)$ folgt leicht

$$\left|\frac{y' - 1}{y}\right| < c_1 \tag{5.3}$$

auf $\tilde{\mathfrak{F}}$. Zwei verschiedene Lösungen von (5. 1) schneiden sich höchstens einmal. Es gibt eine und nur eine Lösung $z(s)$ von (5. 1), die den Randbedingungen $z(0) = 1$, $z(+\infty) = 0$ genügt. Man betrachte nämlich die eindeutig bestimmte Lösung $z(s; A)$ mit $z(0) = 1$, $z(A) = 0$. Wendet man (5. 2) auf den Halbstrahl an, der bei $s = A$ beginnt und die umgekehrte Richtung hat, so folgt

$$0 < z(s) \leqq \frac{1}{C} e^{-cs}, \quad s \geqq 0, \tag{5.4}$$

mit $z = z(s; A)$, $s < A$. Diese Größe nimmt nun bei festem s mit $A \to \infty$ monoton zu. Die Grenzfunktion existiert und erfüllt (5. 4). Offenbar ist $z(s)$ eindeutig bestimmt. z kann durch y ausgedrückt werden. Aus $zy' - z'y = \text{const} = 1$ folgt

$$z(s; A) = y(s) \int_s^A \frac{dt}{y^2(t)}, \quad z(s) = z(s; \infty). \tag{5. 5}$$

Weiter folgt $z'(A; A)\, y(A) = -1$ und daher

$$z'(0; A) = -\frac{1}{y(A)} + \int_0^A z(s; A)\, K\, ds. \tag{5. 6}$$

Nun hängt y stetig von s und dem Anfangselement des Halbstrahls ab. Also ist es auch bei $z(s; A)$ der Fall. Aus leicht ersichtlichen Gleichmäßigkeitseigenschaften der Konvergenz für $A \to \infty$ folgt Ähnliches für $z(s)$ und

$$z'(0) = \int_0^\infty z(s)\, K\, ds. \tag{5. 7}$$

Die von einem Punkte von $\tilde{\mathfrak{F}}$ ausgehenden Halbstrahlen bilden nun in $\tilde{\mathfrak{F}}$ ein Feld. In geodätischen Polarkoordinaten r = Entfernung vom Pol und ϑ = Winkel mit fester Richtung im Pol lautet das Bogenelement

$$ds^2 = dr^2 + y^2(r, \vartheta)\, d\vartheta^2, \tag{5. 8}$$

wo für $\vartheta = \text{const}$ das $y(r, \vartheta)$ die oben betrachtete Lösung von (5. 1) mit r statt s ist. Die Kreise schneiden die Radien senkrecht.

S sei ein Strahl, der die Radien schneidet. Der Schnittwinkel α = Winkel zwischen der positiven Richtung von S und der negativen Richtung des Radius genügt dann längs S der Gleichung

$$\frac{d\alpha}{ds} = \frac{y_r'}{y} \sin \alpha. \tag{5. 9}$$ [4]

---

[4]) Für das Bogenelement $ds^2 = du^2 + E dv^2$ ist nach Gauß längs einer Geodätischen G

$$\frac{d\alpha}{dv} = -\frac{\partial \sqrt{E}}{\partial u}.$$

Vgl. etwa Bianchi, Vorlesungen über Differentialgeometrie, S. 155. Ist ds das Bogenelement längs G, so wird wegen $\sqrt{E}\, dv = \sin \alpha\, ds$

$$\frac{d\alpha}{ds} = -\frac{1}{\sqrt{E}} \frac{\partial \sqrt{E}}{\partial u} \sin \alpha.$$

Diese Gleichung gilt für den Winkel α zwischen ds und der Richtung wachsender u auf $v = \text{const}$. Daher das umgekehrte Vorzeichen in (5, 9).

Beweis von Satz 3. 1. Wir erinnern an die am Anfang von § 3 gegebene Definition des Symbols p_t. H sei nun der gegebene Halbstrahl in $\mathfrak{F}$ mit dem Anfangspunkt p, p' der außerdem vorgegebene Punkt. $H(T)$ sei der Strahl, welcher p' mit dem Punkte p_T auf H verbindet. Wir betrachten p' als Anfangspunkt. $q^{\cdot}(T)$ sei auf $H\,(T)$ vor p_T derjenige Punkt, der von p_T ebensoweit entfernt ist wie p,

$$s\left(q\,(T), p_T\right) = s\,(p, p_T) = T.$$

Dann ist, $q = q\,(T)$,

$$\begin{aligned} s\,(p', q) &= \left|s\,(p', p_T) - s\,(q, p_T)\right| \\ &= \left|s\,(p', p_T) - s\,(p, p_T)\right| \leqq s\,(p, p'). \end{aligned} \tag{5.10}$$

Für $T \to \infty$ haben die Punkte q daher mindestens einen Häufungspunkt q'. Man darf $q' \neq p'$ annehmen, denn sonst könnte man statt p den Punkt p_s, $s > 0$ fest, auf H als Anfangspunkt von H betrachten, und die neuen Punkte q wären dann die Punkte q_s auf $H\,(T)$. Für eine passende Folge von Werten $T \to \infty$ gilt nun $q\,(T) \to q'$, und $H\,(T)$ konvergiert gegen den Halbstrahl H' von p' durch q' [5]). Gleichzeitig konvergiert $q\,(T)_t$ auf $H\,(T)$ gegen q'_t auf H'. Wir zeigen, daß H' zu H asymptotisch ist.

Wir können annehmen, daß p' nicht auf H oder seiner Verlängerung liegt, da sonst alles trivial wäre. Der Bogen $p\,p'$ schneidet dann H in p (und nur in p). Wir betrachten nun alle Strahlen von den Punkten dieses Bogens zum Hilfspunkt p_T auf H. Sie erfüllen bei p_T einen Winkelraum einfach. Wir betrachten ferner alle in diesem Raum enthaltenen geodätischen Kreisbögen um p_T und bezeichnen mit $l(t)$ die Länge desjenigen von ihnen, welcher im Punkte p_t auf H beginnt. Er endet in $q\,(T)_t$ auf $H\,(T)$. Betrachtet man p_T als Pol, so ist

$$l\,(t) = \int y\,(r, \vartheta)\, d\,\vartheta, \qquad r = T - t,$$

wo über jenen Winkelraum integriert wird. Wegen (5. 2) ist nun

$$l\,(t) < \frac{1}{C}\, e^{-c\,(t-t')}\, l\,(t'), \qquad 0 < t' < t. \tag{5.11}$$

Hieraus kann man leicht schließen, daß H' zu H asymptotisch ist, wenn gezeigt werden kann, daß $l\,(t')$ etwa für $t' = s\,(p, p')$ bei dem Grenzübergang $T \to \infty$ beschränkt bleibt. Verschiebt man p' längs $p\,p'$ und wendet man (5. 10) an, so sieht man, daß der besagte Bogen ganz im geodätischen Dreieck $p\,p'\,p_T$ liegt. Die Behauptung ist dann eine Folge des Lemmas:

Die Länge eines um eine Ecke eines geodätischen Dreiecks gezogenen und innerhalb desselben liegenden geodätischen Kreisbogens ist kleiner als C^{-1} mal der gegenüberliegenden Seite (C ist das C von (5. 2)).

---

[5]) Man sieht übrigens sofort, daß $\sphericalangle\, p\,p'\,p_T$ mit wachsendem T zunimmt und kleiner als π bleibt.

Der Beweis ergibt sich leicht mit Hilfe von (5. 2), wenn man den Bogen vermittels der von der Ecke ausgehenden Strahlen auf die Seite abbildet.

Anschließend beweisen wir (3. 2). Eine obere Schranke von $\sigma(P_{t+a}, P'_t)$ bei festem t erhält man in derselben Weise wie bei der Ableitung von (2. 2) durch Parallelverschiebung der ersten Richtung längs der geodätischen Strecke $p_{t+a}\, p'_t$ und eine nachfolgende Drehung. Der Winkel $\Delta\alpha$ dieser Drehung ist nicht größer als das Streckenintegral für den Winkel α mit der Asymptotenschar

$$\int \left| \frac{d\alpha}{ds} \right| ds.$$

Hier kann man vor dem obigen Grenzübergang $T \to \infty$ (5. 9) und (5. 3) anwenden. Nach erfolgtem $T \to \infty$ folgt dann

$$\sigma(P_{t+a}, P'_t) \leqq (1 + c_1)\, s\,(p_{t+a}, p'_t),$$

und damit (3. 2).

Es bleibt noch die Eindeutigkeit von H' zu beweisen. Sie folgt aus der schärferen Aussage: Haben zwei verschiedene Halbstrahlen H', H'' denselben Anfangspunkt $p' = p''$ und beziehen sich die Symbole p'_t bzw. p''_t auf H' bzw. H'', so wächst $s\,(p'_t, p''_t)$ für $t \to \infty$ über alle Grenzen[6]). Wäre dies nicht der Fall, so würde auf Grund des obigen Lemmas folgen, daß es geodätische Kreisbögen von H' nach H'' um p' gäbe, die bei beliebig großem Radius beschränkte Länge hätten. Dies würde der Voraussetzung (5. 2) widersprechen. Damit ist Satz 3. 1 vollständig bewiesen.

§ 6.

Beweis von Satz 3, 2.

(5. 9) gilt längs eines Strahles S in einem Felde von Halbstrahlen mit gemeinsamem Anfangspunkt. α war der Schnittwinkel mit der negativen Richtung des Halbstrahls. Ist A die jeweilige Distanz des Pols vom Schnittpunkt und setzt man $t = A - r$, so folgt

$$\left.\frac{y'_r}{y}\right|_{r=A} = \left.\frac{\partial}{\partial r} \frac{y(r, \vartheta)}{y(A, \vartheta)}\right|_{r=A} = -\, z'_t(t; A)\,|_{t=0}.$$

Verlegt man nun den Pol längs einem Feldstrahl rückwärts ins Unendliche, so wird aus (5. 9) formal

$$\frac{d\alpha}{ds} = -\, z'_0 \sin\alpha \tag{6. 1}$$

---

[6]) Bei der Anwendung auf die Eindeutigkeit berücksichtige man: Für zwei asymptotische Halbstrahlen mit gleichem Anfangspunkt ist notwendig $a = 0$.

für den Schnittwinkel mit der positiven Richtung in einem Felde positiv asymptotischer Strahlen; $z_0' = z'(0)$ wurde am Anfang von § 5 definiert und bezieht sich hier auf (5. 1) längs dem im Schnittpunkt beginnenden (positiven) Feldhalbstrahl. Die strenge Rechtfertigung des Grenzüberganges gelingt leicht bei der integrierten Form von (5. 9), wenn man die erwähnte Gleichmäßigkeit bezüglich des Anfangselementes in $z'(0; A) \to z'(0)$ berücksichtigt [7]).

Aus (5. 3) folgt

$$|z_0'| \leqq c_1 \tag{6. 2}$$

auf $\tilde{\mathfrak{F}}$. Ist die schärfere Voraussetzung A' erfüllt, so ist darüber hinaus

$$A': \ -z_0' \geqq c > 0 \tag{6. 3}$$

auf $\tilde{\tilde{\mathfrak{F}}}$.

Beweis von Satz 3. 2. Man kann die beiden gegebenen Halbstrahlen durch solche, H_1 und H_2, mit gemeinsamem Anfangspunkt p ersetzen. Nach Voraussetzung ist dann $H_1 \neq H_2$. Man betrachte einen Strahl S durch p derart, daß H_1 und H_2 auf verschiedenen Seiten von S liegen. Ferner betrachte man die beiden Asymptotenfelder, welchen H_1 bzw. H_2 angehören, und die Änderung der beiden Schnittwinkel α_1 und α_2 längs S, $\alpha_i = \alpha_i(s)$. Wegen (6. 1) und (6. 3) wächst α_i mit wachsendem s von $\alpha_i(-\infty) = 0$ bis $\alpha_i(+\infty) = \pi$. Genau einmal tritt daher $\alpha_1 + \alpha_2 = \pi$ ein. Dann ergänzen sich aber die beiden Feldhalbstrahlen zu einem Strahl mit den gewünschten Eigenschaften.

Wir wollen die Eindeutigkeit noch unter der schwächeren Bedingung beweisen. Sind zwei Strahlen S_1 und S_2 zueinander sowohl positiv wie negativ asymptotisch, so ist notwendig $S_1 = S_2$. Dies folgt aus dem folgenden Satz.

$b > 0$ sei beliebig vorgegeben. Zu jedem $\varepsilon > 0$ gibt es dann ein $L > 0$ mit folgender Eigenschaft. Sind B_1, B_2 zwei geodätische Bögen der Länge $> L$, deren Anfangspunkte ebenso wie ihre Endpunkte um weniger als b voneinander entfernt sind, so hat B_2 Punkte in der ε-Umgebung des Mittelpunktes von B_1.

Wie leicht zu sehen, kann man sich beim Beweise mit dem Fall eines gemeinsamen Anfangspunktes p begnügen. Die Endpunkte seien p_1 und p_2. Der geodätische Kreis vom Radius $l - b$, $l =$ Länge von B_1, um p wird durch B_1 und B_2 in zwei Bögen zerlegt, von denen einer ganz im Dreieck $p\, p_0\, p_1$ liegt. Seine Länge ist nach dem Lemma von § 5 kleiner als $C^{-1} b$. Die Länge des Bogens durch den Mittelpunkt von B_1 ist daher wegen (5. 11) kleiner als

$$C^{-2} b \exp\{-c(l - b - \tfrac{1}{2} l)\} < C^{-2} b \exp\{-c(\tfrac{1}{2} L - b)\}.$$

Daraus folgt der Satz.

[7]) Man berücksichtige ferner, daß bei dem Grenzübergang jedes α bei festem s sich monoton ändert und kleiner als π bleibt.

§ 7.

Beweis der Hilfssätze von § 3.

Wir betrachten noch einmal die in § 5 definierten Lösungen $z(s; A) = z(s, \alpha; A)$ und $z(s) = z(s, \alpha)$ von (5. 1) längs eines Halbstrahls mit festem Anfangspunkt, aber beliebigem Anfangswinkel α mit einer festen Richtung. Es war bereits bewiesen worden, daß $z'_0 = z'(0)$ eine auf $\mathfrak{F}$ stetige Funktion des Linienelementes ist. Es ist nun entscheidend für den Beweis der Hilfssätze, daß

$$\frac{d z'_0}{d \alpha}$$

existiert. Wir behaupten, daß diese Ableitung ebenfalls stetig vom Linienelement abhängt.

Bisher wurde nur von der zweimaligen stetigen Differenzierbarkeit von $\mathfrak{F}$ und allein von Vor. A') Gebrauch gemacht. Wir benötigen zum Beweis der Behauptung auch die dritten Ableitungen und Vor. B'). Daß

$$u = u(s, \alpha; A) = z'_\alpha(s, \alpha; A)$$

vorhanden ist und stetig von s, α abhängt, folgt aus klassischen Sätzen. u genügt der formal differenzierten Variationsgleichung

$$\frac{d^2 u}{d s^2} + K u = - K'_\alpha z(s, \alpha; A) \tag{7.1}$$

und verschwindet für $s = 0, A$. Nach (5. 8) kann unter Verwendung der Richtungsableitung normal zum Strahl $K'_\alpha = y(s, \alpha)\, dK/d\bar{s}$ geschrieben werden. Aus (5. 1), mit z statt y, und (7. 1) folgt

$$(u' z - u z')' = - \frac{d K}{d \bar{s}} z^2 y,$$

und durch Integration wegen der Randbedingungen für u und $z = z(s, \alpha; A)$

$$u'_s(0, \alpha; A) = \frac{\partial}{\partial \alpha} z'_s(0, \alpha; A) = \int_0^A \frac{d K}{d \bar{s}} z^2 y \, d s. \tag{7.2}$$

Zum Beweise unserer Behauptung genügt der Nachweis, daß der Integrand

$$\frac{d K}{d \bar{s}}(s, \alpha)\, z^2(s, \alpha; A)\, y(s, \alpha), \tag{7.3}$$

der für beliebiges festes A stetig von s, α abhängt, $0 \leqq s \leqq A$, in allen diesen Intervallen absolut unter einer festen integrablen Funktion von s allein liegt[8]). Die Behauptung folgt dann durch Grenzübergang $A \to \infty$. Nun

[8]) An der betreffenden Stelle in Hopf I war beim Beweise ein Versehen unterlaufen. Seine Berichtigung ist leicht, und ich nehme an, daß der Leser sie selbst ausführen konnte. — Bei dieser Gelegenheit sei noch bemerkt, daß der Passus in der dritten Zeile auf S. 293 falsch ist, aber ohne weiteres weggelassen werden kann.

ist nach (5. 5) die Wurzel $z y^{\frac{1}{2}}$ aus dem Produkt des zweiten und dritten Faktors gleich

$$y^{\frac{3}{2}}(s) \int_s^A \frac{d\,t}{y^2(t)} = y^{-\frac{1}{2}}(s) \int_s^A \left(\frac{y(s)}{y(t)}\right)^2 d\,t.$$

Wegen (5. 2) ist für $0 < s < A$ das Integral rechts kleiner als

$$C^{-2} \int_s^\infty e^{-2c(t-s)}\,d\,t = \frac{1}{2\,c\,C^2}.$$

Für $s > 1$ hat also (7. 3) die Majorante

$$\frac{C^*}{y(s)} = \frac{C^*}{y(1)} \cdot \frac{y(1)}{y(s)} < C^{**} e^{-cs}.$$

Für $s \leqq 1$ ist die gleichmäßige Beschränktheit von (7. 3) direkt zu erkennen. Aus alledem folgt

$$\frac{d\,z_0'}{d\,\alpha} = \int_0^\infty \frac{d\,K}{d\,s}\, z^2\, y\, d\,s,$$

$z = z(s, \alpha)$, und man sieht leicht, daß diese Größe eine stetige Funktion des Linienelementes ist.

Beweis von Hilfssatz 3. 4. Man verbinde beide Punkte durch eine geodätische Strecke S, $0 \leqq s \leqq s^*$. β, γ seien von S an gezählt. Der Schnittwinkel $\alpha(s)$ in dem jeweiligen Asymptotenfelde, welchem die beiden Halbstrahlen angehören, erfüllt längs S die Differentialgleichung

$$\frac{d\,\alpha}{d\,s} = -z_0' \sin\alpha, \qquad z_0' = f(s, \alpha).$$

Dabei ist $\alpha(0) = \beta$, $\alpha(s^*) = \gamma$. Nach obigem wissen wir, daß f und $\partial f/\partial\alpha$ stetige Funktionen von s, α sind. Nach einem wohlbekannten Satz ist also die Ableitung von $\alpha(s; \beta)$ nach dem Anfangswert β vorhanden und in s, β stetig. Damit ist der Hilfssatz vollständig bewiesen. Aus Symmetriegründen ist natürlich $d\gamma/d\beta \neq 0$.

Beweis von Hilfssatz 3. 3. Aus der vorangehenden Betrachtung folgt in Verbindung mit den Sätzen über stetige Abhängigkeit von Anfangswerten, daß $\varphi_1(s, \alpha)$, $\varphi_2(s, \alpha)$ stetige Funktionen sind. Ebenso $\alpha(s, \varphi_1)$ und $\alpha(s, \varphi_2)$.

$$\left.\frac{\partial\,\varphi_1}{\partial\,\alpha}\right|_s = \frac{1}{\left.\frac{\partial\,\alpha}{\partial\,\varphi_1}\right|_s}$$

(bei festem s) hängt nach dem vorangehend bewiesenen Satze stetig von s, φ_1, also auch stetig von s, α ab. Analoges gilt von φ_2. Nach dem Theorem über implizite Funktionen, angewandt auf $\alpha(\varphi_1, s) = \text{const}$, ist

$$\left.\frac{\partial \varphi_1}{\partial s}\right|_\alpha = -\frac{\left.\frac{\partial \alpha}{\partial s}\right|_{\varphi_1}}{\left.\frac{\partial \alpha}{\partial \varphi_1}\right|_s}.$$

Die Anwendung ist berechtigt, denn vom Nenner gilt Hilfssatz 3. 4, d. h. Stetigkeit in s, φ_1 und Nichtverschwinden, und der Zähler ist nach (6. 1) gleich $-z_0' \sin\alpha$, was ebenfalls stetig von s, φ_1 abhängt. Die linke Seite hängt damit stetig von s, α ab. Gleiches gilt von φ_2. Der erste Teil des Hilfssatzes ist also bewiesen. Die Funktionaldeterminante ist nun nach obigem gleich

$$\frac{\left.\frac{\partial \alpha}{\partial s}\right|_{\varphi_2} - \left.\frac{\partial \alpha}{\partial s}\right|_{\varphi_1}}{\left.\frac{\partial \alpha}{\partial \varphi_1}\right|_s \left.\frac{\partial \alpha}{\partial \varphi_2}\right|_s}.$$

Der Zähler ist nach (6. 1) gleich $\sin\alpha\,[(z_0')_1 + (z_0')_2]$; die Vorzeichenänderung rührt daher, daß bei dem Subtrahend im Zähler $\pi - \alpha$ der Winkel ist, von dem (6. 1) gilt. Das Verschwinden der eckigen Klammer ist aber unmöglich. Sonst hätte man nämlich eine Lösung $z(s)$ der Gleichung (5. 1) längs eines ganzen Strahles, welche den Bedingungen $z(\pm\infty) = 0$, $z(0) = 1$ genügen würde. Die Unmöglichkeit hiervon folgt leicht aus (5. 2), wenn man den Nullpunkt auf dem Strahl nach $-\infty$ verlegt. Damit ist der Hilfssatz bewiesen.

Beweis des Haupthilfssatzes 3. 5. Die Menge von Stromlinien hat dann und nur dann das Maß $m = 0$ in Ω, wenn es bei der Menge aller entsprechenden Stromlinien in $\widetilde{\Omega}$ der Fall ist. Oder: Wenn es bei derjenigen Teilmenge dieser Menge zutrifft, bei welcher die Strahlen einen festen Halbstrahl H_0 in $\widetilde{\mathfrak{F}}$ schneiden; denn man kann abzählbar viele H_0 so angeben, daß jeder Strahl in $\widetilde{\mathfrak{F}}$ mindestens einen von ihnen schneidet. Für festes H_0 sind die Schnittstrahlen durch s, α gekennzeichnet. Die Linienelemente auf ihnen lassen sich durch die Koordinaten τ, s, α charakterisieren, wo τ die Maßzahl der Strecke vom Schnittpunkt zum Trägerpunkt bedeutet. Man überlegt sich leicht, daß $dm = \sin\alpha\, d\alpha\, ds\, d\tau$ ist. Daraus ergibt sich: Für eine Menge der erwähnten Art ist dann und nur dann $m = 0$, wenn die Menge der Schnittelemente das Maß $\iint ds\, d\alpha = 0$ besitzt. Hieraus und aus Hilfssatz 3. 3 folgt der Haupthilfssatz.

Zusatz bei der Korrektur.

Beweis der Hauptsätze auf Grund der schwächeren Unstabilitätsbedingung (5. 2). Es braucht nur noch die Existenzbehauptung des Satzes 3. 2 aus dieser Bedingung hergeleitet zu werden. Man gehe wieder von zwei verschiedenen Halbstrahlen H', H'' auf $\mathfrak{F}$ mit gemeinsamem Anfangspunkt p_0 aus und betrachte die geodätische Verbindungsstrecke eines Punktes p' auf H' mit einem Punkte p'' auf H''. Zum Beweise der Behauptung genügt der Nachweis, daß für $p' \to \infty$ und $p'' \to \infty$ die Minimalentfernung h zwischen p_0 und $p'p''$ beschränkt bleibt.

Der Flächeninhalt des Dreiecks $p_0 p' p''$ ist wegen (5. 8) mit p_0 als Pol

$$F = \iint y(r, \vartheta)\, dr\, d\vartheta. \tag{1}$$

Es sei $h > 1$. Dann folgt mit Rücksicht auf (5. 2)

$$\int y\, dr > \int_1^h y\, dr > C y(1, \vartheta) \int_1^h e^{c(r-1)} dr,$$

also

$$F > C^* \vartheta_0 (e^{c(h-1)} - 1), \tag{2}$$

unter ϑ_0 den Winkel bei p_0 verstanden. (2) gilt natürlich auch für $h < 1$. Andererseits zerlege man das Dreieck durch die kürzeste Strecke von p_0 zur Strecke $p'p''$ in zwei Teildreiecke D', D''. Auf sie wende man (1) mit p' bzw. p'' als Pol an. Nach (5. 2) ist nun in jedem Falle

$$\int y\, dr < C^{**} \bar{y},$$

wo $\bar{y}$ der Wert im Endpunkt der Integrationsstrecke — der letztere liegt auf der gemeinsamen Seite der Teildreiecke — ist. Wegen $\bar{y}\, d\vartheta \leqq ds =$ Bogenelement auf jener Seite gilt von den Dreiecksinhalten

$$F' < C^{**} h, \quad F'' < C^{**} h.$$

Hieraus folgt in Verbindung mit (2) die Behauptung. Zugleich folgt, daß die Fläche des Dreiecks beschränkt bleibt.

Topologische Natur der Flächen $\mathfrak{F}$. Die geschlossenen Flächen der betrachteten Art stellen gestaltlich keine anderen Typen als die speziellen Flächen negativer Krümmung dar. Die curvatura integra einer Fläche $\mathfrak{F}$ ist nämlich negativ. Zum Beweise dieser Tatsache beachte man, daß die in § 5 und 6 betrachtete Größe z_0', die im Elementraum Ω eine stetige Funktion $w(P)$ darstellt ($w \neq 0$), längs einer Stromlinie die Riccatische Gleichung $\frac{dw}{dt} = -K - w^2$ befriedigt. Wegen

$$\int_\Omega w(\mathfrak{T}_t P)\, dm = \int_\Omega w(P)\, dm$$

ergibt sich

$$2\pi \int_{\mathfrak{F}} K\, do = \int_\Omega K\, dm = -\int_\Omega w^2 dm.$$

(Eingegangen am 19. 4. 1940.)

Statistik der geodätischen Linien in Mannigfaltigkeiten Negativer Krümmung, *Berlin Verh. Sachs. Akad., Wiss, Leipzig, Math-Nat.*, **K1, 261-304, (1939),**
and

Statistik der Lösungen geodätisher problem von unstabilen Typus II, *Mathematische Annalen*, **117, pp. 590-608, (1940)**

Commentary

Ya. G. Sinai

The first paper played an important role in the development of ergodic theory. There E. Hopf proved that for surfaces of negative curvature and finite area, the geodesic flow is ergodic and mixing. In the final part of the paper he proved ergodicity for geodesic flows on compact manifolds of constant negative curvature. The second paper can be considered as a natural continuation of the first. Hopf considered geodesic flows on surfaces whose curvature could be positive and negative. In addition, he imposed an important condition that for each geodesic, any interval of time which the geodesic spends in a domain of positive curvature is followed by a sufficiently long interval when the geodesic lies in a domain of negative curvature. In both papers E. Hopf used very intensively the notions of horocycles and horospheres which appeared earlier in the works of Morse (1924) and Hedlund (1939). In order to prove ergodicity Hopf invented the object which is now called Hopf's chain. He remarked that time averages of any continuous function take constant values along any stable or unstable horocycles or horospheres. Thus, in order to prove ergodicity at least locally, one has to connect any two points by a chain consisting of local pieces of stable and unstable horocycles or horospheres. Hopf understood that this argument works if stable and unstable horocycles satisfy the property which is now called absolute continuity of foliations (see the Encyclopedia of Mathematical Sciences). He proved this property in the two-dimensional case.

This work of Hopf was an inspiration for Smale and Anosov and other mathematicians in the studies of hyperbolic dynamical systems. Smale (1967) introduced an important class of so-called A-systems. Anosov introduced (1967) the class of dynamical systems which bear his name (see pages 257-263 of the encylopedia, which has the references to basic papers on the whole subject). Each Anosov system has so-called stable and unstable invariant foliations which are horocycle flows in the case of geodesic flows on surfaces of negative curvature. These geodesic flows are a particular and important example of Anosov systems. Anosov proved in full generality the ergodicity of Anosov systems which, in particular, gives ergodicity of geodesic flows on n-dimensional compact manifolds of negative curvature. In the work of Anosov and others, the concept of absolute continuity was deeply investigated and all the needed results for the validity of Hopf arguments were proven. Another important result of Anosov is his theorem about structure stability of Anosov systems. A similar theorem was proven in greater generality by Smale and is now considered a central result in chaos theory.

The second paper by Hopf had an interesting continuation in the theory of billards. L.A. Bunimovich proved (see again the references in the encyclopedia) that the motion of a billiard in a domain having the form of a stadium is ergodic. The behavior of trajectories here resembles the behavior outlined in the second paper by Hopf. Each trajectory spends some time in a round domain after which it has enough time to recover its instability. Later, V. Donnay and L. Bunimovich extended these results to a wider class of domains. Using similar ideas V. Donnay constructed examples of two-dimensional topological spheres carrying ergodic geodesic flows (see the most recent results in the papers by Donnay and Pugh (2001)).

References

ANOSOV, D., "Geodesic Flows on Closed Riemannian Manifolds of Negative Curvature," *Proc. Steklov Institute*, **vol. 90**, p.210, (1967).

DONNAY, V. AND C. PUGH, "Finite Horizon Riemann Structures and Ergodicity," Preprint, (2001).

DONNAY, V. AND C. PUGH, "Anosov Geodesic Flows for Embedded Surfaces," Preprint, (2001).

ENCYCLOPEDIA OF MATHEMATICAL SCIENCES, **vol. 100**, Springer-Verlag, p.459, (2000).

HEDLUND, G.A., "The Dynamics of Geodesic Flows," *Bull. Amer. Math. Soc.*, **45**, p.241, (1939).

MORSE, M., "A fundamental class of geodesics on any closed surface of genus greater than one," *Trans. Amer. Math. Soc.*, **26**, p.25-60, (1924).

SMALE, S., "Differentiable dynamical systems," *BAMS*, **73**, p.747-817, (1967).

Reprinted from the Proceedings of the NATIONAL ACADEMY OF SCIENCES,
Vol. 34, No. 2, pp. 47–51. February, 1948

*CLOSED SURFACES WITHOUT CONJUGATE POINTS**

BY EBERHARD HOPF

N.Y.U. INSTITUTE FOR MATHEMATICS AND MECHANICS

Communicated by Marston Morse, December 8, 1947

The present note contains the proof of the following

THEOREM. *Let S be a closed surface of class C''' in the sense of Riemannian geometry. If no geodesically conjugate points exist on S the total curvature of S must be negative or zero. In the latter case the Gaussian curvature must vanish everywhere on S.*

From the second part of the theorem and from known facts one infers: if S is the topological image of the torus or of the Klein bottle and if F contains no geodesically conjugate points S must be the one-to-one and isometric image of a Euclidean model of such a surface.

This second part of the theorem also forms the subject of a recent paper by Morse and Hedlund.[1] These authors prove the theorem under an additional hypothesis about S (non-existence of focal points on S) and raise the question if the theorem holds true without this restriction.

The idea underlying the proof for the first part of the theorem had been outlined in a previous paper by the author.[2] It was found that only little additional remarks are required to make the proof cover the entire theorem. In the present note the proof will be developed ab ovo.

Proof of the Theorem. Consider the Jacobi differential equation of normal variation

$$y''(s) + K(s)y(s) = 0 \tag{1}$$

(K = curvature) along an arbitrary oriented geodesic on S. The direction of increasing arc length is always chosen as to coincide with the direction of the geodesic. The solutions of (1) are well known to exist on the whole axis of s. Non-existence of conjugate points means: any non-trivial solution of (1) possesses at most one zero, any two non-identical solutions intersect at most once. Let

$$y(s;\ a,\ b)$$

be the (according to this intersection property unique) solution satisfying

$$y(a;\ a,\ b) = 1, \qquad y(b;\ a,\ b) = 0. \tag{2}$$

These functions satisfy the identity

$$y(s{:}\ a,\ b) = y(\alpha;\ a,\ b)y(s;\ \alpha,\ \beta) + y(\beta;\ a,\ b)y(s;\ \beta,\ \alpha). \tag{3}$$

Both sides represent solutions of (1) which, according to (2), intersect at $s = \alpha,\ \beta$ and which therefore coincide. In the special case $\alpha = a'$ and $\beta = b$ (3) becomes

$$y(s;\ a,\ b) = y(a';\ a,\ b)y(s;\ a',\ b). \tag{3'}$$

From (2) and from the intersection property we infer that

$$y(s;\ a,\ b') > 0 \text{ for } s < b' \text{ and } a < b'. \tag{4}$$

The two solutions $y(s;\ a,\ b)$ and $y(s;\ a,\ b')$, $b < b'$, intersect at $s = a$ but nowhere else. Hence, on account of (4),

$$y(s;\ a,\ b) \geqq y(s;\ a,\ b') \text{ for } s \leqq a < b < b'. \tag{4'}$$

(4) and (4′) imply the existence of the limit

$$y(s;\ a) = \lim_{b=+\infty} y(s;\ a,\ b) \tag{5}$$

at any $s \leqq a$.

If in (3) α, β are both chosen less than a it becomes obvious that (5) exists at any s and that $y(s;\ a)$ is a solution of (1). It is also inferred that

$$y'(s;\ a) = \lim_{b=+\infty} y'(s;\ a,\ b) \tag{5'}$$

holds at any s. From (2) and (4) we get

$$y(a;\ a) = 1,\ y(s;\ a) \geqq 0$$

for any s. As $y(s;\ a)$ is a solution of (1) we even have $y > 0$ everywhere. The function

$$u(s) = \frac{y'(s;\ a)}{y(s;\ a)}$$

which, according to (3′), does not depend on a is thus continuous at every s. It is a solution of the Riccati equation

$$u'(s) + u^2(s) + K(s) = 0. \tag{6}$$

In this manner, a perfectly well-defined function u is obtained on any oriented geodesic of S. The value of this function is perfectly well determined at every point of the geodesic and is obviously independent of the choice of the point where s is counted from. u is an everywhere continuous solution of (6) along the geodesic.

Let us note for later use that $u(s)$ is the limit of

$$\frac{y'(s;\ a,\ b)}{y\ (s;\ a,\ b)}$$

as $b \to \infty$. This fraction equals

$$\frac{Y'(s;\ b)}{Y\ (s;\ b)}$$

where $Y(s;\ b)$ is the solution of (1) satisfying $Y = 0,\ Y' = 1$ at $s = b$, and thus we may write

$$u(s) = \lim_{n=\infty} \frac{Y'_s(s;\ s+n)}{Y(s;\ s+n)}. \tag{7}$$

We next prove the boundedness of u for all geodesics on S. From the closedness of S one infers that

$$K > -A^2 \tag{8}$$

on S with a suitable constant $A > 0$. A Sturmian argument will show that

$$|u| \leqq A. \tag{9}$$

Let

$$z(s;\ a,\ b) = \frac{\sinh A(b-s)}{\sinh A(b-a)}$$

be the solution of

$$z''(s) - A^2 z(s) = 0 \tag{10}$$

satisfying

$$z(a;\ a,\ b) = 1, \qquad z(b;\ a,\ b) = 0.$$

From (1) and (10) we obtain

$$(zy' - yz')' = -(K + A^2)zy. \tag{11}$$

For $y = y(s;\ a,\ b)$ and $z = z(s;\ a,\ b)$, $a < b$, z and y are both positive for $s < b$. As $zy' - yz' = 0$ for $s = b$ we infer that $zy' - yz' > 0$ for $s < b$, i.e., that

$$u(s;\ b) = \frac{y'(s;\ a,\ b)}{y(s;\ a,\ b)} > \frac{z'(s;\ a,\ b)}{z(s;\ a,\ b)}$$

and therefore, on passing to the limit $b \to +\infty$, that

$$u(s) \geqq -A.$$

On the other hand, $y = y(s;\ a)$ and $z = z(s;\ a, c)$, $c < a$, are both positive for $s > c$, and it follows from (11) that

$$(zy' - yz')_s < (zy' - yz')_c = -y(c)z'(c) < 0 \text{ for } s > c.$$

We therefore have, for $s > c$,

$$u(s) = \frac{y'(s;\ a)}{y(s;\ a)} < \frac{z'(s;\ a, c)}{z(s;\ a, c)}$$

and finally, on letting $c \to -\infty$, $u \leqq A$.

We now consider the three-dimensional space Ω of the oriented line elements P situated on S. Let p denote the bearer point of P on S. Let, furthermore, P_t denote the line element obtained by moving P by arc length t tangentially along the geodesic through P. P_t depends continuously on P and t. For arbitrary fixed t, the transition from P to P_t is a one-to-one mapping of Ω into itself that is well known to leave the volume differential

$$dm = dod\varphi$$

invariant where do and $d\varphi$ denote the differentials of area and angle on S, respectively. Now, the function u constructed above is a bounded function of P, $u = u(P)$. For P fixed, $u(t) = u(P_t)$ is differentiable and satisfies the Riccati equation

$$\frac{du(P_t)}{dt} + u^2(P_t) + K(P_t) = 0 \qquad (K(P) = K(p)). \tag{12}$$

$u(P)$ is easily shown to be measurable (by a somewhat longer argument one can even prove its continuity). Equation (7) can be written

$$u(P) = \lim_{n=\infty} \left. \frac{\dfrac{dY_n(s;\ P)}{ds}}{Y_n(s;\ P)} \right/_{s=0}, \tag{13}$$

where $Y_n(s;\ P)$ is the solution of the Jacobi equation

$$Y''(s) + K(P_s)Y(s) = 0$$

that vanishes at P_n, i.e., at $s = n$, with the derivative one. It follows from the well-known continuity theorem for the solution of the initial value problem (which applies here as S is of class C''') that $Y_n(s;\ P)$ and its derivative with respect to s are continuous functions of P for s fixed. As $Y_n < 0$ for $s < n$ the fraction on the right in (13) depends continuously on P. As a limit function of a sequence of continuous functions $u(P)$ must be measurable. Together with the boundedness this implies the summability of $u(P)$ over Ω in the sense of the measure m.

From (12) we obtain on integrating with respect to t

$$u(P_1) - u(P) + \int_0^1 K(P_t)dt = \int_0^1 u^2(P_t)dt \tag{14}$$

for every P. Now, the invariance of the measure m under the mapping $P \rightarrow P_t$ implies the equality

$$\int_\Omega f(P_t)dm_P = \int_\Omega f(P)dm_P$$

for any summable function $f(P)$. By integration of (14) with respect to P and by using the last identity one obtains on the left-hand side

$$\int_\Omega [\int_0^1 K(P_t)dt]dm_p = \int_0^1 [\int_\Omega K(P_t)dm_P]dt = \int_\Omega K(P)dm$$
$$= 2\pi \int_S K(p)do.$$

The order of integration may be changed since $K(P_t)$ depends continuously on P and t. The right-hand side of (14) must be a summable function of P. The resulting equality

$$2\pi \int_S K(p)do = -\int_\Omega [\int_0^1 u^2(P_t)dt]dm_P$$

immediately proves the validity of the theorem. If the total curvature is zero the right-hand side of (14) must be zero for almost all P. For every such P, $u(P_t)$ must vanish at every t, $0 \leqq t \leqq 1$, because it depends continuously on t. From (12) one infers that $K(P) = 0$ for such a P. As a continuous function K must therefore vanish everywhere on S.

* The German original of this note had been dedicated and presented to C. Caratheodory on his seventieth birthday on September 13, 1943. Though accepted for subsequent publication it never appeared in print. Its loss by an air raid did not become known to the author until long after the end of the war.

1 Morse, M., and Hedlund, G. A., "Manifolds Without Conjugate Points," *Trans. Am. Math. Soc.*, **51**, 362–386 (1942).

2 Hopf, E., "Statistik der Loesungen geodaetischer Probleme vom unstabilen Typus," *Math. Annalen*, **117**, 590–608, in particular p. 608 (1940/41).

Closed Surfaces Without Conjugate Points, *Proc. Nat. Acad. Sci.*, **U.S.A. 34, pp. 47-51, (1948).**

Commentary

Ya. G. Sinai

In this paper, E. Hopf proved a remarkable theorem which states that on the two-dimensional torus any Riemannian metric without conjugate points is a flat metric. This proof uses some arguments from ergodic theory.

Extending Hopf's result to the multi-dimensional case was an open problem for many years. This was fully solved by D. Burago and S. Ivanov (1994).

REFERENCES

BURAGO, D. AND S. IVANOV, "Riemannian tori without conjugate points are flat," *Geom. Funct. Anal.*, **4**, n3, p.259-269, (1994).

Reprinted from Journal of Rational Mechanics and Analysis
Vol. 1, No. 1, January, 1952
Printed in U.S.A.

Statistical Hydromechanics and Functional Calculus

EBERHARD HOPF

Graduate Institute for Applied Mathematics, Indiana University, Bloomington, Indiana*

1. Introduction. The differential law which governs the motion of a deterministic mechanical system has the symbolic form

$$\frac{du}{dt} = \mathfrak{F}(u)$$

where u is an instantaneous phase of the system and where the right hand side is completely determined by the phase. We consider only the case in which $\mathfrak{F}$ does not contain t explicitly. In general the law completely determines the future history u^t of the system if its initial phase u^0 is exactly known at some moment of time t^0. There are mechanical systems the phases of which are characterized by a very large number of independent parameters and the phase motions of which are tremendously complicated. Two examples are the classical model of a gas with its very large number of degrees of freedom and the flow of a viscous incompressible fluid at a very large value of the overall Reynolds number. In both cases the important task is not the determination of the exact phase motion with an exactly given initial phase but the determination of the statistical properties of the "typical" phase motion. In order to achieve this goal statistical mechanics studies probability distributions of simultaneous phases and their evolution in time resulting from the individual phase motions. It is particularly interested in stationary phase distributions because only these can describe statistical equilibrium of the system. But not all of these stationary distributions are of importance. Statistical mechanics constructs certain "relevant" phase distributions which characterize the "typical" phase motions and which must be used for all statistical predictions. Classical statistics of a gas has reached this goal to a large extent. The relevant distributions are the canonical ones if one confines oneself to a scale in gas space on which the interactions of the

* This paper was written under an ONR research contract. Parts of it were presented in several lectures which the author gave during the last four years at New York University, Indiana University, and University of Michigan.

molecules are negligible, and it is just this scale on which the important macroscopic properties of a gas are based. On the other hand, in statistical hydromechanics—the theory of highly turbulent fluid flow—the small scale on which the "fluid elements" interact seems to be of decisive importance.* The relevant phase distributions based on this scale—the hypothetical Kolmogoroff distributions—must be mathematically very different from canonical distributions. So far all attempts to determine the relevant hydrodynamical phase distributions, at least to a sufficient degree of approximation, have met with considerable mathematical difficulties.† In 1940 I made a hitherto unpublished attempt to overcome these difficulties or, at least, to formulate the problem mathematically by introducing the characteristic functional or Fourier transform Φ of an arbitrary hydrodynamical phase distribution. I derived from the Navier-Stokes equations a functional differential equation which expresses $\partial\Phi/\partial t$ directly in terms of the first and second functional derivatives of Φ in the sense of Volterrá. The correlations of simultaneous velocities at n space points are n-th functional derivatives of Φ. Formally, the Φ-equation is completely equivalent to the infinite system of partial differential equations for these correlation functions, but it describes the differential evolution of a phase distribution in a more concise way. Of course, the mathematical level of the problem is raised into the domain of functional analysis, but this is natural and even unavoidable if one views the problem *in toto*. My attempts at finding the "relevant" solutious of the Φ-equation have been unsuccessful even in the simplest case of boundary-free flow, but I believe that the mathematical difficulties of the problem arise from the fact that it is unprecedented and not from any intrinsic complexity. I publish the mere beginnings of the theory with some hesitation but also with the hope that they will contribute to the ultimate solution of the problem. In order to deal with correlations of velocities at different times one needs a simple extension of the theory. This and other matters will be taken up in a second paper.

2. The phase space of a flow problem. We consider an incompressible fluid of constant density $\rho = 1$ and of viscosity μ which moves between given material walls. The boundary condition is that the fluid adheres to the walls or, in mathematical terms, that the velocity vector

$$u = (u_1, u_2, u_3)$$

* The analogy between fluid elements and molecules is a merely formal one. It is known that in many applications the size of the smallest eddies for highly turbulent flow is still very large in comparison to the molecular scale. In other words, for most purposes of statistical hydromechanics a fluid may be regarded as a strictly continuous medium. Statistical hydrodynamics is the theory of the "typical" solutions of the Navier-Stokes equations.

† We refer the reader to the following monograph in which the important recent advances in the statistical theory of turbulence—the theories of Kolmogoroff, Onsager, von Weizsäcker, Heisenberg, Batchelor and others—are described. *Les théories de la turbulence*, par L. Agostini et J. Bass. Publ. scientif. et techn. du ministère de l'air. Paris 1950.

of the fluid coincides on a wall with the velocity of that wall. In order to allow for strictly stationary or statistically stationary flows we assume that the relations which express the boundary conditions do not explicitly contain the time. This means that the boundary configuration remains the same geometrical whole at all times although the bounding walls may move, and that the velocity vector is at all times the same in every geometrical boundary point. Stationariness of the boundary data restricts, of course, the shape of the walls and also their type of motion if they move rigidly. At any rate, we find it convenient to confine ourselves to the following type of boundary condition. Time-independent values of u are prescribed on the boundary* of a region which does not change with time. We speak of a definite flow problem if its boundary data are definitely fixed. Such a definite flow problem has an infinity of possible solutions, and a solution will not be determined unless initial data are prescribed at a given moment of time.

Let R denote the region of x-space, $x = (x_1, x_2, x_3)$, which is occupied by the fluid at all times and let B denote its boundary. Any sufficiently smooth velocity field $u = u(x)$ defined in $R + B$ may be considered as a possible instantaneous state or phrase of the fluid provided that it is solenoidal in R,

$$\text{(1)} \qquad \operatorname{div} u \equiv u_{\alpha,\alpha} = 0,$$

and that it assumes the boundary values given on B. The basic equations of flow are the Navier-Stokes equations

$$\text{(2)} \qquad u_{\alpha,t} + u_\beta u_{\alpha,\beta} = -p_{,\alpha} + \mu u_{\alpha,\beta\beta} \qquad (\alpha = 1, 2, 3)$$

together with (1). The notation

$$f_{\alpha,\beta} = \frac{\partial f_\alpha}{\partial x_\beta}, \qquad f_{\alpha,\beta\gamma} = \frac{\partial^2 f_\alpha}{\partial x_\beta \partial x_\gamma}, \qquad f_{\alpha,t} = \frac{\partial f_\alpha}{\partial t},$$

is employed and also the usual summation convention is made use of. The differential system (2) + (1) if taken in conjunction with the boundary conditions (u given on B) uniquely determines the rate of change of a phase

$$\frac{\partial u(x, t)}{\partial t} = u_{,t} = (u_{1,t}, u_{2,t}, u_{3,t})$$

in terms of the instantaneous phase $u = u(x, t)$ at the same moment of time t. The idea is that one can write (2) in the form

$$u_{\alpha,t} + p_{,\alpha} = -u_\beta u_{\alpha,\beta} + \mu u_{\alpha,\beta\beta}$$

or vectorially,

$$u_{,t} + \operatorname{grad} p = -u_\beta u_{,\beta} + \mu u_{,\beta\beta},$$

* Of course, boundary values must be compatible with the condition of incompressibility.

where the right hand vector field is completely determined by the instantaneous phase u, and the problem is to represent a given vector field in R as a sum of a gradient field and a solenoidal field which assumes given values on B (we must have div $(u_{,t}) = (\text{div } u)_{,t} = 0$). This decomposition is well known to be unique.* In other words, the laws of flow plus the boundary conditions uniquely determine the pressure distribution as well as the rate of change of the phase in terms of the instantaneous phase $u = u(x, t)$. Thus they furnish the differential description of the time-succession of phases $u^t = u(x, t)$ in $R + B$. We presume that they determine this succession of phases uniquely† for all $t > 0$ once the initial phase $u^0 = u(x, 0)$ is given.

The possible phases $u = u(x)$ in $R + B$ may be regarded as points of a function space, the u-space or phase space associated with the flow problem. We denote it by Ω. The flow of the fluid in R carries a phase u, given at $t = 0$, into another well determined phase u^t at time t. This transition from a given phase to another one may be regarded as a functional operation T^t which is defined and single-valued in Ω,

$$u^t = T^t u, \qquad u^t = u(x, t), \qquad u = u(x) = u(x, 0).$$

For u fixed and t variable, $t \geq 0$, T^t describes the flow of the fluid in R which starts from the initial velocity field $u = u(x)$. For t fixed and u variable, u in Ω, T^t defines a transformation of the phase space into itself or part of itself. Under our assumption that the boundary data of the flow problem be stationary the operation which carries a phase at time t_0 into another one at time $t_0 + t$ is independent of t_0 and, therefore, identical with T^t. Hence the group T^t has the property that

$$T^t T^s = T^{t+s}, \qquad T^0 = \text{identity}.$$

Needless to say, for a definite flow problem the phase motions depend also upon the parameter μ, the viscosity of the fluid.

3. Phase distributions, averages and characteristic functionals. An instantaneous phase distribution (or ensemble) is completely characterized by its

* Actually the decomposition is already determined if only the normal component of the solenoidal field is given on B. This apparent overdeterminacy is but an instance of a phenomenon generally observed in mixed boundary and initial value problems for parabolic differential systems. The solution of this problem does exist in the hydrodynamical case for all times $t > 0$ and it does satisfy those apparently superfluous boundary conditions. Other methods have been used to prove the existence. See, for instance, my paper: *Über die Anfangswertaufgabe für die hydrodynamischen Grundgleichungen*, Math. Nachrichten **4** (1950), p. 213–231.

† Uniqueness has not yet been proved. The customary methods furnish this proof in the case of plane flow but, curiously enough, they fail to do so in the general three-dimensional case even with viscosity present. I find it hard to see something serious behind these mathematical difficulties. Moreover, it seems to me that the final conclusions of this paper—the Φ-equation—are not affected if uniqueness fails to hold.

probability distribution (or relative frequency distribution) in Ω. Let $P(A)$ denote the probability that the point of the phase distribution falls into (the relative frequency with which the ensemble occurs in) the part A of the phase space Ω. $P(A)$ is, mathematically speaking, a completely additive set function which is defined for practically all subsets $A, B, \cdots$ of Ω such that

$$P(A) \geq 0, \qquad P(\Omega) = 1.$$

Consider such a phase distribution at time $t = 0$. After elapse of time t all its individual phases u have moved into their new positions T_u^t. If $P(A)$ denotes the distribution at $t = 0$ then the distribution $P^t(B)$ at time $t > 0$ is evidently determined by the formula

$$P^t(B) = P^0(T^{-t}B), \qquad P^0(A) = P(A) \tag{3}$$

where $T^{-t}B$ denotes the subset of precisely those phases u for which T_u^t comes to lie within the part B of Ω. From now on we mean by a "phase distribution" a whole time succession P^t of probability distributions in Ω provided that they are interconnected by the laws of flow, i.e. by (3). Thus phase distributions are to be distinguished from instantaneous phase distributions.

A phase distribution is called stationary (ergodic theory calls this an invariant measure) if

$$P^t(A) = P(A)$$

holds for all t and for all parts A of Ω.

The average of a phase function $F(u)$ with respect to a given phase distribution is defined by the Lebesgue-Stieltjes-integral

$$\bar{F} = \int_\Omega F(u)P^t(du) \tag{4}$$

where $P^t(du)$ denotes the "differential" of the probability distribution P^t. Note and remember later on that the average is only defined with respect to a definite phase distribution. In general it changes with time t. If, however, the underlying phase distribution is stationary then the average of any phase function stays constant in time. Note also that the following identity holds,

$$\bar{F} = \int_\Omega F(u)P^t(du) = \int_\Omega F(T_u^t)P(du). \tag{5}$$

It is easily proved if one observes that, by virtue of (3), the Lebesgue-sums of approximation are the same for both integrals. It is important to grasp fully the scope of the concept of phase function. It is a law which attaches a (real or complex) value to every (or practically every) phase u. Remember that u is a symbol for an admissible vector field $u(x)$ defined in the part $R + B$ of x-space occupied by the fluid. A phase function is a functional whose argument is a whole vector field $u(x)$ in $R + B$ which satisfies (1) and the given boundary conditions. Examples of such phase functionals are

$$F(u) = \frac{1}{2}\int_R u^2\,dx = \frac{1}{2}\int_R u_\alpha u_\alpha\,dx,$$

which is the kinetic energy of the velocity field $u = u(x)$;

$$F(u) = \mu\int_R u_{\alpha,\beta}\,u_{\alpha,\beta}\,dx,$$

which is the energy dissipation associated with the solenoidal field $u(x)$;

$$F(u) = \int_R p_{,\alpha}\,p_{,\alpha}\,dx,$$

which is the integral of the square of the pressure gradient associated with the field. Remember here that the laws of flow uniquely determine the instantaneous pressure gradient in terms of the instantaneous phase $u = u(x)$. Simpler examples are

$$F(u) = u_\alpha(x^1).$$

which is the value of the α-th component of the field $u(x)$ in a given geometric point x^1 of R;

$$F(u) = u_\alpha(x^1)u_\beta(x^2)\ \cdots\ u_\omega(x^n);$$

and

$$F(u) = p_{,\alpha}(x^1).$$

Let

$$(y, u) = \int_R y\cdot u\,dx = \int_R y_\alpha u_\alpha\,dx$$

denote the scalar product of two real vector fields $y = y(x)$ and $u = u(x)$ and consider the expression

$$e^{i(y,u)} = \exp\,(i(y, u)) \tag{6}$$

where $y(x)$ is an arbitrary continuous field in $R + B$ while $u(x)$ is an arbitrary fluid phase in $R + B$ (admissible vector field). The average of this functional of with respect to a phase distribution $P^t(A)$ is denoted by

$$\Phi(y, t) = \overline{e^{i(y,u)}} = \int_\Omega e^{i(y,u)}P^t(du). \tag{7}$$

We call Φ which is a functional of an arbitrary vector field $y = y(x)$ in $R + B$

and a function of t the "characteristic functional" of the phase distribution P^t. For t fixed, say $t = 0$, it is the functional analogue of what probability theory calls the characteristic function of a finite-dimensional probability distribution $P(A)$. If P is a probability distribution in an n-dimensional u-space,

$$u = (u_1, u_2, \cdots, u_n),$$

then its characteristic function

$$\int e^{i(y,u)} P(du), \qquad (y, u) = y_\nu u_\nu,$$

is a function of an arbitrary set y of n numbers $y_1, y_2, \cdots, y_n$.

Since y is to be a real field

$$\Phi^*(y) = \Phi(-y)$$

must hold for the conjugate value.

(3) implies that

$$\Phi(y, t) = \int_\Omega \exp\,(i(y, T^t u))P(du). \tag{8}$$

The incompressibility (1) of the fluid mirrors itself in the following property of the characteristic functional:

$$\Phi(y + \operatorname{grad} \varphi) = \Phi(y) \tag{9}$$

holds for every single-valued scalar function $\varphi(X)$ in $R + B$ which vanishes on B.* This follows, upon using (1), from

$$\begin{aligned}(\operatorname{grad} \varphi, u) &= \int_R \varphi_{,\alpha}\, u_\alpha\, dx \\ &= \int_R (\varphi u_\alpha)_{,\alpha}\, dx = -\int_B \varphi u_n\, dS = 0.\end{aligned}$$

4. The functional derivatives of Φ. A function $\Phi(y)$ of a finite set of variables $y_1, \cdots, y_n$ is said to be differentiable for a particular value of the argument y

* We admitted mathematically very general values of u on the boundary B subject only to the condition

$$\int_B u_n\, dS = 0$$

(incompressibility). Hence (9) holds for all functions φ with a constant value on B. But if we restrict the boundary values of u to physically realizable ones — B consists of impenetrable walls, $u_n = 0$ — then (9) holds for all smooth functions in $R + B$ without restriction of ϕ on B.

if the change $\Delta\Phi$ resulting from a slight change $\delta y = (\delta y_1, \cdots, \delta y_n)$ of its argument is predominantly linear in δy,

$$\Delta\Phi = \delta\Phi + \eta, \qquad \delta\Phi = A_\nu \delta y_\nu$$

where $\eta/\sqrt{\delta y_\nu \delta y_\nu} \to 0$ as $\delta y \to 0$. The linear form $\delta\Phi$ is the differential or variation, and its coefficients A_ν are the partial derivatives of Φ,

$$A_\nu = \frac{\partial\Phi}{\partial y_\nu}, \qquad \delta\Phi = \frac{\partial\Phi}{\partial y_\nu}\,\delta y_\nu.$$

Suppose now that Φ is a functional whose argument is an arbitrary number-valued function of a single real variable x, $0 \leq x \leq 1$, $y = y(x)$. Φ is said to be differentiable (for a particular "value" of the argument, i.e. for a particular function $y = y(x)$) if its change $\Delta\Phi$ resulting from a slight change $\delta y = \delta y(x)$ of its argument $y(x)$ is predominantly linear in $\delta y(x)$, in other words, if

$$\Delta\Phi = \delta\Phi + \eta, \qquad \delta\Phi = \int_0^1 A(x)\,\delta y(x) dx$$

where

$$\eta \Big/ \sqrt{\int \delta^2 y(x) dx} \to 0 \text{ as the denominator} \to 0.^*$$

If we would strictly define the partial derivative by the coefficient of $\delta y(x)$ in the linear form $\delta\Phi$ this partial derivative would have to be not $A(x)$ but $A(x)dx$. It is customary, however, to call $A(x)$ itself the functional or Volterra derivative of Φ with respect to $y(x)$ for the value x†. Note that the point x is the analogue of the discrete index ν in the finite case. We choose the somewhat cumbersome but very suggestive notation

$$A(x) = \frac{\partial\Phi}{\partial y(x) dx}$$

* In the functional case there are also other types of linear forms and distances which may be used to define differentiability differently. The types chosen above are, however, those which are of import in our context.

† The rule to define the partial derivative as the coefficient of the variation of the argument in $\delta\Phi$ would no longer be violated by this choice if we look at the situation in the following way. The argument of Φ is a point function $y(x)$ but we may consider this as the density of a mass distribution (set function) on $0 \leq x \leq 1$. Φ then becomes a functional of an arbitrary mass distribution. If we denote by $Y(S)$ the mass contained in the subset S of the interval (0,1) the differential $\delta\Phi$ has the form

$$\int_0^1 A(x)\,\delta Y(dx).$$

Since now Y is the argument and δY its variation, $A(x)$ as defined by the rule mentioned above is actually the partial derivative with respect to its argument.

for the functional derivative. The relation

$$\delta\Phi = \int_0^1 \frac{\partial\Phi}{\partial y(x)dx}\,\delta y(x)dx$$

becomes then the complete analogue of the relation

$$\delta\Phi = \sum \frac{\partial\Phi}{\partial y_\nu}\,\delta y_\nu$$

of the discrete case.

The hydrodynamical functional $\Phi(y)$ introduced above is a functional of a vector field $y = y(x) = (y_1(x), y_2(x), y_3(x))$ defined in some region R of the three dimensional x-space. In this case the differential has the form

$$\delta\Phi = \int_R A_\alpha(x)\delta y_\alpha(x)dx \tag{10}$$

where

$$dx = dx_1\, dx_2\, dx_3$$

is the element of volume in x-space. $A_\alpha(x)$ is again called the functional derivative of Φ and we denote it by

$$A_\alpha(x) = \frac{\partial\Phi}{\partial y_\alpha(x)dx}. \tag{11}$$

Note that the present analogue of the discrete index ν is a pair (x, α) of a point x in R and a component-subscript $\alpha = 1, 2, 3$. Again we have

$$\delta\Phi = \int_R \frac{\partial\Phi}{\partial y_\alpha(x)dx}\,\delta y_\alpha(x)dx.$$

The first of the two functionals

$$\Phi = \int_R a_\alpha(x)y_\alpha(x)dx, \qquad \Phi = y_\alpha(x^1)$$

is differentiable with the derivative

$$\frac{\partial\Phi}{\partial y_\alpha(x)dx} = a_\alpha(x).$$

The second one is not differentiable because the differential has not the required integral form (10).

Since the first order derivative is again a functional of the field $y = y(x)$ one may consider second order derivatives

$$\frac{\partial}{\partial y_\beta(x')dx'}\left(\frac{\partial\Phi}{\partial y_\alpha(x)dx}\right) = \frac{\partial^2\Phi}{\partial y_\beta(x')dx'\partial y_\alpha(x)dx}.$$

Thus the second derivative of the first example above is zero. The quadratic functional

$$\Phi = \int_R\int_R K_{\alpha\beta}(x, x')y_\alpha(x)y_\beta(x')dxdx', \qquad K_{\alpha\beta}(x, x') = K_{\beta\alpha}(x', x),$$

has the derivatives

$$\frac{\partial\Phi}{\partial y_\alpha(x)dx} = 2\int_R K_{\alpha\beta}(x, x')y_\beta(x')dx'$$

and

$$\frac{\partial^2\Phi}{\partial y_\beta(x')dx'\ \partial y_\alpha(x)dx} = 2\ K_{\alpha\beta}(x, x')$$

while the third derivative is zero. The quadratic functional

$$\Phi = \int_R K_{\alpha\beta}(x)y_\alpha(x)y_\beta(x)dx, \qquad K_{\alpha\beta}(x) = K_{\beta\alpha}(x),$$

however, behaves differently. It has a first derivative

$$\frac{\partial\Phi}{\partial y_\alpha(x')dx'} = 2K_{\alpha\beta}(x')y_\beta(x')$$

but none of the second order.

The successive derivatives of the characteristic function $\Phi(y)$ of a probability distribution, evaluated at $y = 0$, are well known to furnish the successive moments or correlations of the distribution. The same is true in our hydrodynamical case where $\Phi = \Phi(y, t)$ is defined by (7). From

$$\begin{aligned}\delta\Phi &= \overline{\delta e^{i(y,u)}} = \overline{i(\delta y, u)e^{i(y,u)}}\\ &= i\overline{\int_R u_\alpha(x)\delta y_\alpha(x)dx\ e^{i(y,u)}}\\ &= i\int_R \overline{u_\alpha(x)e^{i(y,u)}}\ \delta y_\alpha(x)dx\end{aligned}$$

we infer that

$$\frac{\partial\Phi}{\partial y_\alpha(x)dx} = i\ \overline{u_\alpha(x)e^{i(y,u)}} \tag{12}$$

and that, generally,

$$\frac{\partial^n \Phi}{\partial y_\alpha(x^1)dx^1 \cdots \partial y_\omega(x^n)dx^n} = i^n \, \overline{u_\alpha(x^1) \cdots u_\omega(x^n)e^{i(y,u)}}. \tag{13}$$

If these derivatives are taken for the vanishing vector field $y = 0$ they evidently furnish the moments of the distribution or, in hydrodynamical language, the spatial correlation functions of the velocities

$$\overline{u_\alpha(x^1)u_\beta(x^2) \cdots u_\omega(x^n)}$$

which depend upon n "index pairs" (x^1, α), (x^2, β), $\cdots$, (x^n, ω). Needless to say that they are defined only with respect to a definite phase distribution P^t and that they vary in general with t.

These correlation functions must satisfy a number of obvious relations in consequence of incompressibility. However, we are not interested in the technique of the correlation functions as something *per se*. Our primary interest is in the functional Φ which embodies the statistical properties of fluid flow in a much more concise form than the complicated infinite set of those functions. Incompressibility is expressed by the property (9) of Φ. It can also be expressed by the differential relation

$$\frac{\partial}{\partial x_\beta}\left(\frac{\partial \Phi}{\partial y_\beta(x)dx}\right) \equiv 0. \tag{14}$$

This is an immediate consequence of (1) and (12). One might, by the way, forget the original definition (7) of Φ and one could then show that a functional Φ of an arbitrary field $y = y(x)$ in $R + B$ which has the property (9) must also have the property (14) and vice versa.

I want to add a remark which will have to be used again and again. The characteristic functional

$$\Phi(y) = \int_\Omega e^{i(y,u)}P(du)$$

of a probability distribution actually characterizes this distribution; in other words, there cannot be two essentially different probability distributions with the same characteristic functional Φ.*

5. The functional differential equation for Φ. To make this subject more clearly understood we begin with a simpler problem of the same sort. The hydrodynamical u-space is replaced by an n-dimensional u-space, $u = (u_1, \cdots, u_n)$, denoted also by Ω, and the differential law of the phase motion becomes an

* We cannot bother here with the precise formulation and proof of this uniqueness theorem which is well known in the finite-dimensional case.

ordinary differential system

$$\frac{du_\nu}{dt} = Q_\nu(u) = Q_\nu(u_1, \cdots, u_n), \qquad \nu = 1, \cdots, n. \tag{15}$$

In the hydrodynamical case the right hand sides of the differential equations of phase motion are functionals of second degree in $u = u(x)$. Correspondingly we assume the functions Q to be polynomials though not necessarily of second degree. It is assumed, too, that the solutions u^t of (15) starting from an arbitrary initial phase u^0 exist for all $t > 0$. Concepts such as T^tu, A, P^t and

$$\Phi(y, t) = \overline{e^{i(y,u)}} = \int_\Omega e^{i(y,u)} P^t(du), \qquad (y, u) = y_\nu u_\nu, \tag{16}$$

are now to be used in the corresponding sense. We assert that the characteristic function Φ of any phase distribution $P^t(A)$ satisfies the linear partial differential equation

$$\frac{\partial \Phi}{\partial t} = i \sum_\nu y_\nu Q_\nu \left(\frac{\partial}{i \partial y_1}, \cdots, \frac{\partial}{i \partial y_n} \right) \Phi \tag{17}$$

in which the expressions Q_ν on the right are to be interpreted as symbolic differential operators. For example, if

$$Q(u_1, u_2) = 2 + 3u_1 - u_2^2$$

then

$$Q\left(\frac{\partial}{i \partial y_1}, \frac{\partial}{i \partial y_2} \right) \Phi = 2\Phi - 3i \frac{\partial \Phi}{\partial y_1} + \frac{\partial^2 \Phi}{\partial y_2^2}.$$

The summation sign is not omitted in (17) because it makes the analogy to the integral sign obtained later in the hydrodynamical case more conspicuous.

The proof of (17) begins with the formula

$$\Phi(y, t) = \int_\Omega \exp\,(i(y, T^tu)) P(du).$$

By differentiation with respect to t,

$$\frac{\partial \Phi}{\partial t} = i \int_\Omega \left(y, \frac{dT^tu}{dt} \right) \exp\,(i(y, T^tu)) P(du). \tag{18}$$

T^tu describes a solution of (17). Hence

$$\left(y, \frac{dT^tu}{dt} \right) = y_\nu Q_\nu(T^tu)$$

and

$$\frac{\partial \Phi}{\partial t} = iy_\nu \int_\Omega Q_\nu(T^t u) \exp(i(y, T^t u))P(du).$$

By virtue of (5) this can be written

$$\frac{\partial \Phi}{\partial t} = iy_\nu \int_\Omega Q_\nu(u) \exp(i(y, u))P^t(du)$$

$$= iy_\nu \overline{Q_\nu(u)e^{i(y,u)}}.$$

Since the $Q(u)$ are polynomials the averages must satisfy the identities

$$\overline{Q_\nu(u)e^{i(y,u)}} \equiv Q_\nu\left(\frac{\partial}{i\partial y_1}, \cdots, \frac{\partial}{i\partial y_n}\right) \overline{e^{i(y,u)}}.$$

This concludes the proof of (17).

(17) is a linear differential equation of an order equal to the maximum degree of the polynomials Q. Its coefficients are linear functions of the independent variables y.

One can apply precisely the same argument to hydrodynamical phase motion $T^t u$. Naturally, the Φ-equation becomes in this case a linear functional differential equation of second order. We do not intend, however, to go on with the theory on this level of generality and so we will merely mention and describe the general Φ-equation. In the following sections we confine ourselves to the important and frequently considered limit case of boundary-free flow. The Φ-equation assumes its simplest form in this case and it will there be fully derived from the Navier-Stokes equations.

The characteristic functional of any phase distribution which moves in accordance with the laws of flow in the x-region R satisfies the equation

$$(19) \quad \frac{\partial \Phi}{\partial t} = \int_R y_\alpha(x) \left[i \frac{\partial}{\partial x_\beta} \frac{\partial^2 \Phi}{\partial y_\beta(x)dx\, \partial y_\alpha(x)dx} + \mu \Delta_x \frac{\partial \Phi}{\partial y_\alpha(x)dx} - \frac{\partial \Pi}{\partial x_\alpha} \right] dx,$$

$$\Delta_x = \partial^2/\partial x_\beta \partial x_\beta,$$

for every "value" of its argument, i.e. for every vector field $y = y(x)$ in $R + B$ and for every value of $t \geq 0$. The space-integration on the right is the analogue of the summation in (17). The three terms between the square brackets clearly arise from the inertia— and friction terms and from the pressure gradient. The essential terms are the functional derivatives of Φ of first and second order, the x-derivatives being merely operations on the "index-point" x. $\Pi = \Pi(y; x, t)$ is an auxiliary quantity which is a functional of the vector field $y = y(x)$ and a function of a point x and of t. It must be determined by the simultaneous functional Φ just as the pressure p is to be determined by the simultaneous velocity field $u(x)$. The determination of Π is this. The expression between the square brackets is the α-th component of a certain vector field, defined in the points x of R. Π must be chosen such that this field is solenoidal in R and that it assumes certain given boundary values on B.*

One can write the Φ-equation in several different forms just as one can do this with the Navier-Stokes equations. One way to eliminate the pressure-functional Π is to use the unique decomposition

$$(20) \qquad y(x) = \tilde{y}(x) + \operatorname{grad} \varphi, \qquad \operatorname{div} \tilde{y} = 0 \text{ in } R, \qquad \tilde{y}_n = 0 \text{ on } B,$$

of an arbitrary field $y(x)$ in $R + B$. One must think here of $\tilde{y}(x)$ as a field which is functionally well determined by the field $y(x)$.† It can then be shown that $\Phi(y, t)$ satisfies the differential equation

$$(21) \qquad \frac{\partial \Phi}{\partial t} = \int_R \tilde{y}_\alpha(x) \left[i \frac{\partial}{\partial x_\beta} \frac{\partial^2 \Phi}{\partial y_\beta(x) dx \, \partial y_\alpha(x) dx} + \mu \Delta_x \frac{\partial \Phi}{\partial y_\alpha(x) dx} \right] dx$$

for all fields $y(x)$ and for all $t \geq 0$. The functional relation (9) and the footnote attached to it show that Φ depends actually only on the field $\tilde{y}(x)$. The relation $\Phi = \Phi(\tilde{y}, t)$ is another expression for the condition of incompressibility. One could use it to bring the Φ-equation into a form where only $\tilde{y}$ appears.

The Φ-equation (19) or (21) furnishes a direct differential description of the evolution of a phase distribution which evolves in accordance with the laws of flow. Naturally, it is better adapted to the needs of statistical hydrodynamics than the equations of flow which describe only individual phase motions.

For a flow problem of regular type (entirely finite region R and sufficiently smooth boundary data) Φ must possess a regular Taylor expansion

$$\Phi = \Phi^0 + \Phi^1 + \Phi^2 + \cdots, \qquad \Phi^0 = 1,$$

where Φ^n is a homogeneous polynomial functional of degree n in $y = y(x)$,

$$\Phi^n = \int_R \cdots \int_R K_{\alpha \cdots \omega}(x^1, \cdots, x^n, t) \, y_\alpha(x^1) \cdots y_\omega(x^n) \, dx^1 \cdots dx^n$$

and where the kernel function K is given by

$$K_{\alpha \cdots \omega}(x^1, \cdots, x^n) = \frac{1}{n!} \frac{\partial^n \Phi}{\partial y_\alpha(x^1) dx^1 \cdots \partial y_\omega(x^n) dx^n} \bigg|_{y=0}.$$

* Under the boundary conditions admitted by us, u given on B independent of t, those boundary values are zero.

† The relation between y and $\tilde{y}$ has a simple geometric meaning if one interprets the totality of all vector fields y in $R + B$ as a Hilbert space H with the scalar product (y, y'). The gradient fields of single-valued functions in $R + B$ form a linear subspace G of H. The linear subspace S formed by the solenoidal fields with vanishing normal component on B is the orthogonal complement of G. $\tilde{y}$ is the projection of y upon S.

We know already that this function is just the correlation function

$$\frac{i^n}{n!}\overline{u_\alpha(x^1)\ \cdots\ u_\omega(x^n)}.$$

The Taylor expansion could have been derived equally well by using the power series for the exponential function in (7).

Upon inserting the power series of Φ into the Φ-equation (21) and upon equating terms of equal degree on both sides one obtains the infinite sequence of differential equations for the correlation functions,

$$\frac{\partial \Phi^n}{\partial t} = \int_R \tilde{y}_\alpha(x)\left[i\frac{\partial}{\partial x_\beta}\left(\frac{\partial^2\Phi^{n+1}}{\partial y_\beta(x)dx\ \partial y_\alpha(x)dx}\right) + \mu\frac{\partial^2}{\partial x_\beta\,\partial x_\beta}\left(\frac{\partial\Phi^n}{\partial y_\alpha(x)dx}\right)\right]dx.$$

Formally, there is of course a one-to-one correspondence between the solutions of the Φ-equation and the solutions Φ^1, Φ^2, $\cdots$ of the infinitely many correlation equations but, obviously, the Φ-equation describes the evolution of place distributions more concisely.*

Let us now briefly examine what uses we can expect of the Φ-equation. For this purpose we envisage a definite flow problem with definitely fixed stationary boundary data and where only the coefficient μ of viscosity is permitted to vary. If μ is sufficiently large there exists a single stationary solution (single point in u-space)—the laminar flow— $u = a(x, \mu)$ which is approached by any other solution as $t \to \infty$. At least, it is entirely within our present mathematical means to prove this rigorously.† If μ is very large the frictional term dominates on the right of the Φ-equation which then becomes of the first order

$$\frac{\partial \Phi}{\partial t} = \mu\int_R \tilde{y}_\alpha(x)\,\frac{\partial^2}{\partial x_\beta\,\partial x_\beta}\,\frac{\partial\Phi}{\partial y_\alpha(x)dx}\,dx.$$

It could be solved by the classical method of characteristics. We shall do this later in the extreme limit use of boundary-free flow. However, the well-explored case of very small Reynolds numbers is analytically simple enough and certainly does not call for a new mathematical technique. What happens if μ takes on smaller values? In exceptional cases of flow problems—for instance, if the bounding walls are all at rest, $u = 0$ on B—the character of general phase motion, namely convergence (as $t \to \infty$) towards a single stationary point in phase space, is preserved down to the smallest values of μ. For a "general" flow problem, how-

* Any attempt to use the correlation equations for purposes of approximation must overcome the difficulty that Φ^{n+1} occurs in the equation for $\partial\Phi^n/\partial t$. For small values of μ or, in other words, for highly turbulent flow, this term is of decisive importance. Besides, only the cases $n = 1, 2$ and at most, perhaps, $n = 3$ appear to be manageable. Therefore one is forced to make a good guess about Φ^{n+1}. This is, of course a perfectly natural procedure and recent progress in this direction is quite remarkable but the question remains how good an approximation one can obtain in this way.

† The stability proof has been known to me for about twelve years.

ever, the picture of phase motion becomes different as soon as μ decreases below a certain critical value.* The extreme case of a very small μ presents in general the following picture. The phase motions or, at least, a majority of them still converge towards something as $t \to \infty$. But this something is far from being a single stationary point in u-space. It must be pictured as a high-dimensional manifold of "typical" phase motions everyone of whicn comes in its course arbitrarily close to every point of this manifold and which all have the same statistical properties. The number of dimensions of this (ergodic) manifold of motions increases beyond limit as $\mu \to 0$ or, physically speaking, the infinitely many degrees of freedom of typical inviscid flow come more and more into play.† The manifold is the bearer of a certain stationary phase distribution $P^t \equiv P$. Its characteristic functional Φ is a stationary solution of the Φ-equation. For intermediate values of μ the determination of Φ is an exceedingly complicated task. It is only when $\mu \to 0$ that Φ should have a simpler asymptotic form. But even the solution of this problem, the problem of highly turbulent flow, is perhaps beyond our present reach. This problem is presumably governed by the case $\mu = 0$ of the Φ-equation (21),

$$\int_R \tilde{y}_\alpha(x) \frac{\partial}{\partial x_\beta} \left[\frac{\partial^2 \Phi}{\partial y_\beta(x) dx \ \partial y_\alpha(x) dx} \right] dx = 0.$$

As to simply setting $\mu = 0$ in the basic equations I may suggest that the problem of the limit passage in the solutions which is so involved in the case of individual flow should become easier if attention is shifted to stationary phase distributions. Note also that the functional differential equation (21) is of second order but that the frictional term contains only functional derivatives of first order which is the opposite of what happens in the individual case.

The presence of bounding walls is a highly complicating factor in all these problems and the statistical problem must become easiest in the extreme case of boundary-free flow. It is this simplest and purest form of the problem which looms in back of the present theories of turbulence. This problem will be considered in the following pages from the point of view of the Φ-equation.

6. Spatially periodic flow. Our subject is from now on the flow of a fluid which occupies the entire x-space. We first consider a very restricted type of such flow. Throughout this section it is supposed that

* I am appealing here to known observed facts and not to proven mathematical theorems on the stability of the laminar flow.

† The picture is described in more detail in the introduction of my paper: *A mathematical example displaying features of turbulence*, Communications on Pure and Applied Math. 1 (1948), pp. 303–322. I designed this example mainly to exhibit an actual mathematical case of such a continuously changing manifold of central phase motions with more and more dimensions. The terms of second degree in this example are, however, very different in character from the hydrodynamical inertia terms, and so the analogy to the hydrodynamical case does not extend much farther.

$$u(x + X) \equiv u(x)$$

where X is an arbitrary lattice vector

$$X = (n_1L, n_2L, n_3L)$$

L being a fixed number. This flow problem is, of course, equivalent to the problem in which the basic region R is a cube C of width L with the boundary condition that the velocities and their first derivatives have the same respective values in equivalent points on opposite faces.

The momentum theorem takes the form

$$\frac{d}{dt}\int_C u\,dx = P$$

where P is the resultant (vectorial) pressure force exerted upon the fluid within C. The initial value problem—to find $u(x, t)$ if $u(x, 0)$ is given—is not a determinate one unless P or, equivalently, the momentum $\int_C u\,dx$ is prescribed for all times $t \geq 0$. We confine ourselves to the case where

$$\int_C u\,dx = 0, \tag{22}$$

in other words, we assume that the fluid is at rest in the mean. The Fourier series for u is

$$u(x) = \sum_k v(k)e^{ik\cdot x} \, * \tag{23}$$

where the vector k runs through all the lattice vectors

$$k = \left(n_1\frac{2\pi}{L}, n_2\frac{2\pi}{L}, n_3\frac{2\pi}{L}\right) \tag{24}$$

of the Fourier space or k-space. $v(k)$ is a vector with complex-valued components v_α. It may be regarded as a vector field which is defined not in the entire k-space but only in the lattice (24) of this space. On denoting by z^* the conjugate of a complex number z (or of a complex vector z) we must have

$$v^*(k) = v(-k). \tag{25}$$

The condition of incompressibility takes the form

$$k\cdot v(k) = k_\alpha v_\alpha(k) = 0. \tag{26}$$

* It is frequently used in turbulence theory. The advantage is that x-differentiations go over into simple algebraic operations.

(22) means that

$$v(0) = 0.$$

Also, $v(k) \to 0$ sufficiently rapidly as $|k| \to \infty$. Upon applying (23) and on eliminating the pressure p by means of (26) the Navier-Stokes-equations are readily seen to go over into ($v(k) = v(k, t)$)

$$\text{(27)} \quad \frac{\partial v_\alpha(k)}{\partial t} = -i \sum_{k'+k''=k} [k''_\beta \cdot v_\beta(k')] \left[v_\alpha(k'') - \frac{k_\gamma v_\gamma(k'')}{|k|^2} k_\alpha \right] - \mu |k|^2 v_\alpha(k)$$

or, vectorially,

$$\text{(28)} \quad \frac{\partial v(k)}{\partial t} = -i \sum_{k'+k''=k} [k'' \cdot v(k')] \left[v(k'') - \frac{k \cdot v(k'')}{|k|^2} k \right] - \mu |k|^2 v(k).$$

For later use we note here another form of (27) obtained by introducing an arbitrary vector $z(k)$. We define

$$\text{(29)} \quad \bar{z}(k) = z(k) - \frac{k \cdot z(k)}{|k|^2} k$$

which is the unique vector of the form $z(k) - \lambda(k)$, λ complex, that is perpendicular to the vector k.* On using the identity

$$z(k) \cdot \left[v(k'') - \frac{k \cdot v(k'')}{|k|^2} k \right] = \bar{z}(k) \cdot v(k'')$$

one readily obtains the following form of (27)

$$\text{(29)} \quad \bar{z}_\alpha(k) \frac{\partial v_\alpha(\cdot k)}{\partial t} = i \sum_{k'+k''=k} [\bar{z}_\alpha(k) v_\alpha(-k'')][k''_\beta v_\beta(-k')] - \mu |k|^2 z_\alpha(k) v_\alpha(-k).$$

A phase of the fluid is now characterized by the infinitely many variables $v_\alpha(k)$ or by a discrete vector field $v = v(k)$ which satisfies the conditions named above, in particular (26). The phase space or v-space is again denoted by Ω. General notation such as $T^t v$ = phase motion, A = part of Ω, P^t = moving phase distribution, $f(v)$ = phase function is used as defined above.

$$\text{(30)} \quad \bar{f} = \int_\Omega f(v)\, P^t(dv)$$

---

* The relation between the fields z and $\bar{z}$ in k-space in the precise Fourier-analogue of the relation (20) between the fields y and $\bar{y}$ in x-space.

denotes the average of a phase functional $f(v)$ with respect to the distribution P^t. Again we have

$$\bar{f} = \int_\Omega f(T^t v) P(dv). \tag{31}$$

In analogy to the former arbitrary real vector field $y = y(x)$ we now introduce an arbitrary complex-valued vector field $z = z(k)$ which is defined only in the lattice points (24) of k-space. Two restrictions only are placed on this field, first, that

$$z^*(k) = z(-k) \tag{32}$$

and, secondly, that $z(k)$ vanishes sufficiently rapidly at $k = \infty$. The scalar product of two such fields is defined by

$$\begin{aligned} [z, v] &= \sum_{k'} z(k') \cdot v^*(k') = \sum_{k'} z_\alpha(k') v_\alpha^*(k') \\ &= \sum_{k'} z(k') \cdot v(-k') = \sum_{k'} z_\alpha(k') v_\alpha(-k'). \end{aligned} \tag{33}$$

The characteristic function of a phase distribution P^t is then given by

$$\Phi(z(k), t) = \overline{e^{i[z,v]}} = \int_\Omega e^{i[z,v]} P^t(dv). \tag{34}$$

Obviously, $[z, v]^* = [z, v]$ and hence

$$\Phi^*(z) = \Phi(-z).$$

Since in the present case Φ is actually a function of the discrete infinity of variables $z_\alpha(k)$ there is no need yet to introduce functional derivatives, and derivatives are understood in the ordinary sense. The property of Φ which expresses incompressibility is now that

$$\Phi(z(k) + \varphi(k)k) = \Phi(z(k)) \tag{35}$$

holds for every field z and for every complex-valued scalar $\varphi(k)$, $\varphi^*(k) = \varphi(-k)$. This follows immediately from (34), (33) and (26). Its differential equivalent is now that

$$k_\alpha \frac{\partial \Phi}{\partial z_\alpha(k)} = 0 \tag{35'}$$

holds for every field z and for every vector k.* The z-derivatives of Φ are obtained from (34) and (33).

---

* The equivalence of (35) and (35′) is readily proved by using the formula

$$\frac{d}{d\tau} \Phi(z(k) + \tau\varphi(k)k) \bigg|_{\tau=0} = \sum_k k_\alpha \frac{\partial \Phi}{\partial z_\alpha(k)} \varphi(k).$$

$$(36)\qquad \frac{\partial \Phi}{\partial z_\alpha(k)} = i\int_\Omega v_\alpha(-k)e^{i[z,v]}P^t(dv) = \overline{iv_\alpha(-k)e^{i[z,v]}}$$

and

$$(37)\qquad \frac{\partial^2 \Phi}{\partial z_\beta(k')\partial z_\alpha(k'')} = -\int_\Omega v_\alpha(-k'')v_\beta(-k')e^{i[z,v]}P^t(dv)$$
$$= -\overline{v_\alpha(-k'')v_\beta(-k')e^{i[z,v]}}.$$

On the other hand, (31) implies that

$$(38)\qquad \Phi = \int_\Omega \exp\,(i[z,\, T^t v])P(dv).$$

Hence

$$(39)\qquad \frac{\partial \Phi}{\partial t} = i\int_\Omega \left[z,\, \frac{dT^t v}{dt}\right] \exp\,(i[z,\, T^t v])P(dv)\,.$$

Since T^t describes a phase motion $T^t v = v(k,\, t)$ and by virtue of (33),

$$(40)\qquad \left[z,\, \frac{dT^t v}{dt}\right] = \sum_k z_\alpha(k)\,\frac{\partial v_\alpha(-k,\, t)}{\partial t}.$$

Hence and from the equations (29′) of phase motion $v(k) = v(k,\, t)$

$$(41)\qquad \frac{\partial \Phi}{\partial t} = -\sum_{k'+k''=k}\sum \tilde{z}_\alpha(k)k''_\beta \int_\Omega v_\alpha(-k'',\, t)v_\beta(-k',\, t)\,\exp\,(i[z,\, T^t v])P(dv)$$
$$-\, i\mu \sum_k |\,k\,|^2\, z_\alpha(k) \int_\Omega v_\alpha(-k,\, t)\,\exp\,(i[z,\, T^t v])\; P(dv).$$

The double sum can just as well be taken over all pairs $(k',\, k'')$ of lattice points if in each term k is replaced by $k' + k''$. All the phase integrals have the form (31) and may therefore be written in the form (30). By virtue of (36) and (37), (41) now becomes

$$(42)\qquad \frac{\partial \Phi}{\partial t} = \sum_{k'}\sum_{k''} z_\alpha(k' + k'')k''_\beta\,\frac{\partial^2 \Phi}{\partial z_\alpha(k'')\partial z_\beta(k')} - \mu \sum_k |\,k\,|^2\, z_\alpha(k)\,\frac{\partial \Phi}{\partial z_\alpha(k)}$$

where

$$(42')\qquad \tilde{z}(k) = z(k) - \frac{k\cdot z(k)}{|\,k\,|^2}\,k.$$

(42) can be written

$$(43)\qquad \frac{\partial \Phi}{\partial t} = \sum_{k'} \frac{\partial}{\partial z_\beta(k')}\left[\sum_{k''} \tilde{z}_\alpha(k' + k'')k''_\beta\,\frac{\partial \Phi}{\partial z_\alpha(k'')}\right] - \mu \sum_k |\,k\,|^2\, z_\alpha(k)\,\frac{\partial \Phi}{\partial z_\alpha(k)}\,.$$

After completion of the limit passage $L \to \infty$ it will turn out that (43) stays meaningful but not (42). The proof of (43) rests upon the fact that

$$\frac{\partial \tilde{z}_\alpha(k + k')}{\partial z_\beta(k')}$$

can be different from zero only if $k'' = 0$ in which case, however, no contribution is made by this term.

(42) is, of course, the special instance of the general Φ-equation (21) and (20) for the case of spatially periodic flow.*

The connection between the two forms $y(x)$ and $z(k)$ of the arbitrary vector field is given by the Fourier-relation

$$y(x) = \frac{1}{L^3} \sum_k z(k) e^{ik \cdot x}. \tag{44}$$

The same relation obtains between the fields $\tilde{y}$ and $\tilde{z}$. (23) and (44) imply that

$$(y, u) = [z, v].$$

7. General boundary-free flow. We must envisage a much broader class of flows in the entire x-space. Some sort of condition at infinity must, however, be imposed in order to insure reasonable behaviour of the flow. This is shown by the solutions

$$u_i = a_{i\nu} x_\nu, \qquad p = -(\dot{a}_{ik} + a_{i\alpha} a_{\alpha k}) x_i x_k$$

of the equations of flow where the a_{ik} are entirely arbitrary functions of t subject to the conditions

$$a_{ik} = a_{ki}, \qquad a_{\alpha\alpha} = 0.$$

Such a solution can become infinite at a finite moment of time. It can also start from the state of rest with $u = p = 0$ without staying at rest. Presumably no such things happen if a condition of the following sort is imposed: u is of smaller order of magnitude than $|x|$ at infinity, spatial averages of certain flow quantities taken over larger and larger spheres have limits. We assume that the average velocity (in the sense of that spatial average) is zero. A clear mathematical understanding of this important flow problem—it is the basis of the present theories of free turbulence—is a matter of great interest. In these theories, boundary-free flow is conceived of as the limit case of spatially periodic flow if the period L becomes infinite. We will proceed along these lines in order to find the correct form of the Φ-equation for general boundary-free flow. The viscosity is kept constant during the limit passage $L \to \infty$. As $L \to \infty$ the cubic

* The decomposition (20) is in this case determined by the following conditions: grad φ is to be periodic but φ need not be periodic. $\tilde{y}(x)$ is to be periodic.

lattice (24) becomes increasingly fine in k-space and the discrete vector field $z(k)$ which is defined in the lattice points only is made to go over into a continuous field $z(k)$ in k-space.

If one attempts the limit passage in the Fourier-form of the equations (27) of individual flow one encounters the difficulty that the limit form of the sum of the right is not an ordinary integral.* This does not concern us, however, since we are not interested in problems of individual flow.

The situation is more fortunate in the statistical theory and the limit form of the Φ-equation (46) as $L \to \infty$ is found in the following way. Φ is a function of an arbitrary discrete vector field $z(k)$ which is defined only in the lattice points (24). One may interpret this field as a stepwise constant vector field in the whole of k-space by defining the value of $z(k')$ constant in each lattice cube

$$n_\nu \frac{2\pi}{L} \leq k'_\nu < (n_\nu + 1)\frac{2\pi}{L}, \qquad \nu = 1, 2, 3$$

and equal to the vector $z(k)$ in the corner point

$$k = \left(n_1 \frac{2\pi}{L}, n_2 \frac{2\pi}{L}, n_3 \frac{2\pi}{L}\right).$$

With this interpretation Φ becomes a functional of a stepwise constant field $z(k)$ and one infers that

$$\begin{aligned} \delta\Phi &= \int_{k'} \frac{\partial \Phi}{\partial z_\alpha(k')dk'} \delta z_\alpha(k')dk' \\ &= \sum_k \left[\int_{\text{cube}} \frac{\partial \Phi}{\partial z_\alpha(k')dk'} dk'\right] \delta z_\alpha(k) \qquad (dk' = dk'_1\, dk'_2\, dk'_3) \end{aligned}$$

* If the vector

$$w(k) = L^3 v(k) = \int_C u(x) e^{ik\cdot x} dx$$

is introduced instead of $v(k)$ then the right hand sum in the equation for $\partial w(k)/\partial t$ has just the right factor $L^{-3} \sim$ volume of the lattice cube in k-space to make this sum go over into

$$\int_{k'+k''=k} (w(k')\cdot k'')\left(w(k'') - \frac{k\cdot w(k'')}{k^2}\, k\right) dk'.$$

This procedure is justified if (and, in general, only if) the limit field $u(x)$ vanishes sufficiently rapidly at infinity to make the limit integral

$$w(k) = \int_\infty u(x) e^{ik\cdot x} dx$$

convergent. Flows with $u(\infty) = 0$, however, are much too restricted to be of interest for statistical hydrodynamics.

where the sum is taken over all lattice points k and where the cube is the one with this lattice point as corner point. The cube-integral is approximately equal to

$$\frac{\partial \Phi}{\partial z_\alpha(k)dk} \Delta k$$

where $\Delta k = (2\pi/L)^3$ is the volume of the cube. We must expect, therefore, that for $L \to \infty$

$$\frac{\partial \Phi}{\partial z_\alpha(k)} \sim \frac{\partial \Phi}{\partial z_\alpha(k)dk} \Delta k \tag{45}$$

holds where the left represents the partial derivative in the discrete case whereas the right is Δk times the functional derivative of the limit functional Φ. Upon application of (45) to the differential equation (43) the right hand sums become at once approximating sums of integrals and the limit form of (43) is plainly the functional differential equation of second order

$$\begin{aligned} \frac{\partial \Phi}{\partial t} = \int_{k'} \frac{\partial}{\partial z_\beta(k')dk'} \left[\int_{k''} \tilde{z}_\alpha(k' + k'') k''_\beta \frac{\partial \Phi}{\partial z_\alpha(k'')dk''} dk'' \right] dk' \\ - \mu \int | k |^2 z_\alpha(k) \frac{\partial \Phi}{\partial z_\alpha(k)dk} dk \end{aligned} \tag{46}$$

where $\tilde{z}$ is given by (42′) or (ϵ_{ik} = Kronecker symbol)

$$z_\alpha(k) = \left(\epsilon_{\alpha\beta} - \frac{k_\alpha k_\beta}{k^2} \right) z_\beta(k). \tag{47}$$

Φ must satisfy (46) for every complex-valued vector field $z(k)$ in k-space for which

$$z^*(k) \equiv z(-k)$$

and which vanishes sufficiently rapidly at $k = \infty$. Incompressibility means that

$$\Phi = \Phi(\tilde{z}(k), t) .$$

Solutions Φ are admitted only if they have the following property. At any given moment of time t Φ is the characteristic functional of a probability distribution in v-space. This property implies, for instance, that

$$\Phi^*(z) = \Phi(-z)$$

and that

$$\Phi(0) = 1, \qquad | \Phi(z) | \leq 1.$$

8. Φ and the double correlations. Homogeneous phase distributions. For spatially periodic flow Φ is a function of a discrete vector field $z(k)$. The terms of its Taylor series

$$\Phi = 1 + \Phi^1 + \Phi^2 + \cdots \tag{48}$$

have the form

$$\Phi^n = \sum \cdots \sum G_{\alpha \cdots \omega}(k^1, \cdots, k^n; t) z_\alpha(k^1) \cdots z_\omega(k^n) \tag{49}$$

where the sum is taken over all systems $(k^1, \cdots, k^n)$ of n in dependent lattice points and where the coefficients G have the values

$$\frac{1}{n!} \frac{\partial^n \Phi}{\partial z_\alpha(k^1) \cdots \partial z_\omega(k^n)}\bigg|_{z=0} = \frac{i^n}{n!} \overline{v_\alpha^*(k^1) \cdots v_\omega^*(k^n)}. \tag{50}$$

In the present limit case $L = \infty$ one might expect Φ^n to have the form

$$\int \cdots \int g_{\alpha \cdots \omega}(k^1, \cdots, k^n) z_\alpha(k^1) \cdots z_\omega(k^n)\, dk^1 \cdots dk^n \tag{51}$$

where the integral is an n-fold k-integral and where the coefficient-function g has the value

$$\frac{1}{n!} \frac{\partial^n \Phi}{\partial z_\alpha(k^1) dk^1 \cdots \partial z_\omega(k^n) dk^n}\bigg|_{z=0} \tag{52}$$

It turns out, however, that (51) is too restricted a form of the n-th degree polynomial Φ^n. Sufficient generality is attained if the differential element of (51)

$$g_{\alpha \cdots \omega}\, dk^1 \cdots dk^n$$

is replaced by the element of a mass distribution in the $3n$-dimensional $(k^1, \cdots, k^n)$-space which need not have a density g in this space. In the case of homogeneous turbulence (see farther below) this mass distribution is actually concentrated on the $3(n - 1)$-dimensional subspace

$$k^1 + \cdots + k^n = 0$$

and the functional derivatives (52) are Dirac functions of the n points k with their singularity on that subspace. Therefore it is important that we first establish a sufficiently general formula which expresses Φ^n in terms of the derivatives of Φ and which stays valid in that case. Only the case $n = 2$ will be dealt with. The case $n = 1$ will turn out to be trivial.

Let $H_{\alpha\beta}(k, k')$ denote a function of two independent and continuously varying points k, k' in k-space, $\alpha, \beta = 1, 2, 3$. We assume that these functions behave, particularly at infinity, in such a way that the following sums and integrals converge. We wish to establish the limit form of the sum

$$\sum_k \sum_{k'} \overline{v_\alpha^*(k) v_\beta^*(k')} H_{\alpha\beta}(k, k') \tag{53}$$

as $L \to \infty$. By virtue of (34), (32) and (33), this sum equals

$$-\sum_k \sum_{k'} \frac{\partial^2 \Phi}{\partial z_\alpha(k) \partial z_\beta(k')}\bigg|_{z=0} H_{\alpha\beta}(k, k'). \tag{53'}$$

It can be written

$$-\sum_k \frac{\partial}{\partial z_\alpha(k)} \left[\sum_{k'} \frac{\partial \Phi}{\partial z_\beta(k')} H_{\alpha\beta}(k, k') \right] \bigg|_{z=0}.$$

On application of (45) to both derivatives this double sum evidently becomes an approximation sum for the integral expression

$$-\int_k \frac{\partial}{\partial z_\alpha(k) dk} \left[\int_{k'} \frac{\partial \Phi}{\partial z_\beta(k') dk'} H_{\alpha\beta}(k, k')\, dk' \right] dk \bigg|_{z=0}. \tag{54}$$

Therefore

$$(53) = (53') \to (54) \qquad \text{as } L \to \infty. \tag{55}$$

Two applications of (55) are made. First we identify (53) with

$$\Phi^2 = -\frac{1}{2!} \sum_k \sum_{k'} \overline{v_\alpha^*(k) v_\beta^*(k')}\, z_\alpha(k) z_\beta(k').$$

On writing $w(k)$ for the argument-field in Φ we obtain the limit form

$$\Phi^2(z) = \frac{1}{2!} \int_k \frac{\partial}{\partial w_\alpha(k) dk} \left[\int_{k'} \frac{\partial \Phi}{\partial w_\beta(k') dk'} z_\alpha(k) z_\beta(k')\, dk' \right] dk \bigg|_{w=0} \tag{56}$$

of Φ^2 as $L \to \infty$. Secondly we identify (53) with

$$\overline{u_\rho(x) u_\sigma(x')} = \sum_k \sum_{k'} \overline{v_\alpha^*(k) v_\beta^*(k')} \epsilon_{\alpha\rho}\, \epsilon_{\beta\rho}\, e^{-i(k\cdot x + k' \cdot x')}$$

and obtain

$$\overline{u_\rho(x) u_\sigma(x')} = -\int_k \frac{\partial}{\partial z_\rho(k) dk} \left[\int_{k'} \frac{\partial \Phi}{\partial z_\sigma(k') dk'} e^{-i(k\cdot x + k'\cdot x')}\, dk' \right] dk \bigg|_{z=0}, \tag{57}$$

which relates the double correlations of the velocities to the functional $\Phi = \Phi(z, t)$.

An instantaneous phase distribution $P(A)$ in u-space is called homogeneous if it is invariant under arbitrary translations of the velocity fields. The precise definition is the following. We interpreted the solenoidal vector fields $u = u(x)$ as points of a u-space. The operation which consists in replacing a field $u(x)$ by the translated one, $u_a = u(x + a)$, a being a fixed vector, may be interpreted as a one-to-one transformation of u-space into itself. A probability distribution $P(A)$ in u-space,

$$P(A) = \text{Prob}\,[u \text{ in } A],$$

A being an arbitrary part of this space, is called homogeneous if

$$\text{Prob}\,[u_a \text{ in } A] = \text{Prob}\,[u \text{ in } A] = P(A)$$

holds for every fixed vector a and for every part A of u-space. Homogeneity of a phase distribution is equivalently expressed by the invariance of its characteristic functional $\Phi = \Phi(y(x))$ under arbitrary translations of the argument-field $y(x)$,

$$\Phi(y_a(x)) \equiv \Phi(y(x)). \tag{58}$$

Indeed, (6) and (7) imply that

$$\Phi(y_a) = \int \exp\,(i(y_a, u))P(du) = \int \exp\,(i(y, u_{-a}))P\,du$$
$$= \int \exp\,(i(y, u))P(du) = \Phi(y),$$

since the probability differential $P(du)$ remains invariant under the transformation $u \to u_a$ of the u-space. The converse implication (58) $\to$ invariance of the phase distribution follows by application of the uniqueness theorem mentioned at the end of section 5.

If Φ is expressed in terms of the argument-field $z(k)$ in k-space, $\Phi = \Phi(z) = \Phi(z(k))$, homogeneity of the phase distribution means that

$$\Phi(e^{ia\cdot k}z(k)) \equiv \Phi(z(k)) \tag{59}$$

holds for arbitrary real vectors a and arbitrary fields z. This follows from (44).

If a phase distribution is homogeneous each term Φ^n of the Taylor series for Φ must have the same property of invariance (59). Concerning the integral representation (Stieljtes-integral!) (51) of Φ^n this means that the mass distribution is concentrated upon the subspace

$$k^1 + \cdots + k^n = 0.$$

For instance, Φ^1 must have the form $-iU_\alpha z_\alpha(0)$, where U is the average velocity of the fluid, which we supposed to be zero. Hence

$$\Phi^1 = 0. \tag{60}$$

Φ^2 must be of the form

$$\Phi^2 = -\tfrac{1}{2}\int \Gamma_{\alpha\beta}(k)z_\alpha(k)z_\beta(-k)\,dk, \qquad \Gamma_{\alpha\beta}(k) = \Gamma_{\beta\alpha}(-k), \tag{61}$$

or, in virtue of the incompressibility, $\Phi = \Phi(\tilde{z})$,

$$\Phi^2 = -\tfrac{1}{2} \int f_{\nu\mu}(k)\tilde{z}_\nu(k)\tilde{z}_\mu(-k)\, dk, \qquad f_{\nu\mu}(k) = f_{\mu\nu}(-k). \tag{62}$$

The relation between Γ and f is obtained from (47),

$$\Gamma_{\alpha\beta}(k) = \left(\epsilon_{\alpha\nu} - \frac{k_\alpha k_\nu}{|k|^2}\right)\left(\epsilon_{\beta\mu} - \frac{k_\beta k_\mu}{|k|^2}\right) f_{\nu\mu}(k). \tag{63}$$

From the original definition of Φ, for instance from (7), it is obvious that

$$\Phi^2 = -\frac{1}{2}\overline{(y, u)^2}$$

can have only real and non-positive values. Reality means that

$$\Gamma^*_{\alpha\beta}(k) = \Gamma_{\alpha\beta}(-k) = \Gamma_{\beta\alpha}(k).$$

Non-positivity of (61) implies that the Hermitean form

$$\Gamma_{\alpha\beta}(k)\xi_\alpha\xi^*_\beta$$

is $\geqq 0$ for every complex vector and at every point k. The connection between Γ or f and the double velocity correlations can readily be established by inserting $\Phi = 1 + \Phi_2 + \cdots$ into (57). Only Φ^2 makes a contribution, and the result is the relation

$$\overline{u_\rho(x)u_\sigma(x')} = \int \Gamma_{\rho\sigma}(k)e^{ik\cdot(x'-x)}\, dk. \tag{64}$$

We repeat the main result which we obtained for homogeneous phase distributions with $\bar{u} = 0$. If the characteristic functional $\Phi = \Phi(z, t)$ is known then the second degree term Φ^2 of the Taylor series

$$\Phi = 1 + \Phi^2 + \Phi^3 + \cdots$$

can be computed from it by means of the formula (56). Φ must have the form (61) or (62). After the functions f and Γ are determined the double velocity correlations are obtained from (64).*

9. Isotropic phase distributions. An instantaneous phase distribution P—again we interpret it first in u-space—is called isotropic if it is invariant not only under any simultaneous translation but also under any simultaneous rotation or reflexion of all the velocity fields $u(x)$. The precise definition is quite analogous to the one for homogeneity.

A functional $\Phi = \Phi(y)$ will be called isotropic if it is invariant under arbitrary

* The functions Γ are the components of the spectral tensor of the double velocity correlation well known in the theory of homogeneous turbulence.

translations, rotations and reflexions of the argument-field. We can now assert that a phase distribution is isotropic if and only if its characteristic functional $\Phi(y)$ is isotropic.

To prove this consider a fixed operation of that kind. Let $u' = u'(x)$ be the field obtained by applying this operation to an arbitrary field $u = u(x)$ and denote by $y^{\backprime}(x)$ the field obtained by applying the inverse operation to an arbitrary field $y(x)$. The transition $u \to u'$ represents again a one-to-one transformation of the u-space into itself which leaves the probability distribution $P(A)$ invariant. From the identity

$$(y^{\backprime}, u) = (y, u')$$

and from (7) it follows exactly as in the homogeneous case that $\Phi(y^{\backprime}) = \Phi(y)$. Since $y \to y^{\backprime}$ can represent any operation of that kind Φ must be isotropic. The converse implication is proved by following the preceding proof backwards and by using the fact that the characteristic functional uniquely characterizes the probability distribution P.

We expressed Φ in terms of vector fields $z(k)$ in k-space. The correct relation between the fields $z(k)$ and $y(x)$ is obtained from (44) by proceeding to the limit $L \to \infty$,

$$y(x) = (2\pi)^{-3} \int z(k) e^{ik \cdot x} \, dk.$$

From this one readily concludes: The necessary and sufficient condition that Φ be isotropic is that $\Phi = \Phi(z)$ satisfies (59) and that $\Phi(z)$ be invariant under arbitrary rotations of $z(k)$ about $k = 0$ and under reflexions at arbitrary planes through $k = 0$.

The restrictions which isotropy imposes upon the form of double velocity correlations and of their spectral tensor are well known. Isotropy of Φ implies isotropy of every term of its Taylor series $1 + \Phi^2 + \cdots$. One can readily show that the most general Φ^2 which is both incompressible and isotropic is given by (62) where

$$f_{\nu\mu}(k) = \epsilon_{\nu\mu} f(|\, k \,|).$$

In other words,

$$\Phi^2 = -\frac{1}{2} \int f(|\, k \,|) \bar{z}(k) \bar{z}(-k) \, dk = -\frac{1}{2} \int f(|\, k \,|) \, |\, \bar{z}(k) \,|^2 \, dk. \tag{65}$$

If this expressed in terms of z,

$$\bar{z}(k) = z(k) - \frac{k \cdot z(k)}{k^2} \, k,$$

Φ^2 is obtained in the form (61). Hence

$$\Gamma_{\alpha\beta}(k) = \left(\epsilon_{\alpha\beta} - \frac{k_\alpha k_\beta}{k^2}\right) f(|\, k\,|). \tag{66}$$

Since the Hermitean form

$$\Gamma_{\alpha\beta}(k)\xi_\alpha \xi_\beta^*$$

is positively definite,

$$\Gamma_{\alpha\alpha}(k) = 2f(|\, k\,|)$$

is $\geqq 0$ for every k. (64) and (66) imply that

$$\frac{1}{2}\overline{u(x)\cdot u(x')} = \int f(|\, k\,|)e^{ik\cdot(x-x')}\, dk \tag{67}$$

and, in particular, that the average kinetic energy per unit mass equals

$$\frac{1}{2}\overline{u^2(x)} = \int f(|\, k\,|)\, dk = 4\pi \int_0^\infty f(|\, k\,|)\,|\, k\,|^2\, d\,|\, k\,|. \tag{67'}$$

The indefinite integral

$$\int f(|\, k\,|)\, dk$$

taken over an arbitrary part of k-space is called the amount of average energy contained in that part. $f(|\, k\,|)$ therefore represents the density of average energy in k-space. (64) and (66) also imply that

$$\frac{1}{2}\overline{(u(x') - u(x))^2} = 2\int f(|\, k\,|)(1 - e^{-ik\cdot(x'-x)})\, dk \tag{68}$$

and that the average dissipation of energy per unit mass ϵ equals

$$\epsilon \equiv \overline{\mu u_{\alpha,\beta}\, u_{\alpha,\beta}} = 2\mu \int f(|\, k\,|)k^2\, dk = 8\pi\mu \int_0^\infty f\,|\, k\,|^4\, d\,|\, k\,|. \tag{69}$$

The integrals (67) and (68) can, of course, be transformed into one-dimensional integrals.

10. Some exact stationary solutions of the Φ-equation in the case $\mu = 0$. We first transform the Φ-equation (48) by using $\tilde{z}(k)$ instead of $z(k)$ as the argument-field. Remember that Φ must be a functional of $\tilde{z}$ only. $\tilde{z}(k)$ is however not an arbitrary field with the property $\tilde{z}^*(k) \equiv \tilde{z}(-k)$ since it must satisfy the orthogonality relation

$$k\cdot\tilde{z}(k) = 0. \tag{70}$$

But this relation is characteristic for fields of the form

$$\tilde{z}(k) = z(k) - \frac{k \cdot z(k)}{k^2} k$$

because every field $w(k)$ which satisfies the relation $k \cdot w(k) = 0$ has the property that $w(k) = \tilde{w}(k)$ holds. In order to perform the transformation we must first imagine Φ as a functional of an entirely arbitrary field $\tilde{z}(k)$ which not necessarily satisfies (70). From

$$\delta\Phi = \int \frac{\partial\Phi}{\partial\tilde{z}_\nu(k)dk} \delta\tilde{z}_\nu(k)dk$$

and from

$$\delta\tilde{z}_\nu(k) = \left(\epsilon_{\nu\alpha} - \frac{k_\nu k_\alpha}{k^2}\right) \delta z_\alpha(k)$$

we infer that

$$\frac{\partial}{\partial z_\alpha(k)dk} = \left(\epsilon_{\nu\alpha} - \frac{k_\nu k_\alpha}{k^2}\right) \frac{\partial}{\partial\tilde{z}_\nu(k)dk}.$$

If this is applied to the Φ-equation (46) the viscosity term becomes

$$-\mu \int k^2 \tilde{z}_\alpha(k) \frac{\partial\Phi}{\partial\tilde{z}_\alpha(k)dk} dk \tag{71}$$

and the inertia term is found to be equal to

$$\int_{k'} \frac{\partial}{\partial\tilde{z}_\alpha(k')dk'} \left[\int_{k''} \tilde{z}_\lambda(k' + k'')K_{\alpha\beta\lambda} \frac{\partial\Phi}{\partial\tilde{z}_\beta(k'')dk''} dk''\right] dk', \tag{72}$$

where

$$K_{\alpha\beta\lambda} = \left(k''_\alpha - \frac{k' \cdot k''}{k'^2} k'_\alpha\right)\left(\epsilon_{\beta\lambda} - \frac{k''_\beta k''_\lambda}{k''^2}\right). \tag{72'}$$

Hence the Φ-equation becomes

$$\frac{\partial\Phi}{\partial t} = (72) + (71), \tag{73}$$

which equation is to be satisfied not for all fields $\tilde{z}(k)$ but only for those which satisfy (70).

We use this form of the Φ-equation to find some stationary solutions in the case $\mu = 0$ or, in other words, to find solutions Φ of the differential equation

$$(72) = 0.$$

First it is shown that an arbitrary quadratic functional

$$\Phi = \int F_{\alpha\beta}(k)\tilde{z}_\alpha(k)\tilde{z}_\beta(-k)\,dk, \qquad F_{\alpha\beta}(k) = F_{\beta\alpha}(-k), \tag{74}$$

is such a solution.

Proof. By virtue of

$$\frac{\partial\Phi}{\partial\tilde{z}_\beta(k'')dk''} = 2F_{\nu\beta}(-k'')\tilde{z}_\nu(-k'')$$

the square bracket in (72) becomes

$$\int_{k''}\left(k''_\alpha - \frac{k'\cdot k''}{k'^2}k'_\alpha\right)\tilde{z}_\lambda(k'+k'')\tilde{z}_\nu(-k'')\cdots dk''.$$

The derivative

$$\frac{\partial}{\partial\tilde{z}_\alpha(k')dk'}$$

of this quadratic functional consists of two terms which are obtained by differentiating the two $\tilde{z}$-factors separately. The first factor contributes only if $k'' = 0$ whereas a contribution of the second one is possible only if $k'' = -k'$. But in both cases

$$k''_\alpha - \frac{k'\cdot k''}{k'^2}k'_\alpha$$

vanishes. Consequently, the expression after the first integral sign in (72) vanishes. Incidentally, it does this for entirely arbitrary fields $\tilde{z}$ and not only for those that satisfy (70).

This solution is however without interest to us because it cannot be a characteristic functional of a phase distribution. A polynomial functional is unbounded whereas a characteristic functional must satisfy the inequality $|\Phi| \leqq 1$.

It is shown next that

$$\Phi = W\left(\int \tilde{z}(k)\cdot\tilde{z}(-k)\,dk\right) = W\left(\int \tilde{z}_\alpha(k)\tilde{z}_\alpha(-k)\,dk\right) \tag{75}$$

is a solution of (72) $= 0$ if $W(\xi)$, $\xi \geqq 0$, is an arbitrary function of a single variable which possesses a continuous second derivative.

Proof. By virtue of

$$\frac{\partial\Phi}{\partial_{\sim\beta}(k'')dk''} = 2W'\tilde{z}_\beta(-k'')$$

and of (70) the square bracket of (72) becomes

$$2W' \int_{k''} z_\lambda(k' + k'')\bar{z}_\lambda(-k'') \left(k''_\alpha - \frac{k' \cdot k''}{k'^2} k'_\alpha\right) dk''. \tag{76}$$

Its functional derivative

$$\frac{\partial}{\partial \bar{z}_\alpha(k')dk'} \tag{76'}$$

is the derivative of the product of two functionals, namely $2W'$ and the integral of (76). The derivative of the integral-factor vanishes. In fact, this derivative is but a special case of a similar derivative dealt with in the preceding example. The derivative (76′) of the first factor $2W'$ is obtained just as the one of $\Phi = W$. Hence (76′) of (76) equals $4W''z_\alpha(-k')$ times the integral of (76), and (72) equals $4W''$ times

$$\iint \bar{z}_\alpha(-k')\bar{z}_\lambda(-k'')\bar{z}_\lambda(k' + k'') \left(k''_\alpha - \frac{k' \cdot k''}{k'^2} k'_\alpha\right) dk'\, dk'.$$

The term

$$-\frac{k' \cdot k''}{k'^2} k'_\alpha$$

may be omitted by virtue of (70). After changing the signs of k' and k'' the integral in question becomes

$$-\iint_{k+k'+k''=0} (\bar{z}(k') \cdot k'')(\bar{z}(k) \cdot \bar{z}(k''))\, dV, \tag{77}$$

where

$$dV = dk\, dk' = dk\, dk'' = dk'\, dk''$$

is the volume element of the six-dimensional subspace $k + k' + k'' = 0$ of nine-dimensional (k, k', k'')-space. The value of (77) is invariant under interchange of k and k''. If these two equal integrals are added the integrand is found to contain the factor

$$\bar{z}(k') \cdot k'' + \bar{z}(k') \cdot k = -\bar{z}(k') \cdot k' = 0.$$

This concludes the proof.

The isotropic solutions (75) are not without interest. For instance,

$$\Phi = \exp\left(-\frac{\kappa^2}{2} \int \bar{z}(k) \cdot \bar{z}(-k)\, dk\right) \tag{78}$$

can be shown to be the characteristic functional of an isotropic Gaussian distribution of solenoidal vector fields. (78) is the particular distribution of this sort in which flows in disjoint regions of x-space are as independent as is possible under the condition of incompressibility. Comparing the second form of the Taylor series for (78),

$$-\frac{\kappa^2}{2}\int \bar{z}(k)\cdot\bar{z}(-k)\,dk,$$

with (65) we see that the spectral energy density $f(|k|)$ has the constant value κ^2. By the way, the same is true for the distributions corresponding to the general functionals (75). Hence (75) leads to equipartition of average energy in k-space. The Gaussian case (78) is the perfect analogue to the classical canonical distribution of the phases of a gas. (75) represents an arbitrary combination of microcanonical distributions.

The characteristic functional of a general homogeneous Gauss-distribution of solenoidal vector fields with $\bar{u} = 0$ can be shown to have the form

$$\exp\left(-\frac{1}{2}\int f_{\alpha\beta}(k)\bar{z}_\alpha(k)\bar{z}_\beta(-k)\,dk\right)^*$$

where

$$f_{\alpha\beta}(k) = f_{\beta\alpha}(-k) = f^*_{\beta\alpha}(k).$$

We have not determined all functionals of this type which are stationary solutions of the case $\mu = 0$ of the Φ-equation.† There must be axisymmetric solutions which correspond to two-dimensional flow. However, one can easily prove that (78) is the only three-dimensional isotropic solution of Gaussian type. This remark will be used later.

11. The basic problem of isotropic turbulence. Current theories of isotropic turbulence presuppose and suggest that stationary isotropic phase distributions actually exist even with viscosity present, $\mu > 0$. Their characteristic functionals $\Phi(z)$ are stationary solutions of the Φ-equation (46), $\mu > 0$‡. Stationariness is not incompatible with the loss of kinetic energy due to viscosity. It obviously implies that the average energy per unit mass (67′) be infinite. There is one exceptional stationary solution, namely $\Phi = 1$. The corresponding phase distribution consists of the single stationary phase $u = v = 0$ (state of rest). The probability $P^t(A) = P(A)$ equals one if the phase region contains this phase $u = v = 0$ and zero if it does not do so. Presumably there exist other stationary isotropic solutions of (46), $\mu > 0$. Among them there should be certain "relevant"

* More generally, a Stieltjes integral should be used. However, this generality is hardly needed in statistical hydromechanics.

† A pupil of mine is working on questions of this nature.

‡ Solutions of (46) which represent typical decaying turbulence are not considered in this paper.

ones which represent the statistical pattern of observed turbulent flow, and the problem of isotropic turbulence amounts to just finding them. Remember that the averages of the physically important quantities can be computed once Φ is known. At the present time it is not possible completely to clarify the problem of "relevance." The points involved are brought out best by a reexamination of a basic idea of the current theory.

We start with the familiar two-parameter-group of simple transformations—coupled changes of scales in x, t and change of μ—which preserve the form of the Navier-Stokes equations (2). Remember that we supposed the density ρ of the fluid to be always equal to one. This group evidently induces a two-parameter-group which leaves the form of the Φ-equation (46) invariant and which we will write down farther below. Naturally, this group carries solutions Φ again into solutions, in general with a different value of μ, and it singly preserves properties of solutions such as homogeneity, axisymmetry, isotropy, stationarity. It is expected that the "relevant" solutions of the Φ-equation, $\mu > 0$ too form a two-parameter family of solutions—more precisely, that the relevant solutions are completely exhausted by applying the group transformations to just one of them. This is clearly a sort of ergodic hypothesis.* The original physical form of this idea is this. One may choose for group parameters two physically basic quantities, μ and the average dissipation ϵ of energy. In statistical equilibrium, and with no preferred direction in space, these two quantities must completely determine the statistical behavior of typical boundary-free flow (Kolmogoroff's first hypothesis). We write

$$\Phi(z(k);\, \mu,\, \epsilon) \tag{79}$$

for the corresponding characteristic functional. The group relation between the different functionals (79) is found to be

$$\Phi(z(k);\, \mu,\, \epsilon) \equiv \Phi(\alpha z(\beta k);\, \mu',\, \epsilon'), \tag{80}$$

where

$$\alpha = \left(\frac{\epsilon}{\epsilon'}\right)^{\frac{1}{4}} \left(\frac{\mu}{\mu'}\right)^{\frac{1}{4}}, \qquad \beta = \left(\frac{\epsilon}{\epsilon'}\right)^{\frac{1}{4}} \left(\frac{\mu}{\mu'}\right)^{\frac{3}{4}}. \tag{80'}$$

Hence the remaining basic problem is to find the single functional

$$\Phi(z(k);\, 1,\, 1).$$

The functional (79) must, as μ varies from 0 to ∞, represent the whole continuous scale of possible degrees of turbulent disorder. As $\mu \to \infty$ it should reduce to the trivial case $\Phi = 1$ mentioned above. This is the turbulence-free

* To be more explicit, this hypothesis is made for three-dimensional flow. It is intuitively clear that general spatial flow can and will produce enough disorder to make this hypothesis valid. However, it cannot be expected to hold for strictly plane flow.

limit case with its exclusive preference for the single "laminar flow" which, in our boundary-free case, is simply the state of rest $u = 0$. What happens if $\mu \to 0$? This is the inviscid limit case in which turbulent disorder reaches its maximum degree. It must be expected that the limit functional

$$\Phi_0\,(z(k);\,\epsilon) = \lim_{\mu=0} \Phi(z(k);\,\mu,\,\epsilon) \tag{81}$$

exists (Kolmogoroff's second hypothesis) and that it is a stationary solution of the case $\mu = 0$ of the Φ-equation (46). (81), (81′) imply that Φ_0 satisfies the following functional equation:

$$\Phi_0(z(k);\,\epsilon) = \Phi_0(\alpha z(\beta k);\,\epsilon'),$$

where

$$\alpha = \left(\frac{\epsilon}{\epsilon'}\right)^{\frac{1}{4}}\gamma, \qquad \beta = \left(\frac{\epsilon}{\epsilon'}\right)^{\frac{1}{4}}\gamma^{-3},$$

with γ being arbitrary. This relation does not determine the full functional, but Kolmogoroff was the first to notice that dimensional arguments do determine the spectral density $f(|\,k\,|)$ in the limit case $\mu \to 0$. In other words, the functional equation does determine the important term Φ_0^2 of the Taylor series for Φ_0. It is worthwhile to consider this matter in more detail.

Denoting the spectral density of average energy by

$$f(|\,k\,|;\,\epsilon,\,\mu)$$

we infer from (80), (80′)—we apply this functional equation to Φ^2—that

$$\int f(|\,k\,|;\,\mu,\,\epsilon)\,|\,\bar{z}(k)\,|^2\,dk = \alpha^2 \int f(|\,k\,|;\,\mu',\,\epsilon')\,|\,\bar{z}(\beta k)\,|^2\,dk$$

holds identically in $z = z(k)$. This implies the functional equation for f

$$f(|\,k\,|;\,\mu,\,\epsilon) = \alpha^2\beta^{-3} f\left(\frac{1}{\beta}\,|\,k\,|;\,\mu',\,\epsilon'\right), \tag{82}$$

where α, β are given by (80′). Its general solution can be obtained if one chooses $\epsilon' = 1$ and μ' such that $\beta = |\,k\,|$,

$$\mu' = \mu\epsilon^{-\frac{1}{3}}\,|\,k\,|^{\frac{4}{3}}.$$

On letting

$$g(\mu') = f(1;\,\mu',\,1) \tag{83}$$

the general solution of (82) is found to be

$$f(|\,k\,|\,;\,\mu,\,\epsilon) = g(\mu\epsilon^{-\frac{1}{3}}|\,k\,|^{\frac{4}{3}})\epsilon^{\frac{2}{3}}|\,k\,|^{\frac{11}{3}}, \tag{84}$$

where $g(\mu')$ is an arbitrary function of a single variable. Of course, the full conditions of our problem must make this function a completely determined one. At present, however, we have nothing but Kolmogoroff's second hypothesis at our disposal. It implies that the limit density

$$f_0(|\,k\,|;\,\epsilon) = \lim_{\mu=0} f(|\,k\,|;\,\mu,\,\epsilon)$$

exists. This means that the limit

$$\lim_{\mu'=0} g(\mu') = g(0)$$

exists. Hence we obtain Kolmogoroff's law

$$f_0(|\,k\,|\,;\,\epsilon) = g(0)\epsilon^{2/3}\,|\,k\,|^{-11/3}.$$

(84) also implies that, for any fixed values $\mu > 0$ and ϵ,

$$f(|\,k\,|\,;\,\epsilon,\,\mu) \sim g(0)\epsilon^{2/3}\,|\,k\,|^{-11/3} \qquad \text{as } k \to 0.^*$$

As far as the quadratic term Φ^2 of Φ is concerned, these are the only conclusions one can draw from the two hypotheses of Kolmogoroff.† The determination of the full functional Φ_0 is an outstanding problem of great interest *per se*. One might consider Φ_0 as representing the statistical structure of the typical flow of an ideal incompressible fluid and one should expect that this same limit case obtains if viscosity is replaced by another agent which similarly cuts off the full turbulent disorder of the ideal case. However, "typical ideal fluid flow" must be very different from the picture one usually has of individual smooth ideal flow with perhaps only a discrete set of slip surfaces and vortex lines. In the strict ideal case phase motions in general tend completely to lose their properties of regularity and only certain statistical quantities (the functional Φ_0!) can be expected to stay meaningful. Loss of regularity in space and time is indicated even by dimensional considerations. Kolmogoroff's law, expressed in x-space, says that the average velocity difference in two points at distance r and at the same time is $\sim r^{1/3}$ and not $\sim r$. One can also deduce that the average displacement of a "fluid particle" in a time interval t is $\sim t^{3/2}$ and not $\sim t$. All this is attenuated by an agent such as viscosity.‡

* This shows, by the way that the integral for Φ^2 converges only if the argument-field $z(k)$ vanishes sufficiently rapidly at $k = 0$. It will therefore be necessary to restrict the range of arbitrary fields $z(k)$, $z^*(k) = z(-k)$ to those which vanish sufficiently rapidly at $k = \infty$ and at $k = 0$.

† Of course, (84) can be translated into a statement about the double velocity correlations.

‡ Mathematically speaking, viscosity introduces a certain measure of compactness into the group T^t of phase motions. It was an attempt at proving this which inspired my paper: *Ein allgemeiner hydrodynamischer Endlichkeitssatz*, Math. Annalen **117** (1941), pp. 764–775. These mathematical investigations should however be carried much further, particularly

From the remark at the end of section 10 we can now draw a negative conclusion. Kolmogoroff's law is incompatible with a Gaussian phase distribution. The Gaussian solution (78) and the hypothetical non-Gaussian functional Φ_0 are two completely different solutions of the case $\mu = 0$ of the Φ-equation (46).*

---

in the important case of boundary-free flow. Those compactness properties are probably much weaker in this extreme case (even with $\mu > 0$) than in regular cases where the fluid is enclosed by entirely finite walls. The contrary remark I made about the boundary-free case at the end of my paper is not justified.

Ergodic theory of deterministic systems confined itself, at least at the time when I was active in this field, to transformation groups which enjoy certain compactness properties. Viscous fluid flow belongs to this category; ideal fluid flow considered as something *per se* (separate from the limit properties $\mu \to 0$ of viscous flow) does not. Here the situation is probably this. The group T_u^t of phase motions is actually defined only for certain not too discontinuous fields u, and it has no reasonable properties of compactness. "Limit fields" of fields T_u^t, $t \to \infty$ in general do not belong to those fields. The group cannot be extended to a larger u-space in which it stays a group of single-valued transformations (single-valued = uniquely determined flow) and in which it becomes compact. However there is the following question. Is an extension possible in which T^t becomes a stochastic process? In other words, is "typical ideal fluid flow" an indeterministic process?

---

* Boundary-free flow in the sense indicated before is approximately realized under very general boundary conditions if the Reynolds number is very large and if attention is confined to sufficiently small x-regions. Furthermore, if these small regions are properly dimensioned and if the flow in them is viewed from a properly moved system of reference the statistical structure of the flow should become a universal one, namely the Kolmogoroff structure corresponding to the functional (79). Viewed in x-regions a certain order larger but still small as compared to the extent of the boundaries, the structure is the one corresponding to Φ_0. It seems feasible to me that the structure becomes the Gaussian one if the flow is viewed on a scale still an order larger than the former one. The difference in the order of magnitude of the latter two scales would explain why in the limit, Reynolds number $= \infty$, the two functionals, Φ_0 and the Gaussian one, appear as completely separate solutions of the case $\mu = 0$ of boundary-free flow.

Statistical Hydromechanics and Functional Calculus, *J. Rational Mech. Anal.* **1, pp. 87-123, (1952).**

Commentary

Ya. G. Sinai

In this paper, E. Hopf considered the dynamics described by the Navier-Stokes equation and similar equations of the hydrodynamical type for which the existence theorems are not strong enough to ensure the existence of solutions on the whole phase space. He proposed to consider the evolution of measures induced by the equations of motion and in some sense, this is more flexible. Then, Hopf used the notion of the characteristic functional of the measure to write down the equations involving functional derivatives for the evolution of the characteristic functional. This equation is now called the Hopf equation.

There is vast literature extending the ideas of Hopf. The existence theorem for the Hopf equation for the Navier-Stokes system was proven by Ladyzhenskaya and Vershik (1977). An extensive bibliography on this subject can be found in the book by Vishik and Fursikov (1980). The ideas of Hopf were very useful in statistical physics. Dobrushin and Fritz (1977) proved the existence theorems for big enough domains in the phase spaces of the infinite-dimensional dynamical systems of statistical mechanics. These domains have full measure with respect to any Gibbs measure satisfying some natural conditions.

References

DOBRUSHIN, R. L. AND J. FRITZ, "Nonequilibrium dynamics of two-dimensional infinite particle systems," *Comm. Math. Phys.*, **57**, p.67-81, (1977).

FURSIKOV, A. V. AND M. I. VISHIK, "Mathematical Problems of Statistical Hydrodynamics," Nauka, Moscow, p. 440, (1980).

LADYZHENSKAYA, O. A. AND A. M. VERSHIK, "Sur ℓ' évolution des mesures déterminés par les éguations de Navier-Stokes et la résolution du probleme de Cauchy pour ℓ' equation statistique de E. Hopf," *Ann. Scuola Norm. Sup. Pisa*, eℓ. Sci (4), n.2, p.209-230, (1977).

On the Ergodic Theorem for Positive Linear Operators.

By *Eberhard Hopf*[1]) in Bloomington (Ind., U.S.A.).

Let X denote a space with points x and let there be given a σ-additive and σ-finite measure defined on all sets belonging to a given σ-field of subsets of X. All integrals $\int f$, $f = f(x)$, $x \in X$, are formed with this measure. If no domain of integration is indicated $\int f$ is extended through all of X. Denote by $L^1 = L^1(X)$ the linear space of all functions $f(x)$ integrable in X and consider a mapping $T = Tf$ of L^1 into L^1 which is linear and positive, $f \geqq 0$ (in X) $\Rightarrow Tf \geqq 0$ (in X). A previous paper[2]) of the author dealt with those positive linear operators from L^1 to L^1 which conserve the integral, $\int Tf \equiv \int f$. They are closely connected with Markoff processes in X, and positivity and conservation of the integral arise simply from the requirement that a probability distribution in X goes over again into such a distribution. The author tried unsuccessfully to prove that the quotients $(n = 1, 2, \ldots)$

$$(1) \qquad Q_n(f, p) = \frac{S_n(f)}{S_n(p)}, \quad S_n(g) \equiv \sum_0^{n-1} T^i g,$$

converge almost everywhere in X if $f \in L^1$, $p \in L^1$, $p > 0$. In the mentioned paper this was proved under the additional hypothesis that the constants are in L^1 and that $T1 = 1$. For the corresponding Markoff process this hypothesis means that X has finite measure and that the measure is invariant under the process. In the special case where T is generated by a measure preserving point transformation $S = Sx$ of X into itself, $Tf(x) = f(Sx)$, this theorem coincides with G. D. Birkhoff's ergodic theorem. The general theorem had remained problematic for five years. In the case of point transformations, it covers the author's generalization of Birkhoff's theorem (infinite invariant measure) and the still more general theorem of W. Hurewicz (no invariant measure assumed) in the form given to it by P. Halmos, $Tf(x) = \omega(x) f(Sx)$ where $\omega(x) > 0$ is a suitably chosen fixed weight function.

Only quite recently R. V. Chacon and D. S. Ornstein succeeded in solving the problem. In their paper[3]) they prove the following somewhat more general

Ergodic theorem. *Let T be a positive linear operator from $L^1(X)$ to $L^1(X)$ with an L^1-norm $\leqq 1$. If $f \in L^1(X)$, $p \in L^1(X)$, $p \geqq 0$, then the quotients $Q_n(f, p)$, $n \to \infty$, have a finite limit in almost every point of the set where $\sum_0^\infty T^i p > 0$.*

It seemed desirable to us to present Chacon's and Ornstein's ingeneous proof in a more transparent form. This is what we do in the following lines. It is, in particular, the subsequent "Basic lemma 1" which seems to us to bring out more clearly the core

1) The author is indebted to the Office of Naval Research for a research grant at Indiana University.

2) *E. Hopf*, The general temporally discrete Markoff process. J. of Ratl. Mechanics and Analysis, **3** (1954), 13—45.

3) *R. V. Chacon* and *D. S. Ornstein*, A general ergodic theorem. Illinois J. of Mathematics, **4** (1960), 153—160.

of their reasoning. Our subsequent proof is very simple. Little touches have been applied in other places too to simplify the presentation. It must be said, however, that the principal ideas are those of Chacon and Ornstein.

The proof makes use of the lattice operations $f \to f^+$, $f \to f^-$ which carry L^1 into L^1,

$$f^+ = \max(f, 0), \quad f^- = -\min(f, 0). \tag{2}$$

There holds

$$f = f^+ - f^-, \quad |f| = f^+ + f^-. \tag{3}$$

f^+, f^- are both $\geqq 0$ but never both > 0. Therefore

$$f = f_2 - f_1, \; f_i \geqq 0 \quad \Rightarrow \quad f^+ \leqq f_2, \; f^- \leqq f_1. \tag{4}$$

From $Tf = Tf^+ - Tf^-$, from the positivity of T and from (4) it follows that

$$(Tf)^+ \leqq Tf^+, \; (Tf)^- \leqq Tf^-, \; |Tf| \leqq T|f|. \tag{5}$$

The assumption that $|T| \leqq 1$ is equivalent to the statement that

$$\varphi \geqq 0 \Rightarrow \int T\varphi \leqq \int \varphi. \tag{6}$$

The proof of the ergodic theorem is carried through in several steps the first of which is the proof of the

Basic lemma 1. *If* $f \in L^1$ *and if*

$$\sup_{n>0} \sum_0^{n-1} T^\nu f > 0$$

holds in each point of a set A *(belonging to the given* σ*-field) then there exist, to each* $\varepsilon > 0$*, functions* $h \in L^1$, $\varphi \in L^1$ *such that*

α) $h^- \leqq f^-$,
β) $h = f + T\varphi - \varphi, \; \varphi \geqq 0$,
γ) $\int h \leqq \int f$,
δ) $\int_A h^- < \varepsilon$.

Proof. We define a sequence of functions $h_i \in L^1$, $i \geqq 0$, $h_0 = f$, such that h_{i+1} is obtained from h_i on applying T only to its positive part,

$$h_{i+1} = Th_i^+ - h_i^-, \; h_0 = f. \tag{7}$$

We show that each $h = h_i$ has the properties α) — γ) and that δ) is satisfied for sufficiently large i. From (7), the positivity of T and from (4),

$$h_{i+1}^- \leqq h_i^- \tag{8}$$

and this implies α). (7) may be written

$$h_{i+1} = h_i + Th_i^+ - h_i^+.$$

Hence, by summation,

$$h_i = f + T\varphi_{i-1} - \varphi_{i-1}, \; \varphi_i = \sum_0^i h_\nu^+, \tag{9}$$

which proves β). But β) in turn implies γ) by virtue of (6). From $\varphi_i - \varphi_{i-1} = h_i^+ \geqq h_i$ and from the first equation (9),

$$\varphi_i \geqq f + T\varphi_{i-1},$$

$i > 0$. As T is order-preserving this implies, by induction, that

$$\varphi_i \geqq \sum_0^{i-1} T^\nu f + T^i \varphi_0,$$

$i > 0$. As $\varphi_0 = h_0^+ = f^+ \geqq f$ it follows finally that

$$\varphi_i = \sum_0^i h_\nu^+ \geqq \sum_0^i T^\nu f,$$

$i > 0$. Evidently this holds for $i = 0$ too. This inequality and the hypothesis of the lemma imply now: In each point of A at least one of the h_ν^+, $\nu \geqq 0$, is positive or, in other words, at least one of the h_ν^- is zero. Hence, from (8), $h_n^- \geqq 0$, it follows that $h_n^- \downarrow 0$ in A and, consequently, that $\int_A h_n^- \to 0$. Lemma 1 is herewith completely proved.

Lemma 2. *If $f \in L^1$ and if*

$$\sup_{n>0} \sum_0^{n-1} T^\nu f \geqq 0$$

holds everywhere in a set A then f satisfies

$$\int_A f + \int_{\bar{A}} f^+ \geqq 0, \quad \bar{A} = X - A.$$

Proof. We prove the lemma first under the more restrictive hypothesis that sup > 0 in A. For any given $\varepsilon > 0$, lemma 1 may be applied. From the statements δ), α), γ) of this lemma it follows that

$$-\varepsilon - \int_{\bar{A}} f^- < -\int_A h^- - \int_{\bar{A}} h^- = -\int h^- \leqq \int h \leqq \int f.$$

This is only another form of the desired inequality, but with $-\varepsilon$ in place of 0 on the right. This inequality is thereby proved since ε was arbitrary.

In order to prove the lemma under the original hypothesis, sup $\geqq 0$ in A, we need merely use a function $p \in L^1$, $p > 0$, (such a p exists as the measure was supposed to be σ-finite) and apply the restricted lemma to $f + \eta p$, $\eta > 0$, in place of f. On letting $\eta \to 0$ we obtain the complete proof of lemma 2.

Remark. Lemma 2 is completely equivalent to the author's lemma (l. c. [2]), the maximal ergodic theorem):

$$\int_B f \geqq 0$$

if B is the set of all points where $\sup_{n>0} \sum_0^{n-1} T^\nu f \geqq 0$ (or the set of all points where this sup is > 0). The equivalence follows from three simple facts. Firstly, the hypothesis of lemma 2 says that $A < B$. Secondly, the left hand side of the final inequality of lemma 2 equals

$$\int_B f + \int_C f^- + \int_{\bar{B}} f^+ \geqq \int_B f$$

where $C = B - A$. Thirdly, in $\bar{B}$ there holds sup $\leqq 0$ and, in particular, $f \leqq 0$ or, in other words, $f^+ = 0$. This makes the equivalence obvious.

Lemma 3. *Let $f \in L^1$, $p \in L^1$, $p \geqq 0$. Then*

$$\sup_{n>0} |Q_n(f, p)| < \infty$$

holds almost everywhere in the set where $p > 0$.

Proof. As $|Q_n(f, p)| \leqq Q_n(|f|, p)$ it suffices to prove the lemma for the case that $f \geqq 0$, and without the absolute value sign. Let A denote the subset of the set $[p > 0]$ where sup $Q_n(f, p) = \infty$. For any number $\omega > 0$, lemma 2 may be applied to $f - \omega p$ and A as

$$Q_n(f - \omega p, p) = Q_n(f, p) - \omega.$$

Therefore,

$$\int_A (f - \omega p) + \int_{\bar A} (f - \omega p)^+ \geqq 0.$$

From $f - \omega p \leqq f$, $f \geqq 0$ it follows that $(f - \omega p)^+ \leqq f$. Hence,

$$\int_A p \leqq \frac{1}{\omega} \int f.$$

As $p > 0$ in A and as ω is arbitrary A must have measure zero. Lemma 3 is thereby proved.

Lemma 4. *If* $f \in L^1$, $p \in L^1$, $p \geqq 0$ *then*

$$\lim_{n \to \infty} T^n f / \sum_0^{n-1} T^\nu p = 0$$

Proof. It suffices to prove this for the case $f \geqq 0$. Choose a number $\varepsilon > 0$ arbitrarily and consider the functions

$$g_n = T^n f - \varepsilon \sum_0^{n-1} T^\nu p, \quad g_0 = f$$

which satisfy the relations

$$g_{n+1} + \varepsilon p = T g_n. \tag{10}$$

It is sufficient to show, for almost every point of the set $[p > 0]$, that $g_n < 0$ for all large n or, in other words, that $\Sigma\, \chi_n$ converges where χ_n is the characteristic function of the set $[g_n \geqq 0]$. From $\chi_n g_n = g_n^+$ and (10),

$$\begin{aligned}\int g_{n+1}^+ + \varepsilon \int \chi_{n+1}\, p &= \int \chi_{n+1}\, (g_{n+1} + \varepsilon p) \\ &= \int \chi_{n+1} T g_n \leqq \int \chi_{n+1} T g_n^+ \leqq \int T g_n^+ \leqq \int g_n^+.\end{aligned}$$

By summation of the resulting inequalities from $n = 0$ on,

$$\int g_n^+ + \varepsilon \int p \sum_1^n \chi_\nu \leqq \int g_0^+$$

and, hence,

$$\int p \sum_1^\infty \chi_\nu \leqq \frac{1}{\varepsilon} \int g_0^+.$$

This obviously implies what we want to show and lemma 4 is herewith completely proved.

The next lemma contains the principal part of the proof of the ergodic theorem. Its proof is based upon lemma 1 but it makes use also of lemma 4.

Lemma 5. *Let* $f \in L^1$, $p \in L^1$, $p \geqq 0$. *If in each point of a set* A *there holds* $p > 0$, $\sum_0^\infty T^\nu p = \infty$ *and*

$$\underline{\lim}\; Q_n(f, p) < a < b < \overline{\lim}\; Q_n(f, p)$$

where a, b *are numbers then* A *has measure zero.*

Proof. Choose a number $\varepsilon > 0$. Part of the hypothesis says that in the set A

$$\sup_{n > 0} Q_n(f - bp) \geqq \overline{\lim}\; Q_n(f - bp) > 0.$$

Therefore we may apply lemma 1 to $f - bp$ in place of f. There exists a function $h \in L^1$ (the h of lemma 1 is now written in the form $h - bp$) such that α)

$$(h - bp)^- \leqq (f - bp)^- \tag{11}$$

and δ)

(12) $$\int_A (h - bp)^- < \varepsilon.$$

(11) is equivalent to the statement that $h < bp$ implies $h \geqq f$. Consequently ($a < b$), $h < ap$ implies $h \geqq f$ which means that

(13) $$(h - ap)^- \leqq (f - ap)^-$$

holds too. Now we make use of β) of lemma 1,

$$Q_n(h, p) = Q_n(f, p) + (T^n\varphi - \varphi) \Big/ \sum_0^{n-1} T^i p,$$

and we apply lemma 4 to φ, p. We also use the hypothesis that $\sum^{\infty} T^i p = \infty$ on A. It follows that $Q_n(h, p)$ has the same upper and the same lower limit as $Q_n(f, p)$ in each point of A (except in the points of a nulset which may be neglected in what follows). In particular,

$$\underline{\lim}\, Q_n(h, p) < a \quad \text{or} \quad \overline{\lim}\, Q_n(ap - h, p) > 0 \text{ in } A.$$

Consequently, lemma 1 may be applied to $ap - h$ in place of the f there. The new auxiliary function h of this lemma is now written in the form $ap - f'$. It follows from γ) that

(14) $$\int (ap - f') \leqq \int (ap - h) \quad \text{or} \quad \int (f' - ap) \geqq \int (h - ap),$$

and from α) that

(15) $$(ap - f')^- \leqq (ap - h)^- \quad \text{or} \quad (f' - ap)^+ \leqq (h - ap)^+,$$

and, finally, from δ) that

(16) $$\int_A (ap - f')^- = \int_A (f' - ap)^+ < \varepsilon.$$

On using the first relation (3) we can write the inequality (14), second form,

(17) $$\int (f' - ap)^- \leqq \int (h - ap)^- + \int [(f' - ap)^+ - (h - ap)^+].$$

As the integrand in the last integral is $\leqq 0$ (see (15)) the inequality stays true if the integral is taken over A. Now,

$$(h - ap)^+ \geqq h - ap = h - bp + (b - a)\, p \geqq -(h - bp)^- + (b - a)\, p.$$

Consequently, the right hand side of (17) does not exceed

$$\int (h - ap)^- + \int_A (f' - ap)^+ + \int_A (h - bp)^- - (b - a) \int_A p.$$

We apply (13) to the first term, (16) to the second term and (12) to the third term. Thereby we obtain the inequality

(18) $$\int (f' - ap)^- \leqq \int (f - ap)^- + 2\varepsilon - (b - a) \int_A p$$

in which h no longer appears. By the same reasoning as above it follows that $Q_n(f', p)$ has the same upper and the same lower limit as $Q_n(h, p)$ in each point of A. The final result is the following. If f satisfies the hypothesis of the present lemma then there exists, to any given $\varepsilon > 0$, a function f' which satisfies the same hypothesis and, in addition, the inequality (18). This result may in turn be applied to f': There exists a function f'' which satisfies the hypothesis of the present lemma and the inequality

$$\int (f'' - ap)^- \leqq \int (f' - ap)^- + 2\varepsilon - (b - a) \int_A p,$$

and so forth. On adding up the first n of these inequalities and on observing that the integral remaining on the left is $\geqq 0$ we find that

$$n[(b-a)\int_A p - 2\varepsilon] \leqq \int (f-ap)^-$$

holds for any integer n. Therefore, the number between the square brackets is $\leqq 0$. Since $\varepsilon > 0$ was arbitrary it follows that $\int_A p = 0$ and, since $p > 0$ was assumed in A, that A has measure zero. Lemma 5 is hereby proved.

Proof of the ergodic theorem. First we prove the theorem for the set where $p > 0$. It suffices to prove it for the case that $f \geqq 0$ (in X). Lemma 3 implies that the (finite or infinite) limit if it exists must be finite almost everywhere. So we need only prove existence. Within the part of the set $[p > 0]$ in which $\sum^{\infty} T^i p < \infty$ this is trivial because then $\overline{\lim}\, Q_n < \infty$ implies that the series $\sum^{\infty} T^i f$ with non-negative terms has a finite sum. In the remaining part $[p > 0,\ \sum T^i p = \infty]$ the positivity of the measure of the set where $\underline{\lim} < \overline{\lim}$ would imply the existence of numbers $a, b,\ a < b$, such that the set $[\underline{\lim} < a < b < \overline{\lim}]$ would have positive measure, in contradiction to lemma 5. Hence, $\underline{\lim} = \overline{\lim}$ holds almost everywhere in that remaining part.

In order to prove the theorem under the hypothesis mentioned there, i. e. for the set where $\sum^{\infty} T^i p > 0$, we observe that this set is the union of the sets $[p > 0]$ and

$$[S_{k+1} p > 0,\ S_k p = 0],\ k > 0. \tag{19}$$

In this set, $T^k p = S_{k+1} p - S_k p > 0$. On letting

$$f' = T^k f,\ p' = T^k p$$

we see that

$$Q_{n-k}(f', p') = \frac{S_n f - S_k f}{S_n p} = Q_n(f, p) - \frac{S_k f}{S_n p}$$

holds in this set and for $n > k$. From this formula and from what we have proved before it is evident that $\lim Q_n(f, p)$ exists and is finite almost everywhere in the set (19). The theorem is hereby completely proved.

Eingegangen 15. Januar 1960.

On the Ergodic Theorem for Positive Linear Operators, *J. Reine Angewandte Mathematik*, **vol. 205, pp. 101-106, (1960).**

Commentary

David Ornstein

The famous Birkhoff ergodic theorem asserts the point-wise convergence of expressions like $\frac{1}{n}\sum_{k=0}^{n-1} f(T^k x)$ when f is in $L^1(M,\mu)$ and T is a positive isometry of $L^1(M,\mu)$ induced by a measure-preserving transformation T. It was Hopf's insight that essentially the same theorem holds for any positive operator of norm less then or equal to 1 on $L^1(M,\mu)$. This was the beginning of the ergodic theory of Markov processes with non-discrete phase state. Now there exists an extensive body of work on this subject; see, for example, the book by U. Krengel (1985).

References

KRENGEL, U., "Ergodic Theorems," *DeGruyter Studies in Mathematics*, vol. 6, Walter de Gruyter and Co., Berlin, (1985).

Credits

The American Mathematical Society gratefully acknowledges the kindness of these institutions in granting the following permissions:

Indiana University

"Hamilton's theory and generalized solutions of the Hamilton-Jacobi equation" (with E.D. Conway), *J. Math. & Mech.* **13** (1964), 939-986.

"Statistical Hydromechanics and Functional Calculus," *J. Rational Mech. Anal.* **1** (1952), 87-123.

John Wiley & Sons, Inc.

"The Partial Differential Equation $u_t + uu_x = \mu u_{xx}$," *Comm. on Pure and Appl. Math.* **3** (1950), 201-230. Reprinted by permission of John Wiley & Sons, Inc.

"A Mathematical Example Displaying Features of Turbulence," *Comm. on Pure and Appl. Math.* **1** (1948), 303-322. Reprinted by permission of John Wiley & Sons, Inc.

S. Hirzel Verlag GmbH & Co.

"Statistik der geodatischen Linien in Mannigfaltigkeiten negativer Krümmung," *Akad. Wiss. Leipzig* **91** (1939), 261-304.

"Abzweigung einer periodischen Lösung von einer stationären Lösung eines Differentialsystems," *Akad. Wiss. Math.- Nateiw. Kl. (Leipzig)* **95** no. 1, (1943), 3-22.

Springer-Verlag GmbH & Co. KG

"Zum analytischen Charakter der Lösungen regulärer zweidimensionaler Variationsprobleme," *Math. Zeitschr.* **30** (1929), 404-413, © 1929 Springer-Verlag GmbH & Co. KG.

"Über den funktionalen, insbesondere den analytischen Charakter der Lösungen elliptischer Differentialgleichungen zweiter Ordung," *Math. Zeitschr.* **34** (1931), 194-233, © 1931 Springer-Verlag GmbH & Co. KG.

"Statistik der Lösungen geodatischer Probleme vom unstabilen Typus II," *Math. Annalen* **117** (1940), 590-608, © 1940 Springer-Verlag GmbH & Co. KG

University of Maryland

"Repeated Branching through Loss of Stability, an Example," *Proc. of the Conference on Differential Equations at the University of Maryland* (1955) 49-56.

Walter de Gruyter GmbH & Co. KG

"Elementare Bemerkungen über die Lösungen partieller Differntialgleichungen zweiter Ordnung vom elliptischen Typus," *Sitzungsberichte Preussische Akad. Wiss. Berlin.* (1927), 147-152.

"Über eine Klasse singularer Integralgleichungen," (with Norbert Wiener), *Sitzungsberichte Preussische Akad. Wiss., Phys.- Math. Kl. Berlin* (1931), 696-706.

"On the ergodic theorem for positive linear operators," *J. Reine Angew. Math.* **205** (1960), 101-106.

Wiley-VCH Verlag Berlin GmbH

"Über die Anfangswertaufgabe für die hydrodynamischen Grundgleichungen," *Math. Nachr.* **4** (1951), 213-231.

Titles in This Series

Volume

17 **Eberhard Hopf:** Selected Works of Eberhard Hopf with Commentary (Cathleen S. Morawetz, James B. Serrin, and Yakov G. Sinai, Editors), 2002

16 **S. A. Amitsur:** Selected Papers of S. A. Amitsur with Commentary, Parts 1 and 2 (Avinoam Mann, Amitai Regev, Louis Rowen, David Saltman, and Lance Small, Editors), 2001

15 **Ilya Piatetski-Shapiro:** Selected Works of Ilya Piatetski-Shapiro (James Cogdell, Simon Gindikin, and Peter Sarnak, Editors), 2000

14 **E. B. Dynkin:** Selected Papers of E. B. Dynkin with Commentary (A. A. Yushkevich, G. M. Seitz, and A. L. Onishchik, Editors), 2000

13 **Frederick J. Almgren, Jr.:** Selected Works of Frederick J. Almgren, Jr. (Jean E. Taylor, Editor), 1999

12 **Ellis Kolchin:** Selected Works of Ellis Kolchin with Commentary (Hyman Bass, Alexandru Buium, and Phyllis J. Cassidy, Editors), 1999

11 **V. S. Varadarajan:** The Selected Works of V. S. Varadarajan, 1999

10 **Maurice Auslander:** Selected Works of Maurice Auslander, Parts 1 and 2 (Idun Reiten, Sverre O. Smalø, and Øyvind Solberg, Editors), 1999

9 **Lipman Bers:** Selected Works of Lipman Bers: Papers on Complex Analysis, Parts 1 and 2 (Irwin Kra and Bernard Maskit, Editors), 1998

8 **Walter E. Thirring:** Selected Papers of Walter E. Thirring with Commentaries, 1998

7 **Robert Steinberg:** Robert Steinberg Collected Papers, 1997

6 **Julia Robinson:** The Collected Works of Julia Robinson (Solomon Feferman, Editor), 1996

5 **Freeman Dyson:** Selected Papers of Freeman Dyson with Commentary, 1996

4 **Witold Hurewicz:** Collected Works of Witold Hurewicz (Krystyna Kuperberg, Editor), 1995

3.2 **A. Adrian Albert:** A. Adrian Albert Collected Mathematical Papers: Nonassociative Algebras and Miscellany, Part 2 (Richard E. Block, Nathan Jacobson, J. Marshall Osborn, David J. Saltman, and Daniel Zelinsky, Editors), 1993

3.1 **A. Adrian Albert:** A. Adrian Albert Collected Mathematical Papers: Associative Algebras and Riemann Matrices, Part 1 (Richard E. Block, Nathan Jacobson, J. Marshall Osborn, David J. Saltman, and Daniel Zelinsky, Editors), 1993

2 **Salomon Bochner:** Collected Papers of Salomon Bochner, Parts 1–4 (Robert C. Gunning, Editor), 1992

1 **R. H. Bing:** The Collected Papers of R. H. Bing, Parts 1 and 2 (Sukhjit Singh, Steve Armentrout, and Robert J. Daverman, Editors), 1988

ISBN 0-8218-2077-X

9 780821 820773

CWORKS/17